Essentials of Engineering Fluid Mechanics

(a)

(b)

Two flow conditions for a liquid flowing from a long smooth tube. The laminar jet (a) has a smooth surface and the turbulent jet (b) shows the effect of eddying motion. Both are at the same flow rate. Exposure time about 1/200 sec.

Essentials of Engineering Fluid Mechanics

FOURTH EDITION

REUBEN M. OLSON

Ohio University

HARPER & ROW, PUBLISHERS, New York
Cambridge, Hagerstown, Philadelphia, San Francisco,
London, Mexico City, São Paulo, Sydney

1817

Sponsoring Editor: Charlie Dresser
Project Editor: Eleanor Castellano
Production Manager: Marion Palen
Compositor: Syntax International Pte. Ltd.
Printer and Binder: Maple Press Company
Art Studio: J & R Technical Services Inc.

Essentials of Engineering Fluid Mechanics, Fourth Edition

Copyright © 1980 by Reuben M. Olson

Library of Congress Cataloging in Publication Data

Olson, Reuben M
  Essentials of engineering fluid mechanics.

  Includes bibliographical references and index.
  1. Fluid mechanics. I. Title.
TA357.047 1980   620.1'06   79-23429
ISBN 0-700-22532-3

Contents

Preface   xiii

Part I
INTRODUCTION   1

1 Definitions and Fluid Properties   3

    1-1.   Definition of a Fluid   *3*
    1-2.   Everyday Experiences with Fluids   *4*
    1-3.   Historical Background   *5*
    1-4.   Symbols and Units   *7*
          THERMODYNAMICS   *10*
    1-5.   Temperature, Heat, and Work   *11*
    1-6.   The First Law of Thermodynamics   *12*
    1-7.   The Second Law of Thermodynamics   *12*
    1-8.   The Perfect Gas   *14*

v

1-9. Flow Processes   *16*
      FLUID PROPERTIES   *17*
1-10. Density, Specific Volume, and Specific Gravity   *17*
1-11. Compressibility or Elasticity   *19*
1-12. Surface Tension and Capillarity   *22*
1-13. Vapor Pressure   *25*
1-14. Viscosity   *25*
      Problems   32
      References   42

2 Fluid Statics   43

2-1. General Differential Equation   *44*
2-2. Hydrostatics   *44*
2-3. Manometry   *47*
2-4. Fluid Forces on Submerged Surfaces   *51*
2-5. Curved Surfaces   *54*
2-6. Aerostatics   *56*
2-7. Isothermal Atmosphere   *57*
2-8. Polytropic Atmosphere   *57*
2-9. The U.S. Standard Atmosphere   *58*
2-10. Static Stability   *60*
      Problems   65
      Reference   76

Part II
CONSERVATION LAWS   77

3 Fluid Kinematics   79

3-1. Types of Flow   *79*
3-2. Fluid Velocity and Acceleration   *83*
3-3. Streamlines and the Stream Function, Pathlines, and
     Streaklines   *85*
3-4. Deformation of a Fluid Element   *88*
3-5. Vorticity and Circulation   *91*
3-6. Conservation of Mass and the Continuity Equation   *95*
     Problems   103
     References   112

4 Fluid Momentum   113

THE MOMENTUM THEOREM OF FLUID MECHANICS   *113*
4-1 The Momentum Theorem   *113*
4-2. Applications of the Momentum Theorem   *116*
EQUATIONS OF MOTION   *124*
4-3. Equations of Motion for Nonviscous Fluids   *125*
4-4. Equations of Motion for Viscous Fluids   *138*
Problems   *142*
References   *157*

5 Fluid Energy   158

5-1. The General Energy Equation   *159*
5-2. Comparison of Equations of Motion with Steady-Flow Energy
Equation   *168*
5-3. Work and Power   *169*
5-4. Viscous Dissipation   *170*
Problems   *172*

Part III
INCOMPRESSIBLE FLOW REGIMES   179

6 Potential Flow   181

6-1. Velocity Potential and Stream Function for Two-Dimensional
Flow   *182*
6-2. Flow Nets for Two-Dimensional Flow   *186*
6-3. Examples of Ideal, Two-Dimensional, Steady Fluid Flows   *189*
6-4. Method of Images   *198*
6-5. Potential Flows from Conformal Transformations in the Complex
Plane   *201*
6-6. Numerical Relaxation Method   *206*
Problems   *211*
References   *217*

7 Viscous Flow: the Boundary Layer   218

7-1. Laminar and Turbulent Flow   *219*
7-2. Description of the Boundary Layer   *223*
7-3. The Prandtl Boundary-Layer Equations   *229*

7-4. The Momentum Equation for the Boundary Layer   *232*
7-5. The Flat Plate in a Uniform Free Stream with No Pressure
Gradients: Approximate Momentum Analysis   *234*
Problems   244
References   249

Part IV
APPLICATIONS
251

8 Dimensionless Numbers and Dynamic Similarity
253

8-1. Introduction   *253*
8-2. Dimensional Analysis of Fluid Systems   *255*
8-3. Force Ratios   *258*
8-4. Normalized Equations of Motion and Energy   *259*
8-5. Dynamic Similitude   *261*
8-6. Modeling Ratios   *265*
8-7. Incomplete Similarity   *267*
8-8. Some General Expressions for Dynamic Similitude   *268*
Problems   273
References   280

9 The Flow of Viscous Fluids in Ducts
281

9-1. Basic Considerations   *281*
9-2. Fully Developed Incompressible Flow in Ducts   *284*
9-3. Fully Developed Laminar Incompressible Flow in Ducts   *286*
9-4. Fully Developed Incompressible Turbulent Flow in Ducts   *289*
9-5. Steady Incompressible Flow in the Entrance Region of
Ducts   *300*
9-6. Contractions, Expansions, and Pipe Fittings; Turbulent
Flow   *305*
9-7. Applications   *311*
9-8. Empirical Pipe-Flow Equations   *315*
9-9. Pipe Networks   *317*
9-10. Flow of Mixtures in Pipes   *320*
Problems   322
References   334

10 The Flow of Compressible Gases
336

10-1. The Velocity of Sound   *337*
10-2. The Mach Number   *339*

10-3. Isentropic Flow of a Perfect Gas   *340*

10-4. Diabatic Flow of a Perfect Gas Without Friction   *349*

10-5. Flow of Compressible Gases in Pipes with Friction   *352*

10-6. Normal Shock Waves   *364*

10-7. Oblique Shocks   *369*

   Problems   *373*

   References   *383*

11 Dynamic Drag and Lift   384

11-1. Fluid Forces on a Body in a Flow Stream   *384*

11-2. Drag   *385*

11-3. Lift   *393*

   Problems   *401*

   References   *406*

12 Open-Channel Flow   407

12-1. Basic Considerations   *407*

12-2. Velocity of an Elementary Wave   *413*

12-3. Uniform Flow   *414*

12-4. Specific Energy and Specific Thrust   *418*

12-5. Gradually Varied Flow   *425*

12-6. Hydraulic Jump   *433*

12-7. Channel Transitions   *438*

12-8. Specialized Examples of Open-Channel Flow   *440*

   Problems   *443*

   References   *450*

13 Flow Measurements   452

13-1. Velocity Measurements   *453*

13-2. Liquid Flow Rates in Pipes   *463*

13-3. Liquid Flow Rates in Open Tanks or Open Channels   *468*

13-4. Subsonic Gas Flow Rates in Pipes   *474*

   Problems   *476*

   References   *481*

14 Turbomachines   483

14-1. General Theory   *483*

14-2. Energy Equation and System Characteristics for Pumps   *488*

14-3. Pump Performance Characteristics   *489*

14.4 Pump and System Combinations   *493*

14-5. Dimensionless Parameters and Dynamic Similarity for Pumps   494

14-6. Pump Cavitation   498

14-7. Hydraulic Turbines   500

14-8. Hydraulic Turbine Characteristics   500

14-9. Dimensionless Parameters and Dynamic Similarity for Turbines   506

  Problems   507

  References   511

15 Varied Flow in Open Channels and Gas Flow in Pipes and Nozzles   512

15-1. Froude and Mach Numbers   513

15-2. Specific Energy and the Fanno Line; Specific Thrust and the Rayleigh Line   514

15-3. Hydraulic Jump and Normal Shock   519

15-4. Some Examples of Similar Flows   523

  Reference   523

16 Convective Heat Transfer   524

16-1. The Thermal Boundary Layer   524

16-2. Flow Over a Flat Plate with No Pressure Gradients   526

16-3. Flow in a Pipe with Uniform Wall Temperature   531

  Problems   534

  References   536

Appendix I
Nomenclature   537

Appendix II
Conversion Factors and Property Values   541

Appendix III
Laminar Flow of a Power-Law Fluid   549

Appendix IV
The Prandtl Boundary Layer Equations   553

Appendix V
Tables for Compressible Gas Flow   **555**

Answers to Selected Problems   571
Index   577

Preface

This text is an outgrowth of experience in hydraulics, thermodynamics, and heat transfer as well as in teaching courses in fluid mechanics for aeronautical, civil, electrical, and mechanical engineers and applied mathematicians. It is based on the premise that fluid mechanics is fluid mechanics, and that the same fundamentals should be taught to all engineering students.

Engineering students should learn much in their courses of study and should try to understand at least some of what is learned. Thus, it is necessary to compromise between knowing and understanding, or between training and education, and to combine physical intuition with rigor at the learning stage. Experience and further study will help understanding. Hence an introductory engineering text should attempt to express the truth, nothing but the truth, but not the whole truth. Engineering educators and researchers seem to be much more concerned with completely understanding what is known than are engineering practitioners. Students may become any one of these, and thus compromises are necessary.

In addition to learning the principles of fluid behavior, the student should develop a good understanding of the properties of fluids which should enable him or her to solve practical problems. One aspect should not be emphasized at the expense of the other. The understanding involves physical ideas and includes the derivation and analysis of resulting equations. Moreover, problem solving involves numerical calculations from working formulas, often with the inclusion of experimentally determined coefficients. Both approaches are included in the text. An attempt has been made in this edition, as in previous ones, to maintain a consistent level of treatment throughout and to emphasize the physical nature of fluid behavior.

The SI system of units is employed throughout, with some exceptions in Chapter 14 on Turbomachines. Particular attention is given to the application of the continuity principle, the steady-flow energy equation (from the first law of thermodynamics), and the momentum theorem and equations of motion (from Newton's second law). Some distinction is made between incompressible and compressible flow, and some chapters are devoted exclusively to one or the other. One chapter is devoted to a discussion of the similarities between open-channel flow and gas flow.

The text is divided into four main parts. Part I, Introduction, includes a review of thermodynamics, a discussion of fluid properties, and a treatment of fluid statics with an introduction to static stability.

In Part II, the conservation laws relating to mass, momentum, and energy are derived in scalar form, basically, with results expressed in vector form when appropriate. Those instructors who believe that students are able to understand and appreciate derivations in vector form, and who prefer to derive them in that way, should do so. The Navier-Stokes equations are subsequently introduced in order to explain viscous dissipation and the Prandtl boundary layer equations.

Part III on incompressible flow regimes includes a chapter on potential flow with an introduction to conformal transformations and relaxation methods. It also contains a chapter on boundary layer flows with a discussion of laminar and turbulent flows, a solution of the Prandtl boundary layer equations, and an extension of the use of pipe-flow results to external boundary layers.

Part IV, Applications, is introduced by a chapter on dimensionless numbers and dynamic similarity. This is followed by a treatment of incompressible flow in ducts—both laminar and turbulent flow in circular and noncircular ducts in both entrances and in the fully-developed flow regime. Next is a chapter on compressible flow and one on open-channel flow, with a later chapter discussing the similarities between the two. A treatment of flow measurements, turbomachines, and a brief introduction to thermal boundary layers completes the text.

Some problems from previous editions have been retained, some have been changed, and some new ones have been added. Many new figures for problems have also been added.

Appreciation is due a number of people for their suggestions about the manuscript. I am indebted to Alvin G. Anderson, Warren E. Ibele, Albert G. Mercer, John F. Ripkin, and Joseph M. Wetzel now or formerly of the University of Minnesota, James R. Steven of the City College of New York, and Edward F. Obert of the University of Wisconsin for comments during preparation of the first edition. Comments of Edward Silberman and Joseph M. Wetzel of the St. Anthony Falls Hydraulic Laboratory of the University of Minnesota were especially helpful in preparing the second edition. For this fourth edition I especially want to thank Valdimar K. Jonsson of the University of Iceland, a former colleague at the University of Minnesota, for his comments. Thanks are also due Steven J. Wright of the University of Michigan and Hans J. Leutheusser of the University of Toronto for their comments upon reviewing the manuscript.

Reuben M. Olson

Essentials of Engineering Fluid Mechanics

Part I
INTRODUCTION

Fluid mechanics is a study of the behavior of fluids at rest and in motion. The study takes into account the various properties of fluids and their effects on the resulting flow patterns, in addition to the forces acting between the fluid and its boundaries. The study includes the application of some fundamental laws—conservation of matter, conservation of momentum, and the first and second laws of thermodynamics—to explain observed fluid behavior and to predict unobserved fluid behavior. An engineer should approach the study of fluid mechanics with the idea of extending his previous contact with physics, mathematics, and mechanics bearing two purposes in mind: (1) to obtain an understanding of and a feeling for the behavior of fluids, and (2) to be able to make calculations for flow situations encountered in engineering practice.

Chapter 1 includes some definitions, a discussion of our own and others' experiences with fluids, a summary of symbols and units, a review of thermodynamics, and a discussion of fluid properties. Chapter 2 deals with fluids at rest, which may have been studied previously in courses in physics, calculus, or statics. The subject of fluid statics is a special case of fluid dynamics. When terms involving fluid motion are omitted, the equations of motion for fluids become the equations of fluid statics.

Chapter 1
Definitions and Fluid Properties

1-1. Definition of a Fluid

Most materials may be designated as being a solid, a liquid, or a gas, although some materials may have properties that suggest a dual designation. A solid generally has a definite shape, while liquids and gases have their shapes governed by their container.

All materials are deformable. Most fluids may be distinguished from most solids on the basis of the magnitude of the deformations or changes in relative position of the material particles that result from external shear forces. These deformations are relatively small for large external shear forces in solids but are relatively large for even small external shear forces in fluids. In addition, a fluid continues to deform under these conditions. There are, however, materials such as jellies, paints, pitch, and polymer solutions that exhibit characteristics of both a solid and a fluid.

The most significant properties which distinguish liquids from gases (both are fluids) are

1. the bulk modulus of elasticity, because gases are more easily compressed than liquids, and
2. the density, because under ordinary conditions gases are much less dense than liquids.

3

A fluid, then, may be defined as a substance that continuously deforms when subjected to shear stresses; a fluid is not capable of sustaining shear stresses at rest. This implies that shear stresses may exist only when a fluid is in motion. For these shear stresses to exist, however, the fluid must be *viscous*, a characteristic exhibited by all *real* fluids. An *ideal* fluid may be defined as *nonviscous*, or *inviscid;* thus no shear stresses exist for this fluid when it is in motion.

Shear stresses are set up in viscous fluids as a result of relative motion between the fluid and its boundaries or between adjacent layers of fluid. Generally, the greater this relative motion, the greater are the shear stresses for a given fluid. It is this viscous property that causes resistance to steady flow, either directly or indirectly—directly for pressure drop in a pipe, and indirectly for drag on a golf ball or a similar situation where the flow separates from the boundary and creates a wake which contributes a significant if not major portion of the drag. Resistance to flow at high speeds in gases, however, may be due to compressibility effects. Also, the drag of surface ships is a result not only of viscous shear but also of the surface waves generated.

Although there are no nonviscous fluids, a study of them is of engineering interest because viscous fluids act like nonviscous fluids in many instances.

The treatment of fluids at rest or without relative motion in Chapter 2 applies to both viscous and nonviscous fluids, since the effects of viscosity do not appear when a fluid is at rest, but only when there is relative motion within the fluid. The basic laws of fluid motion introduced in Chapters 3–5 apply to both types of fluid. An introduction to the irrotational flow of non-viscous incompressible fluids is given in Chapter 6, and high-speed flow of compressible nonviscous gases is introduced in Chapter 9. Essentially all the rest of the book deals with viscous fluids.

1-2. Everyday Experiences with Fluids

Prehistoric man's contact with fluids was probably confined largely to the air he breathed and the water he drank and swam in. Later he made use of hydrostatic principles not only in swimming but in making rafts, and he used dynamic principles of fluid motion (or motion in a fluid) when he developed crude boats or canoes and spears and arrows. Throughout history man has increased his everyday contact with fluids.

In the home we are all familiar with water piping and the water hammering which results when a faucet is closed rapidly. Hydraulic engineers must make mathematically elaborate designs for surge tanks in water turbine systems in order to prevent water-hammer damage. The vortex we see when a bathtub is drained is fundamentally the same as a tornado and the swirls in the wake downstream of a bridge pier or behind a canoe. Hot water or steam radiators to heat homes and the radiators for cooling in an automobile depend on the growth of thermal and fluid boundary layers for their effectiveness as heat convectors.

When we say that something or someone is "slower than molasses in January," we are using a qualitative measure of fluid viscosity because we are aware that molasses flows slowly when cold. Lubricating oils for our automobiles are purchased according to a viscosity rating, the SAE number (10 or 20, for example), though we may request a thick or thin oil or a light or heavy oil instead. Few of us are aware that cold water is twice as viscous as warm water (this variation exists in tap water in northern states between winter and summer).

The windmill on a farm has its counterpart in ship, boat, and airplane propellers, in pumps, blowers, fans, turbines, kitchen blenders and malted milk mixers. In each, a torque and thrust is applied either to a fluid or by a fluid, and all are examples of a lifting vane.

The shifting of snow or sand in a high wind, both along the ground and in the air above it, has a counterpart in rivers, where sediment is carried to them by smaller streams and then in them to be deposited as sand bars or carried to the sea to form deltas.

Deposits of snow behind snow fences and the hollow regions around a tree trunk or pole following a blizzard are due to the same type of flow pattern that exists behind a golf ball in flight. Incidentally, ship hulls and aircraft wings and fuselages are made *smooth* in order to reduce drag, but golf balls are made *rough* in order to reduce their drag.

Even a physiologist is concerned with fluid mechanics. The heart is a pump which pumps a fluid (blood) through a piping system (vessels and arteries).

Rockets sent skyward by small boys on the Fourth of July are essentially the same in principle as the rockets used to send manmade satellites into interstellar space. Rocket nozzles have the same general shape as nozzles in gas and steam turbines and the aspirators used in a garden hose to draw liquid fertilizer into the water stream for sprinkling.

We are all aware of aerodynamic drag if we have ever walked or cycled against or with a high wind. In rowing a boat or paddling a canoe we find that we must row or paddle faster and harder in order to go faster, not only to accelerate but also to maintain a higher speed.

The crack of a whip in air induces a shock wave because the tip of the whip travels at supersonic speed. A hunter shooting a rifle sends a supersonic missile to hit a target. Many of us have seen shooting stars disappear because they were traveling so fast they burned up because of "friction."

Thus we all are continually dealing with fluids at rest and in motion although we are seldom aware of them quantitatively, if even qualitatively.

1-3. Historical Background [1]

The application of fluid mechanics began in connection with the motion of stones, spears, and arrows. Ships with oars and sails were used as early as 3000 B.C. Irrigation systems have been found in prehistoric ruins in both

Egypt and Mesopotamia. The early Greeks recognized air and water as two of the four forms of matter (the others were fire and earth). Aristotle (fourth century B.C.) studied the motions of bodies in thin media and in voids. Archimedes (third century B.C.) formulated the well-known laws of floating bodies.

The Roman aqueducts were built in the fourth century B.C., although written evidence indicates the builders did not understand pipe resistance. Da Vinci (1452–1519) hinted he advocated an experimental approach to science, saying, "Remember when discoursing on the flow of water to adduce first experience and then reason." He correctly described many flow phenomena. Galileo (1564–1642) contributed much to the science of mechanics.

The Italian school of hydraulics included Castelli (1577–1644), Torricelli (1608–1647), and Guglielmini (1655–1710), and ideas concerning the steady flow continuity equation in rivers, flow from a container, the barometer, and some qualitative concepts of flow resistance in rivers came from them. A Frenchman, Mariotte (1620–1684), made experiments in which he measured forces of jets and of the wind. In addition to his well-known laws of motion, Newton (1642–1727) proposed that fluid resistance is proportional to what we now call velocity gradient, and he also made experiments on the drag of spheres.

The mathematical science of fluid mechanics—hydrodynamics—was due to four eighteenth-century mathematicians: Daniel Bernoulli and Loenhard Euler (Swiss) and Clairaut and d'Alembert (French). These were followed by Lagrange (1736–1813), Laplace (1749–1827), and an engineer, Gerstner (1756–1832), who contributed ideas on surface waves.

Experimentalists of the eighteenth century added much. These men included Poleni, who derived an equation for weir flow; de Pitot, who developed a tube for measuring velocities; Chézy, who developed a resistance formula for open channels; Borda, who performed experiments on resistance with rotating arms and analyzed flow through orifices; Bossut, who built a towing tank; Du Buat, who pioneered the French school of hydraulics; and Venturi, who experimented with flow in changing cross sections.

In the nineteenth century, the Frenchmen Coulomb (1736–1806) and Prony (1755–1839) conducted tests and drew conclusions regarding flow resistance; the German brothers Ernst (1795–1878) and Wilhelm Weber (1804–1891) conducted tests on wave motion; the French engineers Burdin (1790–1873), Fourneyman (1802–1867), Coriolis (1792–1843), and the American engineer Francis (1815–1892) contributed toward the development and analysis of hydraulic turbines; the Scotsman Russell (1808–1882) and the Alsatian Reech (1805–1880) conducted tests on waves and towed ship models; the Englishman Smith (1808–1874) and the Swede Ericsson (1803–1889) developed the screw propeller; the German Hagen (1797–1889), the Frenchman Poiseuille (1799–1869), and the Saxon Weisbach (1806–1871) did extensive work on pipe flow; the Frenchman Saint-Venant (1797–1886) analyzed the sonic orifice and contributed to open channel hydraulics; the

Frenchmen Dupuit (1804–1866), Bresse (1822–1883), and Bazin (1829–1917) and the Irishman Manning (1816–1897) did extensive work in open channel hydraulics; the Frenchman Darcy (1803–1858) did work on pipe flow and percolating flow; the German Lilienthal (1848–1896) and the Englishmen Phillips (1845–1912) and Lanchester (1868–1946) did extensive pioneering work on the lift of vanes, Lanchester introducing a theory of lift; and the Englishmen William Froude (1810–1879) and his son Robert (1846–1924) did extensive ship model testing.

Classical and applied hydrodynamics were advanced during the nineteenth century by Navier (1785–1836), Cauchy (1789–1857), Poisson (1781–1840), Saint-Venant, and Boussinesq (1842–1929) in France; Stokes (1819–1903), Airy (1801–1892), Reynolds (1842–1912), Lord Kelvin (1824–1907), Lord Rayleigh (1842–1919), and Lamb (1849–1934) in Great Britain; Helmholtz (1821–1894) and Kirchoff (1824–1887) in Germany; and Joukowsky (1847–1921) in Russia.

At the end of the nineteenth century, theoretical hydrodynamics, based on Euler's equations of motion for an ideal (nonviscous) fluid, had reached a comparatively high level of development. It did not explain, however, many observed effects, such as the pressure drop in pipes, and thus practicing engineers developed their own empirical science of hydraulics. These two fields—hydrodynamics and hydraulics—had much too little in common at that time. In 1904 Prandtl (1875–1953) in Germany introduced the concept of a boundary layer, a thin region adjacent to a boundary, in which the viscous effects were concentrated. This proved to be the concept which unified modern fluid mechanics—aerodynamics, hydraulics, gas dynamics, and convective heat transfer. It explained the differences in the behavior of real fluids as observed by hydraulicians and the predictions from the theory of nonviscous fluids by classical hydrodynamicists. Prandtl is properly considered to be the father of modern fluid mechanics.

Progress during the present century has included both analytical and experimental studies of boundary layer flow and control, turbulence structure, flow stability, multiphase flows, heat transfer to and from flowing fluids, flow of rarified gases, magnetohydrodynamics and plasmas, and many applied problems. It is the duty of engineers, even though they may not be engaged in research, to keep abreast of current developments in order that they may be applied expeditiously in engineering designs.

1-4. Symbols and Units

Fluid mechanics encompasses the fields of classical hydrodynamics, hydraulics, boundary layer theory, and gas dynamics, and it forms integral parts of thermodynamics, heat transfer, chemical engineering, and theoretical and applied mechanics. No single system of symbols or units is common to all these fields.

A list of symbols which is a composite of those recommended by the

American Standards Association (ASA) in various fields is given in Appendix I. The only exception involves dimensionless numbers named in honor of individuals who have distinguished themselves in their respective fields. For example, the ASA recommends that N_R be used for the Reynolds number. But when subscripts (and in some instances, subscripts *on* subscripts) are needed this becomes cumbersome. More commonly used are (usually) the first two letters of the individual's name, such as Re for the Reynolds number and Pr for the Prandtl number. The sole exception is M for the Mach number in gas dynamics.

Physical quantities may be designated in terms of a few primary dimensions such as force, length, time, and temperature (F, L, T, and θ), or as mass, length, time, and temperature (M, L, T, and θ).[†] Many physical quantities require more than one primary dimension in order that they may be described. Each one of these primary dimensions, in turn, may be specified by any one of a variety of units. The dimension of length, for example, may be expressed in units of inches, feet, meters, fathoms, rods, or miles, to cite just a few possibilities.

Newton's second law states that force is proportional to time rate of change of momentum. Thus we may write $F = ma/g_c$, where g_c is a factor of proportionality. In this equation a coherent system of units is one in which a unit force produces a unit acceleration on a unit mass. A number of coherent systems of units is tabulated in Table 1-1.

Table 1-1. COHERENT SYSTEM OF UNITS

SYSTEM	FORCE	MASS	LENGTH	TIME	g_c
Absolute English	poundal (pdl)	pound (lb_m)	ft	s	1 lb_m ft/pdl s^2
Absolute metric (cgs)	dyne	gram (g)	cm	s	1 g cm/dyn s^2
Technical English	pound (lb_f)	slug	ft	s	1 slug ft/lb_f s^2
SI	newton (N)	kilogram (kg)	meter (m)	s	1 kg m/N s^2

None of these coherent systems of units involves the acceleration of gravity. The technical English system has been used for many years in fluid mechanics, hydraulics, aerodynamics, and engineering mechanics in North America. The fourth system has been used in university physics courses in North America and is part of the *Système International d' Unités* (SI) and has been adopted by the International Organization for Standardization. It has been adopted in many countries, and it may well become generally accepted throughout the world.

The fourth coherent system, the SI system, will be used for numerical examples and problems in this text (exceptions occur in Chapter 14). The g_c may be omitted when writing equations, since its numerical value is unity in a coherent system. This is equivalent to defining

$$1 \text{ N} = 1 \text{ kg m/s}^2 \qquad \text{or} \qquad 1 \text{ kg} = 1 \text{ N s}^2/\text{m}$$

[†] See Chapter 8 for further details.

It will be noted that many equations, tables, and graphs in fluid mechanics contain nondimensional parameters (peruse through Chapter 7 and on throughout the rest of the text). In these instances the system of units used is of no consequence and the equations, tables, and graphs are valid for any consistent system of units.

Equations relating physical quantities should be correct regardless of the system of units used to specify numerical values of the parameters in the equations. If a coherent system of units is used, conversion factors such as g_c or J (the mechanical equivalent of heat) need not appear in the equations. Thus, numerical calculations may be made directly. If known quantities are expressed in N-kg-m-s units in the SI system, the unknown quantity will be in those units.

Some conversion factors are given near the end of the book in Table A-1 (Appendix II). A list of frequently used metric units and unit symbols is given in Table 1-2.

Example 1-1

The temperature rise when a high-speed gas is brought to rest without heat transfer is $\Delta T = V^2/2c_p$, where c_p is the specific heat capacity at constant pressure and V is the speed. For air at 457 m/s, what is the temperature rise when it is brought to rest?

SOLUTION

In the SI system $c_p = 1005$ J/kg $^\circ$K $= 1005$ m^2/s^2 $^\circ$K and

$$\Delta T = \frac{(457 \text{ m/s})^2}{2(1005 \text{ m}^2/\text{s}^2 \text{ }^\circ\text{K})} = 104^\circ\text{C}$$

Example 1-2

Fluid velocity often is measured indirectly by measuring the pressure rise when the fluid is brought to rest. Then $V = \sqrt{2\Delta p/\rho}$, where Δp is the pressure rise and ρ is the fluid density. The measured pressure rise for water is 7200 Pa. What is the velocity being measured?

SOLUTION

In the SI system, $\rho = 1000$ kg/m^3. Then

$$V = \sqrt{\frac{2(7200 \text{ N/m}^2)}{1000 \text{ kg/m}^3}} = 3.79\sqrt{\text{m N/kg}}$$

$$= 3.79\sqrt{\text{kg m}^2/\text{s}^2 \text{ kg}} = 3.79 \text{ m/s}$$

Table 1-2.   UNITS AND UNIT SYMBOLS

QUANTITY	UNIT	SYMBOL
Acceleration	meter per second squared[a]	m/s^2
Area	square centimeter	cm^2
	square meter[a]	m^2
Density	kilogram per cubic meter[a]	kg/m^3
	gram per liter	g/L
Energy, work, heat	joule[a]	J
	kilojoule	kJ
	megajoule	MJ
	kilowatt-hour (3.6 MJ)	kW h
Force	newton[a]	N
	kilonewton	kN
Length	millimeter	mm
	centimeter	cm
	meter[a]	m
	kilometer	km
Mass	kilogram[a]	kg
	gram	g
	milligram	mg
Plane angle	degree	°
Power or heat flow rate	watt[a]	W
	kilowatt	kW
Pressure	pascal[a] (N/m^2)	Pa
	kilopascal	kPa
Rotational speed	radians per second	rad/s
	revolution per second	r/s
	revolution per minute	r/min
Temperature	degree Celsius	°C
	degree Kelvin[a]	°K
Time	second[a]	s
	minute	min
	hour	h
	day	d
Volume	cubic meter[a]	m^3
	liter	L
	cubic centimeter	cm^3
	milliliter	mL

[a] SI units.

THERMODYNAMICS

A treatment of fluid mechanics, especially one which includes gas flows, requires an inclusion of thermodynamic principles. Only those concepts and equations which apply to fluid mechanics, or which will simplify the treatment, will be reviewed.

Fundamental in this discussion is the concept of a *system* and of a *control volume*. A system contains a fixed quantity of matter upon which we wish to focus our attention. The region outside the system boundaries is called its surroundings and could be a part of another system. In dealing with fluid flows it is generally more convenient to deal with a fixed region in space rather than with fixed particles of matter. This fixed region is called a *control volume*; it is bounded by a *control surface*. Fluid in motion passes through the control surface so that the fluid particles that occupy the control volume are in general different from one instant to another.

The laws of conservation of mass, Newton's laws of motion, and the first and second laws of thermodynamics are generally formulated for a system of fixed mass particles, and it will be necessary to reformulate them to apply to a control volume.

1-5. Temperature, Heat, and Work

If two systems or bodies, one which feels hot and one cold, are brought into contact, after some time they will feel less hot and less cold, respectively, and eventually they will be in a state of equilibrium where no further changes take place. They are then said to be at the same *temperature*. Arbitrary scales of temperature may be defined. One example is the Fahrenheit scale, defined as 32°F at the temperature of an equilibrium mixture of pure ice and air-saturated water (ice point) and 212°F at the temperature of an equilibrium mixture of pure water and water vapor at a pressure of one atmosphere (the steam point). These points on the Celsius scale are 0°C and 100°C, respectively. Absolute temperatures are defined by

degrees Kelvin = degrees Celsius + 273.15

or

$$°K \approx °C + 273 \tag{1.1}$$

The temperature of a substance is a property of that substance; that is, it is a characteristic of the equilibrium state.

If a system or fluid in a control volume is not in thermal equilibrium with the surroundings, then energy will pass through the boundaries of the system or through the control surface. The energy which is being transferred is called *heat*. We shall consider heat transfer to be positive if it is added *to* the system or *to* fluid as it passes through a control volume under consideration, and negative if heat is removed. Heat is *not* a property of a substance since it is energy in transition. The quantity transferred will be designated by Q, and in terms of energy per unit mass, by q.

Work flows from one system or a control volume if the only external effect is equivalent to the raising of a weight (exerting a force through a distance). We will consider work to be positive if it is done *by* the system or *by* fluid as it passes through a control volume, and negative if work is done on it. Examples of positive work are an expansion of the boundaries of a system, and work done by a fluid as it passes through a turbine runner. Work done on fluid as it passes through a control volume such as a pump or compressor impeller is considered negative work. Work, like heat, is *not* a property of a substance, since work, too, is energy in transition. Energy may be defined as the capacity to do work. Work transferred will be designated by W; if measured in terms of energy per unit mass, w will be used.

1-6. The First Law of Thermodynamics

The first law of thermodynamics is an expression of the conservation of energy. Whenever energy is transferred across the boundaries of a system, the energy of the system changes by an equal amount. The increase in energy of the system is equal to the decrease in energy of its surroundings, and vice versa.

One form of the first law states that if work is added to a substance in a closed system with no heat transfer to or from the system, the amount of work added depends only upon the end states of the process. Thus we may define a property E whose change represents this work. For a unit mass of substance,

$$e_1 - e_2 = w'_{1-2}$$

The quantity e may be called the *energy content*. It consists of kinetic energy, potential energy, and internal energy.

If the process is accompanied by a transfer of heat, then the heat added to the substance within the system is the difference between the actual work added and the work added during a no-heat-transfer process between the same two end states. This may be expressed (per unit mass) as

$$q_{1-2} = w_{1-2} - w'_{1-2}$$

It follows that

$$q_{1-2} = e_2 - e_1 + w_{1-2} \tag{1.2}$$

For a closed system (the mass of the system is constant), the energy content consists of kinetic energy, potential energy, and internal energy u. For the closed system with negligible kinetic and potential energy

$$q_{1-2} = u_2 - u_1 + w_{1-2} \tag{1.3}$$

We will be concerned only with energy conversions between the various forms of mechanical energy and thermal energy and will not consider energy involved in chemical and nuclear reactions, or work done as a result of capillary, electric, or magnetic effects.

Equations (1.2) and (1.3) express the first law of thermodynamics, relating heat, work, and energy content. A more complete treatment of energy content is given in the development of the steady-flow equation for a control volume in Chapter 5.

1-7. The Second Law of Thermodynamics

The first law of thermodynamics contains no restrictions regarding the direction in which changes in the form of energy may take place. Thus mechanical energy may be transformed into thermal energy, and thermal energy may be transformed into mechanical energy, according to the first law. The second law of thermodynamics, however, places restrictions on the direction of these energy transformations. For example, mechanical energy dissipated into thermal energy as a result of viscous shear in a real fluid cannot be recovered as mechanical energy; the process is irreversible according to the second law.

Of the many ways of expressing the second law of thermodynamics, the most useful in fluid mechanics is expressed as an axiom known as the inequality of Clausius. This states that if a system undergoes a complete cyclic process (the substance within the system is brought back to its original equilibrium state), the integral of dQ/T is equal to or less than zero:

$$\oint \frac{dQ}{T} \leq 0$$

If the cycle is performed reversibly, the integral equals zero. A cyclic reversible process leaves no trace of its occurrence either within the system or in its surroundings—both are brought back to their respective initial states. No real or actual process is reversible because of friction, viscous shear, temperature difference, and so forth, although some real processes are very nearly reversible and may be considered to be so for purposes of analysis. For real processes, then, the integral is less than zero.

Another second law axiom states that the integral of dQ/T for a reversible process between a reference condition or state and some final condition or state defines the change in a new property, and this property is called *entropy*. If entropy per unit mass is designated as s,

$$ds = \frac{dq_{rev}}{T}$$

From the first law for a closed system [Eq. (1.3)]

$$dq = du + dw$$

so that

$$ds = \frac{du + p\,dv}{T} \tag{1.4}$$

when only $p\,dv$ work is done. It follows that for a reversible process wherein no heat is transferred to or from the system (an *adiabatic* process), $ds = 0$ and for an irreversible adiabatic process, $ds > 0$.

Another property called the *enthalpy* H (or h per unit mass) is defined as the sum of the internal energy and the product of pressure and specific volume:

$$h = u + pv \tag{1.5}$$

In differential form, $dh = du + p\,dv + v\,dp$ so that Eq. (1.4) may be written as

$$T\,ds = dh - v\,dp \tag{1.6}$$

Equation (1.6) is perhaps the most useful form or expression of the first and second laws of thermodynamics for our purpose.

1-8. The Perfect Gas

A perfect gas is one which obeys the equation of state given by

$$pv = RT \quad \text{or} \quad p = \rho RT \tag{1.7}$$

in which R is a constant for any particular gas and is independent of pressure and temperature, p the absolute pressure, v the specific volume, and ρ the density. For many engineering applications, it is sufficiently accurate to assume many gases to behave according to the perfect gas equation. [The term "perfect gas" should not be confused with "ideal fluid." A perfect gas obeys Eq. (1.7); an ideal fluid is nonviscous.] Values of the gas constant together with other properties are listed for a number of gases in Table A-2 (Appendix II).

An indication of the accuracy of the perfect gas equation when applied to the mixture of gases known as air is shown in Fig. 1-1, where values of the compressibility factor Z ($Z = pv/RT$) are shown as a function of pressure and temperature. The shaded area indicates the range of pressures and temperatures in which the perfect gas equation is accurate within 1 percent.

The so-called specific heat capacities at constant pressure c_p and at constant volume c_v are defined as

$$c_p = \left(\frac{\partial h}{\partial T}\right)_p$$

or the change of enthalpy with respect to temperature at constant pressure and

$$c_v = \left(\frac{\partial u}{\partial T}\right)_v$$

or the change of internal energy with respect to temperature at constant volume. For a perfect gas, c_v is a function of temperature,

$$du = c_v\,dT$$

and if c_v is a constant,

$$u_2 - u_1 = c_v(T_2 - T_1) \tag{1.8}$$

Thus for *any* process or path between states 1 and 2 the internal energy change per unit mass equals c_v times the temperature change.

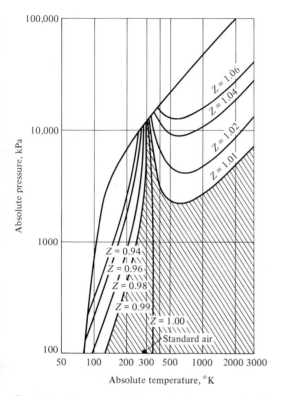

Fig. 1-1.   Accuracy of the perfect gas equation for air. [Source: Data from Newman A. Hall and Warren E. Ibele, "The Tabulation of Imperfect-Gas Properties of Air, Nitrogen, and Oxygen," *Trans, ASME*, Vol. 76 (Oct. 1954), pp. 1039–1056. Used with permission of the publishers, The American Society of Mechanical Engineers.]

Also, $dh = c_p\, dT$ and if c_p is a constant,

$$h_2 - h_1 = c_p(T_2 - T_1) \tag{1.9}$$

Thus for *any* process or path between states 1 and 2 the enthalpy change per unit mass equals c_p times the temperature change.

From the definition of enthalpy it can be shown that, for a perfect gas,

$$c_p - c_v = R \tag{1.10}$$

We also may define a ratio for all substances as

$$k = \frac{c_p}{c_v} \tag{1.11}$$

so that for a perfect gas

$$c_v = \frac{R}{k - 1} \tag{1.12}$$

and

$$c_p = \frac{Rk}{k-1} \tag{1.13}$$

Some typical values of the gas constant, specific heats, and the specific heat ratio are listed in Table A-2 (Appendix II).

1-9. Flow Processes

Substances (fluids) may undergo changes in state via a number of different paths or processes. Those already mentioned and some additional ones will be listed and defined for ready reference:

A system is *isothermal* if its temperature remains constant. The equation of state then indicates that $pv = $ constant for a perfect gas.

A system is *isobaric* if its pressure remains constant.

A system or process is *adiabatic* if no heat is transferred to or from the system to its surroundings. For this process, $dq = 0$.

If the system or process is adiabatic and changes occur reversibly (often called a frictionless adiabatic process), then the process is called *isentropic*. The entropy during this process remains constant, and $ds = 0$. In special situations, the flow may be isentropic but neither adiabatic nor reversible. For example, if a real fluid flows through a pipe, there is an increase in internal entropy due to viscous shear or turbulence. If heat were removed in a manner such that the decrease in fluid entropy from the heat removed just balanced the increase in fluid entropy from internal causes, the flow could occur at constant entropy and be isentropic. However, it would be neither adiabatic (heat is transferred) nor reversible (internal friction exists). From Eqs. (1.6), (1.7), and (1.13) for an *isentropic* process or flow,

$$p_1 v_1^k = p_2 v_2^k = \text{constant} \tag{1.14a}$$

$$\frac{T_2}{T_1} = \left(\frac{p_2}{p_1}\right)^{(k-1)/k} \tag{1.14b}$$

and

$$\frac{\rho_2}{\rho_1} = \left(\frac{p_2}{p_1}\right)^{1/k} \tag{1.14c}$$

A *polytropic* process is a general one for which

$$p_1 v_1^n = p_2 v_2^n = \text{constant} \tag{1.15}$$

where $n = 1$ for an isothermal process, $n = 0$ for a constant-pressure process, and $n = k$ for an isentropic (reversible or frictionless adiabatic) process.

Equations (1.14) apply to polytropic processes with k replaced by $n \neq 0$.

FLUID PROPERTIES

All real fluids have or exhibit certain measurable characteristics or properties that are of engineering importance. Fluid density, compressibility, capillarity, and vapor pressure may be of interest for fluids at rest; and in addition to these, viscosity is significant for real fluids in motion. The thermal properties, such as gas constant, internal energy, enthalpy and entropy for gases, and specific heat capacities and conductivity for both gases and liquids, are important in heat transfer and gas dynamics. Pressure and temperature are considered to be thermodynamic properties, though they might be considered as independent properties upon which the (other) tabular properties depend. Some fluid properties are actually combinations of the properties already mentioned. Thermal diffusivity, for example, involves thermal conductivity, density, and specific heat capacity at constant pressure; the Prandtl number involves viscosity, specific heat capacity at constant pressure, and thermal conductivity (or fluid viscosity and thermal diffusivity); and kinematic viscosity involves dynamic viscosity and density.

Even though all fluids are composed of discrete particles, we shall consider them to have properties of a continuum. The gross properties of a continuum depend, of course, on the molecular structure of the fluid and on the nature of the intermolecular forces. Molecular spacing in gases is an order of magnitude larger than the size of the individual molecules, whereas in liquids it is of the same order as the size of the molecules. Analyses based on this molecular structure are studied in the statistical mechanics of fluids. The behavior of fluids, however, generally may be examined on a *macroscopic scale*, which is much larger than a microscopic or molecular scale. Truesdell states [2], "By considering a fluid or a solid as an assembly of molecules, it is impossible to derive anything in contradiction to the view of matter as continuous. Conversely, there is nothing in the continuum view of matter that is not also accessible to a molecular picture." Thus we may assume that fluids behave as though they were continuous in their structure.

1-10. Density, Specific Volume, and Specific Gravity

The *density* ρ of a substance is a measure of the concentration of matter and is expressed in terms of mass per unit volume; it is determined by taking the ratio of the mass of a substance contained within a particular region divided by the volume of this region. The region should be both small enough and yet large enough so that there are no significant variations in density in subregions within it. Near the nose of a high-speed missile, for example, the concentration of matter may vary tenfold within a region much less than 1 cm in linear dimension. Thus the measured region should not be too large. If the measured region is too small, however, it may contain a different number of molecules at different instances.

This is illustrated in Fig. 1-2. As the volume δV containing a fluid mass δm is reduced in size about some point P, the ratio $\delta m / \delta V$ reaches a limiting

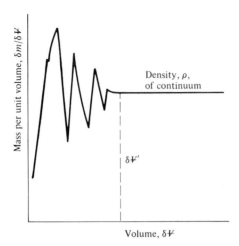

Fig. 1-2. Fluid density about a point.

value ρ. If the volume δV is further reduced, molecular effects appear and the volume may contain a different total mass of molecules δm at various instants. Thus in Fig. 1-2

$$\rho = \lim_{\delta V \to \delta V'} \frac{\delta m}{\delta V}$$

The density of water at room temperature is about 1000 kg/m³. The density of standard air, defined at an absolute pressure of 1.013×10^5 Pa and a temperature of 15°C, is 1.225 kg/m³.

Temperature and pressure have a slight effect on the density of liquids, and a pronounced effect on the density of gases.

The *specific volume v* is the volume occupied by a unit mass of a substance and thus is the reciprocal of density:

$$v = \frac{1}{\rho} \tag{1.16}$$

Specific weight γ is the force of gravity on the mass contained in a unit volume of a substance. Thus

$$\gamma = \rho g \tag{1.17}$$

For a water density of 1000 kg/m³ and the acceleration of gravity $g = 9.81$ m/s², the specific weight of water is

$$\gamma = (1000 \text{ kg/m}^3)(9.81 \text{ m/s}^2) = 9810 \text{ N/m}^3$$

Strictly speaking, specific weight is not a true fluid property, since it depends on the value of the local gravitational acceleration. Hydrostatic forces exerted by fluids do depend on gravity, however, and it is customary to use the specific weight in calculations which involve them.

Specific gravity s is a term used to compare the density of a substance with that of water. Since the density of all liquids depends on temperature as well as pressure, the temperature of the liquid in question, as well as the reference temperature of water, should be stated in giving precise values of specific gravity:

$$s = \frac{\rho}{\rho_w} \qquad (1.18)$$

Values of density or specific gravity of liquids may be found in numerous references [3]. Some typical values for various liquids are given in Tables A-3 and A-4 (Appendix II). Detailed data for water are given in Table A-5 (Appendix II).

The densities of gases may be calculated from any one of various equations of state as a function of pressure and temperature. For a perfect gas,

$$\rho = \frac{p}{RT} \qquad [1.7]$$

Typical values of the gas constant R for various gases are given in Table A-2 (Appendix II).

Example 1-3

Compute the density of air at a pressure of 14.0×10^5 Pa abs and a temperature of 40°C.

SOLUTION

$$\rho = \frac{14.0 \times 10^5 \text{ Pa abs}}{(287.1 \text{ J/kg } °\text{K})(313°\text{K})}$$

$$= 15.58 \text{ kg/m}^3$$

Figure 1-1 indicates that $Z = 1.00$ and thus the perfect gas equation is essentially exact at the given pressure and temperature.

1-11. Compressibility or Elasticity

Fluids may be deformed by viscous shear or compressed by an external pressure applied to a volume of fluid. All fluids are compressible by this method, liquids to a much smaller degree, however, than gases.

The compressibility is defined in terms of an average bulk modulus of elasticity

$$\bar{K} = -\frac{p_2 - p_1}{(V_2 - V_1)/V_1} = \frac{\Delta p}{\Delta V/V} \qquad (1.19)$$

where V_2 and V_1 are the volumes of the substance at pressure p_2 and p_1, respectively. The bulk modulus varies with the pressure for gases, and with both pressure and temperature (though but slightly) for liquids. Thus, the true bulk modulus of elasticity is the limiting value of Eq. (1.19) when the pressure and volume changes become infinitesimal.

$$K = -\frac{dp}{dV/V} \tag{1.20a}$$

If a unit mass of substance is considered,

$$K = -\frac{dp}{dv/v} \tag{1.20b}$$

and

$$K = +\frac{dp}{d\rho/\rho} \tag{1.20c}$$

The denominators of Eqs. (1.19) and (1.20) are dimensionless, so that K has the dimensions of pressure, or force per unit area. Equation (1.19) is used for liquids, and Eq. (1.20) for gases, though there are special forms for isothermal and for isentropic compression, as will be shown.

The value of K for water at $20°$C is about 2.18×10^9 Pa at atmospheric pressure, and it increases essentially linearly to about 2.86×10^9 Pa at a pressure of 1000 atm [3]. Thus in this range at $20°$C

$$K = (2.18 \times 10^9 + 6.7p)\text{Pa} \tag{1.21}$$

where p is the gage pressure in Pa. The value of K for water varies a few percent with temperature changes over the range from $0°$ to $100°$C.

The change in volume of a gas for a change in pressure depends on the compression process. If *isothermal* (constant temperature), the gas equation may be expressed in logarithmic form as $\ln p = \ln \rho + \ln(RT)$ and differentiated to obtain $dp/p = d\rho/\rho$. Thus

$$K_{\text{isothermal}} = \frac{dp}{d\rho/\rho} = \frac{dp}{dp/p} = p \tag{1.22}$$

and the elastic modulus equals the absolute pressure during an isothermal compression.

If the compression is adiabatic and is carried out slowly so that equilibrium conditions exist, the compression may be considered reversible and adiabatic, or *isentropic* (see Sec. 1-9). For this process, $p/\rho^k = $ constant (k is the ratio of specific heat capacities). The logarithmic form, $\ln p - k \ln \rho = \ln C$, may be differentiated to obtain $dp/p = k \, d\rho/\rho$. Thus

$$K_{\text{isoentropic}} = \frac{dp}{d\rho/\rho} = \frac{dp}{dp/kp} = kp \tag{1.23}$$

and the elastic modulus equals the absolute pressure times the ratio of specific heat capacities ($k = c_p/c_v = 1.4$ for air) during an isentropic compression.

The bulk modulus of elasticity K is of interest in acoustics as well as in fluid mechanics. The velocity of sound in any medium is

$$c = \sqrt{\frac{K}{\rho}}$$

(1.24)

and for a gas, sound waves are transmitted essentially isentropically (see Sec. 10-1), so that the velocity of sound in a perfect gas is

$$c = \sqrt{\frac{kp}{\rho}} = \sqrt{kRT}$$

(1.25a)

Example 1-4

What is the speed of sound in water at 20°C and at atmospheric pressure?

SOLUTION
In Eq. (1.24), $K = 2.18 \times 10^9$ Pa and $\rho = 998$ kg/m^3 from Table A-5 (Appendix II) so that

$$c = \sqrt{(2.18 \times 10^9)/998} = 1478 \text{ m/s}$$

Example 1-5

What is the speed of sound in air at 20°C at sea level and at an altitude where the pressure is less than that at sea level? Refer to Table A-2 (Appendix II) for gas properties.

SOLUTION
Equation (1.25a) indicates that the speed of sound is independent of pressure, and thus

$$c = \sqrt{kRT} = \sqrt{(1.4)(287.1)(293)} = 343 \text{ m/s}$$

The speed of sound in a mixture of liquid and tiny gas bubbles may be calculated from Eq. (1.24). Results are valid for nonresonant sound frequencies. Let subscripts m, 1, and g refer to property values for the mixture, liquid, and gas, respectively. For a nonresonant mixture, sound is considered to be propagated at constant temperature at low gas concentrations, and the sound speed is given by $c_m = \sqrt{K_m/\rho_m}$. If x is the proportion of gas by volume, the density of the mixture is

$$\rho_m = x\rho_g + (1 - x)\rho_l$$

and the elastic modulus K_m for the mixture is given by

$$\frac{1}{K_m} = \frac{x}{K_g} + \frac{1-x}{K_l}$$

and thus

$$c_m = \sqrt{\frac{p_g K_l}{[xK_l + (1-x)p_g][x\rho_g + (1-x)\rho_l]}} \qquad (1.25b)$$

1-12. Surface Tension and Capillarity

Very small drops of liquid in a gas and very small gas bubbles in a liquid assume a spherical shape in the absence of extraneous forces such as viscous shear. If a spoon is held under a dripping faucet, the water may rise much above the upper edge of the spoon before spilling. Similarly, water may be poured into a clean glass to a level above the lip of the glass. If a small-bore clean glass tube is placed vertically into a free water surface, the water will rise in the tube. If the liquid is mercury the mercury will be depressed within the tube. Clean plastic tubing similarly placed in water may result in a rise of water within the tube, a depression within the tube, or no rise or depression of the water column, depending on the type of plastic tubing used. Hysteresis effects may also be exhibited—a falling liquid column may come to rest at a different position than a rising liquid column. In manometers (Sec. 2-3) mercury under water results in a well-defined, clean interface or meniscus, whereas carbon tetrachloride under water produces a poorly defined, unclear interface or meniscus. All these are examples of the effects of the *surface tension* of liquids.

This property called surface tension is actually a result of a difference in mutual attraction between liquid molecules near a surface as compared with that farther within the mass of liquid. Thus work is done in bringing molecules to this surface, and the creation of a free surface requires an expenditure of energy. This energy per unit area of surface is called the *surface tension*, designated by σ. Surface tension σ has dimensions of energy per unit area or force per unit length. This force may be considered to be in a direction normal to any line drawn on the interface and in the plane of the interface, as shown in Fig. 1-3.

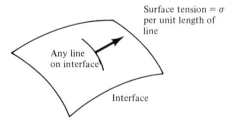

Fig. 1-3.   Surface tension force on an interface.

Consider a spherical drop of liquid or a spherical gas bubble in a liquid, having a radius r. If the bubble expands to a radius $(r + dr)$ from an addition of mass, the increase in energy is $\sigma\, dA = \sigma\, d(4\pi r^2) = \sigma(8\pi r)\, dr$, when there is no change in temperature or density of the phases at the interface. This is equal to the work done by the force from the difference in pressure inside the bubble and the pressure outside times the distance dr the force moves, or $\Delta p(4\pi r^2)\, dr$. Equating the increase in energy to the work done gives $\Delta p = 2\sigma/r$.

For a general curved surface where r_1 and r_2 are the principal radii of the surface, it may be shown that

$$\Delta p = \sigma\left(\frac{1}{r_1} + \frac{1}{r_2}\right) \tag{1.26}$$

For a sphere, $r_1 = r_2$, and from Eq. (1.26), $\Delta p = 2\sigma/r$, as above. For a cylinder one radius of curvature is infinite, so that $\Delta p = \sigma/r$. (Both these results easily may be obtained by making a simple force balance.)

When a liquid–gas interface is in contact with a solid surface, three interface forces exist: between gas and liquid, between gas and solid, and between liquid and solid. Thus in Fig. 1-4, equilibrium gives the following scalar condition:

$$\sigma_{gs} = \sigma_{sl} + \sigma_{gl} \cos \theta$$

from which the contact angle θ may be determined. A liquid in air is said to wet a surface when $\theta < \pi/2$, and the degree of wetting increases as θ decreases toward zero. The contact angle θ for water, air, and clean glass is essentially zero degrees. When $\theta > \pi/2$, the liquid is said to be nonwetting. Mercury, for example, has a contact angle θ of about 130–150° for many surfaces. Water drops on waxed surfaces or on surfaces treated to be non-wetting (some fabrics are so treated) have contact angles greater than $\pi/2$.

The rise of a column of liquid in a small tube is due to surface tension and is referred to as *capillary* action. In Fig. 1-5, for example, the weight of the liquid column in the tube, the force from the pressure difference across the liquid–gas interface times the tube area, and the peripheral force around the tube circumference are all equal. This may be written as

$$\rho g h(\pi r^2) = \Delta p \pi r^2 = \sigma 2\pi r \cos \theta$$

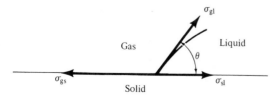

Fig. 1-4.   Contact angle for gas–liquid–solid interface.

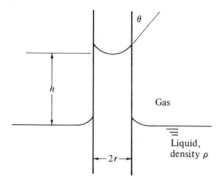

Fig. 1-5. Capillary rise of a liquid in a small circular tube.

from which the capillary rise is

$$h = \frac{2\sigma}{gr\rho} \cos \theta \qquad (1.27)$$

When $\theta < \pi/2$, a capillary rise occurs; when $\theta = \pi/2$, there is no rise or depression of the liquid inside the tube; and when $\theta > \pi/2$, there is a depression of the liquid inside the tube. These are illustrated in Fig. 1-6.

Attempts are generally made to avoid effects of surface tension. Manometers (Sec. 2-3) contain liquids that are chosen to have a readable interface. Liquids having a high surface tension between them are best (for mercury under water $\sigma = 0.375$ N/m, and for carbon tetrachloride under water $\sigma = 0.045$ N/m). Hydraulic models are made large enough so that shallow depths, which otherwise might be affected by surface tension, are avoided. Wetting agents may be used to reduce these effects if the size of the model is made as large as possible, yet smaller than would be desirable. Surface tension plays a role in deterring the growth of small gas nuclei in liquids when they pass through low-pressure regions, yet the precise role is not understood. This rapid growth and collapse of bubbles in liquids is one form of cavitation.

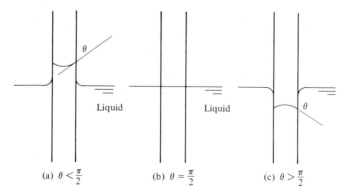

Fig. 1-6. Effect of contact angle θ on capillarity in a small circular tube.

Forces due to surface tension are, with the preceding exceptions, generally small compared with the forces due to gravity, viscosity, and pressure in engineering practice.

Typical values of surface tension for some liquids are given in Table A-6 (Appendix II). The surface tension of water in air (about 0.073 N/m) may be reduced to about one-half by the addition of wetting agents.

Example 1-6

To what height h (Fig. 1-5) would water at room temperature rise in a clean glass tube 2.5 mm in diameter?

SOLUTION

$$h = \frac{2\sigma \cos \theta}{gr\rho}, \quad \text{where} \quad \theta = 0°$$

$$= \frac{(2)(0.073)(1)}{(9.81)(0.00125)(1000)}$$

$$= 0.012 \text{ m} = 12 \text{ mm}$$

1-13. Vapor Pressure

If a liquid and its vapor coexist in equilibrium, the vapor is called a saturated vapor, and the pressure exerted by this saturated vapor is called the *vapor pressure*. The vapor pressure is a function of temperature for a given substance. The vapor pressure of liquids is of practical importance in barometers, pump-piping systems, and from an elementary point of view, in the formation of cavities in low-pressure regions within a liquid. These are not uncommon in pumps, hydraulic turbines, and on boat and ship propellers.

Values of vapor pressure for some liquids at various temperatures are shown in Table A-7 (Appendix II).

1-14. Viscosity

The viscosity of a fluid is a measure of its resistance to flow. It is this property of all real fluids that distinguishes them from ideal, or nonviscous, fluids. Shear resistance is measured as total shear force, a unit shear stress being the shear per unit area subjected to shear.

The viscosity of a gas increases with temperature because of the greater molecular activity as the temperature increases. The kinetic theory of gases (no intermolecular forces are considered) shows that as gas molecules move in random directions superimposed on the mean fluid motion, they collide with other molecules in adjacent fluid layers. The molecules with which they collide move (on the average) at a different fluid velocity, and the collisions will increase or decrease their mean fluid motion depending on whether they collide with faster or slower molecules. This interchange of molecular momentum is manifested as fluid viscosity. A perfect gas, then, may have viscosity and would be called a perfect real gas, as contrasted with a perfect ideal, or nonviscous, gas. It may be shown that the viscosity of a perfect gas is linearly related to the mean free path of the gas molecules. A measurement of gas viscosity, then, may be used to determine the molecular mean free path.

For liquids, molecular spacings are much smaller than for gases, and molecular cohesion is very strong. Increased temperatures decrease this molecular cohesion, and this is manifested as a reduction in liquid viscosity.

Newton postulated that the shear stress within a fluid is proportional to the spatial rate of change of velocity normal to the flow. This spatial rate of change of velocity is called the *velocity gradient*, which is also the time rate of angular deformation. In Fig. 1-7, the velocity u varies with the distance y from the boundary at location A, and the curve connecting the tips of the vectors representing the velocities is called the *velocity profile*. The velocity gradient at any value of y is defined as

$$\frac{du}{dy} = \lim_{\Delta y \to 0} \left(\frac{\Delta u}{\Delta y} \right)$$

and represents the reciprocal of the slope of the velocity profile curve shown in Fig. 1-7.

For a highly viscous fluid at low velocities, the fluid flows in parallel layers, and for this type of flow the shear stress τ at any value of y is

$$\tau = \mu \left(\frac{du}{dy} \right) \tag{1.28}$$

where μ is the factor of proportionality known as the *dynamic viscosity*.

Dynamic viscosity is seen to be the ratio of a shear stress to a velocity gradient and thus its dimensions are force times time per unit area or mass per unit length and time.

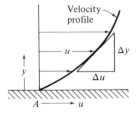

Fig. 1-7. Velocity profile and velocity gradient.

In the SI system of units, shear stress is expressed as N/m^2 (Pa) and velocity gradient as $(m/s)/m$ and thus the SI units for dynamic viscosity are

$$\mu = \frac{N/m^2}{(m/s)/m} = \frac{N\ s}{m^2} = \frac{kg}{m\ s}$$

since $1\ N = 1\ kg\ m/s^2$ as explained in Sec. 1-4.

Kinematic viscosity v is defined as the ratio of dynamic viscosity to density:

$$v = \frac{\mu}{\rho} \tag{1.29}$$

and has dimensions of area per unit time, and units of m^2/s in the SI system.

Conversion factors for viscosity in other units are listed in Table A-1 (Appendix II).

The kinematic viscosity of liquids is largely a function of temperature since this is the case for both dynamic viscosity and density. Thus values of both dynamic viscosity μ and kinematic viscosity v may be given in graphical or tabular form as a function of temperature. The kinematic viscosity of gases, however, is a function of pressure as well as temperature. The dynamic viscosity of gases is essentially dependent on temperature only, but the density of a gas is a function of both pressure and temperature and is calculated from a gas equation of state.

The dynamic viscosity of various liquids and gases is shown as a function of temperature in Fig. A-1 (Appendix II), and the kinematic viscosity of liquids (gas density depends to a greater extent on pressure) as a function of temperature in Fig. A-2 (Appendix II). More detailed data for water are given in Table A-8 (Appendix II).

Example 1-7

An oil (dynamic viscosity $\mu = 0.080\ kg/m\ s$ and density $\rho = 825\ kg/m^3$) flows along a surface with a velocity profile given by the equation $u = 50y - 10^4 y^2$ m/s, where y is the distance from the boundary surface in meters. What is the shear stress at the boundary surface?

SOLUTION

The velocity gradient at $y = 0$ is $(du/dy)_{y=0} = 50\ (m/s)/m$. Thus

$$\tau = \mu(du/dy)_{y=0} = (0.080)(50) = 4.0\ Pa$$

When a velocity profile is given, the velocity gradient easily may be obtained by differentiation.

For a linear velocity profile in a fluid between parallel surfaces a distance h apart, one of which is at rest and the other moving with a velocity V, the velocity gradient is uniform at V/h, as shown in Fig. 1-8a.

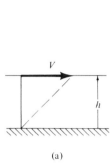 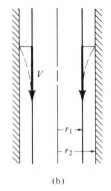 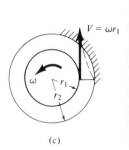

(a)　　　　　　　　　　　(b)　　　　　　　　　　　(c)

Fig. 1-8.　Velocity profiles for laminar motion. (a) Parallel plates. (b) Axial motion of shaft. (c) Rotating shaft.

When there is a relative motion in an axial direction between a circular shaft of radius r_1 and a sleeve of radius r_2 (uniform clearance of $r_2 - r_1$), the shearing force over the shaft is equal in magnitude to that inside the sleeve. For a sleeve of length L,

$$\mu\left(\frac{dV}{dr}\right)_1 (2\pi r_1)L = \mu\left(\frac{dV}{dr}\right)_2 (2\pi r_2)L$$

from which

$$\frac{(dV/dr)_1}{(dV/dr)_2} = \frac{r_2}{r_1}$$

This indicates that the velocity gradient at the shaft surface (at r_1) is greater than that on the surface of the sleeve (at r_2). This situation is shown in Fig. 1-8b.

If a shaft of radius r_1 rotates at an angular velocity ω in a bearing of radius r_2 with a uniform clearance, the torque on the shaft is equal in magnitude to that on the bearing. Thus for a bearing of length L

$$\mu\left(\frac{dV}{dr}\right)_1 (2\pi r_1)(r_1 L) = \mu\left(\frac{dV}{dr}\right)_2 (2\pi r_2)(r_2 L)$$

from which

$$\frac{(dV/dr)_1}{(dV/dr)_2} = \frac{r_2^2}{r_1^2}$$

which indicates, as before, that the velocity gradient at the shaft surface is greater than that on the surface of the bearing. This is shown in Fig. 1-8c.

In both situations shown in Fig. 1-8b and c, if the clearance is much less than the shaft radius such that the gap relatively is very small, that is, if $(r_2 - r_1) \ll r_1$, the velocity gradient may be considered to be $V/(r_2 - r_1)$, a straight line, with little error, generally no more than the uncertainty in the fluid viscosity.

NEWTONIAN AND NON-NEWTONIAN FLUIDS

A fluid for which the dynamic viscosity μ depends on temperature (and slightly on pressure) and is independent of the shear rate is called a *Newtonian* fluid. A graph relating shear stress and shear rate (velocity gradient) is a straight line through the origin whose slope is the dynamic viscosity, $\mu = \tau/(du/dy)$. This graph is often called the flow curve and is shown in Fig. 1-9.

Fluids whose viscous behavior is not described by Eq. (1.28) are called *non-Newtonian* fluids. Although these non-Newtonian fluids are not uncommon, their viscous behavior is not yet completely understood.

Fluids have been classified according to their viscous behavior in different ways. Wilkinson [4] classifies them in two groups as Newtonian and non-Newtonian. The non-Newtonian group has three subgroups:

1. fluids for which the shear stress depends only on the shear rate, and although the relation between them is not linear it is independent of the time of application of the shear stress;
2. fluids for which the shear stress depends not only on the shear rate, but also on the time the fluid has been sheared or on its previous history; and
3. viscoelastic fluids which exhibit characteristics of both elastic solids and viscous fluids.

Metzner [5,6] classifies fluids into four general categories:

I. *Purely Viscous Fluids.* These include Newtonian fluids and non-Newtonian fluids for which the shear stress depends only on shear rate and is time independent. Air and water are Newtonian. Gases and low-molecular-weight liquids are almost always Newtonian, and for them $\tau = \mu(du/dy)$ for parallel laminar flow (only one nonzero velocity component). A number of nonlinear constitutive equations have been used to describe the viscous behavior of non-Newtonian viscous fluids. For parallel flow they include:

a. The power-law equation

$$\tau = K\left(\frac{du}{dy}\right)^n \tag{1.30}$$

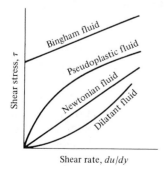

Fig. 1-9. Viscous behavior of fluids: flow curves.

where K is a consistency index and n is a flow behavior index, which become μ and 1, respectively, for a Newtonian fluid.

    b. The Ellis equation

$$\frac{du}{dy} = \frac{\tau_1}{\mu_0} + \left(\frac{\tau}{K}\right)^{1/n}$$

which corrects for the inaccuracy of the power-law equation at low shear rates.

    c. The Bingham equation

$$\tau = \tau_1 + \mu_B\left(\frac{du}{dy}\right)$$

which describes the behavior of a fluid which acts as a solid for shear stresses less than τ_1 and as a Newtonian fluid for shear stresses greater than τ_1 (Fig. 1-9).

    d. The three-parameter Eyring–Powell equation

$$\tau = \mu\left(\frac{du}{dy}\right) + C_1 \sinh^{-1}\left(\frac{1}{C_2}\frac{du}{dy}\right)$$

which is not explicit in du/dy, but is more accurate over a larger range of shear rates than the preceding equations.

    Power-law fluids such as most non-Newtonian slurries, polymer solutions, and molten polymers (including cellulose derivatives) have a flow behavior index n less than unity. These fluids are called *pseudoplastic* because the apparent viscosity decreases with increasing shear rate—the flow curve becomes flatter as the shear rate increases (Fig. 1-9). Suspensions of solids with high solid concentrations and highly saturated polymer solutions have a flow behavior index n greater than unity. These fluids are called *dilatant* because the apparent viscosity increases with increasing shear rate—the flow curve steepens with increasing shear rate (Fig. 1-9). Different values of flow index n result in different shapes of velocity profiles for laminar flow in circular ducts (Appendix III) and exhibit different effects of flow rate on pressure drop for these flows.

    Typical Bingham fluids include some slurries, drilling muds, oil paints, toothpaste, and sewage sludges.

II. *Time-Dependent Fluids.* Fluids for which the apparent viscosity decreases with time under a constant shear rate are called *thixotropic;* those for which the apparent viscosity increases with time are called *rheopectic.* This behavior is characteristic of gypsum pastes, slurries, and suspensions of solids in liquids. The time dependency is often significant for only short time periods of shear stress application. Thus the time-dependent fluids are often treated as purely viscous fluids. Thixotropic fluids such as printers' ink often exhibit hysteresis effects.

III. *Viscoelastic Fluids.* Materials such as pitch, flour dough, and some solid and molten polymers exhibit characteristics of both elastic solids and viscous fluids.

IV. *More Complex Rheological Systems.* Some fluids exhibit characteristics of all three categories I, II, and III. In a superimposed magnetic field, a fluid may have a shear-rate–shear-stress relation which includes magnetohydrodynamic effects. For low-density gases, the fluid may have to be considered as made up of discrete particles and not be treated as a continuum. These are typical of fluids in this fourth category.

Most of the work on non-Newtonian fluids during the past decade has been done by people in applied mathematics and mechanics and by chemical engineers, but interest is being shown by people in other engineering areas. An informative discussion of the civil engineering aspects of non-Newtonian fluids is given by Bugliarello *et al.* [7]. Engineers in many areas are becoming increasingly confronted with the flow of non-Newtonian fluids in pipes as well as in the design and selection of pumps for these fluids [8]. Experiments have shown that frictional effects for some non-Newtonian fluids are less than for a Newtonian fluid under equivalent turbulent flow conditions, resulting in a lower pressure drop in pipes and reduced drag on bodies submerged in liquids. This latter situation is of interest in naval hydrodynamics.

The study of non-Newtonian fluids is a part of the general science of rheology [9].

Example 1-8

The space between two parallel plates 1.5 cm apart is filled with an oil of viscosity $\mu = 0.050$ kg/m s. A thin 30- $\times$ 60-cm rectangular plate is pulled through the oil 0.50 cm from one plate and 1.00 cm from the other. What force is needed to pull the plate at 0.40 m/s?

SOLUTION
The total force overcomes the viscous shear over both the upper and the lower surface of the moving plate. Thus,

$$
\begin{aligned}
F_{total} &= F_{upper} + F_{lower} \\
&= \text{(upper shear stress)(area)} + \text{(lower shear stress)(area)} \\
&= \mu(V/h_{upper})(A) + \mu(V/h_{lower})(A) \\
&= (0.050)(0.40/0.005)(0.180) + (0.050)(0.40/0.010)(0.180) \\
&= 0.72 + 0.36 \\
&= 1.08 \text{ N}
\end{aligned}
$$

Example 1-9

A sleeve 30 cm long encases a vertical rod 2.50 cm in diameter with a uniform radial clearance of 0.005 cm. When the sleeve and rod are immersed in olive oil at 15°C the effective weight of the sleeve is 4.50 N. How fast will the sleeve slide down the rod? Neglect the drag on the outer portion of the sleeve (it is about 0.01 N). Refer to Fig. 1-8b.

SOLUTION
The radial clearance is much less than the rod diameter, and thus the velocity gradient in the oil may be considered constant at $V/\Delta r$. From Fig. A-1 (Appendix II) the dynamic viscosity is about 9.5×10^{-2} kg/m s. Thus,

$$F = \tau A_{\text{sheared}} = \mu(V/\Delta r)\pi DL$$

and

$$V = \frac{F\,\Delta r}{\mu\pi\,DL} = \frac{(4.50)(0.005 \times 10^{-2})}{(9.5 \times 10^{-2})(\pi)(0.025)(0.30)}$$

$$= 0.100 \text{ m/s}$$

Example 1-10

A shaft 15.00 cm in diameter rotates at 1800 r/min inside a bearing 15.05 cm in diameter and 30.0 cm long. The uniform space between them is filled with an oil of viscosity $\mu = 0.018$ kg/m s. What power is required to overcome viscous resistance in the bearing? Refer to Fig. 1-8c.

SOLUTION
The radial clearance is much less than the shaft diameter, and thus the velocity gradient may be considered constant at $V/\Delta r$. Thus,

$$\text{power} = \text{total shear force times peripheral shaft speed}$$
$$= \tau(A_{\text{sheared}})(V), \qquad \text{where} \quad V = \omega r = 60\pi r = 4.5\pi \text{ m/s}$$
$$P = \mu(V/\Delta r)(\pi\,DL)(V)$$
$$= (0.018)(4.5\pi/0.00025)(\pi)(0.150)(0.300)(4.5\pi)$$
$$= 2034 \text{ W} = 2.034 \text{ kW}$$

PROBLEMS

1-1. The dynamic viscosity of water at 20°C is 1.00 kg/m s. Show that this is equal to 1.00 N s/m².

1-2. Show that a kinematic viscosity of 100 ft²/s is equivalent to 9.29 m²/s.

1-3. Kinematic viscosity v is defined as the ratio of dynamic viscosity μ to density ρ ($v = \mu/\rho$). Show that kinematic viscosity has units of m^2/s when dynamic viscosity has units of N s/m^2 and density has units of kg/m^3.

1-4. What is the kinematic viscosity of water in units of m^2/s for a dynamic viscosity of 1.20×10^{-3} N s/m^2 and a density of 999 kg/m^3?

1-5. Convert the following quantities (refer to Table A-1 in Appendix II):
(a) A flow of 500 gal/min to m^3/s.
(b) A flow rate of 1 acre ft/day to m^3/s (1 acre = 43,560 ft^2).
(c) A pressure of 2000 lb_f/ft^2 to kN/m^2.
(d) A speed of 60 mi/h to m/s.
(e) A density of 50 lb_m/ft^3 to kg/m^3.

1-6. Convert a standard atmosphere at 14.696 $lb_f/in.^2$ to units of pascals (newtons per square meters) absolute.

1-7. Show that the expression $\rho V^2/2$ (called the dynamic pressure) has units of pressure in Pa when ρ is in kg/m^3 and V is the velocity in m/s.

1-8. Show that the expression $V^2/2g$ (called the velocity, kinetic or dynamic head) has units of meters head, which represents energy per unit weight (J/N), when V is velocity in m/s and g is the acceleration of gravity in m/s^2.

1-9. The expression $V^2/2$ represents energy per unit mass, where V is velocity. For V in m/s, show that $V^2/2$ has units of J/kg.

1-10. Show that $\rho V^2/2$ represents pressure as well as energy per unit volume.

1-11. What is the kinetic energy per unit mass ($V^2/2$) for a velocity $V = 10$ m/s for
(a) water of density 1000 kg/m^3,
(b) air of density 1.225 kg/m^3, and
(c) mercury of density 13,560 kg/m^3?

1-12. For each of the following flow systems state whether a significant amount of work is done *by* or *on* the fluid, and whether a significant amount of heat is *added to* or *removed from* the fluid.
(a) Water flowing through a centrifugal pump
(b) Water flowing through a hydraulic turbine
(c) Water flowing over the spillway of a large dam
(d) Water flowing through the radiator of an automobile
(e) Water flowing through the entire cooling system of an automobile
(f) Air flowing through an automobile radiator
(g) Air flowing through an automobile radiator and past the engine
(h) Air flowing through the slipstream of an airplane propeller
(i) Air passing through an air compressor
(j) Steam passing through a steam turbine

1-13. Temperatures may be expressed as absolute or relative (degrees Kelvin or Celsius, for example). Which of these *may* or *should* be used for making calculations involving
(a) temperature differences,

(b) temperature ratios, and

(c) calculations of gas densities from the perfect gas equation?

1-14. Determine the volume occupied by a mass of 1 kg of each of the gases listed in Table A-2 (Appendix II) at a temperature of 15°C and a pressure of 1.013×10^5 Pa abs.

1-15. Because differences in internal energy and differences in enthalpy are of more general interest than absolute values, numerical values of u and of h for vapors and gases, as well as for liquids, are tabulated with respect to a reference state other than absolute zero temperature. May both u and h be assumed zero at this reference state? Explain.

1-16. An automobile tire is filled with air at a gage pressure of 200 kPa inside a garage at 25°C. What will be the gage pressure in the tire at

(a) $-15°C$ and

(b) 40°C assuming no change in volume? Assume a standard atmospheric pressure of 1.013×10^5 Pa abs.

1-17. An automobile tire has a constant volume of 0.070 m³ and contains 0.25 kg of air. At an atmospheric pressure of 100 kPa abs, what will be the range of gage pressures in the tire for a temperature range from 15° to 75°C?

1-18. Calculate the change in enthalpy for the air in the tire for the temperature variation in Prob. 1-17.

1-19. Show from Eq. (1.4) or Eq. (1.6) that the entropy change of the air in the tire of Prob. 1-17, as the temperature increases from the garage temperature T_1 to the maximum temperature T_2, is given by

$$s_2 - s_1 = c_v \ln \frac{T_2}{T_1}$$

Does the entropy increase or does it decrease as the temperature increases?

1-20. From Eqs. (1.6) and (1.14c) show that, for an isentropic flow or process of a perfect gas, the change in enthalpy is given by

$$\Delta h = \frac{k}{k-1} \Delta \left(\frac{p}{\rho} \right)$$

From Eqs. (1.9) and (1.13) show that the isentropic requirement is needlessly restrictive and that the change in enthalpy may be expressed by this equation for any type of flow or process.

1-21. What is the increase in enthalpy of a unit mass of air, methane, and xenon for a temperature increase of 50°C by means of

(a) a constant-pressure process,

(b) an adiabatic process, and

(c) a constant-volume process?

1-22. What is the density of air at a 10,000-m altitude in kg/m³? The pressure is 2.64×10^4 Pa abs and the temperature is $-50°C$.

1-23. What is the density of air in the standard atmosphere at 11,000-m altitude, where the pressure is 2.263×10^4 Pa abs and the temperature is $-56.5°C$?

1-24. A balloon is to filled with helium and is to expand to a sphere 18 m in diameter at an altitude of 11,000 m where $p = 22.63$ kPa abs and $T = -56.5°C$. There is to be no stress in the balloon fabric (gas pressure inside equals air pressure outside). What volume of helium must be added at ground level where $p = 101.3$ kPa abs and $T = 15°C$?

1-25. Air at 7.0×10^5 Pa abs and $50°C$ expands isentropically to a temperature of $5°C$.
(a) What is the final pressure?
(b) What is the final air density?

1-26. Methane at 3.0×10^5 Pa abs and $50°C$ expands isentropically to a pressure of 1.6×10^5 Pa abs.
(a) What is the final temperature?
(b) What is the final gas density?

1-27. What is the density of air at 10,000-kPa abs pressure and $130°C$
(a) considering air to be a perfect gas and
(b) for actual air?

1-28. What is the density of air at 1000-kPa abs pressure and $200°K$
(a) considering air to be a perfect gas and
(b) for actual air?

1-29. From Table A-4 (Appendix II) calculate the density of alcohol, seawater, carbon tetrachloride, acetylene tetrabromide, and mercury in kg/m^3.

1-30. What is the specific weight γ of water at $20°C$ in units of N/m^3? Refer to Table A-3 (Appendix II).

1-31. Calculate the density of each gas listed in Table A-2 (Appendix II) in a standard atmosphere at sea level ($T = 15°C$ and $p = 1.013 \times 10^5$ Pa abs).

1-32. Does the equation $p = \rho RT$ apply to
(a) only a nonviscous perfect gas,
(b) only a viscous perfect gas, or
(c) both viscous and nonviscous perfect gases?

1-33. Nitrogen in a high-pressure container is at 1400-kPa abs pressure and $25°C$ temperature. What is its density?

1-34. A gas at 140 kPa abs and $4°C$ has a density of 2.65 kg/m^3, and at 1400 kPa abs and $60°C$ the density is 21.50 kg/m^3. Is the gas a perfect gas?

1-35. The elastic modulus is defined as $K = -dp/(dv/v)$. Show that this is equivalent to $K = +dp/(d\rho/\rho)$.

1-36. Estimate the external pressure required to reduce a given volume of water by 1 percent near atmospheric pressure.

1-37. Estimate the bulk elastic modulus for water at $20°C$ at a pressure of 60×10^5 Pa and at 600×10^5 Pa.

1-38. The pressure at a 6-km depth in the ocean is 61×10^6 Pa. What is the specific weight of seawater at that depth when the specific weight is 10.0 kN/m^3 at the surface and the average bulk modulus of elasticity is 2.4×10^9 Pa? Assume that gravity does not vary appreciably.

1-39. The *Trieste* submerged to a depth of 11,030 m in the Marianas Trench early in 1960. The pressure at that depth is about 113.6 MPa.
   (a) What is the average elastic modulus K throughout this range of depth, assuming it to vary linearly and to be the same as for fresh water?
   (b) What is the increase in density of sea water from the surface to a depth of 11,030 m?

1-40. What is the speed of sound in sea water at 20°C
   (a) at the surface of the sea and
   (b) at a depth of 8000 m?

1-41. Show that Eq. (1.25b) follows from the discussion preceding it.

1-42. Calculate the speed of sound in water at 20°C at atmospheric pressure containing
   (a) 0.1 percent gas nuclei by volume and
   (b) 1 percent gas nuclei by volume and compare with Example 1-4. The results apply at frequencies below bubble resonance.

1-43. Calculate the speed of sound in water at 20°C at normal atmospheric pressure containing
   (a) 0.2 percent and
   (b) 0.8 percent gas nuclei by volume and compare with Example 1-4. The results apply at frequencies below bubble resonance.

1-44. The speed of sound in any medium is $c = \sqrt{K/\rho}$. Suppose water contains 100 parts per million by volume of minute gas nuclei evenly distributed. How does the speed of sound in this mixture compare with the speed of sound in pure water at the same temperature and pressure? Explain.

1-45. The speed of sound in any medium is $c = \sqrt{K/\rho}$. Suppose minute water droplets are sprayed into dry air at a low concentration. How would the speed of sound in this mixture compare with the speed of sound in dry air at the same pressure and temperature? Explain.

1-46. The elastic modulus K is evaluated at two different pressures for a given gas, p_2 being higher than p_1. Is K_2 equal to K_1, or greater than K_1? Explain.

1-47. Air expands from 400 to 200 kPa abs. What is the elastic modulus K
   (a) at the beginning of expansion for an isothermal expansion,
   (b) at the end of the expansion for a reversible adiabatic expansion, and
   (c) at the end of a polytropic expansion for which $k = c_p/c_v$ is replaced by $n = 1.2$?

1-48. The added pressure due to surface tension in a small bubble over the ambient pressure may be found by equating the pressure difference times the great circle area of the bubble to the surface tension times bubble circumference. Show that, for a small bubble of radius R, $\Delta p = 2\sigma/R$.

1-49. Calculate the excess pressure inside a 0.001-cm radius bubble over the ambient pressure for an air bubble in
(a) carbon tetrachloride,
(b) water, and
(c) mercury.

1-50. To what height would water at room temperature rise in a clean glass tube 4 mm in diameter?

1-51. Calculate the excess pressure in a water jet 1 mm in diameter resulting from surface tension.

Prob. 1-51

1-52. Repeat Prob. 1-51 for a water jet 3 mm in diameter.

1-53. Define fluid viscosity as though you were writing for *Scientific American* magazine.

1-54. From kinetic theory, the dynamic viscosity of a perfect gas increases with temperature and is independent of pressure. Under what conditions does the kinematic viscosity of a perfect gas also increase with temperature? Under what conditions does it decrease?

1-55. Show that the term *velocity gradient* is synonymous with *rate of angular deformation* and *time rate of shear strain*.

1-56. Verify the conversion factors for viscosity in Table A-1 of Appendix II.

1-57. In Fig. 1-8b and 1-8c, is the shear stress greater at r_1 or at r_2? Explain.

1-58. In Fig. 1-7, is the shear stress greater at the boundary or out in the flow? Explain.

1-59. It is assumed that bubble nuclei may expand and produce a form of cavitation known as intermittent bubble cavitation when the ambient pressure is reduced or the nuclei pass through a low-pressure region in a liquid. Would a given reduction in pressure have a greater effect on the growth of a large or on a small bubble? Explain.

1-60. A fluid of viscosity $\mu = 1.2 \times 10^{-3}$ kg/m s flows along a surface with a velocity profile given by

$$u = 750y - 2.5 \times 10^6 y^3 \quad \text{m/s}$$

where y is in meters. What is the boundary shear stress?

1-61. Air at 10°C, a pressure of 8×10^4 N/m² absolute and viscosity $\mu = 1.77 \times 10^{-5}$ Ns/m² flows along a smooth surface with a velocity profile given by

$$u = 1000y - (10^9 y^3/9) \quad \text{m/s}$$

where y is in meters. What is the shear stress at the smooth boundary surface?

1-62. The velocity profile for laminar flow in a round pipe is given by the equation

$$u = u_{max}[1 - (r/R)^2]$$

where u_{max} is the centerline velocity, R is the pipe radius, and r is the radial distance from the pipe centerline.

(a) What is the velocity gradient at the pipe wall? Note that the normal distance from the pipe wall is $y = R - r$; thus, $dy = -dr$, and $du/dy = -du/dr$.

(b) What is the shear stress at the pipe centerline?

(c) Show that the velocity gradient, and hence shear stress, varies linearly from zero at the pipe centerline to a maximum at the pipe wall.

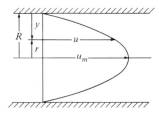

Prob. 1-62

1-63. What is the wall shear stress when fuel oil at 10°C flows with an average velocity of 2.00 m/s ($u_{max} = 4.0$ m/s) in a pipe 20 cm in diameter? Refer to Fig. A-2 (Appendix II) for viscosity and also to Prob. 1-62.

1-64. Oil with a density of 850 kg/m³ flows through a 10-cm diameter pipe. The shear stress at the pipe wall is measured as 3.2 Pa, and the velocity profile is given by

$$u = 2 - 800r^2 \quad \text{m/s}$$

where r is the radial distance from the pipe centerline in meters. What is the kinematic viscosity of the oil?

1-65. A fluid at uniform pressure is contained between two parallel plates 0.04 cm apart. One plate moves at a velocity of 0.90 m/s relative to the other.

(a) What is the rate of angular deformation within the fluid?

(b) What is the shear stress in the fluid for a dynamic viscosity of 0.10 kg/m s?

1-66. Two parallel plates 1 cm apart are at rest. A thin third plate is pulled through an oil which fills the space between plates at rest, at 0.40 m/s. The plate in motion is 0.40×0.80 m and is $\frac{1}{3}$ cm from one plate and $\frac{2}{3}$ cm from the other. For an oil of viscosity 0.120 kg/m s, what force is required to pull the plate?

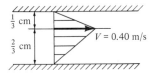

$\frac{1}{3}$ cm

$V = 0.40$ m/s

$\frac{2}{3}$ cm

Prob. 1-66

1-67. A square flat plate 0.60×0.60 m in size and weighing $W = 60$ N slides down a flat slope at a $10°$ angle on an oil film 1.00 mm thick. The oil viscosity is 0.50 kg/m s. At what speed does the plate slide? Neglect effects at the edges of the plate.

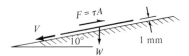

$F = \tau A$

V

$10°$

1 mm

W

Prob. 1-67

1-68. A rod 150 mm in diameter moves along its axis at 0.15 m/s inside a concentric cylinder 150.4 mm in diameter and 1.10 m long. The space between them is filled with an oil of specific gravity 0.82 and kinematic viscosity $v = 10^{-4}$ m²/s. What is the viscous force resisting the motion?

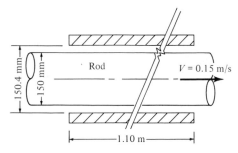

150.4 mm

150 mm

Rod

$V = 0.15$ m/s

1.10 m

Prob. 1-68

1-69. A sleeve 0.4 m long encases a vertical rod 2.00 cm in diameter with a uniform radial clearance of 0.004 cm. When the sleeve and rod are immersed in an oil of viscosity 0.10 kg/m s, the effective weight of the sleeve is 4.0 N. How fast will the sleeve slide down the rod? Consider the drag on the outer portion of the sleeve to be negligible.

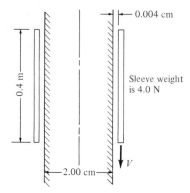

Prob. 1-69

1-70. A viscous fluid of viscosity μ occupies the annular space between a fixed and a rotating cylinder. For laminar motion, is the shape of the velocity profile similar to curve A, B(linear), or C for the two situations shown? Prove your conclusions.

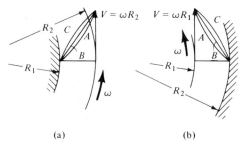

(a)   (b)

Prob. 1-70

1-71. A shaft 7.50 cm in diameter rotates concentrically in a bearing 7.55 cm in diameter and 30 cm long. The space between them contains oil. A torque of 1.10 N m is required to maintain the shaft at a constant rotation of 18 r/s. What is the viscosity of the oil?

1-72. A thin-walled cylinder 25 cm long rotates at 60 r/min midway between two fixed concentric cylinders 25 cm and 26 cm diameter, respectively. A torque of 3.50 N m is required to maintain the motion. What is the viscosity of the oil which fills the space between the two fixed cylinders?

1-73. A 15.0-cm-diameter shaft rotates at 1800 r/min in a stationary journal bearing 40 cm long and 15.1 cm in diameter inside. The oil viscosity is 2.0×10^{-2} kg/m s. What power is required to overcome viscous resistance in the bearing?

1-74. Two coaxial flat circular plates 40 cm in diameter are spaced 0.030 cm apart. The space between them is filled with an oil of viscosity $\mu = 1.00$ kg/m s. The plates are horizontal; the bottom one is at rest, and the upper one rotates at 2 r/s. What torque is required to maintain the given speed?

1-75. An 18.90-cm-diameter cylinder rotates concentrically outside a fixed cylinder 18.30 cm in diameter and 30 cm long. The clearance at both ends is 0.30 cm. A torque of 1.25 N m maintains a rotational speed of 1 r/s. What is the viscosity of the liquid between cylinders?

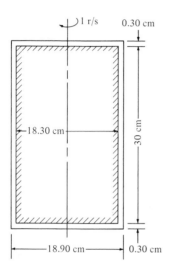

Prob. 1-75

1-76. A uniform oil film 0.12 mm thick of viscosity 0.12 kg/m s separates two disks 220 mm in diameter. One disk rotates at 8 r/s coaxially and relative to the other. What torque is required to maintain this motion? Neglect edge effects.

1-77. Given the fluid drive shown. What is the torque T transmitted for constant ω_1 and ω_2? Express in terms of μ, R, h, and the slip $(\omega_1 - \omega_2)$.

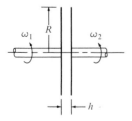

Prob. 1-77

1-78. A torque T is required to rotate the cone at an angular speed ω. The gap h is small. What is the viscosity μ of the fluid which fills the gap?

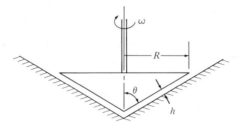

Prob. 1-78

References

1. H. Rouse and S. Ince, *History of Hydraulics* (Iowa City: Iowa Institute of Hydraulic Reasearch, State University of Iowa, 1957).
2. C. Truesdell, *Essays in the History of Mechanics* (New York: Springer-Verlag, 1968), p. 364.
3. *Handbook of Chemistry and Physics*, 59th ed. (West Palm Beach: CRC Press, Inc., 1978–1979).
4. W. L. Wilkinson, *Non-Newtonian Fluids* (Elmsford, New York: Pergamon Press, 1960).
5. A. B. Metzner, "Flow of Non-Newtonian Fluids," Sec. 7 of *Handbook of Fluid Dynamics*, edited by V. L. Streeter (New York: McGraw-Hill Book Company, Inc., 1961).
6. A. B. Metzner, "Heat Transfer in Non-Newtonian Fluids," in *Advances in Heat Transfer*, Vol. 2 (New York: Academic Press, 1965), pp. 357–397.
7. G. Bugliarello, V. C. Behn, C. E. Carver, E. M. Krokosky, J. F. Ripken, and R. L. Schiffman, "Non-Newtonian Flows," *Civil Eng.*, Vol. 35 (1965), pp. 68–70.
8. G. Bugliarello, "Some Considerations on the Analysis and Design of Hydraulic Machinery for Non-Newtonian Fluids," *Proc. Tenth Congress Intl. Assoc. for Hyd. Research*, Vol. 4 (1963), Paper No. 4.1.
9. A. G. Fredrickson, *Principles and Applications of Rheology* (Englewood Cliffs, New Jersey: Prentice-Hall, Inc., 1964).
10. For a brief history of fluid mechanics, see also G. A. Tokaty, *Fluidmechanics* (Henley-on-Thames, Oxfordshire, England: G. T. Foulis and Co. Ltd., 1971).

Chapter 2
Fluid Statics

This chapter deals primarily with the variation in fluid pressure with elevation in a gravitational field such as the earth's and the result of these pressure variations on surfaces submerged in fluids at rest. A study of these variations will enable us to determine, for example, pressure differences as measured with manometers, hydrostatic forces on dams and spillway gates, buoyancy on submerged bodies, the pressure and density variations with altitude in the atmosphere, and criteria for static stability of submerged and floating bodies.

The forces acting on the surfaces of a given fluid element isolated within a body of fluid at rest are due to pressure. It should be emphasized that pressure (or pressure intensity) is a scalar quantity and therefore the pressure acts equally in any and all directions at a point in a fluid. Areas and forces are vector quantities. The area vector always points in a direction normal to the area, and its magnitude is equal to the magnitude of the area. Hence a force resulting from pressure is a vector whose magnitude is the product of the pressure intensity and the magnitude of the area and which points in a direction normal to the area. Forces on surfaces resulting from pressure always act normal to that surface.

Forces due to surface tension are not included since none of the surfaces of the fluid element within the fluid is a free surface or an interface with a

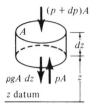

Fig. 2-1. Elementary cylinder of fluid at rest.

different fluid. Tangential shear stresses or forces exist only when there is relative motion between adjacent fluid layers; they cannot exist when a fluid is at rest.

2-1. General Differential Equation

A simple differential equation will be derived which can then be integrated both for liquids of constant density and for gases of variable density.

Consider a right cylinder, circular or otherwise. Equilibrium of external forces shown in Fig. 2-1 gives[†]

$$pA - (p + dp)A - \rho g A \, dz = 0$$

from which

$$dp = -\rho g \, dz = -\gamma \, dz \tag{2.1}$$

(In vector form this equation is grad $p = \rho \mathbf{g}$ and since $\partial p/\partial x = \partial p/\partial y = 0$, the vector equation becomes $\partial p/\partial z = -\rho g$ because z is measured positively upwards and g is directed vertically downwards.) This equation expresses the familiar facts that (1) the pressure intensity decreases with elevation (dp is negative when dz is positive), and (2) the pressure intensity remains the same if there is no change in elevation (dz is zero, and thus dp is zero). Thus, for example, (a) the pressure intensity decreases as we go from a given depth towards the surface of a lake, or as we go upwards in the atmosphere; and (b) the pressure intensity at a 20-m depth in a lake at one point is the same as that at a 20-m depth at some other point in the lake, but pressures at equal elevations in an open container accelerated linearly or rotationally are not the same because the fluids are not at rest.

2-2. Hydrostatics

Equation (2.1) can be integrated for a constant-density fluid,

$$\int_{p_1}^{p_2} dp = -\rho g \int_{z_1}^{z_2} dz$$

[†] Recall that the specific weight $\gamma = \rho g$ and is thus not a true property of a fluid since it depends on the local gravitational acceleration.

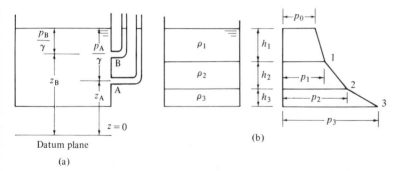

Fig. 2-2.   Piezometric heads in liquids at rest. (a) Homogeneous liquid.
(b) Several liquids of different densities.

from which

$$p_2 - p_1 = -\rho g(z_2 - z_1) = -\gamma(z_2 - z_1) \qquad (2.2)$$

where the pressure change is in newtons per square meter, the specific weight[†]
is in newtons per cubic meter, and the elevation change is in meters. This
equation is commonly expressed as

$$\Delta p = \gamma h \qquad (2.3)$$

where h is the height difference between points for which the pressure
difference is to be calculated. Note that the equation gives only the magnitude of Δp; whether the pressure increases or decreases is determined from
Eq. (2.1.).

The term p/γ is defined as the *pressure head*, z is defined as the *potential
head* with respect to an arbitrary datum, and $(p/\gamma) + z$ is defined as the
piezometric head. Thus if we write $(p_2 - p_1)/\gamma = -(z_2 - z_1)$, we find that
the increase in pressure head equals the decrease in potential head. If we
write $(p_1/\gamma) + z_1 = (p_2/\gamma) + z_2$, we find that the piezometric head is constant
in a homogeneous liquid at rest.[‡] This is illustrated in Fig. 2-2.

The pressure head at B is p_B/γ and at A it is p_A/γ. It is obvious that
$(p_B/\gamma) + z_B = (p_A/\gamma) + z_A$. The pressure intensity p or the pressure head p/γ
is, of course, greater at A than at B. Also, the potential head z is greater
at B than at A. But the sum of the pressure head and potential head, called
the piezometric head (measured by piezometer taps at B and A and indicated
by the height of the liquid columns connected to them), is the same for both.

If there are several liquids of different densities in a container at rest,
the various liquids will form horizontal layers with the liquid of greatest
density at the bottom and the liquid of lowest density at the top, if they are

---

[†] Recall that the specific weight $\gamma = \rho g$ and is thus not a true property of a fluid since it depends
on the local gravitational acceleration.
[‡] Note that this is a special form of the Bernoulli equation (Sec. 4-3) when the velocity terms
are zero.

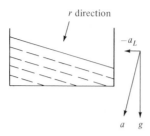

Fig. 2-3. Constant linear acceleration of a liquid in a container.

insoluble in one another. Horizontal planes are planes of constant pressure. If the density of the lightest liquid is ρ_1 and its depth h_1, the next ρ_2 and h_2, and so on, the pressure at the first interface is

$$p_1 = p_0 + \rho_1 g h_1$$

where p_0 is the pressure at the surface of the upper liquid. The pressure at the second interface is

$$p_2 = p_1 + \rho_2 g h_2 = p_0 + \rho_1 g h_1 + \rho_2 g h_2$$

and so on. In this situation the piezometric head as defined here is *not* constant at all levels within the various liquids.

The preceding results apply to body forces (forces which act on the volume or mass of a fluid element) resulting from gravity. If there are body forces due to uniform acceleration other than gravity, such as linear accelerations or centrifugal effects in rigid-body-type rotations, points equidistant below a free liquid surface are all at the same pressure. These are shown in Fig. 2-3 for liquid in an open container accelerated in a horizontal plane at a uniform linear acceleration, and in Fig. 2-4 for a liquid rotating as a solid body (with uniform rotation) in an open container with centrifugal forces due to acceleration in a radial direction. The dashed lines represent surfaces of constant pressure below the free surfaces shown as solid lines. Pressure changes along a vertical line may be determined from Eq. (2.2) or (2.3).

Pressure intensity p may be given in a variety of units, all expressing a force per unit area. In addition, it may be stated as an absolute pressure, referred to a perfect vacuum; or it may be stated as a gage pressure, referred

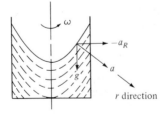

Fig. 2-4. Constant centrifugal acceleration of a liquid in a container.

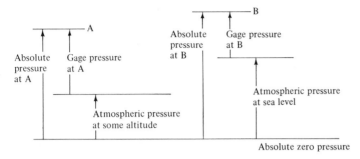

Fig. 2-5. Absolute and gage pressures.

to the surroundings or ambient environment. The difference between the absolute pressure and the corresponding gage pressure is the atmospheric or ambient pressure. This is indicated in Fig. 2-5.

The absolute pressure at B is greater than the absolute pressure at A. Yet the gage pressure at B is less than the gage pressure at A, since the gage pressure at B is referred to a different (higher) atmospheric or ambient pressure than at A.

Example-2-1

The practical depth limit for a free diver is about 50 m. What is the pressure intensity at that depth in (a) fresh water, and (b) seawater?

SOLUTION

(a) From Eq. (2.3).

$$p = \rho g h = (1000)(9.81)(50) = 4.91 \times 10^5 \quad \text{Pa gage}$$

(b) The specific gravity of seawater is about 1.025, thus

$$p = (1.025)(1000)(9.81)(50) = 5.03 \times 10^5 \quad \text{Pa gage}$$

2-3. Manometry

In order to determine pressures or pressure differences from manometer measurements, we apply Eqs. (2.1) and (2.3), starting at one end of the manometer system and writing pressure differences between successive fluid interfaces.

The simplest form of a manometer is a barometer (Fig. 2-6), which is used to determine the absolute atmospheric pressure intensity. Equations (2.1) and (2.3) for the barometer become

$$p_v + \gamma_b h_b = p_a$$

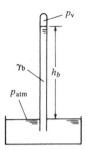

Fig. 2-6. Barometer.

The advantages of using mercury are apparent from a comparison (Table 2-1) of the vapor pressure p_v and the barometer height h_b for mercury, water, and benzene, for example. Not only is the specific weight of mercury such that a more reasonable height of barometer is possible, but the vapor pressure is low (it can be ignored). A standard atmosphere at 15°C and a pressure of 1.013×10^5 Pa abs is assumed.

Water barometers are used, however, in connection with cavitation tests on propellers, water pumps, and water turbines. Note that liquid heights in barometers are *not* directly related by the specific weights of the liquids, nor are they the same for a given liquid, because of the variation in vapor pressure with different liquids or with temperature for a given liquid. The temperature correction applied to a common mercury barometer corrects for variations in specific weight of mercury, the expansion of the glass tube containing the mercury, and the expansion of the (usually) brass measuring scale.

Differential manometers are used to measure the difference in pressure intensity between a given point and the atmosphere, or between two points neither of which is at atmospheric pressure. Some manometers are so arranged that a relatively large pressure difference is measured in terms of a relatively small manometer deflection, while in others a relatively small pressure difference is measured in terms of a relatively large manometer deflection. (The term *manometer deflection* refers to the distance between pertinent liquid–liquid or liquid–gas interfaces.) Manometers of the latter type are often called *micromanometers*.

The manometer deflection generally depends on the relative specific weights of the two or more fluids used in the manometer. Not only should the specific weights be different, but adjacent liquids must be compatible.

Table 2-1. COMPARISON OF BAROMETER LIQUIDS

BAROMETER FLUID	p_v (Pa abs)	γ_b (N/m³)	$\gamma_b h_b$ (kPa)	h_b (m)
Mercury	0.18	133,200	101.3	0.760
Water	2340	9,810	99.0	10.09
Benzene	9990	8,670	91.3	10.53

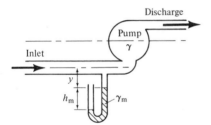

Fig. 2-7.   See Example 2-2. Open manometer.

Their surface tension properties should be such that a good meniscus (interface) is obtained, and they should not mix or dissolve together.
    Some examples of manometer installations follow.

Example 2-2

What is the suction pressure on a centrifugal pump as measured with a mercury manometer as shown in Fig. 2-7? Water is flowing.

SOLUTION
In this instance the hydrostatic equations [Eqs. (2.2) and (2.3)] are applied from the open end of the manometer to the inlet pipe centerline, in terms of gage pressure:

$$0 - \gamma_m h_m - \gamma y = p_{inlet}$$

Let $h_m = 10$ cm and $y = 20$ cm. Then

$$p_{inlet} = -(13{,}560)(9.81)(0.10) - (1000)(9.81)(0.20)$$
$$= -15.26 \text{ kPa gage}$$

Densities are obtained from Table A-3 and A-5 (Appendix II).

Example 2-3

The pressure drop across an orifice in a pipeline conveying oil can be measured with an acetylene tetrabromide manometer as shown in Fig. 2-8. What is the magnitude of $p_1 - p_2$?

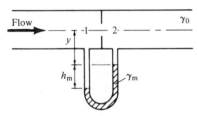

Fig. 2-8.   See Example 2-3. Differential manometer.

SOLUTION

The hydrostatic equations are applied from point 1 to point 2 to obtain

$$p_1 + \gamma_o(y + h_m) - \gamma_m h_m - \gamma_o y = p_2$$

so that

$$p_1 - p_2 = h_m(\gamma_m - \gamma_o)$$

Let $h_m = 36$ cm and the specific weight of the oil be $s_o = 0.80$. Then $\gamma_m = (2.96)(9810) = 29.0$ kN/m^3 from Table A-4 (Appendix II), $\gamma_o = (9810)(0.80) = 7850$ N/m^3, and

$$p_1 - p_2 = (0.36)(29,000 - 7850) = 7.61 \text{ kPa}$$

The piezometric head difference between points 1 and 2 is

$$h_1 - h_2 = (p_1 - p_2)/\gamma_o = 0.97 \text{ m of oil}$$

Note that the distance y does not appear in the solution and thus a differential manometer of this type may be placed at any elevation with respect to the pipe, whereas in Example 2-2, the position of the manometer y was important.

It can be shown that a manometer connected as in Fig. 2-8 indicates the difference in piezometric pressure $(p_1 + \gamma_o z_1) - (p_2 + \gamma_o z_2)$, or the difference in piezometric head, $[(p_1/\gamma_o) + z_1] - [(p_2/\gamma_o) + z_2]$, between points 1 and 2, and indicates the difference in pressure intensity or pressure head only for a horizontal system.

The manometers shown in Examples 2-2 and 2-3 are of a type used to measure moderate or even large pressure differences with reasonably small manometer deflections. The deflection is a function of the pressure difference being measured and the relative densities of the fluids. It is often necessary to select an appropriate manometer liquid to measure a given pressure difference with a maximum deflection limited by the size of manometer available.

Manometers also are designed to produce a relatively large deflection for a very small pressure difference or change in pressure difference. These are called micromanometers and are available commercially in many forms. One type is shown in Example 2-4.

Example 2-4

A sensitive manometer containing water and oil is shown in Fig. 2-9. What pressure difference is indicated when the liquid levels in the upper chambers were initially at the same level and the manometer deflection was zero when $p_A = p_B$? The level in A is $1/450$ cm higher and that in B is $1/450$ cm lower than initially.

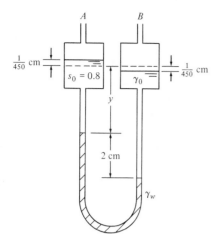

Fig. 2-9.   See Example 2-4. Micromanometer.

SOLUTION

The hydrostatic equations give

$$p_A + \gamma_o[y + (1/450)(0.01)] + \gamma_w(0.02) - \gamma_o[(0.02)$$
$$+ y - (1/450)(0.01)] = p_B$$

$$p_B - p_A = (0.02)(\gamma_w - \gamma_o) + \frac{2\gamma_o}{45,000}$$

$$= 39.6 \text{ Pa}, \qquad \text{which is less than } 1/2500 \text{ atm}$$

Note that the relative cross-sectional areas of chambers A and B and the manometer tubing determine the magnitude of the second term in the final expression for $p_B - p_A$, and this term often may be neglected.

2-4. Fluid Forces on Submerged Surfaces

A flat surface submerged horizontally at a depth h has the same pressure intensity imposed on every point of that surface (Fig. 2-10). Then the total

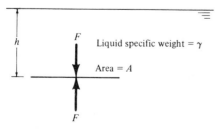

Fig. 2-10.   Flat horizontal surface submerged in a liquid.

force F acting downward on the top surface is

$$F = \bar{p}A = \gamma h A$$

This is also equal to the total force F acting upward on the bottom surface, since the pressure intensity is the same on both the top and bottom surfaces.

If the flat surface is inclined from the horizontal at some arbitrary angle θ, the resultant force (acting normal to the flat surface) is equal in magnitude to the product of the pressure intensity at the centroid of the surface and the surface area. This is shown in Fig. 2-11.

The total force F acting on one side (equal in magnitude but oppositely directed to that on the opposite side) is the integral of the forces acting on each infinitesimal area dA:

$$dF = p\,dA = \gamma h\,dA = \gamma y \sin \theta\, dA$$

$$F = \int_A \gamma y \sin \theta\, dA = \gamma \sin \theta \int_A y\, dA$$

But $\int y\, dA$ is the first moment of the area about O, defined in terms of y, the distance from O to the centroid, as

$$\int_A y\, dA = \bar{y}A$$

$$F = \gamma(\sin \theta)\bar{y}A = \gamma \bar{h} A = \bar{p}A \tag{2.4}$$

since the presence $\bar{p}$ at the centroid is equal to $\gamma \bar{h}$.

The force F is *not* applied at the centroid of the area but is always below it by an amount which diminishes with depth. If, in Fig. 2-11, h_F is the depth at which the resultant force F is applied at the center of pressure CP, we can determine a value of y_F, measured parallel to the flat surface, from which h_F may be determined. We defined y_F in such a way that Fy_F, the first moment of the force F about the point O, is expressed in terms of the first moment of the forces on the infinitesimal area dA. Then

$$Fy_F = \int_A \gamma y^2 \sin \theta\, dA$$

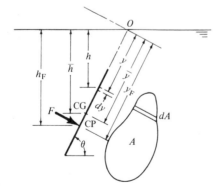

Fig. 2-11.   Flat inclined surface submerged in a liquid.

so that

$$y_F = \frac{\gamma \sin \theta \int_A y^2 \, dA}{\gamma \sin \theta \int_A y \, dA} = \frac{\int_A y^2 \, dA}{\int_A y \, dA}$$

The numerator is the moment of inertia of the total area of the flat surface about O, which can be expressed in terms of the moment of inertia about the centroid of the flat area:

$$\int_A y^2 \, dA = I_{CG} + \bar{y}^2 A$$

Table 2-2. MOMENTS OF INERTIA FOR VARIOUS PLANE SURFACES ABOUT THEIR CENTER OF GRAVITY

SURFACE		I_{CG}
Rectangle or square		$\dfrac{1}{12} Ah^2$
Triangle		$\dfrac{1}{18} Ah^2$
Quadrant of circle (or semicircle)		$\left(\dfrac{1}{4} - \dfrac{16}{9\pi^2}\right) Ar^2 = 0.0699 \, Ar^2$
Quadrant of ellipse (or semiellipse)		$\left(\dfrac{1}{4} - \dfrac{16}{9\pi^2}\right) Aa^2 = 0.0699 \, Aa^2$
Parabola		$\left(\dfrac{3}{7} - \dfrac{9}{25}\right) Ah^2 = 0.0686 \, Ah^2$
Circle		$\dfrac{1}{16} Ad^2$
Ellipse		$\dfrac{1}{16} Ah^2$

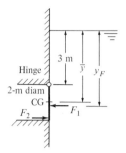

Fig. 2-12.   Example 2-5. Vertical gate.

The denominator is simply $\bar{y}A$. Thus

$$y_F = \bar{y} + \frac{I_{CG}}{A\bar{y}} \tag{2.5}$$

It is convenient to express I_{CG} for any flat area in terms of that area, and this is done in Table 2-2.

Example 2-5

What is (a) the hydrostatic force and (b) the center of pressure on a round vertical gate 2 m in diameter when its top is submerged 3 m in water as shown in Fig. 2-12? (c) What minimum force F_2 will open the gate?

SOLUTION

The force on the gate is the pressure at its center times its area. The center of pressure is below the center of the gate by an amount given by Eq. (2.5). The force F_2 is found by taking moments about the gate hinge.

(a) $F_1 = \bar{p}A = \gamma\bar{y}A = (9810)(4)(\pi) = 1.23 \times 10^5$ N

(b) $y_F = \bar{y} + I_{CG}/A\bar{y} = 4 + (\frac{1}{16})(A)(2)^2/A(4) = 4 + \frac{1}{16} = 4.063$ m

(c) Moments about the hinge gives

$$(1 + \tfrac{1}{16})F_1 = 2F_2 \qquad \text{and} \qquad F_2 = \tfrac{17}{32}F_1 = 6.53 \times 10^4 \text{ N}$$

2.5 Curved Surfaces

The static forces on a curved submerged surface are expressed in terms of the horizontal and vertical components (Fig. 2-13).

1. The horizontal component is equal (in magnitude and point of application) to the force exerted on a projection of the curved surface in a vertical plane.

2. The vertical component is equal to the weight of fluid directly above the surface and is applied at the centroid of this fluid.

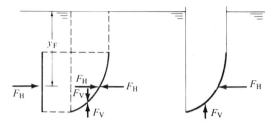

Fig. 2-13. Identical curved surfaces submerged in a liquid.

To be in equilibrium, the horizontal and vertical components on opposite sides of the surface are equal and opposite, respectively. The resultant force on either side is the vector sum of the two components and, again, are equal but opposite for the two sides of the curved surface. Note that the horizontal component F_H and the vertical component F_V are the same for each of the two similar curved surfaces shown in Fig. 2-13.

These generalizations regarding horizontal and vertical force components on curved surfaces are directly applicable to the horizontal and inclined flat surfaces discussed in Sec. 2-4. They can also be used to determine buoyant forces.

Buoyancy is the resultant of the surface forces due to pressure on a submerged body and is equal to the weight of fluid displaced.

In Fig. 2-14, the horizontal force on surface 1–2–3 is equal and opposite to that on surface 1–4–3, so that the net horizontal force is zero. The vertical force on the top surface 2–1–4 (F_{VT}) is equal to the weight of liquid above that surface (diagonally hatched "volume"), and that on the bottom surface 2–3–4 (F_{VB}) is equal to the liquid above that surface (horizontally hatched "volume"), *even though liquid does not occupy all that volume* (see Fig. 2-13). The net vertical force is, then, represented by the weight of liquid in the "volume" 1–4–3–2, the displaced volume. This is known as Archimedes' principle.

Example 2-6

What is the resultant force per meter length on a horizontal circular cylinder 2 m in diameter when water is 2 m deep on one side and 1 m deep on the other (Fig. 2-15)?

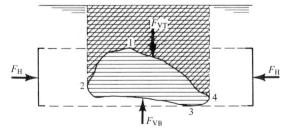

Fig. 2-14. Submerged volume in a liquid.

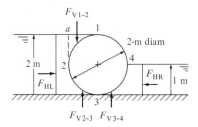

Fig. 2-15. See Example 2-6. Cylindrical dam.

SOLUTION

The horizontal force on the left side is the force on a vertical projection:

$$F_{HL} = \bar{p}A = \gamma \bar{h}A = (9810)(1)(2) = 19.62 \text{ kN}$$

From Eq. (2.5) the center of pressure on a rectangle which is in contact with a free surface is two-thirds the height down from the free surface. Thus F_{HL} acts $\frac{2}{3}$ m above the bottom. The horizontal force on the right side is

$$F_{HR} = \bar{p}A = \gamma \bar{h}A = (9810)(0.5)(1) = 4.90 \text{ kN}$$

and it acts $\frac{1}{3}$ m above the bottom. The vertical force on surface 1–2 is equal to the weight of water in the region 1–a–2–1:

$$F_{V\,1-2} = \gamma \left(r^2 - \frac{\pi r^2}{4} \right) = 9810 \left(1 - \frac{\pi}{4} \right) = 2.10 \text{ kN}$$

The vertical force on surface 2–3 is equal to the weight of water in the region 1–a–2–3–1:

$$F_{V\,2-3} = \gamma \left(r^2 + \frac{\pi r^2}{4} \right) = 17.51 \text{ kN}$$

The vertical force on surface 3–4 is equal to the weight of water in region 3–4–0–3.

$$F_{V\,3-4} = \frac{\gamma \pi r^2}{4} = 7.70 \text{ kN}$$

The net vertical force is 23.11 kN upward, and the net horizontal force is 14.72 kN toward the right. The resultant force is $\sqrt{23.11^2 + 14.72^2} = 27.40$ kN at an angle arctan $1.57 = 57.5°$ from the horizontal.

───

2-6. Aerostatics

Aerostatics is a subject of interest in meteorology and aeronautics. We will integrate Eq. (2.1) for an isothermal atmosphere and for a polytropic atmosphere, and then show that the resulting equations for an isothermal atmosphere combined with special conditions for the polytropic atmosphere will describe the U.S. standard atmosphere. Equation (1.14) for a reversible adiabatic system will apply for an atmosphere with no heat transfer, since the air will be essentially at rest and no entropy changes due to irreversibilities such as turbulence will occur. The atmosphere will be considered to be dry air with no water vapor present and to be a perfect gas.

2-7. Isothermal Atmosphere

If we assume a specified temperature that does not vary with altitude, the variable specific weight can be expressed in terms of the varying pressure from the gas equation of state $\rho = p/RT$, and the hydrostatic equation $dp = -\rho g\, dz$ becomes[†]

$$\int_{p_1}^{p_2} \frac{dp}{p} = -\frac{g}{RT} \int_{z_1}^{z_2} dz$$

Integration gives

$$\frac{p_2}{p_1} = e^{-g(z_2 - z_1)/RT} \tag{2.6}$$

This equation indicates an exponential decrease in pressure with an increase in altitude, and that at an infinite altitude the pressure decreases to zero. Note that the pressures in Eq. (2.6) are absolute and may be expressed in any absolute units, such as pounds per square inch, pounds per square foot, atmospheres, inches of mercury, newtons per square meter, and so forth.

The density ratio is

$$\frac{\rho_2}{\rho_1} = e^{-g(z_2 - z_1)/RT}$$

2-8. Polytropic Atmosphere

Again, the hydrostatic equation $dp = -\gamma\, dz = -\rho g\, dz$ can be integrated by expressing the variable density in terms of the variable pressure. For a polytropic atmosphere, from Eq. (1.15),

$$\frac{p}{\rho^n} = \frac{p_1}{\rho_1^n}$$

where subscript 1 refers to any point where conditions are known. Then

$$\rho = \rho_1 \left(\frac{p}{p_1}\right)^{1/n}$$

[†] The acceleration of gravity g varies with both latitude and altitude (about 0.0031 m/s² per 1000 m increase in altitude) but is considered constant at some average value for the range of elevations involved.

and Eq. (2.1) becomes

$$\int_{p_1}^{p_2} p^{-1/n}\, dp = -\rho_1 p_1^{-1/n}\, g \int_{z_1}^{z_2} dz$$

Integrating and simplifying yields

$$z_2 - z_1 = \frac{n}{(n-1)} \frac{RT_1}{g} \left[1 - \left(\frac{p_2}{p_1} \right)^{(n-1)/n} \right] \tag{2.7a}$$

or

$$\frac{p_2}{p_1} = \left[1 - g \frac{(n-1)}{n} \frac{(z_2 - z_1)}{RT_1} \right]^{n/(n-1)} \tag{2.7b}$$

Equations (2-7) give the pressure or elevation in terms of the other for a given set of conditions at some known point 1. From Eqs. (1.14) and (2.7b)

$$\frac{T_2}{T_1} = \left[1 - \frac{(n-1)}{n} \frac{g}{RT_1} (z_2 - z_1) \right] \tag{2.8a}$$

or

$$\frac{T_2 - T_1}{z_2 - z_1} = -\frac{g(n-1)}{nR} \tag{2.8b}$$

indicating a linear variation of temperature with altitude. Equations (2.8) imply a lapse rate (temperature change per unit increase in altitude or elevation) of $-0.00976°C/m$ for a dry adiabatic atmosphere with $n = k$.[†]

2-9. The U.S. Standard Atmosphere [1]

Reference [1] "depicts idealized middle-latitude year-round mean conditions for the range of solar activity that occurs between sunspot minimum and sunspot maximum." Data were obtained from balloon, rocket, and satellite flights over a number of years.

The U.S. standard atmosphere is defined as 1.01325×10^5 Pa abs pressure and $15°C$ ($288.15°K$) temperature at sea level; it is further defined with regions of constant temperature and other regions of various linear temperature variations up to an altitude of 700 km.

Appropriate values of the polytropic exponent n in Eqs. (2.7) and (2.8) are obtained by solving Eq. (2.8) for n. This gives

$$n = \frac{1}{1 + (R/g)(dt/dz)} \tag{2.9}$$

[†] Combining the differential form of the first law of thermodynamics ($dq = c_p dT - dp/\rho$) and the differential form of the hydrostatic equation ($dp = -\rho g\, dz$) also shows that $dT/dz = -g/c_p = -g(k-1)/kR$ for a dry adiabatic atmosphere.

Table 2-3.   DEFINING PROPERTIES OF THE U.S. STANDARD ATMOSPHERE

ALTITUDE (m)	TEMPERATURE (°C)	TYPE OF ATMOSPHERE	LAPSE RATE (°C/km)	$\bar{g}$ (m/s²)	n	PRESSURE p(Pa)	DENSITY ρ(kg/m³)
0	15.0					1.013×10^5	1.225
		Polytropic	−6.5	9.790	1.235		
11,000	−56.5					2.263×10^4	3.639×10^{-1}
		Isothermal	0.0	9.759			
20,000	−56.5					5.475×10^3	8.804×10^{-2}
		Polytropic	+1.0	9.727	0.972		
32,000	−44.5					8.680×10^2	1.323×10^{-2}
		Polytropic	+2.8	9.685	0.924		
47,000	−2.5					1.109×10^2	1.427×10^{-3}
		Isothermal	0.0	9.654			
52,000	−2.5					5.900×10^1	7.594×10^{-4}
		Polytropic	−2.0	9.633	1.063		
61,000	−20.5					1.821×10^1	2.511×10^{-4}
		Polytropic	−4.0	9.592	1.136		
79,000	−92.5					1.038	2.001×10^{-5}
		Isothermal	0.0	9.549			
88,743	−92.5					1.644×10^{-1}	3.170×10^{-6}

For temperature gradients specified in the U.S. standard atmosphere, values of n are listed in Table 2-3 and Fig. 2-16 together with other pertinent properties of the standard atmosphere up to an altitude of 88,743 m.

Example 2-7

What are the air temperature, pressure, and density at 10 km altitude for the U.S. standard atmosphere?

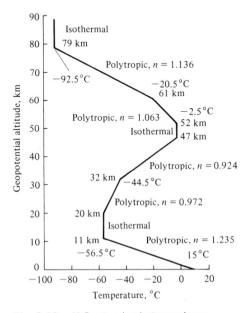

Fig. 2-16.   U.S. standard atmosphere.

SOLUTION

This is in the first polytropic region for which $n = 1.235$, $\bar{g} = 9.79$ m/s^2, and $R = 287.1$ mN/kg $°$K. The temperature is $T_2 = T_1 + (dT/dz)(z_2 - z_1) = 15 + (-6.5)(10) = -50°$C. From Eq. (2.7b) the pressure ratio is

$$\frac{p_2}{p_1} = \left[1 - \frac{(9.79)(0.235)(10,000)}{(1.235)(287.1)(288)} \right]^{1.235/0.235} = 0.262$$

$$p_2 = (0.262)(1.013 \times 10^5) = 2.64 \times 10^4 \quad \text{N/m}^2$$

$$\rho_2 = \frac{p_2}{RT_2} = \frac{2.64 \times 10^4}{(287.1)(223)} = 0.413 \quad \text{kg/m}^3$$

2-10. Static Stability

THE ATMOSPHERE

If a certain air mass is displaced vertically and, as a result of buoyancy due to differences in temperature, the displaced air tends to return to its original position, the atmosphere is considered to be stable. If a displacement results in no tendency to return to the original position or to be displaced further, the atmosphere is considered to be neutral. If a displacement results in a tendency for further displacement, the atmosphere is considered to be unstable.

An air mass moved vertically will be at the same pressure as the ambient air, and since there will be no heat transfer, the temperature will vary according to Eq. (1.14b):

$$\frac{T_2}{T_1} = \left(\frac{p_2}{p_1} \right)^{(k-1)/k}$$

If the atmosphere itself is truly adiabatic, the displaced air will always be at the same pressure and temperature as the ambient air, and there will be no buoyant forces tending to move the air mass. In order that $n < k$, the lapse rate would have to be greater algebraically than that for an adiabatic atmosphere ($> -0.00976°$C/m); thus when an air mass is moved vertically, its temperature would be less than ambient if moved upward and greater than ambient if moved downward, and buoyancy would tend to return it to its original level. Finally, if $n > k$, the lapse rate would have to be less algebraically than that for an adiabatic atmosphere; when an air mass is moved vertically, its temperature would be greater than ambient if moved upward and less if moved downward, and buoyancy would tend to displace the air mass further.

Thus, stability in the atmosphere may be defined in terms of the lapse rate or the polytropic exponent. If $n < k$, the atmosphere is stable; if $n = k$, the atmosphere is neutral; and if $n > k$, the atmosphere is unstable.

The U.S. standard atmosphere is used for determining altitude from altimeters (pressure gages) and, since the altimeters do not correct for varia-

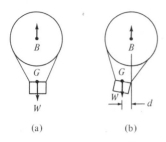

(a)   (b)

Fig. 2-17.   Stability of a submerged body.

tions in the temperature of the air beneath them, they read too high in cold weather and too low in warm weather.

SUBMERGED BODIES

The stability of a submerged body in a fluid at rest depends not only on the possibility of its rising or falling but also on whether, when tipped, the body will return to its initial position or will move farther from it. In *stable equilibrium* a body will return to its initial position if tipped. If *unstable*, the body will continue to move farther. In *neutral equilibrium* the body will remain at rest in its new position.

The position of the center of gravity of the body is designated G and the centroid of the displaced fluid (the center of buoyancy) is designated B. Thus in Fig. 2-17, if the gondola is displaced from position (a) to (b), a couple Wd tends to restore the balloon and gondola to the initial position (a), indicating stable equilibrium. If G were above B, a slight displacement would cause the system to continue tipping until B would be above G. Thus, stability requires that B be above G. If B and G coincide, neutral equilibrium exists.

FLOATING BODIES

Floating bodies may be in a condition of stable equilibrium even though the center of gravity G is above the center of buoyancy B. The position of B generally changes when a floating body tips about a horizontal axis because the shape of the volume of displaced liquid generally changes.

In Fig. 2-18a the floating body is in a state of equilibrium; the net force on the body is zero and there are no net moments on the body. In

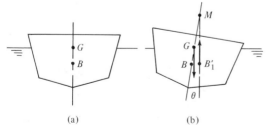

(a)   (b)

Fig. 2-18.   Metacentric height for a floating body.

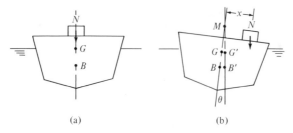

Fig. 2-19. Determination of metacentric height experimentally.

Fig. 2-18b the body is tipped through a small angle θ and the new center of buoyancy is at B_1. Since the weight of the body acts downward through G and the buoyant force acts upward through B_1, there is a restoring couple which tends to turn the body back to its original position shown in Fig. 2-18a. A vertical line through B_1 intersects the line through BG (extended) at M, called the *metacenter*. When M is above G the body is stable, but when M is below G the system is unstable because an overturning couple would tend to overturn the floating body. The distance from G to M is called the *metacentric height*, and it must be positive (M above G) for stability.

Therefore, the relative position of G and M determines stability of a floating body, whereas the relative position of G and B determines stability for a submerged body.

If a ship's cargo is an unconfined liquid, the center of gravity G would shift towards the center of buoyancy B_1 when the ship is tipped as in Fig. 2-18. Therefore, liquid cargo or ballast should be contained in compartments in the ship's hold to improve stability.

The metacentric height $\overline{GM}$ for tilting about a horizontal axis may be determined experimentally. In Fig. 2-19 let a weight N be moved athwartship from the center of an initially horizontal deck by a distance x. The ship will tilt through a small angle θ which can be measured by a plumb bob or other suitable means. The new center of gravity G' and the new center of buoyancy B' are then in a new vertical line. The center of gravity shifts from G to G', and equating moments give

$$Nx = WGG' = W\overline{GM} \tan \theta$$

and thus

$$\overline{GM} = \frac{Nx}{W} \cot \theta \qquad (2.10)$$

Here W is the weight of the floating body (including N) and N is the weight of the movable body.

The metacentric height $\overline{GM}$ is defined as the limiting value as θ approaches zero and may be obtained from a plot of $\overline{GM}$ versus θ for various values of x (or N).

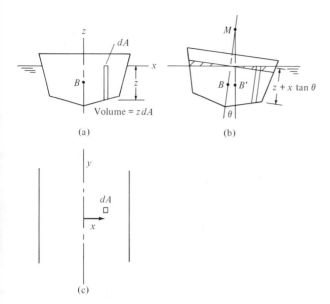

Fig. 2-20. Calculation of metacentric height.

The position of the metacenter or metacentric height also may be calculated. In Fig. 2-20, the x coordinate of the center of buoyancy is $\bar{x}_0$ such that with the total submerged volume $V\!\!\!/$

$$V\!\!\!/\bar{x}_0 = \int (z\,dA)x$$

where $(z\,dA)x$ is the moment of an infinitesimal submerged column. The value of $\bar{x}_0$ should be zero if the submerged portion is symmetrical about the yz plane. After a tilt or roll through a small angle θ, the new $\bar{x}$ coordinate is obtained from

$$V\!\!\!/\bar{x} = \int [(z + x \tan \theta)\,dA]x$$

The difference between these expressions is

$$V\!\!\!/(\bar{x} - \bar{x}_0) = \int x^2 \tan \theta \, dA = I \tan \theta$$

where I is the moment of inertia of the area of the floating body in the plane of the displaced surface. If $\bar{x}_0 = 0$, the last expression gives

$$\frac{V\!\!\!/\bar{x}}{\tan \theta} = V\!\!\!/\overline{BM} = I$$

and from Fig. 2-19, $\overline{GM} = \overline{BM} - \overline{BG}$ so that

$$\overline{GM} = \frac{I}{V\!\!\!/} - \overline{BG} > 0 \text{ for stability} \tag{2.11}$$

$$= 0 \text{ for neutral equilibrium}$$
$$< 0 \text{ for instability}$$

Equation (2.11) gives the position of the metacenter for rolling (tilting about the longitudinal axis) through small angles. For pitching (tilting about a transverse horizontal axis) the metacenter generally is at a different location.

Example 2-8

A long cylindrical log of specific gravity $\frac{2}{3}$ will float in water with its axis horizontal. As its length gradually is reduced it will float in the same position. If its length is reduced so that it is only a fraction of its diameter, it will float with its axis vertical. What is its diameter-to-length ratio for which it will just float with its axis vertical?

SOLUTION

Let the diameter be D and the length L. The required condition is when the metacentric height is zero in Eq. (2.11). Then $I\!\!\not{V} = \overline{BG}$, where $I = AD^2/16$ from Table 2-1, V is the submerged volume and equals $\frac{2}{3}AL$. The center of gravity is $L/2$ from the bottom of the short log and the center of buoyancy is $L/3$ from the bottom, since for a specific gravity of $\frac{2}{3}$ the log is two-thirds submerged. Then $\overline{BG} = L/6$ and setting Eq. (2.11) equal to zero gives

$$\frac{\frac{1}{16}AD^2}{2L/3} - \frac{L}{6} = 0 \qquad \text{from which} \qquad \frac{D}{L} = 1.33$$

Example 2-9

A ship has a displacement of 6000 metric tons. A body of 30 metric tons mass is moved laterally on the deck 12 m, and the end of a 1.80-m plumb bob moves 92 mm. What is the transverse metacentric height?

SOLUTION

In Eq. (2.10) $N = 30$ tons, $W = 6000$ tons, $x = 12$ m, and $\cot \theta = \frac{1800}{92}$. Thus the metacentric height is

$$\overline{GM} = \frac{(30)(12)}{6000}\left(\frac{1800}{92}\right) = 1.17 \text{ m}$$

Example 2-10

A rectangular freshwater barge is 15 m long, 6.8 m wide, and 2.2 m high and has a total mass of 80,000 kg. Assume the center of mass and the center of buoyancy to be at the center of the appropriate volumes. What is the transverse metacentric height?

SOLUTION

The center of gravity G is 1.1 m above the bottom of the barge, and the center of buoyancy B is 0.39 m above the bottom. This is one-half the submergence, which is equal to the volume of water displaced divided by the barge area of $(15)(6.8) = 102$ m². The submergence is

$$\frac{80,000}{(102)(1000)} = 0.784 \text{ m}$$

From Eq. (2.11),

$$\overline{GM} = \frac{I}{\Psi} - \overline{BG} = \frac{\frac{1}{12}A(6.8)^2}{0.78A} - (1.10 - 0.39)$$

$$= 4.23 \text{ m}$$

■■■■

PROBLEMS

2-1. Explain why the equations relating pressure changes with elevation [Eqs. (2.1), (2.6), and (2.7b)] all involve acceleration of gravity g, in addition to fluid properties.

2-2. Lake Baikal in southern Siberia is reported to be 1620 m deep. Estimate the pressure at the bottom of this lake.

2-3. The practical depth limit for a suit diver is about 185 m. What is the pressure intensity in seawater at that depth?

2-4. The Ewing camera has been used at a 5500-m depth in the Atlantic Ocean. What is the pressure intensity at that depth?
  (a) Assume constant density equal to that at the surface.
  (b) Use answer to part (a) and calculate the density at the 5500-m depth; then calculate the pressure at that depth using the average density.

2-5. A mercury barometer has a column 750 mm high.
  (a) What atmospheric pressure in Pa abs does this indicate?
  (b) What height of water column would produce this pressure?

2-6. Seals are known to dive to depths of about 60 m, whalebone whales to 350 m, and sperm whales to 950 m. What are the corresponding pressures these mammals are exposed to during dives?

2-7. What is the difference in pressure between the surface of a lake and a point 20 m below the surface? What is the difference in piezometric head between these points?

2-8. A spherical bubble migrates upward in water. At a depth of 15 m its diameter is 5 mm. What is its diameter just as it reaches the water surface? Assume a standard atmospheric pressure above the water.

2-9. A mercury barometer reads 760 mm. A pressure gage on a steam turbine indicates a pressure of 380 Pa gage. What is the absolute pressure of the steam?

2-10. Explain whether absolute or gage pressures *may* or *should* be used in making calculations involving
(a) pressure differences,
(b) pressure ratios, and
(c) the density of a gas from the perfect gas equation.

2-11. How would the height of a water barometer on a warm summer day at 40°C compare with that on a cold fall day at 4°C, assuming the same atmospheric pressure exists on both days? Explain.

2-12. What is the specific gravity of fluid *A*?

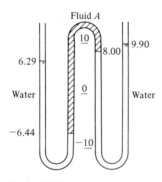

Prob. 2-12

2-13. What is the specific gravity of fluid *B*?

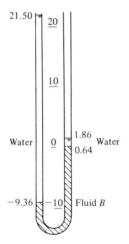

Prob. 2-13

2-14. What difference in piezometric head in meters of flowing fluid is indicated by a deflection of 40 cm of mercury (see Fig. 2-8) when the

flowing fluid is
(a) water,
(b) fuel oil, and
(c) carbon tetrachloride?
Let the specific gravity of fuel oil be 0.97 and carbon tetrachloride be 1.59.

2-15. In Fig. 2-7, acetylene tetrabromide is used in the manometer to measure the pump inlet pressure. The manometer deflection is 45.0 cm and $y = 20.0$ cm. What is the pump inlet pressure
(a) in Pa gage and
(b) in centimeters of mercury vacuum?

2-16. A mercury manometer is to be used to measure the pressure drop between two points along a horizontal pipe in which water flows. The maximum pressure drop to be measured is 140 kPa. What is the corresponding manometer deflection?

2-17. Two small tanks containing alcohol ($s = 0.80$) are connected to a manometer containing mercury ($s = 13.56$). The high-pressure tank is 1.50 m below the low-pressure tank.
(a) What is the pressure difference between the tanks for a manometer deflection of 21.0 cm?
(b) What piezometric head does this represent?

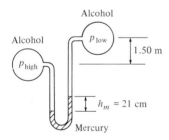

Prob. 2-17

2-18. A micromanometer similar to that in Fig. 2-9 has reservoirs at the top which have an area 60 times that of the tubes below. Oil is in one side and has a specific gravity of $s = 0.86$, and water is in the other side; the meniscus separating them is in one leg of the U-tube below. When a small pressure difference is applied between the two reservoirs, the meniscus rises 137 mm in the tube. What is the pressure difference which is applied?

2-19. Water flows downward in the duct. What is the difference in piezometric head between points A and B when the space at the top of the inverted U-tube contains
(a) air and
(b) an oil of specific gravity 0.85?

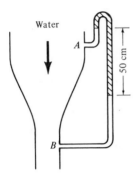

Prob. 2-19

2-20. What is the air pressure in the tank?

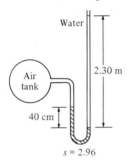

Prob. 2-20

2-21. In a pipe in which oil flows, what is
(a) the pressure drop and
(b) the drop in piezometric head across an orifice if it is measured with a mercury manometer which indicates a deflection of 20 cm? Assume $s = 0.92$ for the oil. See Fig. 2-8.

2-22. An oil ($s = 0.9$) is to be used in a manometer to measure the pressure drop across a length of horizontal pipe. Draw a sketch showing the manometer arrangement, the relative position of the oil–water meniscus in each leg of the manometer, and the flow direction. The pressure drops in the direction of flow.

2-23. The deflection of a differential manometer is a direct indication of the difference in *piezometric* pressure or head between two piezometer holes to which the manometer is connected. It indicates the difference in pressure intensity or pressure head only for a horizontal flow system. Show that these statements are true by writing expressions for the difference in piezometric pressure or head and the difference in pressure intensity or pressure head for the system shown in Fig. 2-8 when the pipe is inclined and when the pipe is horizontal.

2-24. A manometer is connected to piezometer taps 10 m apart in a pipe inclined 30° from the horizontal. Oil of specific gravity $s = 0.9$ flows in the pipe. The pressure drop due to viscous shear is to be measured.

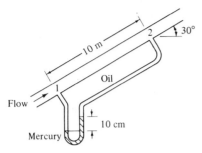

Prob. 2-24

The manometer contains mercury and shows a deflection of 10.0 cm.
(a) What is $p_1 - p_2$?
(b) What is the pressure drop due to viscous shear?
(c) Explain the difference in answers for parts (a) and (b).

2-25. Show that the resultant force on any rectangular flat surface whose base is horizontal and whose top coincides with a free liquid surface at rest is located $2H/3$ from the liquid surface, H being the height of the rectangle.

2-26. A vertical plate has dimensions as shown. It is submerged in a liquid so that the upper edge coincides with the free surface of the liquid. Is the total force on the square portion greater, the same, or less than the total force on the semicircular portion?

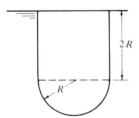

Prob. 2-26

2-27. A flat plate consists of a semicircle of radius R attached to a rectangle of width $2R$ and height L (instead of the $2R$ dimension in Prob. 2-26). What is the ratio of L/R such that the force of the liquid on the semicircle is the same as the force on the rectangle when the plate is submerged vertically in the liquid with the upper edge of the rectangle coinciding with the free surface of the liquid?

2-28. What is the total force on one side of vertical forms for concrete 2.5 m long and 1.20 m high when fresh mortar (specific gravity 2.70) is poured into it? Assume the fresh mortar acts as a liquid.

2-29. An open cubical tank 3 m on each side is filled with oil of specific gravity 0.80.
(a) What is the force of oil on each side of the tank?

(b) What is the total force on the bottom?

(c) Locate the depth to the center of pressure on each side.

2-30. An open cubical tank 6 m on each side is filled with 2 m of water, 2 m of carbon tetrachloride, and 2 m of oil of specific gravity 0.80.

(a) What is the liquid force on each side of the tank?

(b) What is the total liquid force on the bottom?

(c) Locate the center of pressure on each side of the tank.

2-31. A large tank with vertical side walls contains water at a 1.40-m depth and a layer of oil with $s = 0.85$ and of depth 0.60 m over the water.

(a) What is the horizontal force on a 5-m width of a side of the tank?

(b) How high above the bottom of the tank is the line of action of this force?

2-32. Determine the magnitude and location of the resultant force on one side of a vertical rectangular gate 2 m wide and 4 m high with the upper horizontal edge submerged 5 m below the free surface of the water. If the specific gravity of the liquid is different from that of water, is there a change in the magnitude of the force? Is there a change in the location of the resultant force?

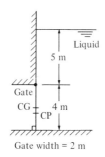

Gate width = 2 m

Prob. 2-32

2-33. A 1.50-m-diameter circular plate is submerged in water, the top and bottom being at depths of 2 and 3 m, respectively.

(a) What is the force of water on one side of the plate?

(b) What is the depth to the center of pressure?

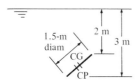

Prob. 2-33

2-34. Given a sea wall as shown,

(a) what is the resultant force per lineal meter of wall?

(b) Determine $\bar{x}$.

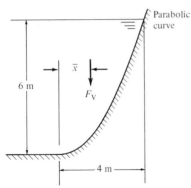

Prob. 2-34

2-35. Show that the moment of F_H about O is a^2/b^2 times the moment of F_V about O, and thus is equal to it for a circular quadrant. For the general case of an elliptical quadrant, what are the conditions for which the resultant of F_H and F_V pass
(a) above O,
(b) through O, and
(c) below O?

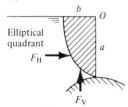

Prob. 2-35

2-36. What is the resultant force on the quarter-circle gate shown? The gate is 3 m wide.

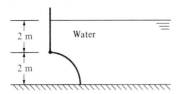

Prob. 2-36

2-37. A circular passage 1.40 m in diameter in the vertical wall of a dam is closed by means of a circular disk which fits the opening of the passage and is held in place by a horizontal shaft passing through its center (a horizontal diameter).
(a) Show that the turning moment required to maintain the disk in a vertical position is independent of the height of water above it if the level of water in the reservoir is above the top of the disk.
(b) What is the magnitude of this turning moment?

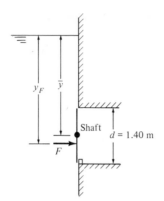

Prob. 2-37

2-38. An open channel has a cross section in the shape of an equilateral triangle with 3.0-m sides and an axis of symmetry which is vertical. A vertical gate at its end is used to close the channel, and it also is an equilateral triangle with 3.0-m sides.
(a) What is the force of water on the triangular gate?
(b) The gate is supported at each of its three corners. What is the force on each support?

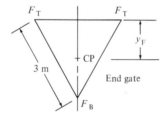

Prob. 2-38

2-39. A 60-cm cube weighing 2000 N is lowered into a large tank containing 1.8 m of water under 1.8 m of oil ($s = 0.8$). If the sides of the cube remain vertical,
(a) how much of the cube protrudes above the oil–water interface, and
(b) what is the total liquid force on one side of the cube?

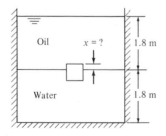

Prob. 2-39

2-40. A hollow sphere is filled with W lb$_f$ of liquid through a small hole in the top. What is the buoyant force on the top half of the sphere?

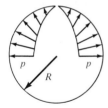

Prob. 2-40

2-41. A hollow right circular cone is filled with W lb$_f$ of liquid through a small hole in its apex. What is the buoyant force on the conical surface of the cone? Its axis is vertical.

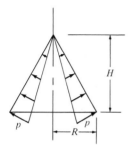

Prob. 2-41

2-42. Let W be the weight of a hydrometer whose stem has a cross-sectional area of A. Show that the distance between specific gravity markings

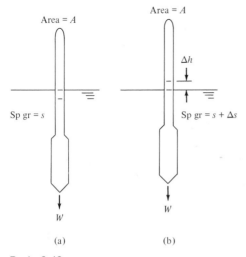

Prob. 2-42

for equal increments of s decreases as the specific gravity increases and thus the marked scale is nonlinear. Let Δh be the stem distance between marks indicating specific gravities of s and $s + \Delta s$. Then show that

$$\Delta h = \frac{W}{A\gamma} \left[\frac{\Delta s}{s(s + \Delta s)} \right]$$

where γ is the specific weight of water.

2-43. A hydrometer weighs 0.40 N in air and has a 5-mm diameter stem. What is the distance between specific gravity marks labeled 1.10 and 1.20?

2-44. One-eighth the total volume of an iceberg projects above the surface of the sea.
(a) What is the average density of the ice in the iceberg?
(b) What is its specific gravity?

2-45. A container of water has a total weight of 40.0 N and rests on a platform scale. A piece of iron of weight 31.4 N and specific gravity 7.80 hangs from a spring scale on a fine wire and is immersed in the water. What is the reading
(a) on the spring scale and
(b) on the platform scale?

2-46. Derive Eqs. (2.7a) and (2.7b) from the polytropic and hydrostatic equations [Eqs. (1.15) and (2.1)].

2-47. Derive Eq. (2.9) from Eq. (2.8b).

2-48. At what height in an isothermal atmosphere would the pressure be one-half that at sea level?

2-49. A 3.00-m constant-diameter balloon is filled with helium. The total weight of the balloon, gas, and payload is 90 N. In an isothermal atmosphere, at what altitude will the balloon come to rest? Ground pressure is 1.00×10^5 Pa abs and temperature is 20°C.

2-50. Verify the lapse rates and polytropic exponents for the U.S. standard atmosphere given in Table 2-3.

2-51. Assume that air in the atmosphere is a perfect gas and that the temperature at a height z above sea level is given by

$$T = T_0 - az$$

where T_0 is the sea level temperature. From Eq. (2.1) show that

$$\frac{p}{p_0} = \left(1 - \frac{az}{T_0} \right)^{g/aR}$$

Show that it is equivalent to Eq. (2.7b) when z_1 is zero.

2-52. An aneroid barometer indicates a pressure of 701 mm mercury at the top of a mountain. What is the corresponding elevation in a U.S. standard atmosphere?

2-53. In a U.S. standard atmosphere, at what elevation is the pressure
(a) 80 percent and
(b) 90 percent of that at sea level?

2-54. What is the pressure in the U.S. standard atmosphere at an altitude of 8500 m?

2-55. A barometer at the foot of a mountain reads 750 mm of mercury and at the top, 545 mm mercury. For a temperature of 13°C at the foot of the mountain, estimate the height of the mountain. Assume a U.S. standard atmosphere polytropic exponent.

2-56. How high will the balloon in Prob. 2-49 rise to an equilibrium condition in a U.S. standard atmosphere? Assume standard sea level conditions at ground level.

2-57. The temperature at the top of a mountain is $-2°C$ and a mercury barometer reads 553 mm. At the foot of the mountain the mercury barometer reads 674 mm. Assume a standard atmosphere and calculate the height of the mountain.

2-58. A diamond mine in South Africa is 2700 m deep. Assume a standard atmosphere at ground level and a polytropic variation of pressure with depth and a linear temperature variation similar to that above ground. What is the air pressure at the bottom of the mine?

2-59. A balloon of 2300 m³ volume is filled with helium at sea level in a U.S. standard atmosphere.
(a) What is the lift capacity at sea level when the balloon weighs 900 N?
(b) What is the lifting capacity at 4600 m in the standard atmosphere? Assume the balloon expands to a volume of 2500 m³ at that altitude.

2-60. From Eq. (2.7a) and (2.9), show that the altitude z may be expressed in terms of known sea level conditions T_0 and p_0 and temperature gradient dT/dz, and a measured pressure p at the altitude z. That is, show that

$$z = \frac{T_0}{-dT/dz}\left[1 - \left(\frac{p}{p_0}\right)^{-\frac{R}{g}\left(\frac{dt}{dz}\right)}\right]$$

for a prescribed temperature gradient.

2-61. Referring to Prob. 2-60, show that

$$p = p_0\left[1 + \left(\frac{dT}{dz}\right)\frac{z}{T_0}\right]^{-\frac{1}{\frac{R}{g}\left(\frac{dt}{dz}\right)}}$$

for a prescribed temperature gradient in a polytropic atmosphere.

2-62. The Empire State Building in New York City is 380 m high. Assume an U.S. standard atmosphere and sea level conditions at the street level. Suppose you have a desk-type aneroid barometer which you carry to the top of the building. If the accuracy of the barometer is within 0.5 mm of mercury, with what accuracy can you measure the height of the building?

2-63. Repeat Example 2-9 for a log of specific gravity $\frac{3}{4}$.

2-64. Repeat Example 2-9 for a body of 25 metric tons, all other data being the same.

2-65. Repeat Example 2-10 for a barge of mass 90,000 kg.

References

1. *U.S. Standard Atmosphere, 1962* (Washington, D.C.: U.S. Government Printing Office, 1962).

Part II
CONSERVATION LAWS

The laws of conservation of mass, conservation of momentum, conservation of energy, and the second law of thermodynamics are expressed in their elementary form for fixed or discrete masses and, in fluid mechanics, for material volumes moving with the fluid and containing the same fluid mass. It is generally more useful in fluid mechanics to write appropriate equations expressing these physical laws for a control volume through which the fluid flows rather than for a fixed mass system. When equations are written for a fixed mass system they are often referred to as Lagrangian equations because they are written following the particular particles of fluid which are of interest. Equations written for a control volume are often referred to as Eulerian equations and relate the flow of mass, momentum, or energy, for example, across the surface of the control volume to changes in these quantities which take place within the control volume.

The relationship between the system and the control-volume methods of analysis is known as the Reynolds transport theorem.

The Reynolds transport theorem states that the rate of change of the integral of a single-valued point function (it could be the mass, momentum, or energy of the fluid) taken throughout a material volume is equal to the time rate of change of this integral taken throughout a control volume

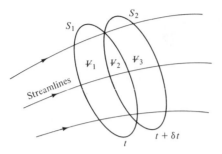

Fig. II-1.   Rate of change of a point function for flow through a control volume.

coinciding with the material volume at a given instant, plus the flux or transport of this point function across the surface of the control volume. In equation form for a point function P, this is

$$\frac{D}{Dt} \int_{\substack{\text{material} \\ \text{volume}}} P \, d\mathcal{V} = \frac{\partial}{\partial t} \int_{\substack{\text{control} \\ \text{volume}}} P \, d\mathcal{V} + \int_{\substack{\text{control} \\ \text{surface}}} P(\mathbf{V} \cdot d\mathbf{S}) \tag{II.1}$$

where t is time, $\mathcal{V}$ is volume, $\mathbf{V}$ is the fluid velocity, $\mathbf{S}$ is the area of the control volume, and D/Dt is the derivative following the particles.

The equation is derived as follows, with reference to Fig. II.1. Let S_1 be the boundary surface of a system at time t. After a time δt the mass of particles has moved to a new space bounded by the surface S_2. The point function P_t designates the mass, momentum, or energy of the system at time t. The region bounded by S_1 is $\mathcal{V}_1 + \mathcal{V}_2$, and the region bounded by S_2 is $\mathcal{V}_2 + \mathcal{V}_3$. The control volume is designated as the region bounded by S_1. Thus

$$P_t = P_{\mathcal{V}_1, t} + P_{\mathcal{V}_2, t}$$

and

$$P_{t+\delta t} = P_{\mathcal{V}_2, t+\delta t} + P_{\mathcal{V}_3, t+\delta t}$$

and thus the change in P for the mass particles in the system in time δt is

$$\delta P = P_{t+\delta t} - P_t = P_{\mathcal{V}_2, t+\delta t} - P_{\mathcal{V}_2, t} + P_{\mathcal{V}_3, t+\delta t} - P_{\mathcal{V}_1, t}$$

and

$$\frac{\delta P}{\delta t} = \frac{\delta P_{\mathcal{V}_2}}{\delta t} + \frac{P_{\mathcal{V}_3, t+\delta t} - P_{\mathcal{V}_1, t}}{\delta t}$$

In the limit as δt approaches zero, the region $\mathcal{V}_2$ approaches the control volume, and then the first term is the rate of change of the point function within the control volume and the second term is the rate at which the function P leaves through the control surface (rate leaving minus rate entering). Integration throughout the material volume, the control volume, and the control surface, respectively, gives Eq. (II.1).

Chapter 3
Fluid Kinematics

In this chapter we will consider the geometry of fluid motions without regard to the forces that produce these motions. These forces will be considered in following chapters. Included here will be a brief description of methods used to describe or classify flows; fluid velocity and acceleration; the concept of streamlines, pathlines, and streaklines; formal statements regarding the conservation of matter as applied to fluid flows; a discussion of fluid displacement and deformation; and vorticity and circulation.

3-1. Types of Flow

There are many criteria according to which fluid flows may be classified. For example, flow may be steady or unsteady, one, two, or three dimensional, uniform or nonuniform, laminar or turbulent, and incompressible or compressible. In addition, gas flows may be subsonic, transonic, supersonic, or hypersonic, and liquids flowing in open channels may be subcritical, critical, or supercritical.

Flow is *steady* when conditions do not vary with *time* or when, in the case of turbulent flow, variations in velocity and pressure at a point are generally small with respect to the mean values and the mean values do not

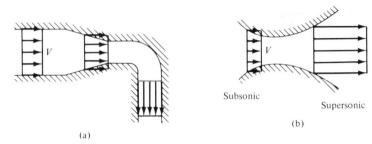

Fig. 3-1. One-dimensional flow. (a) Incompressible flow in a duct. (b) Compressible gas flow in a nozzle. Velocity, pressure, and temperature are assumed uniform throughout any section.

vary with time. If the flow is not steady, it is *unsteady*. Constant flow of water in a pipe is steady, but during opening or closing of a valve the flow is unsteady. Air flow through a blow-down type of wind tunnel would be essentially unsteady, but during a testing time of some 10–30 s the flow would be steady. A given flow may be steady with respect to one observer and unsteady with respect to another. For example, flow around the upstream portion of a bridge pier would appear steady to an observer on the pier, and it would appear unsteady to an observer floating by on the water.

Flow may be classified as one, two, or three dimensional. Various definitions of flow on this basis are found in technical literature.

Mathematical or hydrodynamical treatments often consider the kinematics of the system, and the type of flow is determined by the number of velocity components that exist.

In this text we will use the following definitions, which are commensurate with current engineering practice. In general the definitions involve the *gradients* of velocity, for example, rather than the *components* of velocity.

One-dimensional flow is flow in which all fluid and flow parameters (velocity, pressure, and temperature—thus density and viscosity) are constant throughout any cross section normal to the flow [1] or are represented by their average values over the cross section. Changes in both flow velocity and area may occur from section to section. Average fluid and flow parameters vary only from section to section. The flow of real fluids cannot be completely one dimensional since the velocity at a boundary must be zero with respect to the boundary.[†] They may, however, be assumed to be one dimensional in many instances. Whenever the flow is one dimensional—actual *or* assumed—a one-dimensional *analysis* may be applied to the flow system, and the average value of the velocity is used to describe the flow at a section. Examples of one-dimensional flow are shown in Fig. 3-1.

Two-dimensional flow is flow which is the same in parallel planes and which is not one-dimensional. It is also necessary that conditions be the

---

[†] There are instances of slip flow at the boundaries of rarefied gas flow, but treatment of these is beyond the scope of this text.

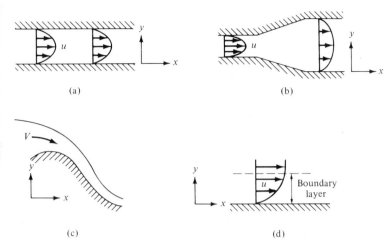

Fig. 3-2.   Two-dimensional flow. (a) Viscous flow between parallel plates;
$u = u(y)$ and $p = p(x)$. (b) Viscous flow between diverging plates; $u = u(x,y)$
and $p = p(x,y)$. (c) Flow over the central part of a wide spillway; $V = V(x,y)$.
(d) Boundary-layer flow past a wide flat plate; $u = u(x,y)$. See also Fig. 7-5.
Flow within boundary layer is two dimensional. Flow outside boundary layer
may be essentially one dimensional.

same along any line normal to the planes. This prevents otherwise identical
flow patterns from being staggered from plane to plane. Usually, two-
dimensional flow may also be defined as flow in which either the fluid or
flow parameters (or both) have spatial gradients in two directions, x and y,
for example. Thus in Fig. 3-2a the velocity varies only in the y direction but
the pressure varies in the x direction, so that gradients exist in two directions.
Note that while the flow is one *directional* since velocity vectors have only
an x component, the flow is two *dimensional*. In Fig. 3-2b the velocity varies
in both the x and the y direction; and the pressure varies in the x direction,
where streamlines are parallel, and in both the x and the y directions, where
streamlines are curved (in the diverging section).

Three-dimensional flow is flow in which the fluid or flow parameters
vary in the x, y, and z directions in a rectangular system of coordinates.
Thus, gradients of the fluid or flow parameters exist in all three directions.
Axisymmetric flow is sometimes considered to be two dimensional, since in
cylindrical coordinates gradients exist in only two directions—axial and
radial.

A special instance in which two-dimensional flow has gradients in two
directions is shown in Fig. 3-3. The flow is not one dimensional although
one directional; but since it is the same in parallel planes, we may consider
it to be two dimensional.

Flow past airfoils of infinite span is two dimensional; flow near the
ends of a finite-span airfoil is three dimensional. Flow past the central portion
of a completely submerged hydrofoil is two dimensional; for a hydrofoil

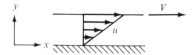

Fig. 3-3. Viscous flow between parallel plates, one at rest and the other moving; $u = u(y)$. Pressure is constant. Flow is not one dimensional, since velocity is not constant across a section. Since flow is the same in parallel planes, the flow is two dimensional.

piercing a free-water surface, the flow is three dimensional near the water surface. Flow over the central portion of a wide spillway in a river is two dimensional, and flow through ducts is usually three dimensional; they may both be considered one dimensional if average values of fluid and flow parameters are assumed to exist throughout any section normal to the mean flow. Many real flow situations which are actually two or three dimensional may be analyzed quite satisfactorily with the assumption of one-dimensional flow.

Flow may be *rotational* or *irrotational* depending on whether or not fluid particles or elements rotate about their own centers as they move throughout space.

Flow may be *uniform* or *nonuniform*, depending on the variation in flow area and the velocity in the direction of flow. If the mean velocity V and the cross-sectional area A are constant in the direction of flow, the flow is uniform. If not, the flow is nonuniform. Liquid flow in a pipe of constant area and in an open channel of constant width and depth exemplifies uniform flow. Liquid flow in varying-area ducts, and all gas flows except low-velocity, constant-area gas flow, are examples of nonuniform flow because the velocity varies from section to section.

Laminar flow exists when fluid layers flow alongside one another either at the same or at slightly different velocities and there is no macroscopic mixing of fluid particles. If a dye (insoluble and of the same density as the flowing fluid) is injected into a laminar flow, the dye will appear as a single thread or filament and will not disperse throughout the fluid. In *turbulent* flow there are velocity fluctuations (both parallel and transverse to the mean velocity) superimposed on the mean motion. There is a mixing of the fluid, and a dye injected into a turbulent flow quickly disperses throughout the fluid. The straight filaments of smoke rising for a few inches from a cigarette held in still air is a well-known example of laminar flow, and the ensuing sinuous or haphazard motion of the smoke above the straight filaments is an example of turbulent motion. Whether the flow is laminar or turbulent depends on the relative magnitudes of the inertia and viscous forces in a fluid system, and there are quantitative values of the ratio of these forces for various flow systems from which the type of flow usually may be determined. There are instances of quasi-laminar-turbulent flow when turbulent flow is not fully established. This is called *transition* flow. This type of flow is usually intermittent at a point or section of the flow—laminar, then turbulent, then laminar, and so on—and is difficult to delineate, analytically as well as experimentally. Laminar flow may be studied analytically, but turbulent flow requires experimental results (combined with analytical) for complete analysis. Laminar flow is associated with low velocities, highly viscous fluids, or small flow passages. Turbulent flow is associated with high velocities, lower viscosity fluids, or large flow passages. In engineering practice, turbulent flow is generally more common than laminar flow.

Flow is considered *incompressible* if the density changes are negligible. All liquid flows and gas flows at low velocities may be considered to be incompressible flow. Gas flow above 60–90 m/s may be considered to be compressible flow. Actually all fluids are somewhat compressible, but generally an incompressible fluid is considered to be one for which the density is independent of pressure up to several hundred atmospheres. It is customary to make a sharp distinction between compressible fluids and compressible flow. Compressible effects in gas flows become important only progressively as the speed increases (see Chapter 10).

Gas flows are considered *subsonic, transonic, supersonic,* or *hypersonic,* depending on whether the velocity is less, about the same, greater, or much greater than the speed of sound (see Fig. 10-2).

Water flowing in an open channel (a river or spillway) is considered *subcitical, critical,* or *supercritical,* depending on whether the velocity is less, the same, or greater than that of an elementary surface wave. A wave generated when a pebble, for example, is dropped into shallow water is considered to be an elementary wave (see Chapter 12).

3-2. Fluid Velocity and Acceleration

Fluid motions may be studied by observing given fluid particles as they move through space or by focusing attention on the motion and properties of different fluid particles as they pass fixed points in space. The first method is called the *Lagrangian method,* and the second the *Eulerian method.* The Eulerian method is generally simpler and will be used here.

Fluid velocity V may vary with position and with time, as may u, v, and w, its components in the x, y, z Cartesian coordinate system, such that

$$u = \frac{dx}{dt} = u(x,y,z,t)$$

$$v = \frac{dy}{dt} = v(x,y,z,t)$$

$$w = \frac{dz}{dt} = w(x,y,z,t)$$

and are also functions of position and time.

Since the velocity of a fluid element or particle is a function of both position and time, we may write for the x component, for example,

$$du = \frac{\partial u}{\partial x} dx + \frac{\partial u}{\partial y} dy + \frac{\partial u}{\partial z} dz + \frac{\partial u}{\partial t} dt$$

with similar expressions for dv and dw. The acceleration in the x direction is

$$a_x = \frac{du}{dt} = \frac{\partial u}{\partial x}\frac{dx}{dt} + \frac{\partial u}{\partial y}\frac{dy}{dt} + \frac{\partial u}{\partial z}\frac{dz}{dt} + \frac{\partial u}{\partial t}$$

$$= \frac{Du}{Dt} = u\frac{\partial u}{\partial x} + v\frac{\partial u}{\partial y} + w\frac{\partial u}{\partial z} + \frac{\partial u}{\partial t}$$

where

$$\frac{D}{Dt} = u\frac{\partial}{\partial x} + v\frac{\partial}{\partial y} + w\frac{\partial}{\partial z} + \frac{\partial}{\partial t} \tag{3.1}$$

is called the material or substantial or particle derivative. The first three terms are associated with the movement of fluid particles and are the *convective* acceleration, and the last term is associated with the change of a property at a fixed point with respect to time and is the *local* acceleration.

In vector form, the acceleration of a fluid particle is

$$\mathbf{a} = \frac{D\mathbf{V}}{Dt} = u\frac{\partial \mathbf{V}}{\partial x} + v\frac{\partial \mathbf{V}}{\partial y} + w\frac{\partial \mathbf{V}}{\partial z} + \frac{\partial \mathbf{V}}{\partial t}$$

$$= (\mathbf{V} \cdot \nabla)\mathbf{V} + \frac{\partial \mathbf{V}}{\partial t} \tag{3.2}$$

In Cartesian coordinates

$$a_x = u\frac{\partial u}{\partial x} + v\frac{\partial u}{\partial y} + w\frac{\partial u}{\partial z} + \frac{\partial u}{\partial t} \tag{3.3a}$$

$$a_y = u\frac{\partial v}{\partial x} + v\frac{\partial v}{\partial y} + w\frac{\partial v}{\partial z} + \frac{\partial v}{\partial t} \tag{3.3b}$$

$$a_z = u\frac{\partial w}{\partial x} + v\frac{\partial w}{\partial y} + w\frac{\partial w}{\partial z} + \frac{\partial w}{\partial t} \tag{3.3c}$$

In cylindrical (r,θ,z) coordinates with velocity components v_r, v_θ, and v_z, the acceleration components are

$$a_r = v_r\frac{\partial v_r}{\partial r} + \frac{v_\theta}{r}\frac{\partial v_r}{\partial \theta} + v_z\frac{\partial v_r}{\partial z} - \frac{v_\theta^2}{r} + \frac{\partial v_r}{\partial t} \tag{3.4a}$$

$$a_\theta = v_r\frac{\partial v_\theta}{\partial r} + \frac{v_\theta}{r}\frac{\partial v_\theta}{\partial \theta} + v_z\frac{\partial v_\theta}{\partial z} + \frac{v_r v_\theta}{r} + \frac{\partial v_\theta}{\partial t} \tag{3.4b}$$

$$a_z = v_r\frac{\partial v_z}{\partial r} + \frac{v_\theta}{r}\frac{\partial v_z}{\partial \theta} + v_z\frac{\partial v_z}{\partial z} + \frac{\partial v_z}{\partial t} \tag{3.4c}$$

There is a natural coordinate system based on a streamline direction s (see the next section), a direction n normal to and in the plane of the streamline, and a meridional direction m normal to the plane of s and n. Expressions for acceleration in this natural coordinate system are given in Sec. 4-3.

Example 3-1

The velocity of a nonviscous incompressible fluid as it steadily approaches the stagnation point at the leading edge of a sphere of radius R is (see the figure for Prob. 3-10)

$$u = u_s\left[1 + \frac{R^3}{x^3}\right]$$

What is the fluid acceleration at (a) $x = -3R$, (b) $x = -2R$, and (c) $x = -R$? (d) When $u_s = 2$ m/s and $R = 2$ cm, what is the magnitude of the acceleration at $x = -2R$?

SOLUTION

Along an x axis extended along a diameter, $a_s = u(\partial u/\partial x)$ where u is as given and $\partial u/\partial x = -3u_s R^3/x^4$. The results are given in the accompanying tabulation.

PART	u	$\partial u/\partial x$	a_x
(a)	$26u_s/27$	$-3u_s/81R$	$-(26/729)u_s^2/R$
(b)	$7u_s/8$	$-3u_s/16R$	$-(21/128)u_s^2/R$
(c)	0	$-3u_s/R$	0
(d)	\multicolumn{3}{l}{$a_x = -(21/128)(2)^2/(0.02) = -32.8$ m/s^2}

It is important to note that if a fluid particle changes velocity as it moves from one point in space to another, there is a *convective* acceleration. If the convective acceleration is zero, the flow is called *uniform* flow. If the velocity of successive fluid particles which pass through a given point in space changes with time, there is a *local* acceleration. If the local acceleration is zero, the flow is called *steady* flow.

3-3. Streamlines and the Stream Function, Pathlines, and Streaklines

A *streamline* is a line to which, at each instant, velocity vectors are tangent. Thus at each instant, $\mathbf{V} = \mathbf{V}(x,y,z)$. From Fig. 3-4, $\mathbf{V}$ is parallel to $d\mathbf{r}$, a segment of the streamline, and thus $\mathbf{V} \times d\mathbf{r} = \mathbf{0}$, where $\mathbf{V}$ and $\mathbf{r}$ are as defined in Sec. 3-2. Expansion of this cross product will show that

$$\frac{dx}{u} = \frac{dy}{v} = \frac{dz}{w} \tag{3.5}$$

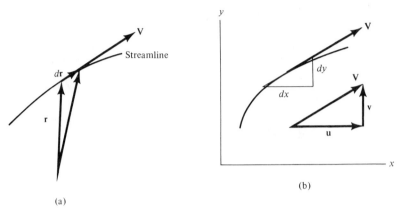

Fig. 3-4. Streamlines and velocity components. (a) Segment $d\mathbf{r}$ of streamline. (b) Velocity components.

is the equation of streamlines, along with $\mathbf{V} \times d\mathbf{r} = \mathbf{0}$. From Fig. 3-4b in the $x-y$ plane, $dy/dx = v/u$, and thus $dz/dy = w/v$ and $dx/dz = u/w$, which are the same as Eq. (3.5). Some streamline patterns are shown in Chapter 6.

A *pathline* is the path or trajectory of a given fluid particle as it moves throughout space. Since the velocity of this particle may be a function of both position and time, $\mathbf{V} = \mathbf{V}(x,y,z,t)$. For *steady* flow, the particle velocity will not depend on time, and thus *pathlines are identical to streamlines for steady flow.*

A *streakline* is an instantaneous locus of all fluid particles that have passed through a given point. If dye is injected into a fluid flow from a given point in the flow, a photograph of the dye streak would be a streakline. If the flow is unsteady, photographs taken at different instances would show different streak lines.

For *unsteady* flow, then, streamlines, pathlines, and streaklines may all be different. For *steady* flow, however, streamlines, pathlines, and streaklines all coincide.[†]

In two-dimensional flow the equations of streamlines may be described by *stream functions,* different values of the stream function ψ designating different streamlines. The x and y components of velocity (u and v) may be obtained from the stream function as follows:

$$u = -\frac{\partial \psi}{\partial y} \tag{3.6a}$$

and

$$v = +\frac{\partial \psi}{\partial x} \tag{3.6b}$$

[†] Film strips which show this very well are available [2].

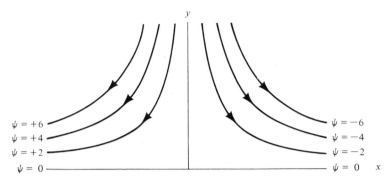

Fig. 3-5.   Streamline pattern for Example 3-2.

Since $\psi = \psi(x,y)$, the total differential of ψ is

$$d\psi = \frac{\partial \psi}{\partial x}\, dx + \frac{\partial \psi}{\partial y}\, dy$$

and from Eq. (3.6), this is

$$d\psi = v\, dx - u\, dy$$

From Eq. (3.5) this is zero along a streamline, and thus streamlines are lines of constant stream function.

In polar (r,θ) coordinates with positive velocity components v_r in the outward radial direction and v_θ in the counterclockwise tangential direction

$$v_r = -\frac{1}{r}\frac{\partial \psi}{\partial \theta} \qquad\qquad (3.7a)$$

and

$$v_\theta = +\frac{\partial \psi}{\partial r} \qquad\qquad (3.7b)$$

Example 3-2

Given $\psi = -2xy$, find the velocity components and describe the flow in the upper half-plane.

SOLUTION

$u = -\partial\psi/\partial y = +2x$ and $v = \partial\psi/\partial x = -2y$. Thus u is to the right in the first quadrant and to the left in the second quadrant; v is downward in the entire upper half-plane. Streamlines are rectangular hyperbolas, obtained by setting $\psi = 0, \pm 2, \pm 4$, etc. The flow pattern is shown in Fig. 3-5.

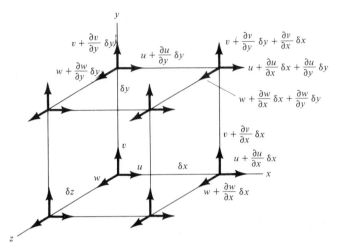

Fig. 3-6. Instantaneous velocity at some corners of an elemental cube of fluid.

3-4. Deformation of a Fluid Element

One method of classifying flows is to describe them as being either irrotational or rotational. In order to have a clearer concept of just what is meant by these terms, it will be informative to examine the types of motion that a fluid element may undergo. Relations are obtained that will be useful in developing expressions for vorticity and circulation in the next section, and in developing the equations of motion in the next chapter.

In general, at a fixed instant the velocity vector which describes a fluid motion will vary throughout space. If we examine the motion of the six faces of an elemental cube of fluid having sides δx, δy, and δz we will find that four types of deformation or movement are possible as a consequence of spatial variations in velocity: (1) translation, (2) linear deformation, (3) angular deformation, and (4) rotation. These may, of course, occur singly or in combination.

Figure 3-6 shows velocity components at each corner of the three faces of the elemental fluid cube which lie in the x–y, y–z, and x–z planes, respectively. During a time dt the faces of the cube will move through space, and the cube will change its form. The four types of movement of the face in the x–y plane are shown in Fig. 3-7.

1. In Fig. 3-7a the *translation* in the x and y directions is $u\,dt$ and $v\,dt$, respectively. There would, of course, be a translation in the z direction also.
2. In Fig. 3-7b the *linear deformations* in the x and y directions are $(\partial u/\partial x)\,\delta x\,dt$ and $(\partial v/\partial y)\,\delta y\,dt$, respectively, with a corresponding linear deformation of $(\partial w/\partial z)\,\delta z\,dt$ in the z direction. Thus the change in length of each of the three sides of the elemental fluid cube in

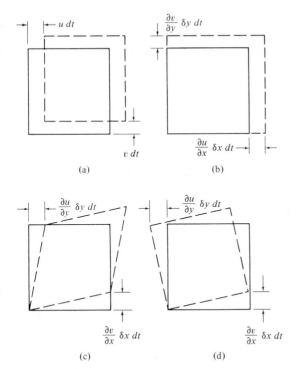

Fig. 3-7. Movements of the face of an elemental cube of sides δx, δy, and δz in the x–y plane in a time dt. (a) Translation. (b) Linear deformation. (c) Angular deformation. (d) Rotation.

Fig. 3-6 will cause a relative change in its volume at a rate of $(V_{dt} - V_0)/V_0\, dt$, or

$$\frac{\left(\delta x + \dfrac{\partial u}{\partial x}\delta x\, dt\right)\left(\delta y + \dfrac{\partial v}{\partial y}\delta y\, dt\right)\left(\delta z + \dfrac{\partial w}{\partial z}\delta z\, dt\right) - \delta x\, \delta y\, \delta z}{\delta x\, \delta y\, \delta z\, dt}$$

$$= \frac{\partial u}{\partial x} + \frac{\partial v}{\partial y} + \frac{\partial w}{\partial z} = \text{div } \mathbf{V} \tag{3.8}$$

For an incompressible fluid this obviously must be zero and is thus an expression of the continuity equation.

3. The net *angular deformation* in Fig. 3-7c, considering positive angles as counterclockwise, is the difference between the angular deformations of the individual sides δx and δy:

$$\left(\frac{\dfrac{\partial v}{\partial x}\delta x\, dt}{\delta x}\right) - \left(\frac{-\dfrac{\partial u}{\partial y}\delta y\, dt}{\delta y}\right) = \left(\frac{\partial v}{\partial x} + \frac{\partial u}{\partial y}\right) dt$$

The rate of distortion is defined as one-half the rate of angular deformation:

$$\dot{\epsilon}_{xy} = \dot{\epsilon}_{yx} = \frac{1}{2}\left(\frac{\partial u}{\partial y} + \frac{\partial v}{\partial x}\right) \tag{3.9a}$$

Similarly, the rate of distortion in shape in the y–z plane is

$$\dot{\epsilon}_{yz} = \dot{\epsilon}_{zy} = \frac{1}{2}\left(\frac{\partial v}{\partial z} + \frac{\partial w}{\partial y}\right) \tag{3.9b}$$

and finally, the rate of distortion in shape in the x–z plane is

$$\dot{\epsilon}_{zx} = \dot{\epsilon}_{xz} = \frac{1}{2}\left(\frac{\partial w}{\partial x} + \frac{\partial u}{\partial z}\right) \tag{3.9c}$$

Also

$$\dot{\epsilon}_{xx} = \frac{\partial u}{\partial x}, \qquad \dot{\epsilon}_{yy} = \frac{\partial v}{\partial y}, \qquad \text{and} \qquad \dot{\epsilon}_{zz} = \frac{\partial w}{\partial z} \tag{3.9d}$$

4. *Rotation* of the elemental face in Fig. 3-7d is the average of the rotation of the sides δx and δy, clockwise again considered positive, and is

$$\frac{1}{2}\left(\frac{\partial v}{\partial x} - \frac{\partial u}{\partial y}\right) dt$$

The rate of rotation is the angular velocity about an axis parallel to the z axis, or

$$\omega_z = \frac{1}{2}\left(\frac{\partial u}{\partial x} - \frac{\partial u}{\partial y}\right) \tag{3.10a}$$

similarly

$$\omega_x = \frac{1}{2}\left(\frac{\partial w}{\partial y} - \frac{\partial v}{\partial z}\right) \qquad \text{in the } y\text{–}z \text{ plane} \tag{3.10b}$$

and

$$\omega_y = \frac{1}{2}\left(\frac{\partial u}{\partial z} - \frac{\partial w}{\partial x}\right) \qquad \text{in the } x\text{–}z \text{ plane} \tag{3.10c}$$

These are components of the resultant angular velocity vector Ω, where

$$\Omega = \frac{1}{2}\operatorname{curl} V = \frac{1}{2}\begin{vmatrix} \mathbf{i} & \mathbf{j} & \mathbf{k} \\ \dfrac{\partial}{\partial x} & \dfrac{\partial}{\partial y} & \dfrac{\partial}{\partial z} \\ u & v & w \end{vmatrix} = \omega_x\mathbf{i} + \omega_y\mathbf{j} + \omega_z\mathbf{k} \tag{3.11}$$

It should be noted that both translation and rotation involve deformation or motion without a change in shape of the fluid element. Linear and angular deformation, however, do involve a change in shape of the fluid element. Only through this linear and angular deformation, but not as a result of translation or rotation, is there heat generated and mechanical energy dissipated as a result of viscous action in a fluid. This will be shown in Sec. 5-3.

If the flow is *irrotational*, the angular velocities in Eqs. (3.10) and (3.11) are zero.

3-5. Vorticity and Circulation

Vorticity ζ is defined as $\zeta = 2\Omega = $ curl $\mathbf{V}$, which is twice the angular velocity or rotation. Angular velocity and rotation are synonymous, and they are often loosely used interchangeably with vorticity, although there is the factor 2 relating them. The vector quantity vorticity is, in general, a function of both position and time in a fluid. A vortex line is a line along which the vorticity vector is tangent, just as a streamline is a line along which the velocity vector is tangent. A vortex tube is bounded by vortex lines, just as a stream tube is bounded by streamlines. A vortex filament is a vortex tube of infinitesimal cross section. The product of vorticity and area is a constant (called *circulation*) along a vortex filament, just as the product of velocity and area is a constant (volumetric flow rate) along a stream filament for incompressible flow. Thus we may write

$$\zeta_1 \, \delta A_1 = \zeta_2 \, \delta A_2$$

and this product is called the strength of the vortex filament. Because of this constancy of strength, vortex filaments (and vortex lines as well) cannot terminate within a fluid. They must form closed curves (a smoke ring is an example) or terminate at a boundary or at infinity if the fluid is imagined to be unbounded.

If the vorticity is zero throughout a fluid in motion, the flow is called irrotational. Then curl $\mathbf{V} = \mathbf{0}$ and

$$\frac{\partial v}{\partial x} = \frac{\partial u}{\partial y}, \qquad \frac{\partial w}{\partial y} = \frac{\partial v}{\partial z}, \qquad \text{and} \qquad \frac{\partial u}{\partial z} = \frac{\partial w}{\partial x} \tag{3.12}$$

which are obtained by setting the coefficients of $\mathbf{i}$, $\mathbf{j}$, and $\mathbf{k}$ in curl $\mathbf{V}$ each equal to zero.

Circulation Γ is defined as the line integral of the tangential component of velocity around a closed curve. It is given by

$$\Gamma = \oint \mathbf{V} \cdot d\mathbf{l} = \oint (u \, dx + v \, dy + w \, dz) \tag{3.13}$$

and is equal to the total strength of all vortex filaments that pass through the closed curve. Thus the component of vorticity at a point within the closed

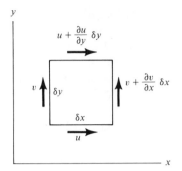

Fig. 3-8. Circulation around face of element in Fig. 3-6 in the x–y plane.

curve is the limit of the circulation per unit area enclosed by the curve:

$$\zeta_A = \lim_{A \to 0} \frac{\Gamma}{A} \qquad (3.14)$$

This can be shown in the x–y plane, for example, by tracing the boundaries of the area in Fig. 3-8 in a counterclockwise direction:

$$\Gamma_z = u\,\delta x + \left(v + \frac{\partial v}{\partial x}\delta x\right)\delta y - \left(u + \frac{\partial u}{\partial y}\delta y\right)\delta x - v\,\delta y$$

$$= \left(\frac{\partial v}{\partial x} - \frac{\partial y}{\partial y}\right)\delta x\,\delta y = \left(\frac{\partial v}{\partial x} - \frac{\partial u}{\partial y}\right)\delta A_z$$

Thus the circulation per unit area is

$$\zeta_z = 2\omega_z = \left(\frac{\partial v}{\partial x} - \frac{\partial u}{\partial y}\right) = \lim_{A_z \to 0} \frac{\Gamma_z}{A_z}$$

Circulation, then, is the counterpart of volumetric flow rate. Flow rate is the product of velocity and area, whereas circulation is the product of vorticity and area. These are illustrated in Fig. 3-9.

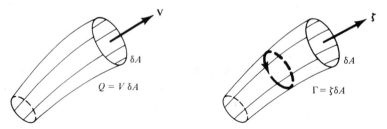

Fig. 3-9. Similarities between stream and vortex filaments. (a) Stream filament. (b) Vortex filament.

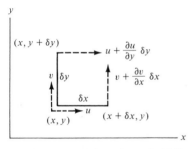

Fig. 3-10.   Rotation of two fluid-line segments.

FLUID ROTATION IN CARTESIAN COORDINATES

Fluid rotation at a point is defined as the instantaneous average rotation of two mutually perpendicular infinitesimal line elements contained in the fluid. In Fig. 3-10 the rotation of each line element is the difference in velocities at the ends divided by the line length. For counterclockwise rotation considered positive, the average rotation of the line elements δx and δy about a z axis normal to the x–y plane is

$$\omega_z = \frac{1}{2}\left[\frac{\left(v + \frac{\partial v}{\partial x}\delta x\right) - v}{\delta x} - \frac{\left(u + \frac{\partial u}{\partial y}\delta y\right) - u}{\delta y}\right]$$

$$= \frac{1}{2}\left(\frac{\partial v}{\partial x} - \frac{\partial u}{\partial y}\right) = \frac{1}{2}\frac{\Gamma_z}{\delta A_z}$$

with similar expressions [Eq. (3.10)] for rotation about the x and y axes.

ROTATION IN NATURAL COORDINATES

In Fig. 3-11 the circulation is obtained by traversing the element counterclockwise, noting that the contribution along the radial sides is zero. Here

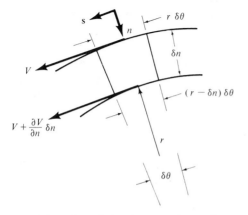

Fig. 3-11.   Circulation in natural coordinates.

V represents V_θ, there being no radial component:

$$\Gamma = Vr\,\delta\theta - \left(V + \frac{\partial V}{\partial n}\,\delta n\right)(r - \delta n)\,\delta\theta$$

$$= \left(\frac{V}{r} - \frac{\partial V}{\partial n}\right)\delta n\,r\,\delta\theta = \left(\frac{V}{r} - \frac{\partial V}{\partial n}\right)\delta A$$

so that

$$\omega = \frac{1}{2}\frac{\Gamma}{\delta A} = \frac{1}{2}\left(\frac{V}{r} - \frac{\partial V}{\partial n}\right) = \frac{1}{2}\left(\frac{V}{r} + \frac{\partial V}{\partial r}\right) \qquad (3.15)$$

For irrotational motion $V/r = \partial V/\partial n$ and the peripheral velocity increases toward the center of curvature of the streamlines, since $\partial V/\partial n$ is positive, and as n increases (toward the center of curvature) V also increases.

To summarize, for *irrotational* motion the curl of the velocity vector is zero and the vorticity is zero. For *rotational* flow, the curl of the velocity vector is *not* zero and there is vorticity.

All fluid flows must satisfy continuity. However, some flows are irrotational and some flows are rotational. These are nicely illustrated in films available for classroom use [3].

Many examples of irrotational flow are given in Chapter 6. They include uniform flow, sources, sinks, vortexes, and various combinations of them. Two types of vortex motion may be worth mentioning here. An irrotational vortex is one for which $V_\theta r = C$. From Eq. (3.15), $\partial V_\theta/\partial r + V_\theta/r = 0$, since $\partial V_\theta/\partial r = -C/r^2$ and $V_\theta/r = +C/r^2$. Streamlines are concentric circles (see Fig. 6-11). An irrotational vortex may exist not only in a nonviscous fluid but also in a viscous fluid. The flow beyond the core of a tornado is a combination of an irrotational vortex and a sink flow (a combination of circular and radially inward streamlines) as shown in Figs. 6-17 and 6-18. Also, if a long vertical circular cylinder is rotated at constant speed in a viscous liquid of very large extent, the laminar motion of the viscous fluid is that of an irrotational vortex. The rotational speed of the fluid at the surface of the cylinder corresponds to the peripheral speed of the cylinder, and it decreases to zero at an infinite distance from the cylinder according to the equation $V_\theta = C/r$.

A rotational vortex may be produced when a fluid rotates about an axis as a solid body, in which case $V_\theta = \omega r$, and from Eq. (3.15), $\omega = \frac{1}{2}(\omega + \omega)$.

An excellent discussion of vorticity and the role of eddies in fluid motion is given by Rouse [4].

Example 3-3

The flow of a nonviscous fluid past a two-dimensional half body (see Fig. 6-12) is described by the x and y components of velocity as

$$u = u_s + \frac{mx}{x^2 + y^2} \qquad \text{and} \qquad v = \frac{my}{x^2 + y^2}$$

where u_s and m are the strengths of the free-stream velocity and source which form the flow pattern. Show that the flow is irrotational.

SOLUTION
To prove the flow is irrotational, it must be shown that $\partial u/\partial y = \partial v/\partial x$:

$$\frac{\partial u}{\partial y} = \frac{-2mxy}{(x^2 + y^2)^2} \qquad \frac{\partial v}{\partial x} = \frac{-2mxy}{(x^2 + y^2)^2}$$

and thus the flow *is* irrotational.

Example 3-4

The flow in a laminar boundary layer is described by the expression

$$u = 1000y - \frac{10^9 y^3}{9} \quad \text{m/s}$$

where y is in meters and $v = 0$. At $y = 1$ mm, is the flow rotational or is it irrotational? Refer to Fig. 1-7.

SOLUTION
It is necessary to determine whether $\partial u/\partial y = \partial v/\partial x$. If so, the flow is irrotational; if not, the flow is rotational:

$$\frac{\partial u}{\partial y} = 1000 - 3 \times \frac{10^9 y^2}{9} = 1000 - 3 \times \frac{10^9 (0.001)^2}{9}$$

$$= 667 \ (\text{m/s})/\text{m}$$

Since $\partial v/\partial x = 0$, $\partial u/\partial y \neq \partial v/\partial x$ and the flow at $y = 1$ mm is rotational.

3-6. Conservation of Mass and the Continuity Equation

The equation of continuity may be written in different forms. It expresses the requirement that a fluid is continuous and that fluid mass is conserved—it is neither created nor destroyed.

Conservation of fluid mass requires that in a *material volume* the mass remains constant, and hence the rate of change of mass is zero. Various forms of the continuity equation for a *control volume* are derived by stating mathematically that the net rate of influx of mass into a given region is equal to the rate of change of mass in that region. The latter obviously will be zero for an incompressible fluid. The general statement may be shown formally from Eq. (II.I). The left-hand side is zero (since the mass remains

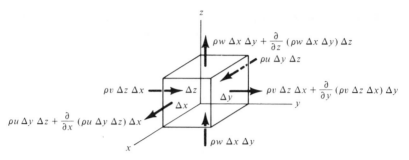

Fig. 3-12. Mass flow rate in and out of an elemental control volume.

constant in a material volume); and with the fluid density ρ as the single-valued point function, the Reynolds transport theorem becomes

$$0 = \frac{\partial}{\partial t} \int_{\substack{\text{control} \\ \text{volume}}} \rho \, dV + \int_{\substack{\text{control} \\ \text{surface}}} \rho(\mathbf{V} \cdot d\mathbf{S}) \tag{3.16}$$

This states that the rate of increase of mass within the control volume plus the net efflux of mass across its control surface is zero. Thus the net rate of influx of mass through the control surface equals the rate of increase of mass within the control volume. Fluid may flow into the control volume either through the control surface or from internal sources (see Chapter 6). Similarly, fluid may flow out of the control volume through the control surface or into internal sinks (see Chapter 6). We will develop forms of the continuity equation in the absence of sources and sinks. When they do occur, they must properly be taken into account, however.

SCALAR DERIVATION OF THE CONTINUITY EQUATION

In Fig. 3-12 the mass flow rate into the region which is a rectangular parallelopiped of sides Δx, Δy, and Δz in the $+x$ direction is $(\rho u \, \Delta y \, \Delta z)$, and out of it, in the $+x$ direction, it is the mass flow rate in, plus the rate of change of the mass flow rate in the $+x$ direction times Δx. This is

$$\rho u \, \Delta y \, \Delta z + \frac{\partial}{\partial x} (\rho u \, \Delta y \, \Delta z) \, \Delta x$$

The net inflow of mass in the $+x$ direction per unit time is the difference between these, which is

$$-\frac{\partial}{\partial x} (\rho u \, \Delta y \, \Delta z) \, \Delta x$$

Similarly, the net rate of mass flow into the region in the $+y$ and the $+z$ directions is

$$-\frac{\partial}{\partial y} (\rho v \, \Delta z \, \Delta x) \, \Delta y \qquad \text{and} \qquad -\frac{\partial}{\partial z} (\rho w \, \Delta x \, \Delta y) \, \Delta z$$

The rate of increase of mass in the region is (if not zero)

$$\frac{\partial}{\partial t}(\rho\,\Delta x\,\Delta y\,\Delta z)$$

and thus

$$-\frac{\partial}{\partial x}(\rho u\,\Delta y\,\Delta z)\,\Delta x - \frac{\partial}{\partial y}(\rho v\,\Delta z\,\Delta x)\,\Delta y - \frac{\partial}{\partial z}(\rho w\,\Delta x\,\Delta y)\,\Delta z$$

$$= \frac{\partial}{\partial t}(\rho\,\Delta x\,\Delta y\,\Delta z)$$

Since Δy and Δz do not vary with x; Δz and Δx do not vary with y; Δx and Δy do not vary with z; and Δx, Δy, and Δz do not vary with t, we can divide through by the quantity $\Delta x\,\Delta y\,\Delta z$, which is the volume of the region considered. We then obtain

$$\frac{\partial(\rho u)}{\partial x} + \frac{\partial(\rho v)}{\partial y} + \frac{\partial(\rho w)}{\partial z} = -\frac{\partial\rho}{\partial t} \tag{3.17}$$

For a constant density fluid ρ is constant, $\partial\rho/\partial t = 0$, and Eq. (3.17) becomes

$$\text{div } \mathbf{V} = \nabla \cdot \mathbf{V} = 0 \tag{3.18a}$$

and

$$\frac{\partial u}{\partial x} + \frac{\partial v}{\partial y} + \frac{\partial w}{\partial z} = 0 \tag{3.18b}$$

for both steady or unsteady flow (the velocity may vary with time as well as position in the fluid).

In cylindrical (r,θ,z) coordinates Eq. (3.17) becomes

$$\frac{\partial(\rho v_r)}{\partial r} + \rho\frac{v_r}{r} + \frac{1}{r}\frac{\partial(\rho v_\theta)}{\partial\theta} + \frac{\partial(\rho v_z)}{\partial z} = -\frac{\partial\rho}{\partial t} \tag{3.19}$$

which for an incompressible fluid becomes

$$\frac{\partial v_r}{\partial r} + \frac{v_r}{r} + \frac{1}{r}\frac{\partial v_\theta}{\partial\theta} + \frac{\partial v_z}{\partial z} = 0 \tag{3.20}$$

for both steady and unsteady flow.

THE CONTINUITY EQUATION FOR A STREAM TUBE

A stream tube is a flow passage bounded by streamlines, and thus no fluid may through the walls of a stream tube (Fig. 3-13). Pipes and nozzles are examples of stream tubes, although they are often considered to be small in cross section so that there is no variation in fluid velocity throughout the cross section. For one-dimensional flow in a stream tube, the continuity equation may be derived by equating the net mass flow rate into a stream tube of elemental length ds and area A to the rate of change of mass within this elemental control volume. Thus in terms of the average flow and fluid

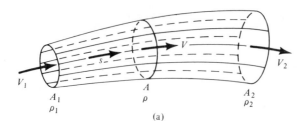

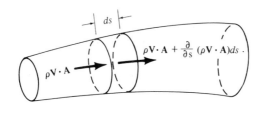

(b)

Fig. 3-13.   A stream tube (bounded by streamlines).

properties

$$\rho \mathbf{V} \cdot \mathbf{A} - \left[\rho \mathbf{V} \cdot \mathbf{A} + \frac{\partial}{\partial s} (\rho \mathbf{V} \cdot \mathbf{A}) \, ds \right] = \frac{\partial}{\partial t} (\rho A \, ds)$$

so that after division by ds (a fixed length)

$$\frac{\partial (\rho A)}{\partial t} + \frac{\partial}{\partial s} (\rho \mathbf{V} \cdot \mathbf{A}) = 0 \tag{3.21}$$

For steady flow, the first term is zero, and thus the second term also is zero, so that

$$\rho \mathbf{V} \cdot \mathbf{A} = \dot{m}, \quad \text{a constant mass flow rate} \tag{3.22}$$

For incompressible flow

$$\mathbf{V} \cdot \mathbf{A} = Q, \quad \text{a constant volumetric flow rate} \tag{3.23}$$

The various forms of the continuity equation are summarized in Table 3-1.

Equations (3.22) and (3.23) may be written in scalar form as

$$\dot{m} = \rho V A \tag{3.24}$$

and

$$Q = V A \tag{3.25}$$

respectively.

The area vector $\mathbf{A}$ is a vector whose magnitude is equal to the magnitude of the area and whose direction is normal to the plane of the area. Thus the

Table 3-1. CONTINUITY EQUATIONS

GENERAL	$\dfrac{\partial \rho}{\partial t} + \nabla \cdot \rho \mathbf{V} = 0$ or $\dfrac{D\rho}{Dt} + \rho \nabla \cdot \mathbf{V} = 0$	VECTOR
Unsteady, compressible	$\dfrac{\partial \rho}{\partial t} + \dfrac{\partial(\rho u)}{\partial x} + \dfrac{\partial(\rho v)}{\partial y} + \dfrac{\partial(\rho w)}{\partial z} = 0$	Cartesian
	$\dfrac{\partial \rho}{\partial t} + \dfrac{\partial(\rho v_r)}{\partial r} + \dfrac{1}{r}\dfrac{\partial(\rho v_\theta)}{\partial \theta} + \dfrac{\partial(\rho v_z)}{\partial z} + \dfrac{\rho v_r}{r} = 0$	Cylindrical
	$\dfrac{\partial(\rho A)}{\partial t} + \dfrac{\partial}{\partial s}(\rho \mathbf{V} \cdot \mathbf{A}) = 0$	Duct
Steady, compressible	$\nabla \cdot \rho \mathbf{V} = 0$	Vector
	$\dfrac{\partial(\rho u)}{\partial x} + \dfrac{\partial(\rho v)}{\partial y} + \dfrac{\partial(\rho w)}{\partial z} = 0$	Cartesian
	$\dfrac{\partial(\rho v_r)}{\partial r} + \dfrac{1}{r}\dfrac{\partial(\rho v_\theta)}{\partial \theta} + \dfrac{\partial(\rho v_z)}{\partial z} + \dfrac{\rho v_r}{r} = 0$	Cylindrical
	$\rho \mathbf{V} \cdot \mathbf{A} = \dot{m}$	
Incompressible	$\nabla \cdot \mathbf{V} = 0$	Vector
Steady or unsteady	$\dfrac{\partial u}{\partial x} + \dfrac{\partial v}{\partial y} + \dfrac{\partial w}{\partial z} = 0$	Cartesian
	$\dfrac{\partial v_r}{\partial r} + \dfrac{1}{r}\dfrac{\partial v_\theta}{\partial \theta} + \dfrac{\partial v_z}{\partial z} + \dfrac{v_r}{r} = 0$	Cylindrical
	$\mathbf{V} \cdot \mathbf{A} = Q$	Duct

dot product of area and velocity indicates that the velocity component normal to the plane of the area and parallel to the vector $\mathbf{A}$ representing the area is multiplied by the magnitude of the area to get a scalar product which represents a volumetric flow rate. Whenever the product of a scalar velocity and a scalar area is given as VA, the dot product $\mathbf{V} \cdot \mathbf{A}$ is implied.

Equation (3.24) states that for a gas flowing in a varying-area duct, the average velocity and density may both change along the direction of flow, but the product of these three quantities must be a constant mass flow rate for steady flow.

Equation (3.25) states that for constant-density flow (flow of liquids or low-velocity gases) the product of average velocity and cross-sectional area is a constant volumetric flow rate.

Example 3-5

Water flows at an average velocity of 3 m/s in the 0.20-m diameter inlet pipe of a pump. What is the average flow velocity in the 0.15-m diameter discharge pipe?

SOLUTION

$$V_2 = V_1(A_1/A_2) = (3)(0.20)^2/(0.15)^2 = 5.33 \text{ m/s}$$

▬▬▬

Example 3-6

Air at 3.50×10^5 Pa abs and $4°C$ flows at the speed of sound in the throat of a nozzle for a supersonic wind tunnel (see Fig. 10-4a, for example). The nozzle exit pressure is 1.978×10^4 Pa abs, and the exit area is four times the throat area. The flow is isentropic from the nozzle throat to the nozzle exit. What is the air velocity at the nozzle exit?

SOLUTION
From continuity, the mass flow rate is the same at the nozzle throat (section 1) as the nozzle exit (section 2). From Eq. (3.24) $V_2 = V_1 A_1 \rho_1/A_2 \rho_2$, where $V_1 = \sqrt{kRT} = \sqrt{(1.4)(287)(277)} = 334$ m/s from Eq. (1.25a):

$$\frac{\rho_1}{\rho_2} = \left(\frac{p_1}{p_2}\right)^{1/k} = \left(\frac{3.50 \times 10^5}{1.978 \times 10^4}\right)^{1/1.4} = 7.79$$

and $A_1/A_2 = 1/4$ as given.

Thus $V_2 = (334)(1/4)(7.79) = 650$ m/s

▬▬▬

Example 3-7

Water flows in a rectangular open irrigation channel 4.50 m wide at a depth of 0.80 m and at an average velocity of 1.25 m/s. The channel branches into two smaller rectangular channels: one is 3.00 m wide, 1.20 m deep, and conveys water at an average velocity of 0.95 m/s; the second branch is 2.50 m wide, and water flows at an average velocity of 0.60 m/s. What is the water depth in the second branch?

SOLUTION
The flow rate in the large channel equals the total flow in the two brances. Thus $Q_0 = Q_1 + Q_2$, or $V_0 A_0 = V_1 A_1 + V_2 A_2$, from which

$$A_2 = \frac{V_0 A_0 - V_1 A_1}{V_2} = \frac{(1.25)(4.5)(0.80) - (0.95)(3.00)(1.20)}{0.60}$$

$$= 1.80 \text{ m}^2$$

and the depth = area/width = $1.80/2.50 = 0.72$ m.

▬▬▬

The differential scalar form of Eq. (3.24) may be obtained by direct differentiation with respect to the stream tube direction s and division by

$VA\rho$, or by differentiating the logarithmic form of the equation with respect to s. We obtain for flow through a conduit (stream tube)

$$\frac{1}{V}\frac{dV}{ds} + \frac{1}{A}\frac{dA}{ds} + \frac{1}{\rho}\frac{d\rho}{ds} = 0$$

which is often written as

$$\frac{dV}{V} + \frac{dA}{A} + \frac{d\rho}{\rho} = 0$$

From this it is apparent that the velocity, area, and density cannot all increase, nor can they all decrease, in the direction of flow (in the direction of positive s).

In these equations $\mathbf{V}$ represents the average velocity vector, and V the average scalar velocity in a direction parallel to the mean streamline direction at a cross section normal to the streamline direction. When the velocity u varies across this section, the average velocity $\mathbf{V}$ is

$$V = \frac{1}{A}\int u\,dA = \frac{1}{n}\sum_{1}^{i=n} u_i \tag{3.26}$$

where u_i are point velocities measured at the centroids of n equal areas throughout a flow cross section.

Example 3-8

The velocity profile for turbulent flow in a circular tube is approximated by $u = u_m(y/R)^{1/7} = u_m(1 - r/R)^{1/7}$, where u_m is the centerline velocity, R is the tube radius, and y is the radial distance from the tube wall ($r = R - y$, thus $dr = -dy$). What is the average flow velocity in the tube in terms of u_m?

SOLUTION

$$V(\pi R^2) = \int_0^R u(2\pi r)\,dr$$

$$V = \frac{2}{R^2}\int_0^R \frac{u_m y^{1/7}}{R^{1/7}}\,r\,dr = \frac{2u_m}{R^{15/7}}\int_0^R y^{1/7}(R - y)\,dy = \frac{49}{60}u_m$$

Streamline patterns for two-dimensional incompressible flow may be used to determine velocity variations in the flow field. From Eq. (3.25), $V_1 n_1 = V_2 n_2$ in Fig. 6-2, and thus velocities vary inversely with streamline spacing. Hence velocity variations may qualitatively be obtained by a simple inspection of a streamline flow pattern. Measurements of streamline spacing may be used to obtain quantitative variations in flow velocity, while exact values may be obtained by differentiation of the stream function which describes the flow. This will be discussed in more detail in Chapter 6.

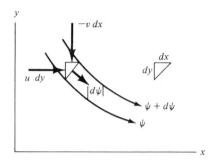

Fig. 3-14.   Flow rate between streamlines for incompressible fluid.

The flow rate between streamlines is equal to the difference between the values of stream functions representing the streamlines. In Fig. 3-14 the flow rate between streamlines ψ and $\psi + d\psi$ toward the right is the sum of the volumetric flow rate in the $+x$ and the $-y$ directions, from continuity. This is

$$u \, dy - v \, dx = -d\psi$$

Thus the flow rate between any two streamlines ψ_1 and ψ_2 is

$$Q = \int_{\psi_1}^{\psi_2} |d\psi| = |\psi_2 - \psi_1|$$

and this may be considered a physical interpretation of the stream function.

Example 3-9

In Example 3-2, velocity components were obtained as $u = +2x$ and $v = -2y$. Show that continuity is satisfied.

SOLUTION

$\partial u / \partial x = 2$ and $\partial v / \partial y = -2$. Thus $\partial u / \partial x + \partial v / \partial y = 0$.

▬▬▬

Example 3-10

Given $u = 6x + xy$. Determine the v component of velocity that will satisfy continuity.

SOLUTION

$\partial u / \partial x = 6 + y = -\partial v / \partial y$. Integration gives $v = -6y - y^2/2 + f(x)$. Note that any $f(x)$ may be used here, since $\partial v / \partial y = -6 - y$ in any instance.

▬▬▬

PROBLEMS

3-1. Classify the following flows as steady or unsteady, uniform or non-uniform. Where there is reason for doubt, give conditions for which flow is as you state.

(a) Water in a garden hose
(b) Water flowing through the nozzles of a rotating sprinkler
(c) Flow through the nozzle attached to the end of a graden hose
(d) Flow of gases through a rocket nozzle
(e) Flow through a blow-down type of supersonic wind tunnel
(f) Flow of water over a wide spillway in a river
(g) Liquid draining from a small tank
(h) Gasoline in the fuel line of an automobile in city traffic; on a superhighway

3-2. Classify the following flows as one, two, or three dimensional:

(a) Flow of water over a wide spillway in a river
(b) Flow in a bend in a river
(c) Flow through the test section of a water tunnel or subsonic wind tunnel. Why is flow which is nearly one-dimensional desirable?
(d) Flow of a nonviscous fluid through a rectangular elbow
(e) Flow of a viscous fluid through an elbow in a round pipe

3-3. Classify the following as laminar or turbulent flow:

(a) Summer-grade lubricating oil flowing from an oil can
(b) Water issuing from a fire nozzle
(c) Flow in a river
(d) Flow through a hypodermic needle
(e) Atmospheric winds
(f) Flow of a viscous liquid at low velocity through a small pipe
(g) Flow of a low-viscosity liquid at a relatively high velocity through a large pipe

3-4. The velocity field in a flow system is given by

$$\mathbf{V} = 5\mathbf{i} + (x + y^2)\mathbf{j} + 3xy\,\mathbf{k}$$

What is the fluid acceleration at $(1,2,3)$ and at $(-1,-2,-3)$?

3-5. A nozzle is shaped such that the axial flow velocity increases linearly from 2 to 18 m/s in a distance of 1.20 m. What is the convective

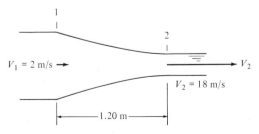

Prob. 3-5

acceleration
(a) at the inlet and
(b) at the exit of the nozzle?

3-6. In Prob. 3-5, the inlet velocity increases 0.40 m/s each second. What is the local acceleration
(a) at the inlet and
(b) at the exit when the velocities are as given in Prob. 3-5?

3-7. In Prob. 3-5, the liquid-flow rate doubles every 10 s such that the velocity at inlet is given by $V/4 = 2^{t/10}$.
(a) What is the local acceleration at inlet at $t = 0$?
(b) What is the local acceleration at the exit at $t = 0$?
(c) What are the corresponding total accelerations at $t = 0$?

3-8. Water flows through a converging nozzle such that its axial velocity increases linearly from 1.00 to 10.00 m/s in a distance of 0.60 m. What is the convective acceleration
(a) at the inlet and
(b) at the exit of the nozzle?

3-9. The nozzle inlet velocity in Prob. 3-8 increases at a rate of 0.2 m/s each second. What is the local acceleration
(a) at inlet and
(b) at the exit of the nozzle when the inlet velocity is 1.00 m/s?

3-10. The velocity of an ideal incompressible fluid as it steadily approaches the stagnation point at the leading edge of a cylinder of radius R held normal to the stream is

$$u = u_s\left(1 - \frac{R^2}{x^2}\right)$$

What is the fluid acceleration at
(a) $x = -3R$,
(b) $x = -2R$, and
(c) $x = -R$?
(d) When $u_s = 2.00$ m/s and $R = 2.0$ cm, what is the magnitude of the acceleration at $x = -3R$?

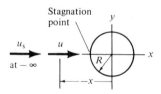

Prob. 3-10

3-11. The tangential velocity V_θ around the surface of a circular cylinder of radius R held normal to a steady irrotational and incompressible flow stream is

$$V_\theta = 2u_s \sin \theta$$

where u_s and θ are defined in Fig. 6-15 and are the free-stream velocity and angular position on the cylinder, respectively. What are the tangential and radial components of acceleration, a_s and a_n respectively, at
(a) $\theta = 0$,
(b) $\theta = \pi/4$, and
(c) $\theta = \pi/2$? Evaluate for $u_s = 2.00$ m/s and $R = 2.0$ cm.

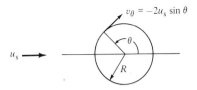

Prob. 3-11

3-12. In Prob. 3-10, $u_s = 3.00$ m/s and $R = 4.0$ cm. What is the acceleration at $x = -2R$?

3-13. In Prob. 3-11, evaluate results in SI units for $u_s = 4.00$ m/s and $R = 3$ cm.

3-14. Water flows through a horizontal conical contraction at a rate of $0.4 \text{ m}^3/\text{s}$. The flow is steady and one dimensional. The diameter of the contraction cone decreases from 0.5 to 0.25 m in a length of 0.6 m. What is
(a) the fluid acceleration and
(b) the pressure gradient at the midsection of the contraction?

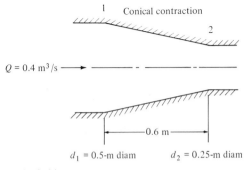

Prob. 3-14

3-15. In Prob. 3-14 the flow rate is increasing at a rate of 60 L/s. What is the total fluid acceleration at the midsection of the contraction?

3-16. A liquid flows through a square duct 4.00 m long with the side dimensions changing linearly from 0.40 m at the entrance to 0.20 m at the exit. For a steady flow rate of 0.36 m^3/s what is the fluid acceleration
  (a) at the entrance,
  (b) at a section one-half the distance between the ends, and
  (c) at the end of the duct?

3-17. Describe the flow given by
  (a) $\psi = -20y$
  (b) $\psi = 10x$
  (c) $\psi = 5x - 8.66y$
  (d) $\psi = x^2$

3-18. Given $u = 2y$ and $v = 2$.
  (a) What is the stream function for this flow?
  (b) Sketch the streamline pattern in the upper half plane for the constant in the stream function set equal to zero.

3-19. Given a stream function $\psi = x^2 + 2x + y^2$. What is the flow rate passing through a straight line connecting the points (2,0) and (0,4) per unit width in the z direction?

3-20. In Fig. 3-5 the streamlines $\psi = 0$ and $\psi = -6$ may be considered boundaries along with the y axis to represent a two-dimensional flow. Determine the flow rate per unit width in the z direction from the discussion of Fig. 3-14 and from integration of the u and v components of velocity along the lines $x = 3$ and $y = 4$, respectively.

3-21. Determine the v component of velocity to within an additive constant which will satisfy continuity for each of the following:

  (a) $u = x^2$ (e) $u = \dfrac{-y}{x^2 + y^2}$

  (b) $u = 6x + xy$ (f) $u = x$

  (c) $u = x^2 + x$ (g) $u = 2xy^2$

  (d) $u = \dfrac{x}{x^2 + y^2}$

3-22. In Prob. 3-21, which flows could be irrotational and for what conditions regarding $f(x)$ would they be?

3-23. Which of the following flows satisfy continuity for an incompressible fluid flow; and of these, which are rotational (typical of a viscous fluid) and which are irrotational (typical of a nonviscous fluid)?

  (a) $u = x^2 \cos y$; and $v = -2x \sin y$
  (b) $u = x + 2$; and $v = 1 - y$
  (c) $u = xyt$; and $v = x^3 - y^2t/2$
  (d) $u = \ln x + y$; and $v = xy - y/x$
  (e) $u = x + y$; and $v = x - y$

3-24. Derive the continuity equation for motion of a steady incompressible fluid in polar coordinates, for two-dimensional flow in the x–y plane.

3-25. Extend the results of Prob. 3-24 to the general case when there is a z component of velocity v_z. This is the continuity equation for steady flow in cylindrical coordinates, Eq. (3.20).

3-26. In Prob. 3-25:
   (a) Describe the flow when $v_\theta = v_z = 0$.
   (b) Describe the flow when $v_r = 0$.
   (c) Describe the flow when $v_r = v_z = 0$.
   (d) Describe the flow when $v_\theta = v_r = 0$.

3-27. A fluid flows radially from a point source with a velocity $v_r(r,t)$. Show that the continuity equation, except for the source itself, is

$$\frac{\partial \rho}{\partial t} + \frac{\rho}{r^2}\frac{\partial}{\partial r}(r^2 v_r) + v_r \frac{\partial \rho}{\partial r} = 0$$

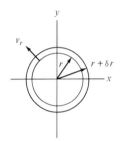

Prob. 3-27

3-28. Let the x and y components of velocity in steady, two-dimensional, incompressible flow be linear functions of x and y such that

$$\mathbf{V} = (ax + by)\mathbf{i} + (cx + dy)\mathbf{j}$$

where a, b, c, and d are constants.
   (a) For what conditions is continuity satisfied?
   (b) What is the vorticity?
   (c) For what conditions is the flow irrotational?

3-29. A velocity field is given by

$$\mathbf{V} = (x^2 + y^2)\mathbf{i} + 2xy\mathbf{j} - 4xz\mathbf{k}$$

What is the angular velocity of the fluid at (2,3,4)?

3-30. Calculate the circulation around the path in the x–y plane bounded by the lines $y = 0$, $x = 6$, $y = 4$, and $x = 0$ for the flow field given by

$$\mathbf{V} = (x^2 + y^2)\mathbf{i} + 2xy\mathbf{j}$$

NOTE: If the flow is irrotational, the circulation should be zero.

3-31. State which of the following flow situations are irrotational and which are rotational:

(a) Laminar flow in a pipe with a velocity profile given by Eq. (9-11a) and Fig. 9-5.

(b) Vortex flow in which the peripheral velocity is $v_\theta = \omega r$.

(c) Vortex flow in which $v_\theta = C/r$.

(d) Flow in a boundary layer in which the velocity profile is given by Eq. (7.17) or Eq. (7.22) and shown in Fig. 7-9.

3-32. Given the stream function $\psi = x^2 + 2y$.

(a) Show that continuity is satisfied.

(b) Is the flow irrotational or is it rotational?

3-33. A flow is described by the stream function

$$\psi = \frac{x^3}{3} - \frac{y^2}{2} - xy$$

What is the circulation around the square enclosed by $x = \pm 2$, $y = \pm 2$?

3-34. Express the fluid vorticity as a function of y for the following laminar boundary-layer velocity profiles (see Fig. 7-9):

(a) $u = \dfrac{2u_s}{\delta} y - \dfrac{u_s}{\delta^2} y^2$

(b) $u = \dfrac{3u_s}{2\delta} y - \dfrac{u_s}{2\delta^3} y^3$

(c) $u = u_s \sin(\pi y/2\delta)$

3-35. What can be said about the vorticity of the flow from the surface to $y = \delta$, the outer edge of the boundary layer, in Prob. 3-34? What is the vorticity at $y > \delta$?

3-36. Water flowing in a 75-mm-diameter pipe discharges into a rectangular tank 1.50×2.50 m in horizontal dimensions. Water rises in the tank at a rate of 0.150 m in 40 s. What is the average flow velocity in the pipe?

3-37. Water flows from a 40-mm-diameter pipe at a rate of 0.150 m³ in 28 s. What is the volumetric flow rate and the average flow velocity in the pipe?

3-38. Water at a flow rate of 0.080 m³/s and alcohol at a flow rate of 0.050 m³/s (specific gravity $s = 0.80$) meet in a Y-connection and flow as a mixture in a 200-mm-diameter pipe. What is the average flow velocity of the mixture?

3-39. A 150-mm-diameter pipe branches into one 100-mm pipe and one 75-mm pipe. The average flow velocity in the 100-mm pipe is 2.50 m/s and in the 75-mm pipe it is 1.95 m/s. What is the flow rate and the average flow velocity in the 150-mm pipe?

3-40. A tank has two water pipes each 100 mm in diameter conveying water into the tank, and one 50-mm and one 120-mm pipe discharging water from the tank. The velocity in one inlet pipe is 2.10 m/s and in the other, 3.50 m/s. The average velocity in the 50-mm discharge pipe is 4.00 m/s. What is the flow rate and average velocity in the 120-mm discharge pipe when the net flow into the tank is zero?

3-41. Standard air enters a porous-wall circular tube 4 cm ID and 80 cm long at an average velocity of 30 m/s. Air enters the main flow stream through the porous pipe wall at 0.60 m/s uniformly over the entire pipe length. For incompressible flow, what is the average flow velocity at the downstream end of the porous pipe?

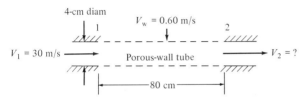

Prob. 3-41

3-42. At one section in a 30-cm-diameter pipe methane flows at an absolute pressure of 8.00×10^5 N/m², a temperature of $40°C$, and an average velocity of 25 m/s.
(a) What is the mass flow rate in kg/s?
(b) What is the average flow velocity at a downstream section where the pressure is 5.00×10^5 N/m² and the temperature is $40°C$?
(c) What is the pressure farther downstream where the average flow velocity is 250 m/s and the temperature is $20°C$?

3-43. Atmospheric air heated at constant pressure to $70°C$ is drawn through a suitable entrance and nozzle to the 0.40×0.40-m square test section of a supersonic wind tunnel. The average velocity in the test section is 1.5 times the speed of sound corresponding to the air temperature in the test section. Assume the air expands isentropically from the outside atmosphere after heating to 2.76×10^4 Pa abs in the test section.
(a) What is the air temperature in the test section?
(b) What is the density of the air in the test section?

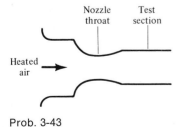

Prob. 3-43

(c) What is the mass flow rate of air through the wind tunnel?

(d) What is the flow rate of air in m³/min of standard air?

3-44. The velocity profile for laminar flow between two parallel plates is given by

$$u = u_m\left(1 - \frac{b^2}{B^2}\right)$$

where u_m is the centerplane velocity, B is the half-spacing between plates, and b is the normal distance from the centerplane. What is the average velocity in terms of u_m?

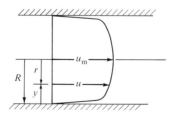

Prob. 3-44

3-45. The velocity profile for laminar flow in a circular tube of radius R is given by

$$u = u_m\left(1 - \frac{r^2}{R^2}\right)$$

where u_m is the centerline velocity and r is the radial location from the centerline. What is the average velocity in terms of u_m?

3-46. The velocity profile for turbulent flow in a circular tube is given approximately by

$$u = u_m\left(\frac{y}{R}\right)^{1/8}$$

$$= u_m\left(1 - \frac{r}{R}\right)^{1/8}$$

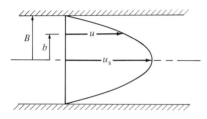

Prob. 3-46

where u_m is the centerline velocity, R is the tube radius, and y is the radial distance from the tube wall ($r = R - y$). What is the average velocity in terms of u_m?

3-47. The velocity u within a turbulent boundary layer at a distance y meters from the boundary surface varies according to the equation

$$u = u_s \left(\frac{y}{\delta}\right)^{1/7}$$

where u_s is the free-stream velocity outside the boundary layer and δ is the boundary layer thickness (see Fig. 7-9). What is the mass flow rate within the boundary layer for standard air in a 1.20-m width of flow for $u_s = 30$ m/s and $\delta = 0.028$ m?

3-48. Repeat Prob. 3-46 with the exponent 1/10 instead of 1/8.

3-49. Velocities were measured at the upstream end of a diffuser in a water tunnel. What is the average velocity?

3-50. Velocities were measured at the downstream end of a diffuser in a water tunnel. What is the average velocity?

3-51. Velocities were measured at the upstream end of a contraction for a water tunnel. What is the average velocity?

3-52. Velocities were measured in a circular tube at 24°C and 9.95 × 10^4 Pa abs. What is the average velocity?

3-53. Velocities were measured in a circular tube at 24°C and 9.86 × 10^4 Pa abs. What is the average velocity?

Velocities measured at the center of equal increments of $(r/R)^2$, representing equal increments of area, in circular ducts of various diameters are as listed here; velocities are in meters per second.

PROBLEM	3-49	3-50	3-51	3-52	3-53
DUCT DIAMETER (cm)	15.2	30.5	45.7	3.53	3.52
$(r/R)^2$			VELOCITY (m/s)		
0.05	14.20	5.55	1.61	41.2	50.7
0.15	14.20	5.12	1.61	39.8	49.4
0.25	14.20	4.54	1.60	38.6	47.2
0.35	14.19	3.89	1.59	37.2	45.1
0.45	14.17	3.32	1.58	35.8	43.2
0.55	14.20	2.87	1.56	34.5	40.8
0.65	14.17	2.41	1.54	32.9	38.6
0.75	14.02	1.98	1.50	31.2	36.0
0.85	13.56	1.71	1.42	29.3	32.6
0.95	13.11	1.37	1.28	25.0	26.5

3-54. Measurements of the velocity distribution with depth in a wide rectangular channel 3.00 m deep are as follows: surface, 1.60 m/s; 0.6-m depth, 1.52 m/s; 1.2-m depth, 1.40 m/s; 1.8-m depth, 1.24 m/s; 2.4-m depth, 1.00 m/s; bottom, 0 m/s. Estimate the flow rate per meter width of channel.

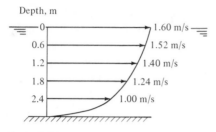

Prob. 3-54

References

1. A. H. Shapiro, *The Dynamics and Thermodynamics of Compressible Fluid Flow*, Vol. I (New York: The Ronald Press Company, 1953), pp. 73–74.
2. FM-47, "Pathlines, Streaklines, Streamlines and Timelines in Steady Flow" and FM-48, "Pathlines, Streaklines, Streamlines in Unsteady Flow," National Committee for Fluid Mechanics Films (Chicago: Encyclopaedia Britannica Educational Corporation).
3. FM-13, "The Bathtub Vortex," FM-14A and FM-14B, "Visualization of Vorticity with Vorticity Meter," Part I and Part II, FM-20, "The Horseshoe Vortex," FM-26, "Tornadoes in Nature and the Laboratory," and FM-70, "The Sink Vortex," National Committee for Fluid Mechanics Films (Chicago: Encyclopedia Britannica Educational Corporation).

Chapter 4
Fluid Momentum

Newton's second law states that the net force acting on a given mass is proportional to the time rate of change of linear momentum of that mass. This law may be applied to a control volume through which fluid flows by the use of the Reynolds transport theorem [Eq. (II.1)] to obtain the momentum theorem for fluid flow. It may also be applied to a fluid element to obtain the equations of motion in differential form, and these equations may be integrated for certain conditions to obtain equations of practical use in solving flow problems.

THE MOMENTUM THEOREM OF FLUID MECHANICS

4-1. The Momentum Theorem

The momentum theorem is concerned only with external forces and provides useful results without requiring a detailed knowledge of the internal processes within the fluid. It may be applied to flows that are steady or unsteady; one, two, or three dimensional; compressible or incompressible.

    As in the case of solids or discrete particles, fluids tend to continue in a state of uniform motion or rest unless acted upon by external forces. If the velocity of a group of fluid particles or of the entire fluid as it passes through the surface of a control volume changes either in magnitude or direction, or both, a net external force acting *on* the fluid is required to produce this change. We will consider

1. normal forces due to pressure,
2. tangential forces due to viscous shear,
3. body forces such as gravity acting in the direction of the gravitational field.

If we let $\Sigma\mathbf{F}$ be the external force acting on the fluid in a nonaccelerating control volume, and $\rho\mathbf{V}$ (which is the momentum flux per unit volume) be the point function P in Eq. (II.1), then the Reynolds transport theorem may be written as

$$\Sigma\mathbf{F} = \frac{D}{Dt} \int_{+\text{ material} \atop \text{volume}} \rho\mathbf{V}\, d\Psi$$

$$= \frac{\partial}{\partial t} \int_{\text{control} \atop \text{volume}} \rho\mathbf{V}\, dV + \int_{\text{control} \atop \text{surface}} \rho\mathbf{V}(\mathbf{V}\cdot d\mathbf{s}) \tag{4.1}$$

The first term on the right represents the change in momentum of the fluid within the control volume when the density or velocity vary with time, and is, of course, zero for steady flow. The second term on the right represents the flux or transport of momentum across the boundaries of the control volume. The $\Sigma\mathbf{F}$ term includes all the surface and body forces mentioned above.

For both steady and unsteady flow, Eq. (4.1) may be expressed as follows:

The net external force acting on the fluid within a prescribed control volume equals the time rate of change of momentum of the fluid within the control volume plus the net rate of momentum flux or transport out of the control volume through its surface.

This is the momentum theorem of fluid mechanics.[†]

[†] *Direct development of the momentum theorem:* In Fig. 4-1 let $\mathbf{M}$ be the linear momentum of the particular fluid within the control volume bounded by the control surface S at time t. At time $t + \delta t$ this same fluid occupies the region bounded by S', and its momentum is

$$\left[\mathbf{M} + \frac{\partial\mathbf{M}}{\partial t}\, dt\right] + [\text{momentum of the fluid in II}] - [\text{momentum of the fluid in I}]$$

The last two terms represent the momentum flow out of the control volume through the control surface S during the time δt. Hence the time rate of change of the momentum of the fluid initially within S is

$$\frac{\partial\mathbf{M}}{\partial t} + \text{momentum flow rate outward through the boundary } S \text{ of the control volume}$$

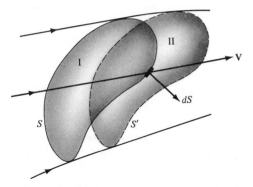

Fig. 4-1.   Control volume for vector derivation of momentum theorem.

For steady flow,

$$\Sigma\mathbf{F} = (\dot{m}\mathbf{V})_{\text{leaving } S} - (\dot{m}\mathbf{V})_{\text{entering } S} \tag{4.3a}$$

or in scalar form

$$\Sigma F_x = (\dot{m}V_x)_{\text{leaving } S} - (\dot{m}V_x)_{\text{entering } S} \tag{4.3b}$$

with similar expressions for the y and z directions, respectively.

Equations (4.3) may also be written as

$$\Sigma\mathbf{F} = \dot{m}(\mathbf{V}_{\text{leaving } S} - \mathbf{V}_{\text{entering } S}) \tag{4.4a}$$

or in scalar form

$$\Sigma F_x = \dot{m}(V_{x \text{ leaving } S} - V_{x \text{ entering } S}) \tag{4.4b}$$

with similar expressions for the y and z directions, respectively.

That part of $\Sigma\mathbf{F}$ which represents forces due to pressure and viscous shear is the integral of these forces over the control surface only, since they are equal and opposite at all locations within the control volume.

The momentum of the fluid within the control volume for the fluid initially within S is

$$\mathbf{M} = \iiint_V \mathbf{V}\rho \, dV$$

and the momentum flow rate (also called the momentum flux or the rate of momentum transport) outward through the control surface S is

$$\iint_S \rho\mathbf{V}(\mathbf{V} \cdot d\mathbf{S})$$

Thus the time rate of change of momentum is equated to the net force acting on the fluid within the control volume.

$$\Sigma\mathbf{F} = \frac{\partial\mathbf{M}}{\partial t} + \iint_S \rho\mathbf{V}(\mathbf{V} \cdot d\mathbf{S}) \tag{4.2}$$

4-2. Applications of the Momentum Theorem

In instances where the fluid velocity changes only in magnitude, the direction along which the momentum equation is written is well defined and a scalar equation is sufficient. When the fluid changes direction, a vector equation may be used or scalar equations must be written in mutually perpendicular directions, and forces and velocities along these directions represent components of the resultant forces and velocities. The velocity change in Eq. (4.4) may be expressed in terms of absolute or relative velocity for *nonaccelerating* control volumes; that is, the fluid velocities may be taken with respect to any nonaccelerating moving reference or to a reference at rest, so long as a consistent reference is used. The earth is usually considered to be a reference at rest.

Forces due to pressure should be based on gage pressure (that is, with respect to the ambient atmospheric pressure) because forces resulting from atmospheric pressure in most cases completely envelop the control volume and thus they cancel and have no resultant effect.

The momentum theorem may be applied to a control volume in which two or more fluids enter and combine within the control volume, such as in the combustion chamber of a jet engine. For steady flow, the net force *on* the fluid is the difference between the flow of momentum of the fluid leaving the control volume and the total flow of momentum of the various fluids entering the control volume [Eq. (4.4)].

The momentum theorem may be applied in the following manner: The region or control volume wherein a momentum change takes place is isolated, and a positive *x* direction (and a positive *y* and *z* direction if necessary) is arbitrarily assigned. The mass flow rate is expressed in terms of either the entrance or exit conditions, since Eqs. (4.3) and (4.4) are valid for steady flow only. Some examples, in which one-dimensional flow is assumed, will illustrate the application of the momentum principle. In all instances, the control volume wherein the momentum change takes place is indicated by dotted lines. Some unsteady flows may be transformed into steady flows by changing the frame of reference. The flow in Example 4-5 is a steady flow when viewed with respect to the rocket. The steady-flow momentum theorem may be applied to a moving wave front (Secs. 10-1 and 12-2) when written with respect to the moving wave front, and to a tidal bore when written with respect to the moving hydraulic jump (Sec. 12-6).

Example 4-1

A water jet strikes a curved blade and is deflected through an angle of 60°. The jet speed is 24 m/s, the jet area is 0.010 m², and the blade is smooth so that the jet speed is constant. What force is exerted by the jet on the vane?

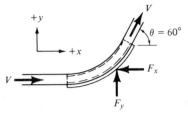

Fig. 4-2. See Example 4-1. Blade force on free jet of liquid.

SOLUTION

In Fig. 4-2,

$$-F_x = V_j A_j \rho (V_j \cos \theta - V_j)$$
$$F_x = (24)(0.010)(1000)[24 - (24)(0.5)] = 2880 \text{ N}$$
$$F_y = V_j A_j \rho (V_j \sin \theta - 0) = (24)(0.010)(1000)(24)(0.866)$$
$$= 4988 \text{ N}$$

The resultant force is $\sqrt{2880^2 + 4988^2} = 5860 \text{ N}$

Example 4-2

In Example 4-1, the blade moves at 6 m/s in the direction of the jet. What force is exerted on the vane by the jet?

SOLUTION

All the water from the jet does not strike the vane per unit time, as before. In Fig. 4-3 the relative speed between the jet and blade is $(V_j - V_b) = 18$ m/s at both the entrance and the exit of the control volume (indicated by dashed lines). Thus with velocity differences with respect to the blade

$$-F_x = (V_j - V_b) A_j \rho [(V_j - V_b) \cos \theta - (V_j - V_b)]$$
$$F_x = (18)(0.010)(1000)(18)(1 - 0.5) = 1620 \text{ N}$$
$$F_y = (V_j - V_b) A_j \rho [(V_j - V_b) \sin \theta - 0]$$
$$= (18)(0.010)(1000)(18)(0.866) = 2806 \text{ N}$$

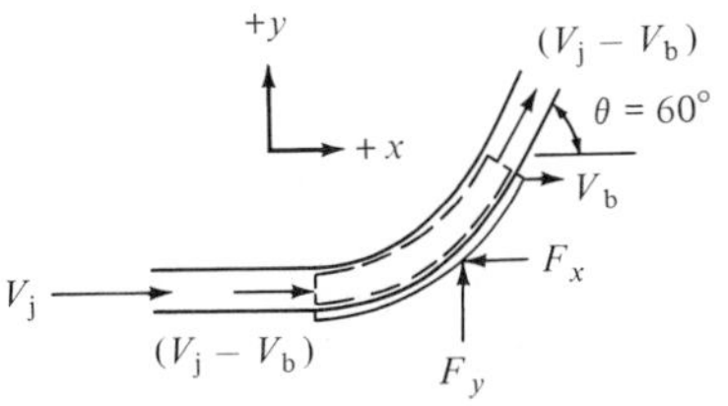

Fig. 4-3. See Example 4-2. Force of moving blade on free jet of liquid.

Results are the same if velocities are taken with respect to the earth. This is left as an exercise.

━━━━━

Example 4-3

Water flows through a reducing elbow at a rate of 0.25 m³/s. The area reduction is from 0.10 m² at inlet to 0.05 m² at exit. The pressure is 1.70×10^5 Pa gage at inlet and 1.60×10^5 Pa gage at exit. What is the resultant force of the water on the elbow? Assume the bend to be in a horizontal plane.

SOLUTION

In Fig. 4-4, let F_x and F_y be the x and y components of the viscous shear and pressure forces of the elbow walls on the fluid as shown. The momentum equations written in the $+x$ and $+y$ directions are

$$p_1 A_1 - F_x - p_2 A_2 \cos \theta = V A \rho (V_2 \cos \theta - V_1)$$

and

$$F_y - p_2 A_2 \sin \theta = (V A \rho)(V_2 \sin \theta - 0)$$

Then

$$F_x = p_1 A_1 - p_2 A_2 \cos \theta - V_1 A_1 \rho (V_2 \cos \theta - V_1)$$
$$= (1.7 \times 10^5)(0.10) - (1.6 \times 10^5)(0.05)(0.5)$$
$$\quad - (2.5)(0.10)(1000)[(5)(0.5) - 2.5]$$
$$= 13.0 \text{ kN on the water}$$

$$F_y = p_2 A_2 \sin \theta + V_1 A_1 \rho (V_2 \sin \theta - 0)$$
$$= (1.6 \times 10^5)(0.05)(0.866) + (0.25)(1000)(5)(0.866)$$
$$= 8.0 \text{ kN on the water}$$

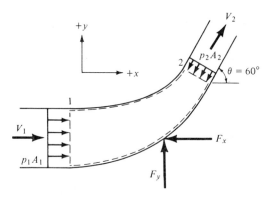

Fig. 4-4. See Example 4-3. Flow through a reducing elbow.

The resultant force of the water on the elbow is $\sqrt{13.0^2 + 8.0^2} = 15.3$ kN at an angle arctan $8.0/13.0$, or $31.6°$ from the x direction, downward and to the right.

Example 4-4

A water jet pump has a jet area of $A_j = 45$ cm^2 and a jet velocity $V_j = 27$ m/s which entrains a secondary stream of water having a velocity $V_s = 3$ m/s in a constant-area pipe of total area $A = 540$ cm^2. At section 2 the water is thoroughly mixed. Assume one-dimensional flow and neglect wall shear. (a) What is the average velocity of the mixed flow at section 2? (b) What is the pressure rise $(p_2 - p_1)$, assuming the pressure of the jet and secondary stream to be the same at section 1? Refer to Fig. 4-5.

SOLUTION

(a) The flow rate of the jet plus secondary equals the flow rate at section 2. Thus,

$$V_j A_j + V_s A_s = V_2 A_2$$

$$V_2 = \frac{V_j A_j + V_s A_s}{A_2} = \frac{(27)(0.0045) + (3)(0.0495)}{0.0540} = 5.0 \text{ m/s}$$

(b) The momentum equation is written between sections 1 and 2. Forces are due solely to pressure. The momentum change is that leaving at section 2 minus that entering at section 1 (the sum of that for the jet and that for the secondary flow). With the flow direction taken as positive,

$$p_1 A_1 - p_2 A_2 = \dot{m}_2 V_2 - (\dot{m}_j V_j + \dot{m}_s V_s)$$

$$p_2 - p_1 = \frac{(\dot{m}_j V_j + \dot{m}_s V_s) - \dot{m}_2 V_2}{A}$$

$$= \frac{(27)^2(0.0045)(1000) + (3)^2(0.0495)(1000) - (5)^2(0.0540)(1000)}{0.0540}$$

$$= 44 \text{ kPa}$$

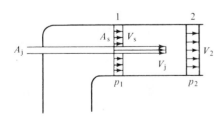

Fig. 4-5.   See Example 4-4. Water jet pump.

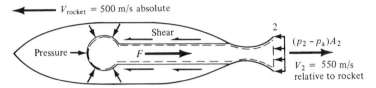

Fig. 4-6.   See Example 4-5. Thrust on rocket.

Example 4-5

A rocket travels at 500 m/s and discharges its exhaust gases at 550 m/s relative to the rocket at a rate of 5.00 kg/s through an exit area of 0.050 m². The absolute pressure at exit is 100 kPa, and the ambient air pressure is 85 kPa abs. What is the thrust on the rocket?

SOLUTION

The gases have a momentum change within the dotted-line region in Fig. 4-6. The external forces acting on the gases are due to pressure and viscous shear, as indicated. Let F be the resultant of these forces. Since the exit pressure is above the ambient pressure, two external forces act. The force F is applied by the rocket walls, and the force $(p_2 - p_a)A_2$ results from the positive gage pressure at exit:

$$F - (p_2 - p_a)A_2 = (VA\rho)(V_2 - 0)$$

Expressing the change in gas velocity as relative to the rocket,

$$F = \dot{m}V_2 + (p_2 - p_a)A_2 = (5.00)(550) + (100 - 85)(1000)(0.050)$$
$$= 3500 \text{ N}$$

The thrust of the gases *on* the rocket is equal and opposite to F. Note that the effect of a positive gage pressure at exit is to increase the thrust, so that a rocket designed to operate at one altitude will produce a higher thrust at a higher altitude, and conversely. This example is rather extreme, and in general a rocket should be designed to expand the gases closer to the ambient pressure. If the gases expand fully, the exit pressure p_2 is fixed by the gas condition as it enters the nozzle and the nozzle design. Lowering the exit pressure will also increase the exit velocity.

Example 4-6

A gas flows through an expanding section in a pipe. The area increases from A_1 to A_2, the velocity decreases from V_1 to V_2, the pressure increases from p_1 to p_2, and the gas density increases from ρ_1 to ρ_2. What is the net force

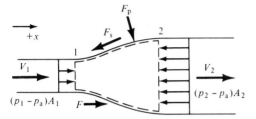

Fig. 4-7.   See Example 4-6. Gas flowing through expansion.

of the walls of the expanding section on the fluid? Pressures p_1 and p_2 are absolute.

SOLUTION
The walls of the expanding section exert both viscous shear (opposite to the flow direction) and a force due to pressure (axial component is in the flow direction). Let the resultant of these be F in the direction shown in Fig. 4-7. The atmospheric pressure is p_a:

$$(p_1 - p_a)A_1 + F - (p_2 - p_a)A_2 = V_1A_1\rho_1(V_2 - V_1) = V_2A_2\rho_2(V_2 - V_1)$$

$$F = p_2A_2 - p_1A_1 + V_1A_1\rho_1(V_2 - V_1) - p_a(A_2 - A_1)$$

The results of Example 4-6 may be used to define the *thrust function*, which is commonly used in gas dynamics. The resulting expression for the net force F may be rearranged to give

$$F = (p_2A_2 + V_2^2A_2\rho_2) - (p_1A_1 + V_1^2A_1\rho_1) - p_a(A_2 - A_1) \qquad (4.5)$$

where the expression $(pA + V^2A\rho)$ is called the thrust function. The net external force is thus equal to the change in the thrust function corrected for the effect of the atmosphere.

We have used the average velocity across a section in determining the momentum flux $V^2A\rho$ through a section in a one-dimensional analysis. If the velocity varies across the section, the true momentum flux is greater than that based on the average velocity. The true momentum flux is found by integrating the product of the mass flow rate and velocity over the area of flow. This is

$$\int_A (u\rho \, dA)u$$

while that based on the average velocity is

$$(V \rho A)V = V^2A\rho$$

The ratio of the true momentum flux to that based on the average velocity is called the *momentum-flux correction factor* β, where

$$\beta = \frac{\int_A u^2 \, dA}{V^2 A} \geq 1 \tag{4.6a}$$

$$\approx \frac{1}{V^2 n} \sum_1^{i=n} u_i^2 \tag{4.6b}$$

where n is the number of equal incremental areas making up the total area A.

Example 4-7

The velocity profile for laminar flow between two parallel plates is given by

$$u = u_m \left(1 - \frac{b^2}{B^2} \right)$$

where u_m is the centerplane velocity, B is the half-spacing between plates, and b is the normal distance from the center plane. (a) What is the average velocity in terms of u_m? (b) What is the momentum correction factor β?

SOLUTION

From Eq. (3.26),

$$\bar{u} = V = \frac{2}{2B} \int_0^B u \, db = \frac{u_m}{B} \int_0^B \left(1 - \frac{b^2}{B^2} \right) db = \left(\frac{2}{3} \right) u_m$$

From Eq. (4.6b),

$$\beta = \frac{2 \int_0^B u^2 \, db}{(2u_m/3)^2 (2B)} = \frac{9}{4B} \int_0^B \left(1 - \frac{b^2}{B^2} \right)^2 db = 1.20$$

Example 4-8

The velocity profile for turbulent flow in a circular tube may be approximated by

$$u = u_m \left(\frac{y}{R} \right)^{1/7} = u_m \left(1 - \frac{r}{R} \right)^{1/7}$$

where u_m is the centerline velocity, R is the tube radius, and y is the radial distance from the tube wall ($r = R - y$). (a) What is the average velocity in terms of u_m? (b) What is the momentum flux factor β?

SOLUTION

From Example 3-8, $\bar{u} = V = \frac{49}{60}u_m$. The momentum flux factor is, from Eq. (4.6a),

$$\beta = \frac{\int_0^R u_m^2(y^{2/7}/R^{2/7})2\pi r\, dr}{(49/60)^2(u_m^2)\pi R^2}$$

$$= \frac{7200}{2401\, R^{16/7}} \int_0^R y^{2/7}(R - y)\, dy = \frac{50}{49} = 1.020$$

For one-dimensional flow, which has a flat velocity profile, $\beta = 1$, and this is an assumption often made. Example 4-8 indicates a value of 1.020; measurements for fully developed turbulent flow in a circular pipe are about 1.03. Thus from a momentum flux point of view, turbulent flow often may be assumed to be one dimensional. Example 4-7 showed $\beta = 1.20$ for laminar flow with a parabolic velocity profile between parallel plates. For fully developed laminar flow in a circular pipe with a parabolic velocity profile, $\beta = 1.333$.

Example 4-9

Velocities measured at the center of equal increments of $(r/R)^2$, representing equal increments of area, at the downstream end of a diffuser in a cavitation-testing water tunnel are as follows: 18.2, 16.8, 14.9, 12.75, 10.9, 9.4, 7.9, 6.5, 5.6, and 4.5 m/s. (a) What is the average flow velocity? (b) What is the momentum flux factor β?

SOLUTION

From Eq. (3.26), the average velocity is $\bar{u} = V = 107.45/10 = 10.75$ m/s. From Eq. (4.6b), the momentum flux factor is

$$\beta = \frac{(18.2)^2 + (16.8)^2 + \cdots + (4.5)^2}{10(10.75)^2} = \frac{1361.4}{(10)(115.56)} = 1.178$$

Example 4-10

An incompressible fluid of density ρ approaches a flat plate with a uniform free-stream velocity u_s and flows past one side of the plate. Because of viscous shear, the fluid motion is retarded within a thin layer of thickness δ. The velocity within this layer varies from zero at the wall to the free stream velocity u_s at $y = \delta$. The fluid within the thin layer has had its momentum

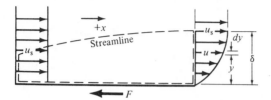

Fig. 4-8. See Example 4-10. Flow along a flat plate.

changed by an external force F applied along the surface or wall of the plate, as shown in Fig. 4-8. The momentum equation is

$$-F = \int_0^\delta (u\rho \, dy)(u - u_s)$$

which can be evaluated if u is expressed as a function of y, and this will be done in Chapter 7.

The steady-state-momentum theorem is a very useful and necessary concept when one is concerned with forces in fluid flows, especially since in using it only body forces and external surface forces are considered. Details of the flow within the control volume are not considered, and need not be. The theorem will be used throughout the text in analyzing many flow situations. Among these are (1) boundary layer analysis in Chapter 7, (2) determination of the sonic speed in a gas in Chapter 10, (3) analysis of shocks in gas flows in Chapter 10, (4) flow in both the entrance and the fully developed region for viscous fluids in ducts in Chapter 9, (5) flow through a sudden expansion in a pipe in Chapter 9, (6) flow through a hydraulic jump in Chapter 12, (7) analysis of impulse turbines in Chapter 14, and (8) comparison of open channel and gas flows in Chapter 15.

EQUATIONS OF MOTION

Newton's second law may be applied to small elements of fluid to develop the equations of motion for a nonviscous fluid and for a viscous fluid. Those for a nonviscous fluid are known as the Euler equations, and those for a viscous fluid are known as the Navier–Stokes equations.

Integration of the Euler equations for incompressible irrotational flow produces the Bernoulli equation which relates velocity, pressure, and elevation changes in nonviscous fluids and is also applicable in many viscous flows when the effects of viscosity are negligible. The Euler equations may be also integrated for incompressible rotational flow, as well as for compressible flow.

The Navier–Stokes equations are nonlinear partial differential equations. Exact solutions have been obtained for only a limited number of laminar flow situations for which some of the terms are zero [1]. Integration is then possible. Approximate solutions have been made for some very slow motions. Finally, many numerical solutions have been made with the use of high-speed computers. Exact solutions include flow between parallel plates with and without pressure gradients, flow in a circular tube, flow between two concentric rotating cylinders, a suddenly accelerated plane wall, flow development between parallel walls, flow in a pipe starting from rest. flow near an oscillating flat plate, flow near a stagnation point, flow near a rotating disk, and flow in convergent and divergent channels. Approximate solutions for very slow motions include parallel flow past a sphere, lubrication in a bearing, and flow between closely spaced parallel plates known as Hele–Shaw flow.

4-3. Equations of Motion for Nonviscous Fluids

We will develop the equations in both natural (streamline) and Cartesian coordinates. When integrated, they are referred to as the Bernoulli equations for both incompressible flow and for compressible flow. The Bernoulli equation for incompressible flow is often referred to as a mechanical-energy equation because of the similarity to the steady-flow energy equation obtained from the first law of thermodynamics for a nonviscous fluid with no heat transfer and no external work. This similarity will be discussed in detail after the steady-flow energy equation is developed in Chapter 5.

Forces acting on a fluid element are of two general types: body forces and surface forces. Body forces are those that act on the volume or mass of the fluid element. They include the force of gravity and the force on a conducting fluid in a magnetic field. Surface forces include both normal forces and tangential forces. Tangential forces generally are due to viscous shear, but they may also include surface tension if the fluid element has a free surface.

For nonviscous fluids there obviously are no viscous forces, and surface tension forces will not be considered since we shall exclude a free surface. Body forces may be of either of the aforementioned types. Surface forces on a nonviscous fluid element submerged within the fluid are normal forces due to pressure. This normal pressure force is due to the thermodynamic pressure used in the equation of state. It can be shown that pressure is a scalar quantity and in this instance is independent of direction. Hence $p_x = p_y = p_z$ and subscripts are not necessary.

It may seem strange to consider the motion of fluids that are nonviscous, since all real fluids are viscous. It is, however, important to understand the flow behavior of a fluid that has only density and no other physical property. In many instances viscous fluids move in a way in which the viscous effects are negligible. Only in a few instances (creeping motion, for example) is the inertia of the fluid not important. Hence, in some instances the fluid inertia is the important characteristic affecting the flow, while in other instances the viscosity of the fluid has a greater effect on the flow than does any other fluid property.

NATURAL COORDINATES

If the velocity V of an elementary fluid particle is a function of position and time, we may write that $V = V(s,t)$. Thus,

$$dV = \frac{\partial V}{\partial s} ds + \frac{\partial V}{\partial t} dt \qquad \text{or} \qquad \frac{dV}{dt} = \frac{\partial V}{\partial s} \frac{ds}{dt} + \frac{\partial V}{\partial t}$$

and since the velocity along a streamline is $V = ds/dt$, the acceleration in that direction is

$$a_s = \frac{dV}{dt} = V \frac{\partial V}{\partial s} + \frac{\partial V}{\partial t}$$

The term $V \partial V/\partial s$ is the convective acceleration, or the variation in velocity along a streamline; $\partial V/\partial t$ is the local acceleration, or the variation of velocity at a point with respect to time.

We will now apply Newton's second law to an element of a nonviscous fluid. From Fig. 4-9, writing $\Sigma F = ma_s$ along the streamline gives

$$p \Delta A - \left(p + \frac{\partial p}{\partial s} \Delta s \right) \Delta A - \rho g \Delta A \Delta s \sin \theta = \rho \Delta A \Delta s \left(V \frac{\partial V}{\partial s} + \frac{\partial V}{\partial t} \right)$$

and since $\sin \theta = \partial z/\partial s$, upon simplifying we get

$$V \frac{\partial V}{\partial s} + \frac{1}{\rho} \frac{\partial p}{\partial s} + g \frac{\partial z}{\partial s} = -\frac{\partial V}{\partial t} \tag{4.7}$$

which is Euler's equation of motion along a streamline for a nonviscous fluid. For steady flow $\partial V/\partial t = 0$, and integration along a streamline for a constant-density fluid gives

$$\frac{V^2}{2} + \frac{p}{\rho} + gz = \text{constant} \tag{4.8a}$$

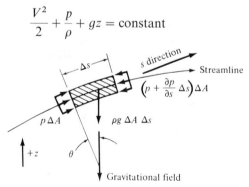

Fig. 4-9. Nonviscous fluid element moving along a streamline.

Equation (4.8a) states that the energy per unit mass of fluid is constant along a streamline for steady incompressible flow of a nonviscous fluid. The pressure may vary with time, however, but it would have no effect on the flow pattern. Equation (4.8a) also may be written as

$$\frac{\rho V_1^2}{2} + p_1 + \rho g z_1 = \frac{\rho V_2^2}{2} + p_2 + \rho g z_2 = \text{constant total pressure} \quad (4.8b)$$

Equation (4.8b) states that the energy per unit volume of fluid remains constant along a streamline. Here $\rho V^2/2$ is called the dynamic pressure, p the static pressure, and $\rho g z$ the potential pressure. Finally, E.q. (4.8a) may be written as

$$\frac{V_1^2}{2g} + \frac{p_1}{g\rho} + z_1 = \frac{V_2^2}{2g} + \frac{p_2}{g\rho} + z_2 = \text{constant total head} \quad (4.8c)$$

where the energy per unit weight is constant along a streamline. Here $V^2/2g$ is called the velocity head, $p/\rho g$ the pressure head, and z the potential head. This is the form of the Bernoulli equation commonly used by hydraulic engineers (with ρg replaced by γ).

Equations (4.8) may be applied to viscous flows in instances when the viscous effects are small, in order to obtain approximate results. Many times these results are very close to the actual results.

For a gas flowing without heat transfer and without viscous effects (frictionless adiabatic or isentropic flow), $p/\rho^k = \text{constant}$ and the equation of motion [Eq. (4.7)] for steady flow is

$$V\frac{\partial V}{\partial s} + \frac{c^{1/k}}{p^{1/k}}\frac{\partial p}{\partial s} + g\frac{\partial z}{\partial s} = 0$$

which, upon integration along a streamline, becomes

$$\frac{V^2}{2} + \frac{k}{k-1}\left(\frac{p}{\rho}\right) + gz = \text{constant}$$

or

$$V_1^2 + \frac{k}{k-1}\left(\frac{p_1}{\rho_1}\right) + gz_1 = \frac{V_2^2}{2} + \frac{k}{k-1}\left(\frac{p_2}{\rho_2}\right) + gz_2 \quad (4.9)$$

If $V_1 = 0$,

$$V_2 = \sqrt{\frac{2k}{k-1}\left[\frac{p_1}{\rho_1} - \frac{p_2}{\rho_2}\right]} \quad (4.10a)$$

$$= \sqrt{\frac{2kR}{k-1}(T_1 - T_2)} \quad (4.10b)$$

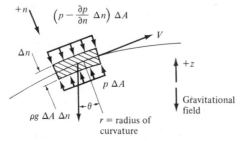

Fig. 4-10.   Normal forces on fluid element moving along a streamline.

This equation was derived for (a) steady flow, (b) a compressible nonviscous perfect gas flowing isentropically, and (c) conditions where no external work is involved.

It can be shown (Sec. 5-2) that the steady-flow energy equation written for *any* adiabatic flow without external work, for both reversible *and* irreversible flow, has the same form as Eq. (4.9), which was derived for reversible adiabatic flow only. This is because the term $[k/(k-1)](p/\rho)$ is equal to $c_p T$, the enthalpy, a property whose changes are independent of path.

If we consider the fluid element of Fig. 4-9 and apply Newton's second law in a direction *normal* to the streamline as in Fig. 4-10, variations in velocity head, pressure head, and potential head (or velocity or dynamic pressure, static pressure, and potential pressure) across streamlines, from one to another, may be determined.

Although the velocity V itself is always tangent to a streamline at any given point, if the streamline is curved, the velocity at an infinitesimal distance from the given point measured along the streamline will have changed direction. The component of this change velocity in a direction normal to the streamline at the given point is designated as V_n. If V_n is a function of position and time, we may write that $V_n = V_n(s,t)$. Thus,

$$dV_n = \frac{\partial V_n}{\partial s}\, ds + \frac{\partial V_n}{\partial t}\, dt$$

or

$$\frac{dV_n}{dt} = \frac{\partial V_n}{\partial s}\frac{ds}{dt} + \frac{\partial V_n}{dt}$$

From Fig. 4-11, $\partial V_n/\partial s = V/r$, and since $ds/dt = V$,

$$a_n = \frac{V^2}{r} + \frac{\partial V_n}{\partial t}$$

The term V^2/r is convective, and $\partial V_n/\partial t$ the local acceleration, respectively.

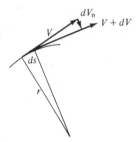

Fig. 4-11.   Flow along a curved streamline.

As before, we will apply Newton's second law to an element of a nonviscous fluid. From Fig. 4-10, writing $\Sigma F = ma_n$ normal to the streamline gives

$$-p\,\Delta A + \left(p - \frac{\partial p}{\partial n}\,\Delta n\right)\Delta A + \rho g\,\Delta A\,\Delta n\cos\theta$$

$$= \rho\,\Delta A\,\Delta n\left(\frac{V^2}{r} + \frac{\partial V_n}{\partial t}\right)$$

and since $\cos\theta = -\partial z/\partial n$, upon simplifying we get

$$\frac{V^2}{r} + \frac{1}{\rho}\frac{\partial p}{\partial n} + g\frac{\partial z}{\partial n} = -\frac{\partial V_n}{\partial t} \qquad (4.11)$$

which is the equation of motion normal to a streamline.

If the streamlines are straight (the radius of curvature r is infinite and V^2/r and V_n are both zero), Eq. (4.11) may be integrated along the direction n to get, for a constant-density fluid,

$$p + \rho g z = \text{constant}$$

which indicates that, in flow of constant-density fluids in which there is no curvature of the streamlines, the piezometric pressure is constant normal to the streamlines. In a horizontal plane under the same conditions, the pressure intensity is constant normal to the streamlines. For steady flow, $\partial V_n/\partial t = 0$ and the last term in Eq. (4.11) drops out.

If streamlines are curved, Eq. (4.11) for steady flow in a horizontal plane becomes

$$\frac{V^2}{r} = -\frac{1}{\rho}\frac{dp}{dn} = +\frac{1}{\rho}\frac{dp}{dr}$$

since the n direction is opposite to the radial r direction (Fig. 4-10). Thus,

$$\frac{dp}{dr} = \frac{\rho V^2}{r}$$

This states that dp/dr is positive for both rotational and irrotational flow. As the radius of curvature increases (positive dr) the pressure also increases (positive dp), and as the radius of curvature decreases the pressure also decreases. Hence the pressure increases away from the center of curvature and decreases toward the center of curvature of curvilinear streamlines.

We will integrate Eq. (4.11) for an irrotational flow and for a rotational flow which rotates as a solid body.

For irrotational flow $\omega = 0$, and from Eq. (3.26)

$$\frac{V}{r} = \frac{\partial V}{\partial n} \qquad \text{and} \qquad \frac{V^2}{r} = V\frac{\partial V}{\partial n}$$

Equation (4.11) may then be written for steady flow as

$$V\frac{\partial V}{\partial n} + \frac{1}{\rho}\frac{\partial p}{\partial n} + g\frac{\partial z}{\partial n} = 0$$

Thus for steady irrotational flow, the equation of motion normal to a streamline is of exactly the same form as the equation of motion along a streamline [Eq. (4.7)]. Integration gives the same result, namely Eqs. (4.8), and thus in an incompressible steady irrotational flow the expression $(\rho V^2/2 + p + \rho g z)$, or its equivalents in Eq. (4.8), is the same *throughout the entire fluid*, not only along individual streamlines but also across streamlines.

If the fluid rotates as a solid body, there is no relative motion between fluid particles and thus no shear, even though the fluid is viscous. The equation of motion normal to a streamline [Eq. (4.11)] for steady flow may therefore be applied. For this, the angular velocity is constant and is given by

$$\omega = \frac{V}{r}$$

Since $\partial V_n/\partial s = V/r$, Eq. (3.26) may be written as

$$\frac{V}{r} - \frac{\partial V}{\partial n} = 2\omega = 2\frac{V}{r}$$

so that

$$-\frac{\partial V}{\partial n} = \frac{V}{r} \qquad \text{and} \qquad \frac{V^2}{r} = -V\frac{\partial V}{\partial n}$$

Substitution of this expression for V^2/r in Eq. (4.11) for steady flow gives, upon integration in a direction normal to the streamlines for a constant-density fluid,

$$-\frac{V^2}{2} + \frac{p}{\rho} + gz = \text{constant} \qquad\qquad (4.12a)$$

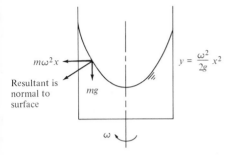

Fig. 4-12. Liquid rotating as a solid body.

or

$$-\frac{V_1^2}{2} + \frac{p_1}{\rho} + gz_1 = -\frac{V_2^2}{2} + \frac{p_2}{\rho} + gz_2 \qquad (4.12b)$$

Note the difference between this equation and Eq. (4.8).

This solid-body rotational vortex may be generated by steadily rotating a cylindrical container about its axis, and the resulting free surface of the liquid in the container will assume the shape of a paraboloid (Fig. 4-12). Pressures at any point within the liquid are determined from the hydrostatic equation ($\Delta p = \rho g \Delta z$) or from Eq. (4.12). The liquid contained within the impeller of a centrifugal pump approximates that of a solid-body rotation. When there is flow through the pump, the pressure rise due to radial flow must be added to that produced by the rotation (tangential flow).

Example 4-11

A 50-mm-diameter nozzle is attached to the end of a 100-mm-diameter pipe. The pressure in the pipe is 365 kPa gage. What is the flow rate through the nozzle? Assume water flows without viscous effects. Refer to Fig. 4-13.

SOLUTION
The continuity and the Bernoulli equations are applied between the nozzle inlet at section 1 and the nozzle exit at section 2. These are

$$V_1 A_1 = V_2 A_2 \qquad \text{and} \qquad \rho\left(\frac{V_1^2}{2}\right) + p_1 = \rho\left(\frac{V_2^2}{2}\right) + p_2$$

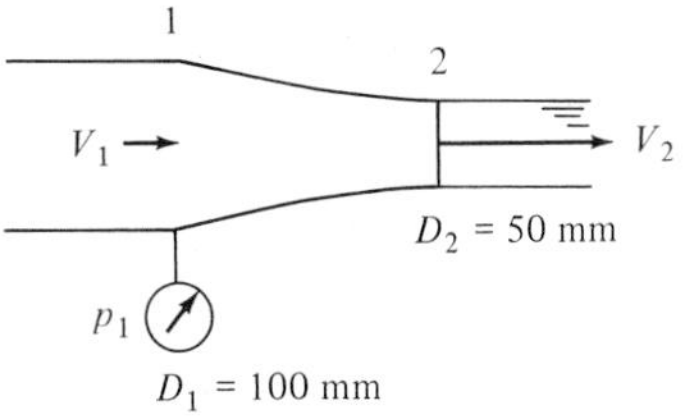

Fig. 4-13. See Example 4-11. Nozzle on the end of a pipe.

from which

$$V_2 = \sqrt{\frac{2p_1}{\rho[1 - (A_2/A_1)^2]}} = \sqrt{\frac{(2)(365,000)}{(1000)[1 - (50/100)^4]}}$$

$$= 27.9 \text{ m/s}$$

$$Q = V_2A_2 = (27.9)\left(\frac{\pi}{4}\right)(0.050)^2 = 0.0548 \text{ m}^3/\text{s}$$

Note that in order to state that the pressure at the nozzle exit is at atmospheric pressure, the streamlines must be parallel. If the nozzle consisted of a straight taper, there would be a contraction of the flow beyond the end of the nozzle, the streamlines would be curved, and the pressure would vary across the exit section from atmospheric at the outer edge of the jet to a value above atmospheric at the jet axis.

Example 4-12

Kerosene flows from a large open tank through a 50-mm-diameter hole in its side. The free surface of the kerosene is 5.20 m above the center line of the hole (see Fig. 4-14). What is the velocity of the jet issuing from the hole? Assume nonviscous flow.

SOLUTION

The Bernoulli equation may be applied from the free surface to a point in the jet where the pressure is known. As in Example 4-11, this is where the streamlines become parallel and the pressure is atmospheric.

The Bernoulli equation is

$$0 + 0 + 5.2 = \frac{V_2^2}{2g} + 0 + 0$$

so that $V_2 = \sqrt{(2g)(5.2)} = 10.10$ m/s. Note that the jet velocity is independent of the fluid density.

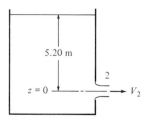

Fig. 4-14.   See Example 4-12. Flow from an open tank.

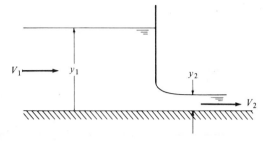

Fig. 4-15.   See Example 4-13. Flow under a sluice gate.

Example 4-13

Water flows under a sluice gate as shown in Fig. 4-15. Assume frictionless one-dimensional flow between sections 1 and 2. What is the flow rate per unit width of channel in terms of the given depths?

SOLUTION

The continuity equation for a unit width of flow is $V_1 y_1 = V_2 y_2$. The Bernoulli equation is $V_1^2/2g + y_1 = V_2^2/2g + y_2$ since the pressures are the same along the free surface. Note that the Bernoulli equation written along the channel bed between sections 1 and 2 is $V_1^2/2g + p_1/\gamma = V_2^2/2g + p_2/\gamma$ since the potential head is constant along the channel bed. But $p_1/\gamma = y_1$ and $p_2/\gamma = y_2$. Combining the continuity and Bernoulli equations gives

$$q = V_2 y_2 = \sqrt{\frac{2g(y_1 - y_2)}{1 - (y_2/y_1)^2}} \, y_2$$

as the volumetric flow rate per unit width.

▬▬

Example 4-14

A submarine travels at a depth of 60 m in the ocean (seawater density is 1025 kg/m³). The pressure on the nose of the submarine is 214 kPa above the atmospheric pressure at the ocean surface. What is the speed of the submarine?

SOLUTION

To make the flow steady, we hold the submarine at rest and let the water approach it at the same relative speed as the submarine was moving through the water. The Bernoulli equation is

$$\frac{\rho V_1^2}{2} + p_1 = \frac{\rho V_1^2}{2} + \gamma h = p_0 = 214{,}000 \text{ N gage}$$

where h is the submerged depth. Thus the water approaches at

$$V_1 = \sqrt{2(p_0 - \gamma h)/\rho} = \sqrt{2[214{,}000 - (1025)(9.81)(60)]/1025}$$
$$= 5.02 \text{ m/s}$$

Example 4-15

An elbow of 30- $\times$ 30-cm cross section is made up of two circular arcs with an inner radius $r_1 = 30$ cm and an outer radius $r_2 = 60$ cm (Fig. 4-16). Assume water flows through this elbow as an irrotational vortex. The pressure difference between the inner and outer walls is 20.0 kPa. What is the flow rate through the elbow?

SOLUTION

For an irrotational vortex, $Vr = $ constant (from Section 3-6) and Eq. 4-8b is used to find either the velocity at any radius or the value of the constant. Then we integrate across the section (since the velocity varies) to find the flow rate: $V_1 r_1 = V_2 r_2 = c$ so that $V_1 = 2V_2$. From the Bernoulli equation (Eq. 4.8b) we obtain

$$p_2 - p_1 = \frac{\rho}{2}(V_1^2 - V_2^2) = \frac{c^2 \rho}{2}\left(\frac{1}{r_1^2} - \frac{1}{r_2^2}\right)$$

Thus,

$$c^2 = \frac{2(p_2 - p_1)}{\rho[(1/r_1^2) - (1/r_2^2)]} = \frac{(2)(20{,}000)}{(1000)[1/(0.3)^2 - 1/(0.6)^2]} = 4.80$$

$$c = 2.191$$

$$Q = \int_{r_1}^{r_2} V(0.3)\, dr = \int_{r_1}^{r_2}\left(\frac{c}{r}\right)(0.3)\, dr = (2.191)(0.3)\ln\left(\frac{r_2}{r_1}\right)$$

$$= (2.191)(0.3)(0.693) = 0.456 \text{ m}^3/\text{s}$$

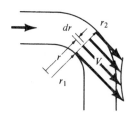

Fig. 4-16. See Example 4-15. Irrotational vortex flow in a square elbow.

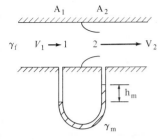

Fig. 4-17.   See Example 4-16. Flow through a nozzle meter.

Example 4-16

A nozzle meter consists of a nozzle throat of circular area A_2 in a pipe of circular area A_1 and is used to measure the volumetric flow rate of a liquid of specific weight γ_f. A manometer connected as shown in Fig. 4-17 uses a heavier liquid of specific weight γ_m and indicates a deflection of h_m meters. What is the flow rate? Neglect viscous effects.

SOLUTION

The continuity equation and the Bernoulli equation written between sections 1 and 2 are solved simultaneously. These are

$$Q = V_1 A_1 = V_2 A_2 \quad \text{and} \quad \frac{V_1^2}{2g} + \frac{p_1}{\gamma_f} = \frac{V_2^2}{2g} + \frac{p_2}{\gamma_f}$$

From the manometer deflection

$$\frac{p_1 - p_2}{\gamma_f} = \frac{h_m(\gamma_m - \gamma_f)}{\gamma_f}$$

Thus

$$Q = V_2 A_2 = \frac{A_2}{\sqrt{1 - (A_2/A_1)^2}} \sqrt{\frac{2gh_m(\gamma_m - \gamma_f)}{\gamma_f}}$$

The neglect of viscous effects for low-viscosity liquids gives quite accurate results in this instance.

Example 4-17

A closed cylindrical drum 1.00 m in diameter and full of water is rotated about its vertical axis at 20 rad/s (Fig. 4-18). What is the difference in pressure between the axis of rotation and the outer edge of the drum?

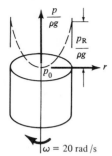

Fig. 4-18. See Example 4-17. Rotational vortex in a rotating drum.

SOLUTION
This is a solid-body rotation and Eq. (4.12b) applies. Let point 1 be the axis and point 2 be the outer edge of the drum. Since $V_1 = 0$,

$$\Delta p = \frac{\rho V_2^2}{2} = \frac{\rho \omega^2 R^2}{2} = \frac{(1000)(20)^2(0.5)^2}{2} = 50 \text{ kPa}$$

CARTESIAN COORDINATES
The normal pressures acting on the six sides of a cubical nonviscous fluid element is shown in Fig. 4-19.

Let X, Y, and Z be the body forces per unit mass in the x, y, and z directions such that the resultant body force is $\mathbf{F} = X\mathbf{i} + Y\mathbf{j} + Z\mathbf{k}$, and let the velocity vector be $\mathbf{V} = u\mathbf{i} + v\mathbf{j} + w\mathbf{k}$. The surface forces on an element of sides dx, dy, and dz and mass $\rho\, dx\, dy\, dz$ are due to pressure only, and the resultant force in the x direction is

$$X(\rho\, dx\, dy\, dz) - \frac{\partial p}{\partial x}\, dx(dy\, dz)$$

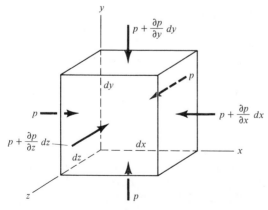

Fig. 4-19. Normal forces on a nonviscous fluid element.

This is equal to the product of mass and acceleration in the x direction, where the acceleration is

$$a_x = \frac{du}{dt} = \frac{\partial u}{\partial t} + u \frac{\partial u}{\partial x} + v \frac{\partial u}{\partial y} + w \frac{\partial u}{\partial z}$$

Equating force per unit mass to acceleration gives

$$X - \frac{1}{\rho} \frac{\partial p}{\partial x} = u \frac{\partial u}{\partial x} + v \frac{\partial u}{\partial y} + w \frac{\partial u}{\partial z} + \frac{\partial u}{\partial t} \tag{4.13a}$$

Similarly for the y and z directions

$$Y - \frac{1}{\rho} \frac{\partial p}{\partial y} = u \frac{\partial v}{\partial x} + v \frac{\partial v}{\partial y} + w \frac{\partial v}{\partial z} + \frac{\partial v}{\partial t} \tag{4.13b}$$

and

$$Z - \frac{1}{\rho} \frac{\partial p}{\partial z} = u \frac{\partial w}{\partial x} + v \frac{\partial w}{\partial y} + w \frac{\partial w}{\partial z} + \frac{\partial w}{\partial t} \tag{4.13c}$$

or, in vector form,

$$\mathbf{F} - \frac{1}{\rho} \nabla p = \frac{D\mathbf{V}}{Dt} \tag{4.14}$$

For irrotational flow, curl $\mathbf{V} = 0$, and $\partial v/\partial x = \partial u/\partial y$, $\partial w/\partial y = \partial v/\partial z$, and $\partial u/\partial z = \partial w/\partial x$. Thus Eqs. (4.13), when multiplied by dx, dy, and dz, respectively, may be written for steady flow as

$$X\,dx - \frac{1}{\rho} \frac{\partial p}{\partial x} dx = u \frac{\partial u}{\partial x} dx + v \frac{\partial v}{\partial x} dx + w \frac{\partial w}{\partial x} dx$$

$$Y\,dy - \frac{1}{\rho} \frac{\partial p}{\partial y} dy = u \frac{\partial u}{\partial y} dy + v \frac{\partial v}{\partial y} dy + w \frac{\partial w}{\partial y} dy \tag{4.15}$$

$$Z\,dz - \frac{1}{\rho} \frac{\partial p}{\partial z} dz = u \frac{\partial u}{\partial z} dz + v \frac{\partial v}{\partial z} dz + w \frac{\partial w}{\partial z} dz$$

If z is measured upward, opposite to gravity, and if only body forces due to gravity are considered, then $X = Y = 0$ and $Z = -g$. Then when Eqs. (4.15) are added together, we get

$$\frac{1}{2} d(V^2) + \frac{dp}{\rho} + g\,dz = 0$$

whose integral is the same as Eq. (4.8), the Bernoulli equation, for a constant-density fluid. Equations (4.14) and (4.15) are the Euler equations in Cartesian coordinates and in vector form.

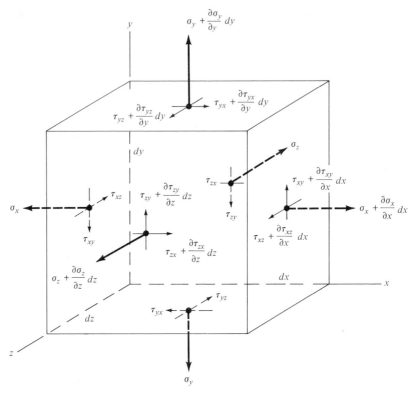

Fig. 4-20. Normal and tangential surface forces on a viscous fluid element in motion.

4-4. Equations of Motion for Viscous Fluids

There were two forces acting on a nonviscous fluid element—a body force and a normal pressure force on its surface. This normal pressure force is the thermodynamic pressure used in the equation of state. A viscous fluid in motion has additional surface forces acting on a fluid element—tangential shear forces and normal forces which are not the same as the normal pressure forces acting on a nonviscous fluid element in motion. An outline of the development of the equations of motion for a viscous fluid will be given; details may be found in books on boundary layer theory by Schlichting [1] and by Rouse [2].

Surface forces on a viscous fluid element in motion are shown in Fig. 4-20. Normal stresses are designated as σ and shear stresses as τ. The first subscript indicates the plane of the area element by giving the direction of the normal to the area, and the second subscript indicates the direction of the stress component. Normal stresses reduce in magnitude to the fluid pressure as the viscosity is reduced to zero, whereas the shear stresses reduce to zero as the viscosity goes to zero.

An inspection of the surface forces in Fig. 4-20 indicates that the net surface forces in the x, y, and z directions are

$$dF_{s(x)} = \left(\frac{\partial \sigma_x}{\partial x} + \frac{\partial \tau_{yx}}{\partial y} + \frac{\partial \tau_{zx}}{\partial z}\right) dx\, dy\, dz$$

$$dF_{s(y)} = \left(\frac{\partial \sigma_y}{\partial y} + \frac{\partial \tau_{zy}}{\partial z} + \frac{\partial \tau_{xy}}{\partial x}\right) dx\, dy\, dz \qquad (4.16)$$

$$dF_{s(z)} = \left(\frac{\partial \sigma_z}{\partial z} + \frac{\partial \tau_{xz}}{\partial x} + \frac{\partial \tau_{yz}}{\partial y}\right) dx\, dy\, dz$$

The body force per unit mass has components X, Y, and Z as before. Then Newton's second law may be written for a unit volume as

$$\rho a_x = \rho X + \frac{\partial \sigma_x}{\partial x} + \frac{\partial \tau_{yx}}{\partial y} + \frac{\partial \tau_{zx}}{\partial z}$$

$$\rho a_y = \rho Y + \frac{\partial \tau_{xy}}{\partial x} + \frac{\partial \sigma_y}{\partial y} + \frac{\partial \tau_{zy}}{\partial z} \qquad (4.17)$$

and

$$\rho a_z = \rho Z + \frac{\partial \tau_{xz}}{\partial x} + \frac{\partial \tau_{yz}}{\partial y} + \frac{\partial \sigma_z}{\partial z}$$

It can be shown that by taking moments about the edges of the elemental volume in Fig. 4-20 corresponding to the axes, $\tau_{xy} = \tau_{yx}$, $\tau_{yz} = \tau_{zy}$ and $\tau_{zx} = \tau_{xz}$.

Phenomenologic relations for the normal and for the shear stresses are necessary in order to express them in terms of fluid properties and flow parameters. For a Newtonian fluid, shear stress is the product of dynamic viscosity and rate of angular deformation. From Section 3-4 we get

$$\tau_{xy} = \mu\left(\frac{\partial u}{\partial y} + \frac{\partial v}{\partial x}\right)$$

$$\tau_{yz} = \mu\left(\frac{\partial v}{\partial z} + \frac{\partial w}{\partial y}\right) \qquad (4.18)$$

and

$$\tau_{zx} = \mu\left(\frac{\partial w}{\partial x} + \frac{\partial u}{\partial z}\right)$$

The normal stresses for an isotropic (properties independent of orientation) Newtonian fluid are also related to the pressure p, the viscosity μ, and velocity gradients from a hypothesis of Stokes. The relationships are

$$\sigma_x = -p + 2\mu\frac{\partial u}{\partial x} - \frac{2}{3}\mu\left(\frac{\partial u}{\partial x} + \frac{\partial v}{\partial y} + \frac{\partial w}{\partial z}\right)$$

$$\sigma_y = -p + 2\mu\frac{\partial v}{\partial y} - \frac{2}{3}\mu\left(\frac{\partial u}{\partial x} + \frac{\partial v}{\partial y} + \frac{\partial w}{\partial z}\right) \qquad (4.19)$$

$$\sigma_z = -p + 2\mu\frac{\partial w}{\partial z} - \frac{2}{3}\mu\left(\frac{\partial u}{\partial x} + \frac{\partial v}{\partial y} + \frac{\partial w}{\partial z}\right)$$

For a nonviscous fluid, $\mu = 0$ and it is thus obvious that $\sigma_x = \sigma_y = \sigma_z = -p$. For a viscous fluid in motion, the arithmetic average of the three normal stresses is called the *pressure*:

$$-p = \tfrac{1}{3}(\sigma_x + \sigma_y + \sigma_z) \qquad (4.20)$$

Substitution of Eq. (3.3) for the accelerations, Eqs. (4.18) for the shear stresses, and Eqs. (4.19) for the normal tresses in Eqs. (4.17) will give the complete equations of motion for viscous fluid motion, the Navier–Stokes equations. For a constant-viscosity fluid and for incompressible flow they become

$$\frac{\partial u}{\partial t} + u\frac{\partial u}{\partial x} + v\frac{\partial u}{\partial y} + w\frac{\partial u}{\partial z}$$

$$= X - \frac{1}{\rho}\frac{\partial p}{\partial x} + v\left(\frac{\partial^2 u}{\partial x^2} + \frac{\partial^2 u}{\partial y^2} + \frac{\partial^2 u}{\partial z^2}\right)$$

$$\frac{\partial v}{\partial t} + u\frac{\partial v}{\partial x} + v\frac{\partial v}{\partial y} + w\frac{\partial v}{\partial z}$$

$$= Y - \frac{1}{\rho}\frac{\partial p}{\partial y} + v\left(\frac{\partial^2 v}{\partial x^2} + \frac{\partial^2 v}{\partial y^2} + \frac{\partial^2 v}{\partial z^2}\right)$$

$$\frac{\partial w}{\partial t} + u\frac{\partial w}{\partial x} + v\frac{\partial w}{\partial y} + w\frac{\partial w}{\partial z}$$

$$= Z - \frac{1}{\rho}\frac{\partial p}{\partial z} + v\left(\frac{\partial^2 w}{\partial x^2} + \frac{\partial^2 w}{\partial y^2} + \frac{\partial^2 w}{\partial z^2}\right) \qquad (4.21)$$

or in vector form

$$\frac{DV}{Dt} = \mathbf{F} - \frac{1}{\rho}\,\text{grad}\,p + v\nabla^2\mathbf{V} \qquad (4.22)$$

Comparison with the Euler equations [Eqs. (4.13) and (4.14)] shows that there is an additional term in the Navier–Stokes equations which accounts for viscous effects.

In cylindrical (r,θ,z) coordinates, the Navier–Stokes equations are

$$\rho\left(\frac{\partial v_r}{\partial t} + v_r\frac{\partial v_r}{\partial r} + \frac{v_\theta}{r}\frac{\partial v_r}{\partial \theta} - \frac{v_\theta^2}{r} + v_z\frac{\partial v_r}{\partial z}\right)$$

$$= -\frac{\partial p}{\partial r}$$

$$+ \mu\left(\frac{\partial^2 v_r}{\partial r^2} + \frac{1}{r}\frac{\partial v_r}{\partial r} - \frac{v_r}{r^2} + \frac{1}{r^2}\frac{\partial^2 v_r}{\partial \theta^2} - \frac{2}{r^2}\frac{\partial v_\theta}{\partial \theta} + \frac{\partial^2 v_r}{\partial z^2}\right) + \rho F_r$$

$$\rho\left(\frac{\partial v_\theta}{\partial t} + v_r\frac{\partial v_\theta}{\partial r} + \frac{v_\theta}{r}\frac{\partial v_\theta}{\partial \theta} + \frac{v_r v_\theta}{r} + v_z\frac{\partial v_\theta}{\partial z}\right)$$

$$= -\frac{1}{r}\frac{\partial p}{\partial \theta}$$

$$+ \mu\left(\frac{\partial^2 v_\theta}{\partial r^2} + \frac{1}{r}\frac{\partial v_\theta}{\partial r} - \frac{v_\theta}{r^2} + \frac{1}{r^2}\frac{\partial^2 v_\theta}{\partial \theta^2} + \frac{2}{r^2}\frac{\partial v_r}{\partial \theta} + \frac{\partial^2 v_\theta}{\partial z^2}\right) + \rho F_\theta$$

$$\rho\left(\frac{\partial v_z}{\partial t} + v_r\frac{\partial v_z}{\partial r} + \frac{v_\theta}{r}\frac{\partial v_z}{\partial \theta} + v_z\frac{\partial v_z}{\partial z}\right)$$

$$= -\frac{\partial p}{\partial z} + \mu\left(\frac{\partial^2 v_z}{\partial r^2} + \frac{1}{r}\frac{\partial v_z}{\partial r} + \frac{1}{r^2}\frac{\partial^2 v_z}{\partial \theta^2} + \frac{\partial^2 v_z}{\partial z^2}\right) + \rho F_z \qquad (4.23)$$

where F_r, F_θ, and F_z are the body force components per unit mass in the three directions indicated. In cylindrical coordinates the stress components for an incompressible fluid are

$$\sigma_r = -p + 2\mu\frac{\partial v_r}{\partial r}; \qquad \tau_{r\theta} = \mu\left[r\frac{\partial}{\partial r}\left(\frac{v_\theta}{r}\right) + \frac{1}{r}\frac{\partial v_r}{\partial \theta}\right]$$

$$\sigma_\theta = -p + 2\mu\left[\frac{1}{r}\frac{\partial v_\theta}{\partial \theta} + \frac{v_r}{r}\right]; \qquad \tau_{\theta z} = \mu\left[\frac{\partial v_\theta}{\partial z} + \frac{1}{r}\frac{\partial v_z}{\partial \theta}\right]$$

$$\sigma_z = -p + 2\mu\frac{\partial v_z}{\partial z}; \qquad \tau_{rz} = \mu\left[\frac{\partial v_r}{\partial z} + \frac{\partial v_z}{\partial r}\right] \qquad (4.24)$$

In 1894 Reynolds developed a modified form of the Navier–Stokes equations for turbulent flow by writing each velocity as an average value plus a fluctuating component ($u = \bar{u} + u'$, and so on). These modified equations indicate shear stresses due to turbulence known as the Reynolds stresses which have been useful in the study of turbulent flow.

PROBLEMS

4-1 A 60-mm water jet impinges on a flat plate normal to the jet. The jet speed is 20.0 m/sec. What is the force on the plate
(a) when it is at rest,
(b) when it moves at 8.0 m/sec against the jet, and
(c) when it moves at 8.0 m/sec in the same direction as the jet?

4-2. Repeat Prob. 4-1 for a U-shaped surface which deflects the jet through 120°.

4-3. A horizontal water jet 10 cm² in area moving at 30 m/sec strikes a smooth plate inclined 45° from the jet direction. What is the magnitude and direction of the resulting force on the plate?

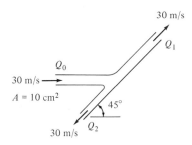

Prob. 4-3

4-4 A 60-mm-diameter water jet with a velocity of 16 m/s strikes a smooth flat plate inclined 35° from the direction of the jet. What is the normal force on the plate
(a) when the plate is at rest,
(b) when the plate moves 3.5 m/s in the direction of the jet, and
(c) when the plate moves 3.5 m/s against the jet?
(d) What is the work done in part (b)? Express in terms of power (force times velocity).

4-5. A water jet 60 mm in diameter moving at 18 m/s strikes a curved vane and is deflected through 150°. What is the resultant force on the vane?

4-6. In Prob. 4-5, the vane is one of a number of vanes on the periphery of a rotor which rotates such that the vanes move away from the jet at 9 m/s. What is the force on the vanes?

Jet

$V_j = 18$ m/s $V_b = 9$ m/s

Prob. 4-6

4-7. A jet-propelled boat moves at 5.2 m/s through fresh water. It takes in water at a rate of 40 m³/min and passes it through pumps and out through ducts at 10.5 m/s. What is the propulsive force?

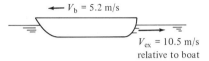

$V_b = 5.2$ m/s

$V_{ex} = 10.5$ m/s
relative to boat

Prob. 4-7

4-8. Water flows through a nozzle at 0.090 m³/s and leaves in a jet 60 mm in diameter. The nozzle entrance is ten times the jet area. What is the momentum change of the water?

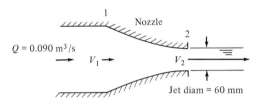

1

Nozzle

2

$Q = 0.090$ m³/s

$V_1 \rightarrow$

V_2

Jet diam = 60 mm

Prob. 4-8

4-9. A 3-cm-diameter nozzle on the end of a 9-cm-diameter fire hose discharges water with a jet velocity of 30 m/s. The pressure in the hose is 350 kPa gage. What is the magnitude and direction of the net force of the water on the nozzle?

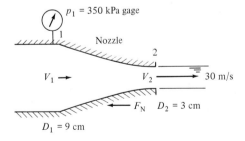

$p_1 = 350$ kPa gage

1

Nozzle

2

$V_1 \rightarrow$

V_2 — 30 m/s

F_N $D_2 = 3$ cm

$D_1 = 9$ cm

Prob. 4-9

4-10. Water enters a conical diffuser at an average velocity of 12 m/s. The diffuser enlarges from a 15-cm diameter at inlet to 30 cm in diameter at exit over a length of 1.40 m. The measured pressure rise is $0.81 \, \rho V_1^2/2$. The pressure at inlet is atmospheric. What is the resultant longitudinal force of the water on the diffuser walls?

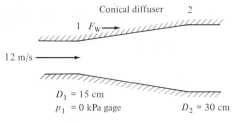

Prob. 4-10

4-11. Water flows through a well-designed contraction cone for a water tunnel at an average velocity at exit of $V_2 = 12.0$ m/s. The contraction inlet is 0.45 m in diameter, and the exit is 0.15 m in diameter. The measured pressure drop is $1.00 \, V_2^2/2$. The exit is at atmospheric pressure. What is the resultant longitudinal force of the water on the contraction walls?

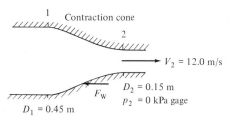

Prob. 4-11

4-12. A reducing elbow turns water through $135°$ in a vertical plane. The inlet diameter is 40 cm, the outlet diameter is 20 cm, and the outlet center is 2.00 m below the horizontal inlet axis. The volume of the elbow is 0.20 m³. What is the net force of the water on the elbow when

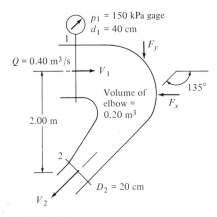

Prob. 4-12

water flows at $0.40 \text{ m}^3/\text{s}$ and the inlet pressure is 150 kPa gage? Neglect viscous effects.

4-13. A water jet pump has a jet area $A_j = 45 \text{ cm}^2$ and a jet velocity $V_j = 27 \text{ m/s}$ which entrains a secondary stream of water having a velocity $V_s = 3 \text{ m/s}$ in a constant-area pipe of total area $A = 540 \text{ cm}^2$. At section 2 the water is thoroughly mixed. Assume one-dimensional flow and neglect wall shear.

(a) What is the average velocity of the mixed flow at section 2?

(b) What is the pressure rise $(p_2 - p_1)$, assuming the pressure of the jet and the secondary stream to be the same at section 1?

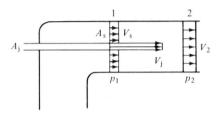

Prob. 4-13

4-14. A rocket burns fuel at a rate of 3.5 kg/s and ejects the exhaust gases at 1400 m/s relative to the rocket and at the same pressure as the surroundings. What is the thrust developed when the rocket travels at a forward speed of

(a) 150 m/s and

(b) 450 m/s?

4-15. Can a rocket travel faster than the relative velocity of the exhaust gases which leave it? Explain.

4-16. A rocket exhausts 45 kg of gas per second at an exit velocity of 1475 m/s relative to the rocket. What is the thrust produced

(a) when the rocket is held at rest on a test stand,

(b) when the rocket travels at 250 m/s, and

(c) if the rocket flies at an altitude where the ambient pressure is 28 kPa abs?

Exit pressure is 100 kPa abs and exit diameter is 30 cm.

4-17. A rocket motor with a 50-cm-diameter nozzle exit develops a thrust of 40 kN at sea level. What thrust would be developed under the same nozzle flow conditions at an altitude of 5 km where the pressure is 54 kPa abs?

4-18. A jet engine takes in air at 70 kg/s and burns fuel at a rate of 4.5 kg/s. The combustion gases leave the engine at ambient pressure and at 1400 m/s relative to the engine.

(a) What thrust is produced when the engine is at rest on a test stand?

(b) What thrust is produced when the engine moves through the air at 250 m/s?

4-19. Calculate the momentum flux factors for the circular duct flows of Probs. 3-49 to 3-53.

4-20. Calculate the momentum flux factor β for
(a) two-dimensional laminar flow between parallel plates in Prob. 3-44,
(b) turbulent flow in a circular tube in Prob. 3-46, and
(c) laminar flow in a circular tube in Prob. 3-45.

4-21. A fluid of density ρ flows along a flat plate. At the beginning the velocity is uniform, but at the end of the plate the fluid velocity is reduced in a region called the *boundary layer*. Within this boundary layer the velocity profile is linear, such that $u = (y/\delta)u_s$ (see Fig. 4-8). What is the total drag force on a unit width of plate?

4-22. What force F is required to prevent the sprinkler from rotating about a vertical axis? The density of the flowing liquid is ρ.

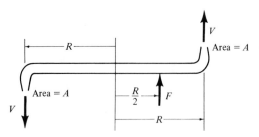

Prob. 4-22

4-23. In Prob. 4-22,
(a) What is the speed of rotation if the sprinkler rotates without friction?
(b) Explain why, when the sprinkler arm rotates, you cannot isolate only the fluid in the reducing elbow in making a momentum analysis but must include the fluid in the arm of the sprinkler as well.

4-24. The drag on a circular cylinder of diameter D may be measured indirectly by determining the change in momentum flux through the

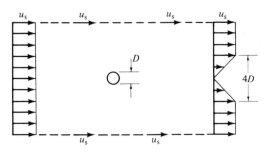

Prob. 4-24

control volume or surface indicated. The pressure is assumed to be constant over the entire control surface.

(a) What is the drag force on the cylinder per unit length of cylinder?

(b) The drag coefficient is defined here as $C_D = F/(\rho u_s^2/2)(D)$. What is the drag coefficient? Compare with that shown in Fig. 11-4 at Re_D between 10^4 and 10^5.

4-25. An incompressible fluid of density ρ enters a circular tube with a uniform velocity profile V and a pressure p_1. When the flow becomes fully developed some distance downstream the velocity profile becomes $u = u_m[1 - (r/R)^2]$, where u is the velocity at radius r in the tube of radius R and u_m is the maximum velocity at the tube centerline. Show that the total wall shear force along the tube wall is $(p_1 - p_2 - \rho V^2/3)\pi R^2$ by using the actual momentum flux for the fully developed condition. Here p_2 is a downstream pressure.

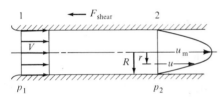

Prob. 4-25

4-26. Water flowing into the vertical inlet of a diversion tunnel for a dam forms an irrotational vortex except near its axis. At a point 1.6 m from the center of rotation, the tangential velocity is 1 m/s.

(a) How much lower is the water surface at a point 0.8 m from the center of rotation than at a point 1.6 m from the center of rotation?

(b) By how much is the water level at the 0.8-m radius below the level of the water far from the axis of the vortex?

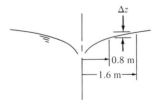

Prob. 4-26

4-27. A 15-cm vertical cylinder is rotated in a viscous liquid. The motion of the liquid is that of an irrotational vortex. The liquid level adjacent to the cylinder is 5 cm lower than the free surface of the liquid far from the cylinder. What is the rotational speed of the cylinder?

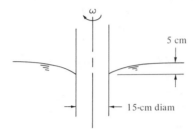

5 cm

15-cm diam

Prob. 4-27

4-28. Suppose particles in a swirling mass of air move in horizontal circles about a fixed vertical axis such that the motion out to a radius R is that of a solid-body rotation with angular velocity ω; beyond R the motion is that of an irrotational vortex. This is known as a cylindrical vortex.

   (a) What is the peripheral velocity V_θ for $r < R$ and for $r > R$?
   (b) Indicate, by a sketch, the variation in pressure with radius in a horizontal plane. Compare with Prob. 4-30a.
   (c) What is the radial pressure gradient at $r = R$? Compare with Prob. 4-30c.
   (d) Why does the irrotational vortex not extend to the axis of rotation in a real fluid? (This is essentially the motion of a tornado, or of dust and leaves or snow often observed on a windy day.)

4-29. A centrifugal pump impeller rotates at 1450 r/min. The inlet radius is 5 cm and the outlet radius is 20 cm. With no flow (pump discharge valve is shut off), the fluid within the impeller rotates as a solid body. What is the pressure difference in newtons per square meter and the pressure head difference in meters of fluid between impeller outlet and inlet

   (a) if the impeller contains water and
   (b) if the pump is not primed, so that the impeller rotates in standard air?

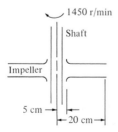

1450 r/min

Shaft

Impeller

5 cm

20 cm

Prob. 4-29

4-30. Water particles in a large reservoir move in horizontal circles about a fixed vertical axis. The motion out to a radius R is that of a solid-body rotation at an angular velocity ω; beyond R the motion is that of an irrotational vortex. This is known as a cylindrical vortex.
  (a) Sketch the shape of the water surface. (It is essentially like that produced by a canoe paddle.)
  (b) What is the peripheral velocity V_θ of the particles for a radius $r > R$ in terms of r, R, and ω?
  (c) What is the slope of the water surface at R in a radial direction?

4-31. The tangential velocity around the surface of a cylinder normal to the flow of a nonviscous fluid at a free-stream velocity u_s and density ρ is $v_\theta = -2u_s \sin\theta$ (see Fig. 6-15).
  (a) Write an expression for the pressure at the surface of the cylinder in terms of the free-stream pressure p_s, u_s, ρ, and θ.
  (b) At what values of θ is the velocity zero? These are the stagnation points.
  (c) At what values of θ is the pressure on the surface of the cylinder equal to the free-stream pressure p_s?

4-32. In Prob. 4-31, what is the difference between the highest and the lowest pressure on the surface of the cylinder for standard air at a free-stream velocity u_s of
  (a) 15 m/s and
  (b) 30 m/s?

4-33. The tangential velocity around the surface of a sphere in a nonviscous fluid flowing past it at a free-stream velocity of u_s is $v_\theta = -\frac{3}{2}u_s \sin\theta$. Answer parts (a), (b), and (c) of Prob. 4-31.

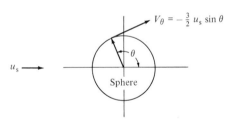

Prob. 4-33

4-34. In Prob. 4-33
  (a) what is the difference between the maximum and the minimum pressure on the surface of the sphere in terms of the fluid ρ and the free-stream velocity u_s?
  What is this pressure difference for $u_s = 3$ m/s and a fluid density 1000 kg/m^3 for
  (b) a billiard ball of diameter 60 mm and
  (c) a basketball of diameter 24 cm?
  (d) Does the size of the sphere affect the answer?

4-35. A circular duct contracts gradually from an area of 0.10 m² to an area of 0.025 m². A differential manometer indicates a 7.50-cm deflection of mercury when water flows through the duct. What is the flow rate for frictionless flow?

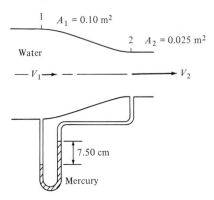

Prob. 4-35

4-36. A 30-cm-diameter vertical, circular duct contracts gradually to 15 cm in diameter. A differential manometer connected between the two sections shows a 10.50-cm deflection of mercury when water flows through the system. Estimate the flow rate for frictionless flow from the larger to the smaller section (significant losses would occur for flow from the smaller to the larger section).

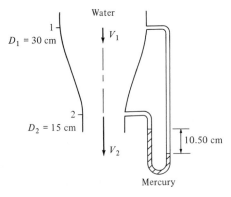

Prob. 4-36

4-37. A liquid flows from a large open tank through a round hole in the side of the tank located H ft below the free liquid surface within the tank.
(a) Where in the flow stream is the velocity equal to $\sqrt{2gH}$, neglecting friction?

(b) Explain why it is necessary to write the Bernoulli equation from a point within the tank to the section of the jet where the streamlines are parallel.

(c) Describe the pressure distribution throughout the flow section at the hole in the tank.

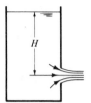

Prob. 4-37

4-38. Write an expression for the volumetric flow rate for a liquid of density ρ flowing through a venturi meter [Fig. 13-11a] in terms of the diameters D_1 and D_2 and the pressure drop $p_1 - p_2$. Assume a nonviscous fluid. Viscous effects may be taken into account by multiplying the result of this problem by a discharge coefficient C_d.

4-39. What is the manometer deflection h_m

(a) when both the wall static pressure tap S and the total pressure tube T are as shown,

(b) when only the total pressure tube T is moved to location B,

(c) when only the static pressure tap S is moved to location A, and

(d) when the total pressure tube T is moved to B and the wall static pressure tap is moved to A?

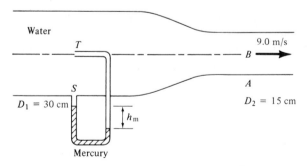

Prob. 4-39

4-40. The pressure is 100 kPa gage near the end of a 20-cm-diameter pipe with an end plate that has a circular opening which produces an 8-cm-diameter jet. What is the flow rate of water and the force of the water on the end plate?

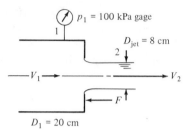

$p_1 = 100$ kPa gage

$D_{jet} = 8$ cm

$D_1 = 20$ cm

Prob. 4-40

4-41. The tank of Example 4-12 has 3 m of an oil of specific gravity 0.85 above 5.2 m of water. What is the jet velocity?

4-42. The tank of Example 4-12 is covered, and the air space above the kerosene is at a gage pressure of 90 kPa. What is the jet velocity?

4-43. A small card may be lifted by placing a spool near it and blowing through the axial hole in the spool. Explain this. A thumb tack or common pin may be needed to prevent relative sliding between the card and spool.

Card

Prob. 4-43

4-44. The width of a rectangular open channel is reduced from 1.8 to 1.5 m, and the bottom is raised 0.30 m at the contracted section. The depth upstream is 1.20 m, and the surface drops 8 cm at the contracted section. What is the flow rate? Neglect viscous effects.

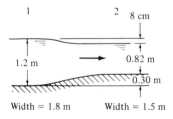

1   2   8 cm

1.2 m   0.82 m

0.30 m

Width = 1.8 m   Width = 1.5 m

Prob. 4-44

4-45. Water flows in a rectangular channel 3.0 m wide at a depth of 9 cm. The channel bottom is gradually raised 6 cm. The water surface rises

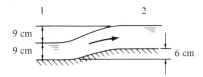

Prob. 4-45

9 cm as the water passes over the raised section of the channel. What is the flow rate? Neglect viscous effects.

4-46. In Prob. 4-45, the channel is 4.0 m wide, the initial depth is 0.10 m, the channel bottom gradually is raised 0.07 m, and the water surface rises 0.1 m. What is the flow rate in m^3/s? Neglect frictional effects.

4-47. Water is 5.40 m deep upstream of a small dam and flows over the downstream face of the dam at a rate of 1.80 m^3/s per meter width of dam.

    (a) What is the water depth at the foot of the dam, assuming no viscous effects?

    (b) What is the horizontal force of the water on the dam per meter width?

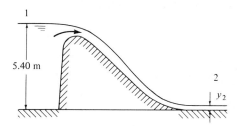

Prob. 4-47

4-48. In the figure for Prob. 4-47, the upstream depth is 3.80 m and the flow rate is 1.14 m^3/s per meter width of dam.

    (a) What is the water depth at the foot of the dam, assuming no losses?

    (b) What is the resultant force of the water on the dam per meter width?

4-49. A small airplane travels through standard air at 70 m/s. What is the gage pressure at a stagnation point on the airplane?

4-50. The stagnation pressure on a small aircraft flying in standard air near sea level is 2.20 kPa gage. What is the speed of the airplane?

4-51. The flow through a fan test stand is controlled by the circular cone as shown. What is the axial force exerted by the air on the cone? Assume frictionless flow and standard air at constant density. *Hint:* Obtain the velocities V_1 and V_2 from the continuity and Bernoulli equations; then write the momentum equation ($p_1 = 6.5$ kPa gage).

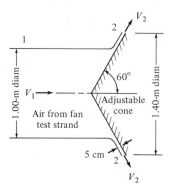

Prob. 4-51

4-52. In Prob. 4-51 the duct diameter is 80 cm, $V_1 = 18$ m/s, $V_2 = 70$ m/s, and the cone half-angle is 45°. For air at a constant density of 1.225 kg/m³ and with no viscous effects, what is the net force of the air stream on the cone?

4-53. Bridge piers 1.60 m wide are on 10.0-m centers and parallel to water flowing in a horizontal river bed. The depth upstream of the piers is 2.40 m and between piers, 2.30 m. Assume one-dimensional, non-viscous flow. What is the flow rate between piers?

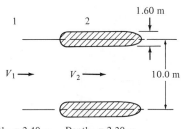

Depth₁ = 2.40 m   Depth₂ = 2.30 m

Prob. 4-53

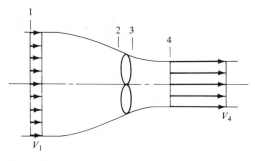

Prob. 4-54

4-54. Fluid is accelerated as it passes through the slipstream of a propeller. Show that the average velocity of the fluid as it passes the plane of the propeller blades is the average of the upstream and the downstream velocities V_1 and V_4. Assume $p_1 = p_4$, $V_2 = V_3$, and neglect viscous effects.

4-55. The figure for Prob. 4-54 represents the slipstream of a ship propeller. The forward speed of the ship is 9.0 m/s in fresh water, the propeller diameter is 50 cm, and 2.36 m³/s of water passes through the propeller blades. What is the thrust developed?

4-56. A helicopter has a mass of 2500 kg and has a 10-m-diameter rotor. What is the average velocity of air through the rotor blades when the helicopter hovers at rest near the ground in a standard atmosphere (density is 1.225 kg/m³)?

4-57. A liquid of density ρ flows steadily without friction in a horizontal duct whose shape is the frustum of a cone at a flow rate of Q. Show that the pressure gradient in the direction of flow $+x$ in terms of the diameter D is

$$\frac{dp}{dx} = \frac{32\rho Q^2}{\pi^2 D^5} \frac{dD}{dx}$$

4-58. A piston in a large tank of cross-sectional area A forces an incompressible ideal fluid of density ρ out of the tank and into the atmosphere through a small pipe of cross-sectional area a. The fluid jet has a cross-sectional area $C_c a$ in the smaller pipe.
   (a) Write the continuity, momentum, and Bernoulli equations for this system to show that the contraction coefficient is

$$C_c = \frac{1}{2 - a/A}$$

   (b) As a becomes very small compared with A, the entrance to the small pipe is called a Borda mouthpiece. What is the contraction coefficient for a Borda mouthpiece?

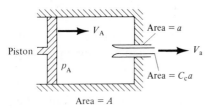

Prob. 4-58

4-59. In Fig. 4-20, show that $\tau_{xy} = \tau_{yx}$.

4-60. Show that a variation in viscous shear in a steady uniform flow is accompanied by a pressure gradient normal to the shear stress gradient.

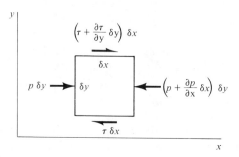

Prob. 4-60

4-61. From the Navier–Stokes equations, derive an expression for the velocity profile for laminar flow between parallel plates, both at rest, a distance of $2b$ apart.

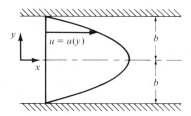

Prob. 4-61

4-62. Determine from the Navier–Stokes equations the velocity profile for plane Couette flow with a pressure gradient dp/dx constant. Couette flow is laminar flow between parallel plates one of which is at rest and the other a distance $y = h$ from the one at rest, which moves at a constant velocity U.

4-63. In Prob. 4-62, there is a special case where there is no net flow and the pressure gradient is zero. What is the velocity profile in the fluid between plates?

4-64. From the Navier–Stokes equations in cylindrical form, derive an expression for the velocity profile for laminar flow in a circular tube. Compare with the derivation of Eq. (9.11a).

4-65. Two viscous fluids occupy the space between two parallel infinite plates. The upper fluid, of viscosity μ_2, occupies one half the space and is adjacent to the upper plate which is moved at a constant velocity U. The lower fluid, of viscosity μ_1, is adjacent to the lower plate which is at rest. Derive expressions for the velocity profile in each fluid.

References

1. H. Schlichting, *Boundary Layer Theory* (translated by J. Kestin), 6th ed. (New York: McGraw-Hill Book Company, Inc., 1968).
2. H. Rouse, *Fluid Mechanics for Hydraulic Engineers* (New York: McGraw-Hill Book Company, Inc., 1938; and New York: Dover Publications, Inc, 1961).

Chapter 5
Fluid Energy

The steady-flow energy equation for fluid flow is obtained from the first law of thermodynamics (given in Sec. 1-6), which is an expression of the conservation of energy; it states that the amount of heat added to a fluid is equal to the change in its energy content plus any work done by the fluid. We will be concerned only with mechanical and thermal energies. For a fluid system of fixed material, the energy content consists of kinetic, internal, and potential energy. For a fluid control volume, the energy content includes displacement energy in addition to these. This displacement energy represents the energy or work required to push the fluid across the boundaries of the control volume.

There are no restrictions on the direction of the interchange of mechanical and thermal energy implied in the first law of thermodynamics. The second law of thermodynamics governs this.

It should be recalled that the Bernoulli equation for constant-density fluids [Eq. (4.8)] is a restricted form of an energy equation; it includes only kinetic, displacement, and potential energy. External work, heat transfer, and internal energy are *not* included. It is derived from purely dynamical principles and is valid only for conditions under which it is derived. The Bernoulli equation for compressible gases is also a restricted form of an energy equation [Eq. (4.9)], which includes kinetic, displacement, internal, and potential energy. External work and heat transfer are *not* included. It was derived for reversible adiabatic flow, yet is valid for both irreversible as well as reversible adiabatic flow for reasons explained near the bottom of p. 127.

The steady-state general energy equation will first be developed and expressed in a form commonly used in hydraulics and in gas dynamics. Then a comparison of the equations of motion with the energy equation will be made. An expression will then be derived that describes viscous dissipation. Finally, expressions for calculating power available or expended in flow systems will be given.

5-1. The General Energy Equation

In the Reynolds transport theorem [Eq. (II.1)] the point function P represents the energy content $e\rho$ per unit volume (e per unit mass in Sec. 1-6), and the energy content may be changed only by adding heat or by doing external work. Thus Eq. (II.1) may be expressed as

$$Q - Wk_{\text{done}} = \frac{\partial}{\partial t} \int_{\substack{\text{control} \\ \text{volume}}} (e\rho)\, d\Psi + \int_{\substack{\text{control} \\ \text{surface}}} e\rho(\mathbf{V} \cdot d\mathbf{S})$$

The energy content e represents kinetic energy, displacement energy and internal energy (the sum is enthalpy), and potential energy per unit mass. For steady flow the first term on the right is zero.

Thus the steady flow energy equation may be written for a control volume as:

heat added $= \Delta$ (kinetic energy) $+ \Delta$ (displacement energy)

$+ \Delta$ (potential energy) $+ \Delta$ (internal energy)

$+$ (work done by the fluid)

The heat added and the internal energy are considered to be forms of thermal energy, and the remaining terms are considered to be forms of mechanical energy.

• *Kinetic Energy.* The kinetic energy of any mass m moving at a velocity V is $mV^2/2$, and in terms of kinetic energy per unit mass it is $V^2/2$. For a fluid passing a section where the velocity is not uniform (two- or three-dimensional flow), the true kinetic energy per second is always greater than that based on the average velocity. The true kinetic energy is found by integrating the product of mass flow rate through an infinitesimal area and the kinetic energy per unit mass over the entire flow area. The ratio of the true kinetic energy per unit time to that based on the average velocity is

System boundary

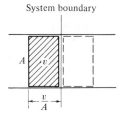

Fig. 5-1. Flow work or displacement energy.

called the *kinetic energy-flux correction factor* α, where

$$\alpha = \frac{\int_A (u^2/2)(u\rho\, dA)}{(V^2/2)(V\rho A)} = \frac{\int_A u^3\, dA}{V^3 A} \geq 1 \tag{5.1a}$$

$$\approx \frac{1}{V^3 n} \sum_1^{i=n} u_i^3 \tag{5.1b}$$

For one-dimensional flow, $\alpha = 1$; for laminar flow in a round pipe with a parabolic velocity profile, $\alpha = 2$; for turbulent flow in a pipe, $\alpha \approx 1.06$.

In the Bernoulli equation the kinetic energy term ($V^2/2$, $\rho V^2/2$, or $V^2/2g$) refers to a point velocity. In the energy equation the kinetic energy term uses the average velocity in a duct, and therefore the true kinetic energy is obtained by integration across the duct area as in Eq. (5.1). The energy equation often is used in a one-dimensional analysis, and α is assumed to be unity.

• *Flow or Displacement Energy.* Flow or displacement energy is the energy or work required to push a unit mass of fluid across the boundary of a system, at entrance or exit. Thus if the shaded area in Fig. 5-1 represents the volume occupied by a unit mass of fluid, its volume is the specific volume v; if the area through which it flows is A, the length of the shaded region is v/A. The work required to displace this volume from the shaded region to the dotted region across the boundary is the force times the distance moved. This is $(pA)(v/A) = pv$. The term pv is the flow or displacement work (or energy) per unit mass of fluid. Although pv has a value for both closed and open systems, only for the open system does it represent energy (flow or displacement work).

• *Potential Energy.* A unit mass of fluid has potential energy dependent upon its elevation above an arbitrary datum elevation where $z = 0$. The work required to bring it to any elevation other than the datum is gz joules per unit mass, and this is its potential energy.

• *Internal Energy.* The internal energy (u per unit mass) is the form in which energy is stored within a substance, and in general it is a function of

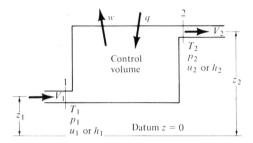

Fig. 5-2. Nomenclature for steady-flow energy equation.

pressure and temperature. For a perfect gas, it is a function only of the temperature [Eq. (1.8)].

The steady-flow energy equation may now be written as (see Fig. 5-2)

$$\alpha_1 \frac{V_1^2}{2} + p_1v_1 + gz_1 + u_1 + q - w = \alpha_2 \frac{V_2^2}{2} + p_2v_2 + gz_2 + u_2$$

or

$$\frac{V_1^2}{2} + p_1v_1 + gz_1 + u_1 + q - w = \frac{V_2^2}{2} + p_2v_2 + gz_2 + u_2 \qquad (5.2)$$

for one-dimensional flow with $\alpha_1 = \alpha_2 = 1$, and this is the steady-flow energy equation applicable to one-dimensional flow of all fluids (compressible or incompressible, with or without friction). In terms of enthalpy ($h = u + pv$), Eq. (5.2) may be written as

$$\frac{V_1^2}{2} + h_1 + gz_1 + q - w = \frac{V_2^2}{2} + h_2 + gz_2 \qquad (5.3a)$$

Recall from Eqs. (1.8) and (1.9) that for a perfect gas

$$u_2 - u_1 = c_v(T_2 - T_1) \qquad \text{and} \qquad h_2 - h_1 = c_p(T_2 - T_1)$$

In general, setting $\alpha_1 = \alpha_2 = 1$ for a one-dimensional analysis is acceptably accurate even though the flow is two- or three-dimensional. Equations (5.2) and (5.3) are the forms of the energy equation generally applicable to gas flows, although they are valid for liquid flows as well. According to the first law of thermodynamics, energy is neither created nor destroyed, so there are actually no energy losses—merely a conversion from one form of energy to another (from thermal to mechanical, or from mechanical to thermal).

COMPRESSIBLE GASES

In thermodynamics, the steady-flow energy equation applied to gases or vapors is generally used in analyses involving heat and work transfer and enthalpy changes in systems such as boilers, compressors, gas or steam turbines, and condensers. In fluid mechanics, we are generally interested in systems involving heat transfer and enthalpy and velocity changes. Thus the work term and potential energy change are of secondary interest and may be considered zero. For open systems with heat transfer, Eq. (5.3) becomes

$$\frac{V_1^2}{2} + h_1 + q = \frac{V_2^2}{2} + h_2 \tag{5.3b}$$

For adiabatic flow (irreversible as well as reversible, that is, with or without friction),

$$\frac{V_1^2}{2} + h_1 = \frac{V_2^2}{2} + h_2 = h_0 \tag{5.3c}$$

where h_0 is the total or stagnation enthalpy, corresponding to a condition of zero velocity reached adiabatically. In general, the stagnation state corresponds to a condition of zero velocity reached isentropically. (See Chapter 10 for a more detailed treatment.)

For isothermal flow, $h_1 = h_2$ for a perfect gas and thus

$$\frac{V_1^2}{2} + q = \frac{V_2^2}{2} \tag{5.3d}$$

Example 5-1

What is the temperature on the nose of a missile moving at 610 m/s in standard air (15°C)?

SOLUTION
This is supersonic flow, and the air must pass through a shock before it reaches the nose of the missile. The flow is *not* isentropic (not reversible), and Eq. (5.3c) may be applied, or $V_1^2/2 + h_1 = h_0$, where we consider the missile to be at rest and the air is approaching it at the speed of the missile in still air. Thus

$$\frac{V_1^2}{2} = c_p(T_0 - T_1) \quad \text{and} \quad T_0 = T_1 + \frac{V_1^2}{2c_p}$$

$$= 15 + \frac{(610)^2}{(2)(1005)}$$

$$= 200°C$$

[Table A-2 (Appendix II) gives $c_p = 1005$ J/kg °K.]

Example 5-2

Methane flows through a horizontal insulated pipeline. Frictional effects reduce pressure and density in the direction of flow such that the velocity is increased from 12 to 120 m/s. What is the corresponding reduction in temperature?

SOLUTION

From Eq. (5.3c), $T_1 - T_2 = (V_2^2 - V_1^2)/2c_p$, where $c_p = 2190$ J/kg $^\circ$K from Table A-2 (Appendix II):

$$T_1 - T_2 = \frac{120^2 - 12^2}{(2)(2190)} = 3.25^\circ C$$

LIQUIDS AND CONSTANT-DENSITY GASES

We neglect the thermal terms, and to account for the conversion of some of the mechanical energy into thermal energy due to viscous dissipation or irreversibilities, this mechanical energy is included in a friction term h_L. Then the energy equation becomes

$$\frac{V_1^2}{2} + \frac{p_1}{\rho} + gz_1 - w = \frac{V_2^2}{2} + \frac{p_2}{\rho} + gz_2 + h_L' \qquad (5.4a)$$

where $h_L' = (u_2 - u_1 - q)$ and represents the amount of mechanical energy converted into thermal energy.

It is easier to measure the available energy dissipation from pressure measurements than to account for it by temperature measurements. For example, if water flows through a 150-m length of 15 cm pipe at a rate of 0.06 m^3/s, the available energy dissipation appears in the form of a pressure drop of about 90 kPa, depending on the pipe roughness. If the pipe were perfectly insulated, there would be no heat loss ($q = 0$) and the increase in internal energy equivalent to the dissipation of mechanical energy would result in a temperature rise of about 0.02°C ($\Delta T = \Delta p/\rho c_v$). The pressure drop can be measured with much more accuracy than the temperature rise, even for the limiting case when no heat is transferred through the pipe walls. If the pipe is not insulated, the temperature rise would be even less than 0.02°C.

Equations (5.2), (5.3), and (5.4a) express the steady-flow energy equation in terms of energy per unit mass of fluid (joules per kilogram). The energy equation for liquids is usually written in a form obtained from Eq. (5.4a) by dividing each term by the acceleration of gravity g. This gives (since the specific weight $\gamma = \rho g$)

$$\frac{V_1^2}{2g} + \frac{p_1}{\gamma} + z_1 - w = \frac{V_2^2}{2g} + \frac{p_2}{\gamma} + z_2 + h_L \qquad (5.4)^\dagger$$

---

† In the energy equation for gases [Eq. (5.2)], pressures must be in absolute units. In the energy equation for liquids or incompressible fluids [Eqs. (5.4a) and (5.4b)], pressures may be either gage or absolute, since the terms involving pressure may be brought to the same side of the equal sign and become $\Delta p/\gamma$. Thus the difference in pressures Δp is the same whether gage or absolute pressures are used.

where each term has the dimensions of energy per unit *weight* of fluid (joules per newton) or head (meters). Thus in hydraulics each term is often referred to as *head*—velocity head, pressure head, potential head, head removed by a turbine (w is positive) or added by a pump (w is negative), and head loss (h_L is always positive in the direction of flow), in the order appearing in Eq. (5.4b).

Equation (5.3a), in terms of energy per unit mass, will be used for compressible gas flow, and Eq. (5.4b), in terms of energy per unit weight, will be used for liquid flow and constant-density gas flow.

Example 5-3

A liquid flows with frictional effects in an insulated horizontal pipe. Show that the viscous dissipation represented by a mechanical energy decrease and a comparable thermal energy increase is related by the expression $\Delta T = \Delta p/\rho c_v$.

SOLUTION

In Eq. (5.4a) $V_1 = V_2$ from continuity, $z_1 = z_2$ since the pipe is horizontal, and $w = 0$ because there is no pump in the line. Thus $h'_L = (p_1 - p_2)/\rho$. For adiabatic flow $h'_L = (u_2 - u_1) = c_v(T_2 - T_1)$ also. Equating these gives $\Delta T = \Delta p/\rho c_v$.

━━━

Example 5-4

Oil ($s = 0.86$) flows in a pipe 75 cm in diameter with an average velocity of 1.52 m/s. The piezometric pressure drop in 300 m of pipe is 10.3 kPa. What is the frictional head loss? (See Fig. 5-3.)

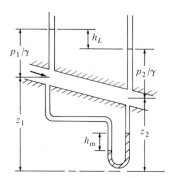

Manometer deflection h_m directly indicates drop in piezometric pressure.

Fig. 5-3.   See Example 5-4. Head loss in a pipe.

SOLUTION

Equation (5.4b) applies, since the flow has viscous effects. Since the velocity remains constant from continuity ($V_1 = V_2$) and $\Delta(p + \gamma z) = 10.3 \text{ kPa}$,

$$\frac{p_1}{\gamma} + z_1 = \frac{p_2}{\gamma} + z_2 + h_L$$

$$h_L = \Delta\left(\frac{p}{\gamma} + z\right) = 10{,}300/(0.86)(9.81)(1000)$$

$$= 1.22 \text{ m}$$

Example 5-5

Two open reservoirs of water (Fig. 5-4) are connected by 1200 m of 250-mm pipe. The level of water in the higher reservoir is 35 m above that in the lower reservoir. The steady flow rate is 0.130 m³/s. (a) What is the total head loss (available energy dissipation)? (b) What is the pressure at the midpoint of the pipe, assuming one-half the head loss occurs upstream and one-half downstream of the midpoint? Assume the midpoint of the pipe to be at the same level or elevation as the lower reservoir.

SOLUTION

(a) Equation (5.4b) may be written between the free surfaces of the reservoirs. The velocity at each point is zero, and the pressures are both atmospheric. Thus Eq. (5.4b) becomes, with the z datum at the lower reservoir,

$$0 + 0 + 35 = 0 + 0 + 0 + h_L$$

$$h_L = 35 \text{ m or J/N}$$

(b) Equation (5.4b) written between the upper reservoir surface and the pipe midpoint is

$$0 + 0 + 35 = \frac{V_3^2}{2g} + \frac{p_3}{\gamma} + 0 + 17.5$$

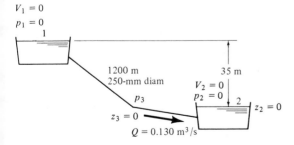

Fig. 5-4. Flow between open reservoirs.

where $V = Q/A = 2.65$ m/s and $V_3^2/2g = 0.36$ m. Thus,

$$p_3 = (35 - 17.5 - 0.36)(1000)(9.81) = 168 \text{ kPa gage}$$

Example 5-6

In Example 5-5 what energy or head must be supplied by a pump to convey 0.130 m^3/s of water from the lower to the higher reservoir?

SOLUTION
For the same flow rate through the same piping system, the head loss through the piping system is again $h_L = 35$ m. The energy equation written from the lower to the higher reservoir surfaces is

$$0 + 0 + 0 - w = 0 + 0 + 35 + 35$$
$$-w = 70 \text{ m} \quad \text{or} \quad \text{J/N}$$

This shows that the pump has to provide for a 35-m increase in potential head and a 35-m head loss in the piping system.

Example 5-7

Air is at rest in a supply reservoir at 80°C ($=353$°K) and at high pressure and exhausts through a converging–diverging nozzle (Fig. 10-4a) to the atmosphere. The air temperature in the nozzle throat is $\frac{5}{6}$ the absolute temperature in the supply reservoir. What is the air speed in the nozzle throat?

SOLUTION
The flow is without heat transfer, so from Eq. (5.3c) $h_1 = V_2^2/2 + h_2$. Then

$$V_2 = \sqrt{2c_p(T_1 - T_2)} = \sqrt{2c_p(T_1/6)}$$
$$= \sqrt{(2)(1005)(353/6)} = 344 \text{ m/s}$$

It is easily verified that the air speed in the nozzle throat is sonic.

Example 5-8

Measurements for performance tests on a diffuser (a gradual expansion) for a cavitation-testing water tunnel are as follows: inlet $V_1 = 14.02$ m/s, $\alpha_1 = 1.004$ and outlet $V_2 = 3.71$ m/s, $\alpha_2 = 1.56$ (inlet velocity profile was very flat and outlet profile peaked). The rise in pressure head is 8.11 m. What is the head loss through the diffuser?

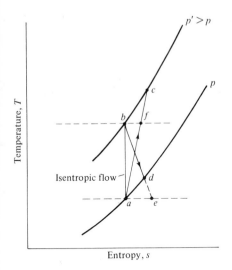

Fig. 5-5.   Reversible and irreversible adiabatic flows.

SOLUTION

From the energy equation

$$h_L = \alpha_1(V_1^2/2g) - \alpha_2(V_2^2/2g) + (p_1 - p_2)/\gamma$$
$$= (1.004)(10.02) - (1.56)(0.70) + (-8.11) = 0.86 \text{ m}$$

Irreversibilities in constant-density flow are accounted for by the term h_L in Eq. (5.4), which is a positive quantity in the equation and increases in the direction of flow.

For compressible gas flows, the second law of thermodynamics requires that if irreversibilities exist from viscous dissipation, the entropy must increase for adiabatic flow. This is shown on a temperature-entropy diagram in Fig. 5-5. An isentropic flow from pressure p to p' or from p' to p is shown by the path $a–b$ or $b–a$ at constant entropy. For a frictional flow in a duct from one pressure to another, the downstream temperature is higher than for frictionless flow. Thus flow from p to p' is from a to c, T_c being higher than T_b; and flow from p' to p is from b to d, T_d being higher than T_a. Since enthalpy is a function of temperature, the downstream enthalpy is also higher for frictional than for frictionless adiabatic flow. From the energy equation [Eq. (5.3c)] a higher downstream enthalpy is associated with a lower downstream velocity. Thus for given initial conditions, a downstream velocity is lower for frictional than for frictionless adiabatic flow. Similarly, if adiabatic flow is from one temperature to another, the downstream pressure is lower for frictional than for frictionless flow (e is lower than d and f is lower than c in Fig. 5-5).

5-2. Comparison of Equations of Motion with Steady-Flow Energy Equation

Both the Bernoulli equation and the steady-flow energy equation are applicable along a stream tube, which constitutes the closed portion of the boundaries of an open system. Thus they may be compared with one another. The Bernoulli equation for steady incompressible flow [Eq. (4.8)] may be written as

$$\Delta\left(\frac{V^2}{2} + \frac{p}{\rho} + gz\right) = 0 \tag{5.5}$$

and the steady-flow energy equation for incompressible flow [Eq. (5.4a)] as

$$-w = \Delta\left(\frac{V^2}{2} + \frac{p}{\rho} + gz\right) + h'_L \tag{5.6}$$

Comparing Eq. (5.5) with Eq. (5.6), we may state that since Eq. (5.5) does not include friction (viscous shear) or external shaft work and Eq.(5.6) does, one difference between the flow of a real and an ideal incompressible fluid is that in order to keep the sum of kinetic, displacement, and potential energy constant, *external shaft work must be done on a real fluid to overcome the so-called head loss due to frictional effects*. If no external shaft work is done on the fluid, then the head loss due to frictional effects decreases the available energy of the fluid.

For gas flow, the Euler equation of motion for steady flow [Eq. (4.7)] with $\partial V/\partial t = 0$ is valid for a *nonviscous fluid only*. The steady-flow energy equation [Eq. (5.2) or Eqs. (5.3)] is valid for *both* viscous and nonviscous fluids (both reversible and irreversible flow). If we compare them in differential form for conditions of *reversible* (frictionless) flow, the Euler equation of motion should remove the results of motion or flow from the steady-flow energy equation, and we should be left with the first law of thermodynamics for a nonflow (closed) system. That this is true is easily verified. Equation (5.2) in differential form with no external work is

$$dq = V\,dV + p\,dv + v\,dp + g\,dz + du$$

and Eq. (4.7) with $\partial V/\partial t = 0$ is

$$V\,dV + \frac{dp}{\rho} + g\,dz = 0$$

Subtraction gives

$$dq = du + p\,dv$$

which is an expression of the first law of thermodynamics for a closed system.

For an adiabatic gas flow (though not necessarily reversible, in which case $p/\rho^k \neq C$) with no external work, the steady-flow energy equation is

$$\frac{V_1^2}{2} + h_1 + gz_1 = \frac{V_2^2}{2} + h_2 + gz_2$$

and since for a perfect gas

$$h = c_p T = \frac{Rk}{k-1} T = \frac{k}{k-1}\left(\frac{p}{\rho}\right)$$

we get

$$\frac{V_1^2}{2} + \frac{k}{k-1}\left(\frac{p_1}{\rho_1}\right) + gz_1 = \frac{V_2^2}{2} + \frac{k}{k-1}\left(\frac{p_2}{\rho_2}\right) + gz_2 \tag{5.7}$$

for *either* reversible *or* irreversible adiabatic flow. This is identical in appearance to Eq. (4.9) obtained from the Euler equation for a *nonviscous* (frictionless) gas.

Equation (5.7) is a highly specialized form of the energy equation and at the same time a specialized integral of the equation of motion. The conditions required to specialize these two equations to produce the same form are not necessarily coincident, since we started from two entirely different initial points of consideration.

We may conclude that the steady-flow energy equation cannot be derived from the equations of motion, nor can the equations of motion be derived from the energy equation.

5-3. Work and Power

Power is the time rate of doing work. Power expended in doing work is expressed as a product of the work done per unit mass of fluid times the mass rate of flow. In terms of the steady-flow energy equation, as written for liquids, it is the product of the work done per unit weight of fluid and the weight rate of flow:

$$P = w(VA\gamma) \quad \text{watts} \tag{5.8}$$

The power available in a jet results from the kinetic energy of the jet, and the power represented is the product of this kinetic energy per unit mass and the mass rate of flow, or kinetic energy per unit weight of fluid times the weight rate of flow:

$$P = \frac{V^2}{2}(VA\rho) = \frac{V^2}{2g}(VA\gamma) \quad \text{watts} \tag{5.9}$$

Power loss in a flow system from viscous dissipation is the product of energy loss per unit mass of fluid and the mass-flow rate, or the energy loss per unit weight of fluid times the weight rate of flow:

$$P = h_L(VA\gamma) \quad \text{watts} \tag{5.10}$$

Power required to overcome a drag force is the product of the drag force and the velocity of the body moving through a fluid. Power is, in reality, a scalar product of two vectors—a force and a velocity. In scalar notation, the drag force is considered to be the component parallel to the velocity:

$$P = FV \quad \text{watts} \tag{5.11}$$

Example 5-9

Water flows from one reservoir at elevation 30 m to another at elevation 15 m through 600 m of 45-cm-diameter cast-iron pipe. The flow rate is 0.25 m³/s, and 30 kW of power are extracted from the water by a turbine. What is the head loss experienced by the water in flowing from the higher to the lower reservoir?

SOLUTION

Apply Eq. (5.4b) between the reservoir surfaces after determining the work extracted from the water. Power = $Q\gamma w$, so that $w = 30,000/(0.25)(9810) = 12.23$ m or joules per unit weight. The velocity and pressure are both zero at each reservoir surface. Thus the energy equation becomes

$$0 + 0 + 30 - (+12.23) = 0 + 0 + 15 + h_L$$

$$h_L = 2.77 \text{ m}$$

∎

Example 5-10

The water jet at an average velocity of 14.02 m/s in the test section of the cavitation-testing water tunnel of Example 5-8 is 15.24 cm in diameter. What is the power in the jet?

SOLUTION
From Eq. (5.9)

$$P = \frac{V^3 A\rho}{2} = \frac{(14.02)^3(\pi/4)(0.1524)^2(1000)}{2} = 25.1 \text{ kW}$$

∎

5-4. Viscous Dissipation

In Sec. 5-2 it was mentioned that external shaft work must be done on a viscous incompressible fluid to overcome mechanical energy dissipation due to viscous effects. These dissipative effects can be expressed in terms of the stresses on the fluid element shown in Fig. 4-20. The rate at which work is done by each stress is the product of the force on the various faces of the element and the corresponding component of velocity parallel to the force. For example, the work done per unit time on the fluid element by τ_{yx} is

$$\left[-\tau_{yx}u + \left(\tau_{yx} + \frac{\partial \tau_{yx}}{\partial y}\,dy\right)\left(u + \frac{\partial u}{\partial y}\,dy\right)\right]dx\,dz = \frac{\partial}{\partial y}(u\tau_{yx})\,dx\,dy\,dz$$

Thus this work rate is the gradient of the product of stress intensity and corresponding velocity component. The total work per unit time per unit volume in a viscous fluid is, from Fig. 4-20,

$$\frac{\partial}{\partial x}(\sigma_x u + \tau_{yx}v + \tau_{zx}w) + \frac{\partial}{\partial y}(\sigma_y v + \tau_{zy}w + \tau_{xy}u)$$

$$+ \frac{\partial}{\partial z}(\sigma_z w + \tau_{xz}u + \tau_{yz}v)$$

Expanding these gives

$$\left[\frac{\partial \sigma_x}{\partial x} + \frac{\partial \tau_{xy}}{\partial y} + \frac{\partial \tau_{x3}}{\partial z}\right]u + \left[\frac{\partial \sigma_y}{\partial y} + \frac{\partial \tau_{yz}}{\partial z} + \frac{\partial \tau_{yx}}{\partial x}\right]v$$

$$+ \left[\frac{\partial \sigma_z}{\partial z} + \frac{\partial \tau_{xz}}{\partial x} + \frac{\partial \tau_{zy}}{\partial x}\right]w + \left(\sigma_x \frac{\partial u}{\partial x} + \sigma_y \frac{\partial v}{\partial y} + \sigma_z \frac{\partial w}{\partial z}\right)$$

$$+ \tau_{yx}\left(\frac{\partial v}{\partial x} + \frac{\partial u}{\partial y}\right) + \tau_{zy}\left(\frac{\partial w}{\partial y} + \frac{\partial v}{\partial z}\right) + \tau_{xz}\left(\frac{\partial u}{\partial z} + \frac{\partial w}{\partial x}\right)$$

The first three terms (in brackets) are the rate of increase of the mechanical energy of the fluid and are not dissipative.

Using Eqs. (4.19) for the shear stresses and Eqs. (4.20) for the normal stresses, the last four terms above may be expressed as

$$-p\left(\frac{\partial u}{\partial x} + \frac{\partial v}{\partial y} + \frac{\partial w}{\partial z}\right) + 2\mu\left[\left(\frac{\partial u}{\partial x}\right)^2 + \left(\frac{\partial v}{\partial y}\right)^2 + \left(\frac{\partial w}{\partial z}\right)^2\right]$$

$$-\frac{2}{3}\mu\left[\frac{\partial u}{\partial x} + \frac{\partial v}{\partial y} + \frac{\partial w}{\partial z}\right]^2$$

$$+ \mu\left[\left(\frac{\partial v}{\partial x} + \frac{\partial u}{\partial y}\right)^2 + \left(\frac{\partial w}{\partial y} + \frac{\partial v}{\partial z}\right)^2 + \left(\frac{\partial u}{\partial z} + \frac{\partial w}{\partial x}\right)^2\right]$$

Here the first term expresses the rate at which fluid is compressed; this is displacement energy and is not dissipative. The last three terms (those multiplied by the viscosity μ) are known collectively as the dissipation function Φ,

$$\Phi = 2\mu\left[\left(\frac{\partial u}{\partial x}\right)^2 + \left(\frac{\partial v}{\partial y}\right)^2 + \left(\frac{\partial w}{\partial z}\right)^2\right]$$
$$+ \mu\left[\left(\frac{\partial v}{\partial x} + \frac{\partial u}{\partial y}\right)^2 + \left(\frac{\partial w}{\partial y} + \frac{\partial v}{\partial z}\right)^2 + \left(\frac{\partial u}{\partial z} + \frac{\partial w}{\partial x}\right)^2\right]$$
$$- \frac{2}{3}\mu\left(\frac{\partial u}{\partial x} + \frac{\partial v}{\partial y} + \frac{\partial w}{\partial z}\right)^2 \tag{5.12}$$

The last term is zero for an incompressible fluid, and the dissipation function then contains only the first two terms with brackets. They indicate that only through linear (first bracketed term) and angular (second bracketed term) deformation is there heat generated as a result of viscous action within the fluid. The second law of thermodynamics places restrictions on the direction of energy transformations and precludes the recovery of the heat generated by viscous action to increase the mechanical energy of the fluid.

PROBLEMS

5-1. Calculate the kinetic energy factor α for the circular duct flows of Probs. 3-49 through 3-53.

5-2. Calculate the kinetic energy correction factor α for
   (a) the two-dimensional laminar flow between parallel plates in Prob. 3-44.
   (b) turbulent flow in a circular tube in Prob. 3-46, and
   (c) laminar flow in a circular tube in Prob. 3-45.

5-3. Air flows through a supersonic wind tunnel. In the nozzle throat of area 0.01 m^2 the temperature is 280°K, the pressure is 50 kPa abs and the air speed is 330 m/s. In the test section of area 0.040 m^2 the temperature is 123.2°K and the pressure is 2.82 kPa abs.
   (a) Calculate the air speed in the test section from continuity.
   (b) Calculate the air speed in the test section from the energy equation.

5-4. Air flows through a horizontal insulated pipe. Frictional effects reduce pressure and density in the direction of flow so that the velocity is increased from 15 to 125 m/s. What is the reduction in temperature?

5-5. The stagnation temperature on the nose of a rocket is 38°C. The rocket travels at an altitude where the air temperature is −12°C and the pressure is 62 kPa abs. What is the speed of the rocket?

5-6. A rocket travels at an altitude where the air temperature is −11°C and the pressure is 61.6 kPa abs. The temperature on the nose of the rocket is 80°C. What is the speed of the rocket?

5-7. Methane at 1200 kPa abs and 40°C enters a rough, insulated 15-cm diameter pipe at 18.0 m/s. At a downstream section the velocity is 180 m/s.

(a) What is the temperature at the downstream section?
(b) What is the pressure at the downstream section?
(c) What is the total wall shear force?

5-8. Air flows at a constant temperature of 40°C in a pipe having a cross-sectional area of 0.010 m². At inlet the pressure is 700 kPa abs and the velocity is 30.0 m/s. At exit the pressure is 600 kPa abs.
(a) What is the exit velocity?
(b) What is the total wall shear force?

5-9. Air flows through a 15-cm diameter horizontal pipe. At inlet the pressure is 825 kPa abs, the temperature is 60°C, and the average velocity is 43.0 m/s. As a result of pipe friction and heat transfer the exit pressure is 550 kPa abs and the exit temperature is 77°C.
(a) What is the exit velocity?
(b) What is the total wall shear?
(c) How much heat must be added to the air or removed from it per unit time?

5-10. Methane at 1200 kPa abs pressure and 30°C enters a rough insulated 15-cm diameter pipe at 20 m/s. At a downstream section the velocity is 200 m/s.
(a) What is the temperature at the downstream section?
(b) What is the pressure at the downstream section?
(c) What is the total wall shear force?

5-11. Air flows at constant temperature (35°C) in a pipe of cross-sectional area of 0.01 m². At inlet the air pressure is 700 kPa abs and the air velocity is 35 m/s. At exit the air pressure is 550 kPa abs.
(a) What is the exit velocity of the air?
(b) What is the total shear on the pipe wall?

5-12. Air flows through a 15-cm-diameter horizontal pipe. At the inlet the pressure is 800 kPa abs, the temperature is 80°C and the average velocity of the air is 45 m/s. As a result of pipe friction and heat transfer the exit pressure is 500 kPa abs and the exit temperature 95°C.
(a) What is the exit velocity?
(b) What is the total wall shear?
(c) What is the heat transfer in joules per kilogram?

5-13. Air flows in a duct. At section A the pressure is 208 kPa abs and the temperature is 115°C. At section B the pressure is 180 kPa abs and the temperature is 102°C. The flow is adiabatic. In what direction is the flow? Refer to Fig. 5-5.

5-14. Air flows in a duct adiabatically. At section A the pressure is 208 kPa abs and the temperature is 115°C. At section B the pressure is 180 kPa abs and the temperature is 93°C. In what direction is the flow? Refer to Fig. 5-5.

5-15. Air at 550 kPa abs and 205°C flows from a reservoir at zero velocity through a converging nozzle and expands adiabatically to 303 kPa abs and 133°C (see Fig. 10-3).

(a) What is the final air velocity?

(b) What is the final air velocity if the expansion to 303 kPa were reversible?

5-16. The first law of thermodynamics states that for steady flow in the absence of heat transfer and work transfer the total energy of a fluid, for example, remains constant as the fluid passes through a control volume. Explain the loss term h_L in the energy equation for liquids in light of this statement.

5-17. Oil of specific gravity $s = 0.90$ flows in a 10-cm pipe. The pressure at section 1 is 345 kPa gage, and the pressure at a downstream section 2, which is 17.0 m lower in elevation than section 1, is 415 kPa gage. What is the head loss?

5-18. At section A in a piping system, the diameter is 15 cm and the pressure is 70 kPa gage. At another section B, which is 6.0 m higher in elevation than section A, the diameter is 30 cm and the pressure is 83 kPa gage. For water flowing at a rate of 0.028 m^3/s, what is the flow direction?

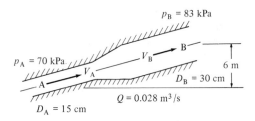

Prob. 5-18

5-19. Kerosene flows in a pipe between points A and B. Point B is 24.5 m higher than point A. The pressure at A is 415 kPa gage, and at B it is 275 kPa gage. In which direction is the flow? Specific weight of kerosene is 8044 N/m^3.

5-20. A 38-mm-diameter jet of water flows from a nozzle connected to a 64-mm-diameter pipe. The jet velocity is 18.3 m/s.

(a) Assuming no viscous effects, what is the pressure in the pipe?

(b) If this pressure in the pipe produces a jet velocity of only 18.0 m/s, what is the head loss due to viscous effects?

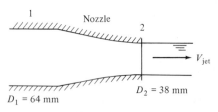

Prob. 5-20

5-21. The average velocity of a jet of water issuing from a round hole in the side of an open tank is 9.70 m/s. The hole is 4.90 m below the free surface of the water in the tank. What is the head loss due to viscous effects?

5-22. At section A in an air duct the pressure is 202 kPa abs and the temperature is 120°C. At section B the pressure is 165 kPa abs and the temperature is 100°C. In what direction is this adiabatic flow? Refer to Fig. 5-5.

5-23. In Prob. 5-22, when the temperature at section A is 125°C, all other data being the same, in what direction is the flow?

5-24. Air at 600 kPa abs and 210°C flows from a reservoir in which the air is at rest, through a converging nozzle and expands adiabatically to 330 kPa abs and 137°C at exit.
(a) What is the air speed at the nozzle exit?
(b) What would be the air speed at the nozzle exit if the flow were to be isentropic to the exit pressure?

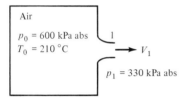

Prob. 5-24

5-25. Kerosene (specific gravity $s = 0.82$) flows in a 15-cm pipe. The pressure at section A is 280 kPa gage, and the pressure at a downstream section B is 350 kPa gage. Section B is 15.0 m lower than section A. What is the head loss?

5-26. At section A in a piping system the pipe diameter is 15 cm and the pressure is 45 kPa gage. At another section B, which is 6.0 m higher than A, the pipe diameter is 30 cm and the pressure is 48 kPa gage. Water flows at 0.030 m³/s.
(a) What is the head loss?
(b) What is the flow direction?

5-27. The velocity of a jet issuing from a circular hole in the side of an open tank is 11.00 m/s. The hole centerline is 6.30 m below the free surface of the water in the tank. What is the head loss due to viscous dissipation?

5-28. An oil flows in a horizontal circular tube such that the flow is laminar with a velocity profile given by

$$u = u_m\left[1 - \left(\frac{r}{R}\right)^2\right]$$

as it leaves the pipe. Here r is the radial position where the velocity is u, u_m being the maximum velocity at the pipe centerline. A short distance away from the end of the pipe the velocity profile becomes uniform (one-dimensional flow). Neglect effects of gravity and air on the jet and calculate

(a) the ratio of jet area to pipe area, and

(b) the energy dissipation within the jet in terms of the pipe velocity. Do not neglect the momentum flux factor β.

5-29. Water flows through a circular reducer which has a 60-cm-diameter entrance and a 30-cm-diameter exit. The inlet velocity is 3.00 m/s, and the inlet pressure is 140 kPa gage. The head loss through the reducer is 0.60 m. What is the net force of the water on the reducer?

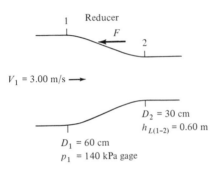

Prob. 5-29

5-30. A turbine operates between a headwater pool 52 m above the tailwater pool. The head loss in the flow system is 6.0 m. The power output of the turbine is 6300 kW at an efficiency of 90 percent. What is the flow rate through the turbine?

5-31. When water flows through a pipeline from one open reservoir to another whose free surface is 15 m below the first, the flow rate is 0.186 m³/s. A pump is installed in the pipeline and 0.186 m³/s are pumped from the lower to the higher reservoir. What power is added to the water by the pump?

5-32. Water flows at an average velocity of 14 m/s in a 15-cm-diameter jet. What is the power in the jet?

5-33. A turbine generates 2600 kW at 200 r/min under a head of 14.3 m when the flow rate is 80 m³/s. What is the turbine efficiency?

5-34. A jet of water issuing from a nozzle at a volumetric flow rate Q with a velocity V_j strikes a series of vanes mounted on a wheel, called a Pelton wheel, which rotates with a peripheral velocity $V_b = \omega R$. The jet is split in half by the vanes, each half being deflected through an angle β. The entire system is at atmospheric pressure.

(a) What is the power in the jet?

(b) What power is given up by the jet to the blades?

(c) Show that the theoretical efficiency, defined as the ratio of power given by the jet to the power in the jet, is independent of the flow rate and depends only on the speed ratio V_b/V_j and the blade angle β.

(d) Show that the theoretical efficiency is a maximum when the speed ratio is $V_b/V_j = 0.5$.

(e) The efficiency would be a maximum for a blade angle of 180°. Why are Pelton wheels made with blade angles of approximately 165°?

(f) What advantage results from splitting the jet in half?

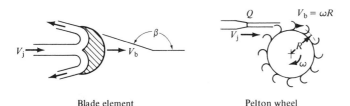

Blade element   Pelton wheel

Prob. 5-34

5-35. A 1.2-m-diameter fan is mounted as shown. Standard air is drawn from the atmosphere and is discharged through a 0.6-m-diameter exit at 26 m/s. Assume one-dimensional, incompressible, nonviscous flow. Note that the exit pressure is atmospheric.

(a) What is the pressure at section 1?

(b) What is the pressure at section 2?

(c) How much power is added to the air by the fan? Use Eq. (5.11).

(d) Use the steady-flow energy equation and Eq. (5.8) to calculate the power added by the fan to the air. Compare with part (c).

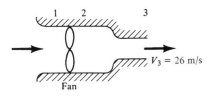

$D_1 = D_2 = 1.2$ m $D_3 = 0.6$ m

Prob. 5-35

5-36. A turbine operates under a total head between headwater and tailwater of 52.0 m with a head loss of 3.5 m in the system. Its power output is 6930 kW at an efficiency of 91 percent. What is the flow rate through the turbine?

5-37. Water flows between two reservoirs 16.4 m different in elevation through a pipeline at a rate of 0.065 m³/s. A pump is installed in the pipeline and pumps 0.065 m³/s from the lower to the higher reservoir. What power is added to the water by the pump?

5-38. A nozzle for an impulse turbine creates a jet 10 cm in diameter with a flow rate of 0.78 m³/s. What power is represented by the kinetic energy of the jet?

5-39. Repeat Prob. 5-35 with the exit velocity 40 m/s, all other parameters being the same.

Part III
INCOMPRESSIBLE
FLOW REGIMES

The equations of motion for nonviscous irrotational fluid flows are attributed to Leonhard Euler, Daniel Bernoulli, and others and have been known since the eighteenth century. The equations of motion for viscous fluids are attributed to, among others, Louis Navier in 1827 and George Stokes in 1845. The viscous effects indicated by the viscous terms in the Navier–Stokes equations were not fully understood until 1904. At that time Ludwig Prandtl explained that the viscous effects are concentrated in a thin region of retarded flow adjacent to a boundary which we now know as the hydrodynamic boundary layer. Outside of this region Prandtl suggested that viscous effects are negligible and the fluid there generally acts as though it has no viscosity.

Chapter 6 gives an introductory treatment to the flow of nonviscous, irrotational, and incompressible fluids, and Chapter 7 introduces the concepts of the incompressible boundary layer.

Chapter 6
Potential Flow

Potential flows are irrotational flows whose velocity components may be derived from velocity potential functions. They apply to incompressible fluids, and since the flows are irrotational, the Bernoulli equation applies throughout the entire flow field. Since velocities are readily obtainable, so then are pressure variations. Many real flows are similar to potential flows, and this in itself justifies a study of potential flows. Many problems in fluid mechanics deal with the relative motion between bodies and fluids of low viscosity—airfoils in air, for example. The viscous fluid may be regarded as consisting of two regions: a thin layer adjacent to the body in which the viscous effects are large, and the rest of the fluid in which the viscous effects are negligible. In this outer region, the flow of the viscous fluid is essentially that of an ideal nonviscous fluid. A knowledge of nonviscous fluid behavior is therefore useful in predicting or analyzing the behavior of real viscous flows.

We will consider the steady flow in two dimensions of an incompressible, irrotational fluid and show mathematically how to describe streamline patterns for various types of flow. Potential flow also exists in three dimensions and for compressible flow [1–3]; but these will not be considered in detail. The differential equations of the velocity potential for incompressible flow are linear, and solutions may be superposed. The corresponding equations for compressible flow are nonlinear, and hence solutions may not be superposed.

From the streamline patterns and the Bernoulli equation the velocity and pressure variations throughout a flow field may be obtained. From the pressure distribution around the surface of a body the drag and lift may be calculated. The drag is generally zero for steady motion in an infinite, continuous, irrotational fluid but is nonzero for free-streamline flows and for some flows with the body near a boundary. The calculated lift for some airfoils is in remarkable agreement with the actual lift experienced by the airfoil in a real fluid. The streamline patterns, together with their associated potential lines, form a flow net which has an exact counterpart in other potential fields such as heat conduction, flow of electricity in a conductor or electrolyte, and electric and magnetic fields.

Different methods of obtaining flow patterns will be shown, and some values of contraction coefficients calculated for potential flow by von Mises will be tabulated.

6-1. Velocity Potential and Stream Function for Two-dimensional Flow

If w is zero and both u and v are independent of z, then streamlines and the paths of particles lie in planes parallel to the $x-y$ plane. If the motion in the $x-y$ plane is known, the motion is everywhere known and the flow is then two dimensional.

It will be shown that a function (or variable) ϕ exists, called the velocity potential, such that the velocity components u and v, as well as their resultant $\mathbf{V}$, are obtainable from

$$u = -\frac{\partial \phi}{\partial x} \tag{6.1a}$$

$$v = -\frac{\partial \phi}{\partial y} \tag{6.1b}$$

and

$$\mathbf{V} = -\nabla\phi = -\operatorname{grad}\phi = -\frac{\partial \phi}{\partial x}\mathbf{i} - \frac{\partial \phi}{\partial y}\mathbf{j} - \frac{\partial \phi}{\partial z}\mathbf{k} \tag{6.1c}$$

The velocity components u and v are also obtainable from a stream function ψ, defined such that

$$u = -\frac{\partial \psi}{\partial y} \tag{6.2a}$$

and

$$v = \frac{\partial \psi}{\partial x} \tag{6.2b}$$

Combining Eqs. (6.1) and the continuity equation $\partial u/\partial x + \partial v/\partial y = 0$, we find that

$$\frac{\partial^2 \phi}{\partial x^2} + \frac{\partial^2 \phi}{\partial y^2} = 0 \tag{6.3}$$

or in vector form

$$\nabla \cdot \nabla \phi = \nabla^2 \phi = 0$$

Equation (6.3) is the Laplace equation.

From Sec. 3-6, $\zeta_z = \partial v/\partial x - \partial u/\partial y$ and with Eq. (6.2)

$$\frac{\partial^2 \psi}{\partial x^2} + \frac{\partial^2 \psi}{\partial y^2} = \zeta_z$$

If the flow is irrotational the vorticity ζ is zero, and

$$\frac{\partial^2 \psi}{\partial x^2} + \frac{\partial^2 \psi}{\partial y^2} = 0 \tag{6.4}$$

or in vector form

$$\nabla \cdot \nabla \psi = \nabla^2 \psi = 0$$

The similarity of Eqs. (6.3) and (6.4) suggests that the potential and stream functions for one flow could represent the potential and stream functions for another flow. That is, given stream and potential functions may be interchanged to produce another flow pattern. An example of this is the source, or sink, and the vortex described in Sec. 6-3.

Equipotential lines are lines of constant ϕ, and streamlines are lines of constant ψ. Equations (6.1) and (6.2) indicate that they always intersect at right angles, and the two families of intersecting lines form a system of curvilinear squares called a *flow net*. A graphical method of drawing flow nets will be described in Sec. 6-2. The derivation of the potential functions and stream functions for a few simple flow patterns will be given in Sec. 6-3.

If ϕ_1 and ϕ_2 are two velocity potentials which satisfy the Laplace equation, then $(\phi_1 + \phi_2)$, which is the result of superimposing the flows represented by ϕ_1 and ϕ_2 individually, will also satisfy the Laplace equation and thus will be the velocity potential for the combined flows. Similarly, the stream function for the resulting combined motion is the sum of the stream functions for each of the superposed flows.

The stream function is a consequence of the continuity equation and thus is applicable to *both* rotational and irrotational flows. The potential function, however, is a consequence of the condition for zero vorticity and thus is applicable to *only* irrotational flows.

DERIVATION OF VELOCITY POTENTIAL AND STREAM FUNCTION

A differential expression $u\,dx + v\,dy$ is called exact if there exists a function, say $-\phi(x, y)$ such that $d\phi = -(u\,dx + v\,dy)$, the test for exactness being that

$$\frac{\partial u}{\partial y} = \frac{\partial v}{\partial x}$$

This is the condition for two-dimensional irrotational flow. Since the total differential of ϕ is

$$d\phi = \frac{\partial \phi}{\partial x}\,dx + \frac{\partial \phi}{\partial y}\,dy$$

it follows that, by equating coefficients of dx and dy in the two expressions for $d\phi$,

$$u = -\frac{\partial \phi}{\partial x} \quad \text{and} \quad v = -\frac{\partial \phi}{\partial y}$$

The function $\phi(x, y)$ is the *velocity potential*, and irrotational flow is called *potential flow*. Use of the negative partial differentials of ϕ produces flow in a direction of decreasing potential.

Similarly, a differential expression $-u\,dy + v\,dx$ is called exact if there exists a function, say, $\psi(x, y)$ such that $d\psi = -u\,dy + v\,dx$, the test for exactness being that

$$-\frac{\partial u}{\partial x} = \frac{\partial v}{\partial y}$$

This is the condition imposed by the continuity equation. Since the total differential of ψ is

$$d\psi = \frac{\partial \psi}{\partial x}\,dx + \frac{\partial \psi}{\partial y}\,dy$$

it also follows that, by equating coefficients of dx and dy in the two expressions for $d\psi$,

$$u = -\frac{\partial \psi}{\partial y} \quad \text{and} \quad v = \frac{\partial \psi}{\partial x}$$

and therefore equipotential lines and streamlines intersect everywhere at right angles. Along a potential line, $u\,dx + v\,dy = 0$, from which

$$\left(\frac{dy}{dx}\right)_\phi = -\frac{u}{v}$$

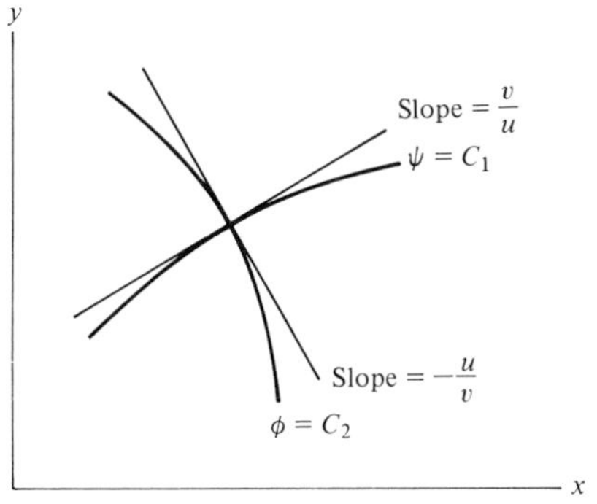

Fig. 6-1. Orthogonal nature of ϕ and ψ.

Along a streamline, $-u\,dy + v\,dx = 0$, or

$$\left(\frac{dy}{dx}\right)_{\psi} = \frac{v}{u}$$

Thus,

$$\left(\frac{dy}{dx}\right)_{\phi} = -\left(\frac{1}{dy/dx}\right)_{\psi}$$

and thus at any point of intersection the equipotential lines are normal to the streamlines. This is shown in Fig. 6-1. In polar form, the radial component of velocity is

$$v_r = -\frac{1}{r}\frac{\partial\psi}{\partial\theta} = -\frac{\partial\phi}{\partial r} \tag{6.5a}$$

and the tangential component is

$$v_{\theta} = \frac{\partial\psi}{\partial r} = -\frac{1}{r}\frac{\partial\phi}{\partial\theta} \tag{6.5b}$$

The vector velocity $\mathbf{V}$ is expressed in terms of the potential function as follows:

$$\mathbf{V} = u\mathbf{i} + v\mathbf{j} = -\left(\frac{\partial\phi}{\partial x}\right)\mathbf{i} - \left(\frac{\partial\phi}{\partial y}\right)\mathbf{j} = -\nabla\phi$$

the negative gradient of ϕ by definition. Since the gradient of a scalar function is always normal to the curve (or the level surface in three dimensions) representing the function, the velocity vectors, and hence streamlines, are always normal to the equipotential lines (or surfaces in three dimensions).

In heat conduction, equipotential lines are lines of constant temperature and streamlines are lines indicating the direction of heat flow. In electrical conduction, equipotential lines are lines of constant voltage and streamlines are lines along which current flows. In fluid flow, equipotential lines are just that—they are *not* lines of constant pressure—and streamlines are lines indicating fluid particle paths for steady flow.

Further differentiation of the preceding partial differential equations gives

$$\frac{\partial^2 \phi}{\partial x^2} = \frac{\partial^2 \psi}{\partial y \, \partial x} \quad \text{and} \quad \frac{\partial^2 \phi}{\partial y^2} = -\frac{\partial^2 \psi}{\partial x \, \partial y}$$

and thus for continuously differentiable functions

$$\frac{\partial^2 \phi}{\partial x^2} + \frac{\partial^2 \phi}{\partial y^2} = 0 \quad \text{and} \quad \frac{\partial^2 \psi}{\partial x^2} + \frac{\partial^2 \psi}{\partial y^2} = 0$$

Also

$$\nabla^2 \phi = \nabla \cdot \nabla \phi = \nabla \cdot \left(\frac{\partial \phi}{\partial x} \mathbf{i} + \frac{\partial \phi}{\partial y} \mathbf{j} \right) = \frac{\partial^2 \phi}{\partial x^2} + \frac{\partial^2 \phi}{\partial y^2} = 0$$

Any function $\phi(x, y)$ which satisfies the Laplace equation is called *harmonic* and is a possible velocity potential which describes some irrotational flow. The velocity components of such a flow may be obtained from Eq. (6.1). There are unlimited solutions to the Laplace equation, and the desired one requires application of appropriate boundary conditions.

All flows must satisfy continuity, and thus it is possible for velocity components to satisfy the continuity requirements of Eq. (3.18) but not satisfy the condition for being irrotational, given by Eq. (3.12). In this instance, the velocity components would *not* be derivable from a potential function ϕ, since the flow would be rotational.

6-2. Flow Nets for Two-dimensional Flow

A set of streamlines and a corresponding set of equipotential lines, with the difference in stream function between streamlines the same as the difference in potential function between equipotential lines, will form a mesh of rectangular or, more often, curvilinear squares. This system of curvilinear squares is known as a flow net. Flow nets for two-dimensional flows may be obtained by various methods.

With a little practice, flow nets for many fluid flows may be drawn freehand with sufficient accuracy for many purposes. They may be obtained by plotting the equations of the streamlines and the equipotential lines. Some of these equations may be obtained by combining simple flows which in themselves are solutions of the Laplace equation. Examples of this method will be given in Sec. 6-3. Flow nets may also be obtained from conformal transformations in the complex plane, as described in Sec. 6-4. They may be obtained by expressing the Laplace equation as a finite-difference equation and solving this numerically by relaxation methods described in Sec. 6-5.

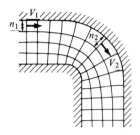

Fig. 6-2. Flow around a bend. (Source: Reproduced, with permission, from *Basic Mechancis of Fluids*, by Hunter Rouse and J. W. Howe. Copyright, 1953, John Wiley & Sons, Inc.)

In the first method streamlines are drawn equally spaced at some section where there is rectilinear flow. The number of intervals or streamlines depends upon the accuracy desired and the effort one is willing to expend— the smaller the increments the greater the accuracy and the greater the effort required to draw the flow net, and conversely. Only in the limit when $\Delta\psi$ and $\Delta\phi$ approach zero will the flow net consist of perfect squares. The larger the intervals, the more the curvilinear squares will depart from perfect squares. Streamlines are drawn by eye, and a system of normal equipotential lines are then added. Successive adjustments are made in both streamlines and equipotential lines until the net appears proper. Diagonal lines through the curvilinear squares should also form a mesh of curvilinear squares; this is often used as an added check on the correctness of the flow net.

From continuity, $V_1 n_1 = V_2 n_2$ in Fig. 6-2, and thus velocities vary inversely with streamline spacing. From the Bernoulli equation for an ideal incompressible fluid when gravity effects are absent, $(\rho V_1^2/2) + p_1 = (\rho V_2^2/2) + p_2$ and pressure variations may also be determined throughout the flow field once velocity variations are known. If gravity effects are present, the complete Bernoulli equation must be used.

For flow around a circular bend, the velocity is greatest ($Vr = C$ for an irrotational vortex) and the pressure least at the inner radius (Fig. 6-2).

The flow into a rounded entrance is shown in Fig. 6-3. Flow from a rounded exit or outlet would be the mirror image of this.

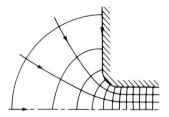

Fig. 6-3. Flow through a rounded inlet. (Source: Reproduced, with permission, from *Basic Mechanics of Fluids*, by Hunter Rouse and J. W. Howe. Copyright, 1953, John Wiley & Sons, Inc.)

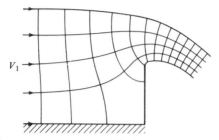

Fig. 6-4. Flow over a weir. (Source: Reproduced, with permission, from *Basic Mechanics of Fluids*, by Hunter Rouse and J. W. Howe. Copyright, 1953, John Wiley & Sons, Inc.)

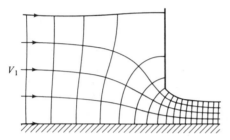

Fig. 6-5. Flow under a sluice gate. (Source: Reproduced, with permission, from *Basic Mechanics of Fluids*, by Hunter Rouse and J. W. Howe. Copyright, 1953, John Wiley & Sons, Inc.)

Flow over a weir under the influence of gravity is shown in Fig. 6-4, and flow under a sluice gate is shown in Fig. 6-5.

Flow nets for flow around corners, shown in Fig. 6-6, indicate infinitesimal velocities at the outer corner in view (a) and infinitesimal streamline spacing with infinite velocities at the inner corners in both views (a) and (b).

Real fluids act like ideal fluids in regions where streamlines converge or diverge only minutely near boundaries. Where divergence of streamlines is appreciable, separation occurs and the flow net will not give a reliable

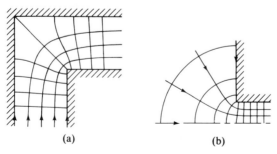

(a)   (b)

Fig. 6-6. Flow around a corner (a) in a duct and (b) into a duct. (Source: Reproduced, with permission, from *Basic Mechanics of Fluids*, by Hunter Rouse and J. W. Howe, Copyright, 1953, John Wiley & Sons, Inc.)

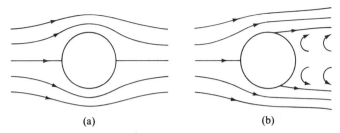

Fig. 6-7. Flow past a circular cylinder. (a) Ideal fluid without separation. (b) Viscous fluid with separation.

picture of the flow if the boundaries of the separation zone are not known. Real fluids tend to separate from a boundary in regions where the pressure increases in the direction of flow. An example of this is shown in Fig. 6-7 for flow past a circular cylinder. The flow of an ideal fluid, view (a), is similar to the flow of a real fluid, view (b), over the upstream portion of a cylinder, but the flow of the ideal fluid and of the real fluid differ over the aft portion of a cylinder. In Figs. 6-4 and 6-5 the flow nets have been drawn by assuming that the fluid would separate from the boundaries at the sharp edges.

6-3. Examples of Ideal, Two-dimensional, Steady Fluid Flows

RECTILINEAR FLOW

Uniform flow in the $+x$ direction is such that $u = u_s$ and $v = 0$. Then $-\partial\phi/\partial x = u_s$ and $\phi = -u_s x + f(y)$ (disregarding the constant), and $v = -\partial\phi/\partial y = 0$ and ϕ is thus not a function of y. The stream function is obtained from $u = -\partial\psi/\partial y = u_s$, from which $\psi = -u_s y + f(x)$ (again disregarding the constant), and $v = \partial\psi/\partial x = 0$ and ψ is thus not a function of x. The flow net for this flow is shown in Fig. 6-8. Since rectangular and polar co-ordinates are related by the expressions $x = r \cos\theta$ and $y = r \sin\theta$, the

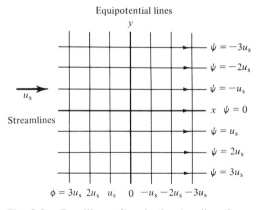

Fig. 6-8. Rectilinear flow in the $+x$ direction.

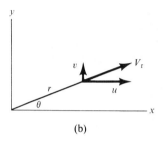

(a)

(b)

Fig. 6-9.   Flow from a two-dimensional, or line, source. (a) Line source. (b) Flow in the x–y plane.

potential and stream functions for uniform flow in the $+x$ direction are

$$\phi = -u_s x = -u_s r \cos \theta$$

$$\psi = -u_s y = -u_s r \sin \theta$$

The velocity potential and stream functions for uniform flow in any direction can always be expressed by first-degree equations in x and y. Thus $\phi = -ux - vy$ and $\psi = -uy + vx$ for the general situation. For example, $\phi = 2x + y$ and $\psi = -x + 2y$ represent flow for which $u = -2$ and $v = -1$ (and $\mathbf{V} = -2\mathbf{i} - 1\mathbf{j}$) throughout the entire x–y plane.

SOURCE

A line source is a line of unit length between two parallel planes, a unit distance apart, from which fluid flows radially in all directions parallel to the planes (Fig. 6-9a). The strength of the source is equal to the volumetric flow rate from it. The potential function and the stream function will be derived. In the x–y plane (Fig. 6-9b) the radial velocity is V_r at a radial distance r from the origin, the location of source. The volumetric flow rate from this line source of unit length is $q = V_r 2\pi r$. From geometry, the definition of the velocity potential, and the stream function we may write for $r \neq 0$

$$V_r = \frac{q}{2\pi r} = -\frac{\partial \phi}{\partial r} = -\frac{1}{r}\frac{\partial \psi}{\partial \theta} \qquad \text{and} \qquad V_\theta = 0$$

Thus

$$\phi_{source} = -\frac{q}{2\pi} \ln r = -\frac{q}{4\pi} \ln(x^2 + y^2) \tag{6.6}$$

and

$$\psi_{source} = -\frac{q}{2\pi} \theta = -\frac{q}{2\pi} \arctan \frac{y}{x} \tag{6.7}$$

where q equals the strength of the source, and $q/2\pi$ the flow rate per radian.

SINK

A sink is a negative source, and the velocity potential and the stream function for a sink are the negative of those for a source:

$$\phi_{sink} = \frac{q}{4\pi} \ln(x^2 + y^2) = \frac{q}{2\pi} \ln r \tag{6.8}$$

$$\psi_{sink} = \frac{q}{2\pi} \tan^{-1} \frac{y}{x} = \frac{q}{2\pi} \theta \tag{6.9}$$

VORTEX

For an irrotational, or free, vortex $Vr = C$ (Sec. 3-5), streamlines are circles about the origin and equipotential lines are radial lines from the origin. The stream function and velocity potential may be described by interchanging ϕ and ψ in the equations for a source. For a vortex rotating in a positive (counterclockwise) direction,

$$\phi_{vortex} = -C \tan^{-1} \frac{y}{x} = -C\theta \tag{6.10}$$

$$\psi_{vortex} = \frac{C}{2} \ln(x^2 + y^2) = C \ln r \tag{6.11}$$

The strength of a vortex depends on the magnitude of C, since at a given radius the greater the value of C the greater the rotational velocity. The *strength* is defined as the magnitude of the *circulation*. Circulation, in turn, is defined as the line integral of the product of the velocity component tangent to an element of the closed path around which the integral is taken times the infinitesimal path length. This is

$$\Gamma = \oint |\mathbf{V}| \cos \theta |d\mathbf{l}| = \oint \mathbf{V} \cdot d\mathbf{l} \tag{6.12}$$

in Fig. 6-10. For the particular case of a path along a circle whose center is at the origin (a vortex streamline), the velocity vector is always constant and tangent to the path of integration, and $dl = r \, d\theta$. Then

$$\Gamma = V \oint dl = V \int_0^{2\pi} r \, d\theta = 2\pi r V = 2\pi C$$

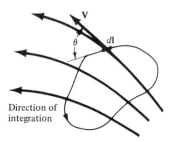

Fig. 6-10. Circulation in a positive direction, with region about which integration is made being to the left of path traversed.

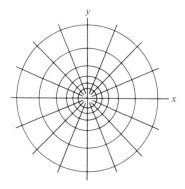

Fig. 6-11. Flow net for a source, sink, and vortex.

For any other closed path which does not include the origin, the circulation for an irrotational vortex is zero.

Figure 6-11 shows the flow net for a source, a sink, and a vortex. The potential functions and stream functions for rectilinear flow, a source, a sink, and a free vortex, may be added in a number of different ways. The results will be solutions of the Laplace equation and will describe different flow situations. Some typical combinations will be illustrated without derivation.

HALF-BODY
The addition of a rectilinear flow and a source flow produces flow around a half-body (Fig. 6-12):

$$\phi = -u_s x - \frac{q}{4\pi} \ln(x^2 + y^2) = -u_s r \cos \theta - \frac{q}{2\pi} \ln r \qquad (6.13)$$

$$\psi = -u_s y - \frac{q}{2\pi} \tan^{-1} \frac{y}{x} = -u_s r \sin \theta - \frac{q}{2\pi} \theta \qquad (6.14)$$

DOUBLET
A source and sink of equal strength, when brought together in such a way that the product of their strength and the distance between them remains

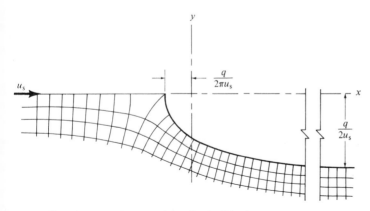

Fig. 6-12.   Flow net for one side of a half-body.

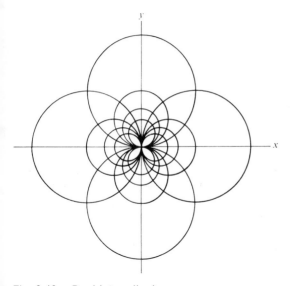

Fig. 6-13.   Doublet or dipole.

constant, produce a doublet, or dipole. The constant C (any C, not necessarily the same as for a vortex) is called the strength of the doublet (Fig. 6-13).

   To form a doublet from a sink and source brought together along the x axis, the potential function at any point P in Fig. 6-14 is that for the source plus that for the sink:

$$\phi_D = -\frac{q}{2\pi} \ln r + \frac{q}{2\pi} \ln(r + dr) = \frac{q}{2\pi} \ln\left(1 + \frac{dr}{r}\right)$$

Expansion of $\ln(1 + dr/r)$ in a power series gives

$$\phi_D = \frac{q}{2\pi}\left[\frac{dr}{r} - \frac{1}{2}\left(\frac{dr}{r}\right)^2 + \cdots\right]$$

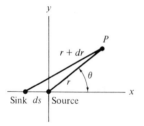

Fig. 6-14. Source–sink combination to form a doublet.

Neglecting higher powers of dr/r gives

$$\phi_D = \frac{q}{2\pi}\frac{dr}{r} = \frac{(q\,ds)\cos\theta}{2\pi r}$$

Thus

$$\phi_D = C\frac{\cos\theta}{r} = \frac{Cx}{x^2 + y^2} \tag{6.15}$$

Similarly

$$\psi_D = -C\frac{\sin\theta}{r} = \frac{-Cy}{x^2 + y^2} \tag{6.16}$$

CYLINDER

Flow past a circular cylinder is obtained by combining a rectilinear flow with a doublet and is shown in Fig. 6-15 for the doublet $C = -u_s R^2$. This is obtained by setting $u = -\partial\phi/\partial x = 0$ at $x = R$ and $y = 0$ in the expression for ϕ a rectilinear flow and for a doublet from Eq. (6.15):

$$\phi = -u_s x - u_s \frac{R^2 x}{x^2 + y^2} = -u_s r \cos\theta - u_s R^2 \frac{\cos\theta}{r} \tag{6.17}$$

$$\psi = -u_s y + u_s \frac{R^2 y}{x^2 + y^2} = -u_s r \sin\theta + u_s R^2 \frac{\sin\theta}{r} \tag{6.18}$$

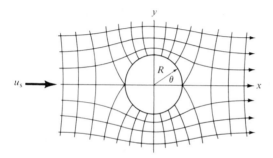

Fig. 6-15. Flow past a circular cylinder.

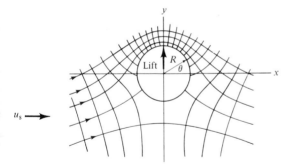

Fig. 6-16.   Rectilinear flow and irrotational vortex around a cylinder produces lift.

At any point in the flow, $\mathbf{V} = -\nabla\phi = -(\partial\phi/\partial x)\mathbf{i} - (\partial\phi/\partial y)\mathbf{j}$, and

$$V = |\mathbf{V}| = \sqrt{\left(\frac{\partial\phi}{\partial x}\right)^2 + \left(\frac{\partial\phi}{\partial y}\right)^2}$$

Along the surface of the cylinder, $V = 2u_s \sin\theta$.

CYLINDER WITH CIRCULATION

Flow around a circular cylinder with circulation is obtained by adding a vortex to the rectilinear flow past the cylinder (a clockwise vortex is used in Fig. 6-16). The strength of the vortex is $C = \Gamma/2\pi$. Then,

$$\phi = -u_s\left(x + \frac{R^2 x}{x^2 + y^2}\right) + \frac{\Gamma}{2\pi}\tan^{-1}\frac{y}{x}$$

$$= -u_s\left(r + \frac{R^2}{r}\right)\cos\theta + \frac{\Gamma}{2\pi}\theta \tag{6.19}$$

$$\psi = -u_s\left(y - \frac{R^2 y}{x^2 + y^2}\right) - \frac{\Gamma}{4\pi}\ln(x^2 + y^2)$$

$$= -u_s\left(r - \frac{R^2}{r}\right)\sin\theta - \frac{\Gamma}{2\pi}\ln r \tag{6.20}$$

With no circulation about the cylinder, the stagnation points (where the velocity is zero) are at $\theta = 0$ and $\theta = \pi$ (Fig. 6-15). As the circulation is increased, the stagnation points move toward each other along the surface of the cylinder until they coincide at $\theta = -\pi/2$. If the circulation is increased more, the stagnation points will be removed from the surface of the cylinder and fluid will rotate completely around the cylinder. The tangential velocity along the surface of the cylinder is

$$V_\theta = -\left(2u_s \sin\theta + \frac{\Gamma}{2\pi R}\right)$$

(negative because in Fig. 6-16 it is clockwise), from which the location of the stagnation points $V_\theta = 0$ can be related to the circulation by the equation

$$\Gamma = -4\pi u_s R \sin \theta$$

A double stagnation point (at $\theta = -\pi/2$) occurs if the velocity due to circulation is twice u_s, from the equation

$$V_{\text{circ}} = \frac{\Gamma}{2\pi R} = -2u_s \sin \theta \bigg|_{V_\theta = 0} = 2u_s \bigg|_{V_\theta = 0}$$

In all of the preceding situations, if the free-stream velocity u_s is known, the velocity at any other point may be obtained from the velocity potential, the stream function, or direct measurement of the streamline spacing on a flow net. If the pressure p_s is known where the velocity u_s is known, the pressure at other points may be obtained from the Bernoulli equation

$$\rho \frac{u_s^2}{2} + p_s = \rho \frac{V^2}{2} + p$$

Drag is defined as the resultant force of a fluid on the surface of a body in a direction parallel to the free-stream velocity of approach u_s. In an ideal fluid, only normal forces due to pressure occur, and for flow past a circular cylinder without circulation, the pressure variation around the surface is given by

$$p - p_s = \rho \frac{u_s^2}{2} - \frac{\rho(2u_s \sin \theta)^2}{2} = \frac{\rho u_s^2}{2} (1 - 4 \sin^2 \theta)$$

For this,

$$\text{drag} = -\int_0^{2\pi} (p - p_s) \cos \theta (R \, d\theta) = 0$$

and

$$\text{lift} = -\int_0^{2\pi} (p - p_s) \sin \theta (R \, d\theta) = 0$$

For flow past a circular cylinder *with* circulation,

$$p - p_s = \frac{\rho}{2} \left[u_s^2 - \left(2u_s \sin \theta + \frac{\Gamma}{2\pi R} \right)^2 \right]$$

and the drag is again zero. The lift for a unit length of cylinder is

$$L = \rho u_s \Gamma \tag{6.21}$$

See Sec. 11-3 for a further discussion of circulation and lift.

TORNADO

A tornado may be approximated by a two-dimensional vortex and a sink, except in a region near the origin, and this is an example of good agreement

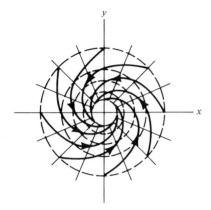

Fig. 6-17.   Irrotational vortex (positive, or counterclockwise) and sink as a tornado model.

between a real and an ideal fluid. The resultant flow is shown in Fig. 6-17. For the ideal fluid

$$\phi = \frac{q}{2\pi} \ln r - \frac{\Gamma}{2\pi} \theta = \frac{q}{4\pi} \ln(x^2 + y^2) - \frac{\Gamma}{2\pi} \tan^{-1} \frac{y}{x}$$

$$\psi = \frac{q}{2\pi} \theta + \frac{\Gamma}{2\pi} \ln r = \frac{q}{2\pi} \tan^{-1} \frac{y}{x} + \frac{\Gamma}{4\pi} \ln(x^2 + y^2)$$

The tangential velocity from the vortex is $V_\theta = \Gamma/2\pi r$, and the radial velocity from the sink is $V_r = -q/2\pi r$. The resultant velocity is $V = \sqrt{\Gamma^2 + q^2}/2\pi r$, which may also be obtained from

$$V = \sqrt{\left(\frac{\partial \phi}{\partial x}\right)^2 + \left(\frac{\partial \phi}{\partial y}\right)^2}$$

Thus $Vr = C = \sqrt{\Gamma^2 + q^2}/2\pi$. Combining this with the energy equation for an ideal fluid gives

$$p_2 - p_1 = \frac{\rho C^2}{2} \left(\frac{1}{r_1^2} - \frac{1}{r_2^2}\right)$$

which indicates a linear relation between p and $1/r^2$.

Measurements made during passage of a tornado on 8 June 1953 at eight barograph stations in the NACA (National Advisory Committee for Aeronautics) Lewis Laboratory near Cleveland, Ohio, are tabulated in Table 6-1. Adjustments in position, but not in direction, of the path of the center of the tornado were made, as well as an adjustment to a standard reference pressure for all stations. From the data, $Vr = C = 7250 \text{ m}^2/\text{s}$, and a graph substantiating the linear relation between p and $1/r^2$ is shown in Fig. 6-18.

Table 6-1. MINIMUM PRESSURES FROM MEASUREMENTS DURING PASSAGE OF A TORNADO

STATION	CORRECTED MINIMUM PRESSURE (kPa abs)	DISTANCE FROM PATH (m)	$\frac{1}{r^2} \times 10^6$
1	97.72	233	18.4
2	97.91	280	12.8
3	98.18	485	4.25
4	97.85	271	13.6
5	98.17	471	4.51
6	97.64	218	21.0
7	98.08	320	9.77
8	98.21	580	2.97

SOURCE: Extracted from W. Lewis and P. J. Perkins, "Recorded Pressure Distribution in the Outer Portion of a Tornado Vortex," *Monthly Weather Review*, Vol. 81 (1953), pp. 379–385, with permission of the Weather Bureau, U.S. Department of Commerce.

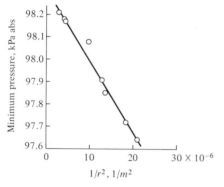

Fig. 6-18. Minimum recorded pressures during passage of a tornado. Theoretical model combining irrotational vortex and a sink indicates linear relation between pressure and $1/r^2$. Measurements verify this.

6-4. Method of Images

Flow patterns for elementary flows (sources, sinks, and vortexes, for example) near boundaries may be obtained by the method of images.

The flow pattern that results from combining two equal sources, two equal sinks, two equal vortexes of opposite sign, or two equal cylinders aligned normal to a uniform flow at right angles to their line of centers will include, in each instance, one straight streamline normal to a line joining them and located midway between them. These flow patterns are shown in Fig. 6-19. The straight streamline midway between the equal sources, sinks, etc. may be considered to be a boundary, and the flow on one side of it is then the flow pattern for a source, sink, vortex, or around a cylinder near a boundary.

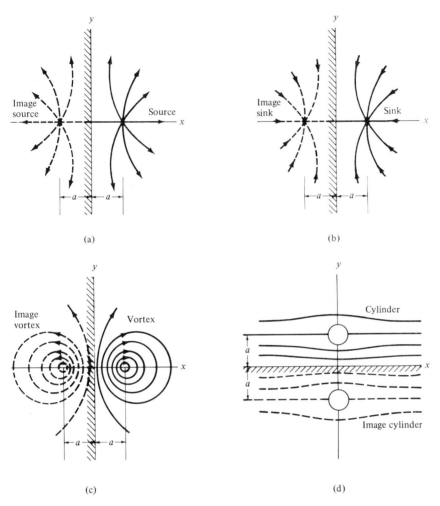

Fig. 6-19.   Method of images. (a) Source near a plane boundary. (b) Sink near a plane boundary. (c) Irrotational vortex near a plane boundary. (d) Cylinder near a plane boundary.

To obtain the flow patterns in these instances it is necessary to consider an *image* on the opposite side of the boundary which is of equal strength and of the proper sign or direction. The boundary is removed, and the flow resulting from the combination is determined.

For a line source of strength q a distance a from a boundary (Fig. 6-19a) the stream function is obtained by adding the stream function for a source at $(a,0)$ to that for a source at $(-a,0)$ to get

$$\psi = -\frac{q}{2\pi} \arctan \frac{y}{x-a} - \frac{q}{2\pi} \arctan \frac{y}{x+a}$$

The flow in either the right or the left half-plane is the resulting flow pattern for a source near a boundary.

In like manner, the stream function for a sink near a boundary (Fig. 6-19b) is

$$\psi = +\frac{q}{2\pi} \arctan \frac{y}{x-a} + \frac{q}{2\pi} \arctan \frac{y}{x+a}$$

and the flow pattern in either the right or the left half-plane represents the flow pattern for a sink near a boundary.

Similarly, a vortex a distance a from a boundary has a stream function given by (Fig. 6-19c)

$$\psi = -\frac{C}{2} \ln[(x-a)^2 + y^2] + \frac{C}{2} \ln[(x+a)^2 + y^2]$$

for a clockwise vortex of strength $C = v_\theta r$ at $(a, 0)$ and an equal counterclockwise vortex at $(-a, 0)$. Note that the vortex at $(a, 0)$ will move along the boundary in the $+y$ direction since it is in the field of the counterclockwise image vortex at $(-a, 0)$. (The image vortex will also move in the $+y$ direction.) Intuitively, one might surmise that a vortex near a boundary would move like a wheel on a track. This is not so, however.

The stream function for two cylinders, one at $(0, +a)$ and the other at $(0, -a)$ as shown in Fig. 6-19d, is obtained by adding the stream functions for a cylinder at each location in a uniform flow (recall that the cylinder results from a doublet in a uniform flow). The result is

$$\psi = -2u_s y + u_s R^2 \left[\frac{(y-a)}{x^2 + (y-a)^2} + \frac{(y+a)}{x^2 + (y+a)^2} \right]$$

A source near a corner produces a flow pattern given by adding three image sources as shown in Fig. 6-20.

A line source between parallel planes a distance a apart produces a flow pattern obtained by having an infinite number of images at $\pm a$, $\pm 2a$, $\pm 3a$,

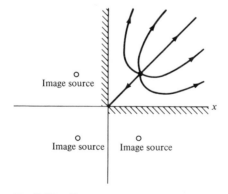

Fig. 6-20. Source near a corner.

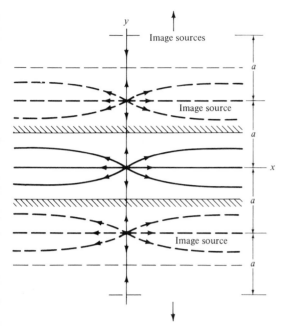

Fig. 6-21. Source between parallel planes.

etc. aligned in a line through the original source and normal to the planes. The stream function for a source of strength q at the origin in Fig. 6-21 would be

$$\psi = -\frac{q}{2\pi}\left[\arctan\frac{y}{x} + \arctan\frac{y-a}{x} + \arctan\frac{y-2a}{x} + \cdots\right.$$

$$\left. + \arctan\frac{y+a}{x} + \arctan\frac{y+2a}{x} + \cdots\right]$$

The method of images may also be used to obtain flow patterns for flow past a cylinder between parallel walls and for source and sink flows near cylindrical boundaries, as well as many other flows near boundaries.

6-5. Potential Flows from Conformal Transformations in the Complex Plane

Potential flow patterns may be obtained by combining simple flow patterns, as was done in Sec. 6-3. This superposition method is valid because of the linearity of the Laplace equation.

A second method of obtaining solutions to the Laplace equation for two-dimensional flow makes use of the functions of a complex variable. By this means a rectangular grid flow net in one complex plane is transformed into the desired flow net in another complex plane. The transformation is

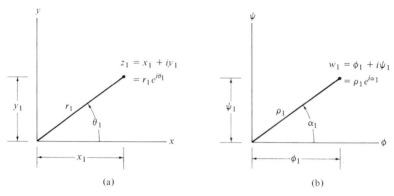

Fig. 6-22. Complex z and w planes.

called *conformal* because infinitesimal squares in the one plane become similar infinitesimal curvilinear squares in the other plane. In some instances, successive conformal transformations are used to obtain a desired flow pattern.

FUNCTIONS OF A COMPLEX VARIABLE

Let (see Fig. 6-22)

$$w = f(z) = f(x + iy) = \phi(x, y) + i\psi(x, y)$$

The function w is said to be a function of a complex variable if

1. $f(z)$ is single-valued and finite;
2. $f(z)$ has single-valued derivatives.

Then within a region for which these two conditions apply, the function is said to be *analytic*.

A complex derivative

$$\lim_{\delta z \to 0} \frac{f(z + \delta z) - f(z)}{\delta z}$$

may approach its limit in an infinite number of ways. But the limit must be the same for all possible ways for the second condition to hold.

Along the x direction

$$\lim_{\substack{\delta y \to 0 \\ \delta x \to 0}} \frac{f(z + \delta z) - f(z)}{\delta x + \delta y} = \lim_{\delta x \to 0} \frac{f(z + \delta x) - f(z)}{\delta x} = \frac{\partial f}{\partial x}$$

Along the y direction

$$\lim_{\substack{\delta x \to 0 \\ \delta y \to 0}} \frac{f(z + \delta z) - f(z)}{\delta x + i \delta y} = \frac{1}{i} \lim_{\delta y \to 0} \frac{f(z + i \delta y) - f(z)}{\delta y} = \frac{1}{i} \frac{\partial f}{\partial y}$$

But condition (2) states that $\partial f/\partial x = (1/i)\partial f/\partial y$, and since $f(z) = \phi + i\psi$,

$$\frac{\partial f}{\partial x} = \frac{\partial \phi}{\partial x} + i\frac{\partial \psi}{\partial x} \quad \text{and} \quad \frac{\partial f}{\partial y} = \frac{\partial \phi}{\partial y} + i\frac{\partial \psi}{\partial y}$$

Combining these and equating real and imaginary parts, we get

$$\frac{\partial \phi}{\partial x} = \frac{\partial \psi}{\partial y} \quad \text{and} \quad \frac{\partial \phi}{\partial y} = -\frac{\partial \psi}{\partial x} \tag{6.22}$$

These are the Cauchy–Riemann equations. Any $f(z) = f(x + iy)$ that satisfies the two conditions initially stated also satisfies the Cauchy–Riemann equations. Where these conditions are not satisfied, we have singular points. From the first part of Eq. (6.22) we obtain $-u$, the negative of the x component of velocity, and from the second part we get $-v$, the negative of the y component of velocity [Eqs. (6.1) and (6.2)].

The functions ϕ and ψ are called conjugate functions, and curves formed by $\phi(x, y) = $ constant and $\psi(x, y) = $ constant form an orthogonal system, a flow net. Lines of constant ϕ are equipotential lines, and lines of constant ψ are streamlines.

If $w = z = \phi + i\psi = x + iy$, $\phi = x$ and $\psi = y$, and potential lines are lines of constant x and streamlines are lines of constant y (Fig. 6-8).

If $w = z^2 = (x + iy)^2 = x^2 - y^2 + i2xy$, then $\phi = x^2 - y^2$ and $\psi = 2xy$, and the streamlines are as shown in Fig. 3-5 (with flow in the opposite direction) and in Fig. 6-23.

The w plane generally is a rectangular grid plane, and the pattern in the z plane is the physical plane showing the flow net being sought (Fig. 6-23). Successive transformations often are made, each step being a translation, a rotation, an inversion, a stretching or a shrinking, for example. The function ϕ is plotted as abscissa and the function ψ as ordinate. Each point in the z plane will correspond to at least one point in the w plane. Curves in the planes may also be transformed, the relation between configurations depending on the relation $w = f(z)$. The transformation will be conformal if corresponding infinitesimal configurations in the two planes are similar.

For example, the rectangular grid in Fig. 6-8 (this is in the so-called w plane) is transformed into the flow of Fig. 6-15 (this is the so-called z plane in which $z = x + iy = r\,e^{i\theta}$) by the complex transformation

$$w = -u_s\left(z + \frac{R^2}{z}\right)$$

$$= -u_s\left(r\,e^{i\theta} + \frac{R^2}{r}e^{-i\theta}\right)$$

which may be separated into two components, ϕ the real component and ψ the imaginary component, such that $w = \phi + i\psi$. Since by definition

$$e^{i\theta} = \cos\theta + i\sin\theta$$

and

$$e^{-i\theta} = \cos(-\theta) + i \sin(-\theta)$$
$$= \cos\theta - i\sin\theta$$

the transformation becomes

$$w = -u_s r(\cos\theta + i\sin\theta) - u_s \frac{R^2}{r}(\cos\theta - i\sin\theta)$$
$$= -u_s\left(r + \frac{R^2}{r}\right)\cos\theta - i\, u_s\left(r - \frac{R^2}{r}\right)\sin\theta$$

so that

$$\phi = -u_s\left(r + \frac{R^2}{r}\right)\cos\theta \tag{6.17}$$

and

$$\psi = -u_s\left(r - \frac{R^2}{r}\right)\sin\theta \tag{6.18}$$

as before in Sec. 6-3.

This particular conformal transformation may be physically described as resulting from plotting a rectangular grid such as in Fig. 6-8 on a rubber sheet, making a cut along the x axis symmetrically about the y axis, and opening the cut to form a circle. The stretched rubber sheet would have the grid lines in the form of the flow net of Fig. 6-15.

If the cut were made along the negative y axis and the two edges of the cut swung 90° to coincide with the x axis, the stretched rubber sheet would have the grid lines in the form of the flow net for Example 3-2. This is shown in Fig. 6-23. Here, $w = z^2/4$.

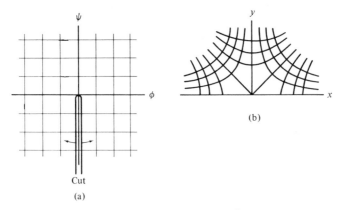

Fig. 6-23.   Conformal transformation $w = z^2/4$.

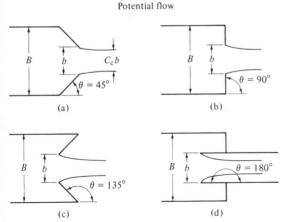

Fig. 6-24. Geometry of two-dimensional jet flow.

A third method of analysis involves the use of the Schwarz–Christoffel theorem, often in conjunction with conformal mapping. This theorem provides a means of mapping the interior of a simple closed polygon into the upper half-plane, with the boundary of the polygon becoming the real axis (the x axis in the complex plane).

Some useful engineering results of this complicated mapping technique were obtained by von Mises for the contraction coefficients for two-dimensional jets of an incompressible fluid in the absence of gravity. The ratio of the resulting cross-sectional area of a jet to the area of the boundary opening is called the *coefficient of contraction*, C_c. For the flow system shown in Fig. 6-24, von Mises calculated C_c values given in Table 6-2.

Values of contraction coefficients in Table 6-2 agree well with experimental values measured for real fluids. The results listed for two-dimensional

Table 6-2. COEFFICIENTS OF CONTRACTION FOR TWO-DIMENSIONAL JETS FOR GEOMETRIES OF FIG. 6-24

b/B	C_c $\theta = 45°$	C_c $\theta = 90°$	C_c $\theta = 135°$	C_c $\theta = 180°$
0.0	0.746	0.611	0.537	0.500
0.1	0.747	0.612	0.546	0.513
0.2	0.747	0.616	0.555	0.528
0.3	0.748	0.622	0.566	0.544
0.4	0.749	0.631	0.580	0.564
0.5	0.752	0.644	0.599	0.586
0.6	0.758	0.662	0.620	0.613
0.7	0.768	0.687	0.652	0.646
0.8	0.789	0.722	0.698	0.691
0.9	0.829	0.781	0.761	0.760
1.0	1.000	1.000	1.000	1.000

flow may be used for axisymmetric jets if the coefficient of contraction is defined by

$$C_c = \frac{b_{jet}}{b} = \left(\frac{d_{jet}}{d}\right)^2$$

and d and D are the diameters corresponding to the widths b and B, respectively. For example, if a small round hole of diameter d is in a large reservoir ($d/D = 0$), the jet diameter would be $\sqrt{0.611} = 0.782$ times the diameter of the hole, for $\theta = 90°$.

6-6. Numerical Relaxation Method

If $f(x, y)$ represents any harmonic function (it could be the stream function ψ, the potential function ϕ, or the temperature T, for example), the Laplace equation is

$$\frac{\partial^2 f}{\partial x^2} + \frac{\partial^2 f}{\partial y^2} = 0$$

This differential equation may be replaced by a difference equation. The solution of the difference equation is not the solution of the differential equation, but with small intervals in the difference equation the solutions may be made sufficiently close to each other to be useful. Values of f at nodes of a grid are used to draw streamlines or equipotential lines by interpolation.

In Fig. 6-25,

$$\frac{\partial^2 f}{\partial x^2}(x, y) = \frac{\dfrac{\partial f}{\partial x}\left(x + \dfrac{h}{2}, y\right) - \dfrac{\partial f}{\partial x}\left(x - \dfrac{h}{2}, y\right)}{h}$$

$$= \frac{\dfrac{f(x + h, y) - f(x, y)}{h} - \dfrac{f(x, y) - f(x - h, y)}{h}}{h}$$

$$= \frac{f(x + h, y) - 2f(x, y) + f(x - h, y)}{h^2}$$

Similarly,

$$\frac{\partial^2 f}{\partial y^2}(x, y) = \frac{f(x, y + h) - 2f(x, y) + f(x, y - h)}{h^2}$$

Adding these, equating them to zero (the Laplace equation), and solving for $f(x, y)$ gives

$$f(x, y) = \frac{f(x + h, y) + f(x - h, y) + f(x, y + h) + f(x, y - h)}{4} \qquad (6.23)$$

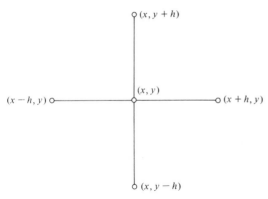

Fig. 6-25.   Interior nodal point.

which indicates that the value of $f(x, y)$ is the average of the function at four neighboring points.

If the value of $f(x, y)$ is needed near a boundary where four neighboring points are not equidistant as in Fig. 6-25, but are as shown in Fig. 6-26, the Laplace equation becomes

$$\frac{\dfrac{f_b - f_0}{bh} - \dfrac{f_0 - f_d}{h}}{(h + bh)/2} + \frac{\dfrac{f_c - f_0}{h} - \dfrac{f_0 - f_a}{ah}}{(h + ah)/2} = 0$$

where $f(x, y) = f(0)$, and ah and bh are the proportionate distances from 0 to a and b, respectively. Solving for f_0 gives

$$f_0 = \frac{ab}{a + b}\left[\frac{1}{b(1 + b)}f_b + \frac{1}{(1 + b)}f_d + \frac{1}{(1 + a)}f_c + \frac{1}{a(1 + a)}f_a\right] \qquad (6.24)$$

This equation prorates the values of the function f at points at unequal distances from the location where the value of the function is sought (f_0).

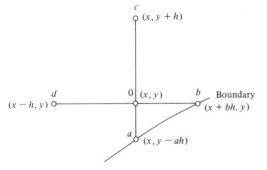

Fig. 6-26.   Nodal point near a boundary.

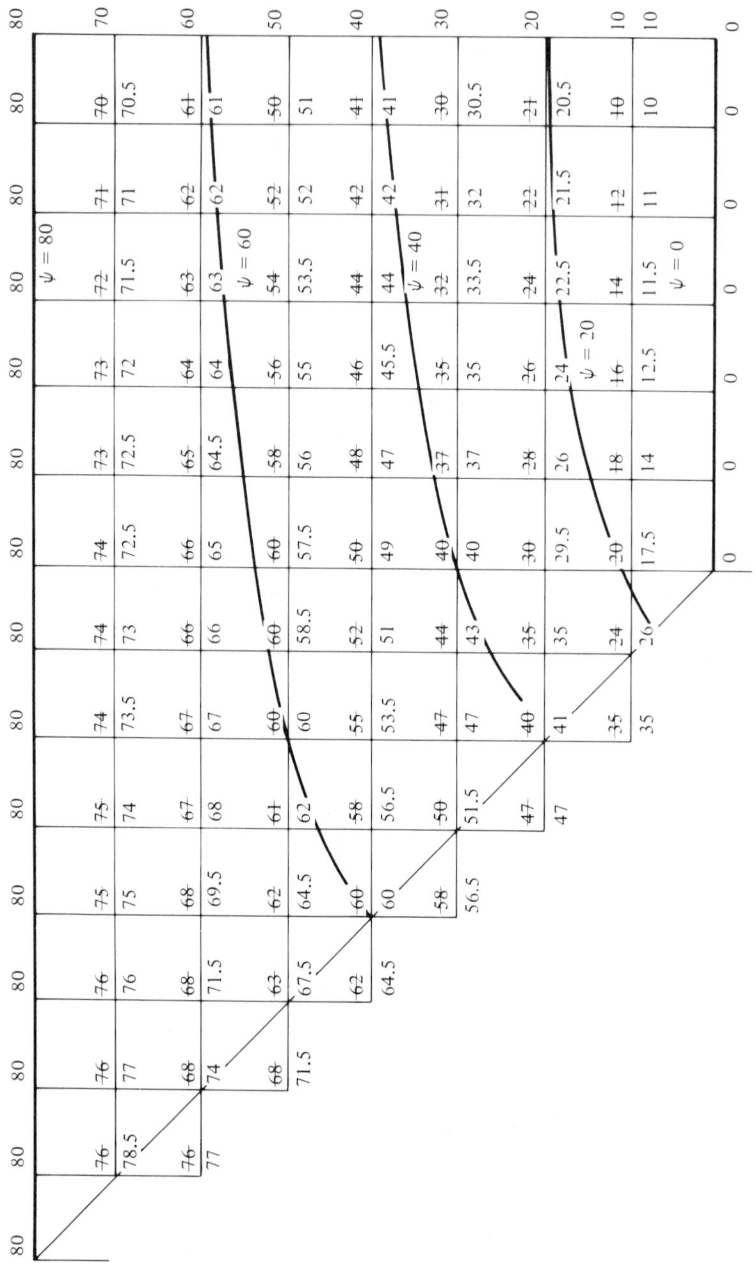

Fig. 6-27. Relaxation network for stream function for flow around a corner.

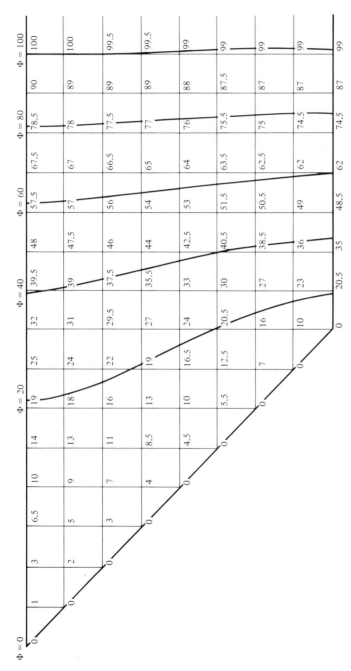

Fig. 6-28.   Relaxation network for potential function for flow around a corner.

If $a = 1$ and $b = \frac{1}{2}$,

$$f_0 = \frac{1}{6}(\tfrac{8}{3}f_b + \tfrac{4}{3}f_d + f_c + f_a)$$

and if both $a = \frac{1}{2}$ and $b = \frac{1}{2}$,

$$f_0 = \frac{1}{6}(2f_b + f_d + f_c + 2f_a)$$

The method is illustrated by solving for values of ψ and ϕ at nodes of a grid for the flow around a corner shown in Fig. 6-6a. The grid for ψ is shown in Fig. 6-27 and for ϕ in Fig. 6-28 together with interpolated streamlines and potential lines, respectively, at intervals of $\Delta\psi = \Delta\phi = 20$. Only one-half the flow duct is shown since the flow is symmetrical about the corner diagonal.

Initial values are chosen and written near each node. In Fig. 6-27 the boundaries of the duct are streamlines. Far to the right the streamlines should be parallel and equally spaced. Nodes are numbered in equal intervals at the right end of the duct as a result. (A smaller grid and longer duct might show a slight nonuniformity at these nodes.) Initial values are shown crossed out; final values to the nearest 0.5 are also given. After assigning initial values, all nodal values are corrected according to Eq. (6.23) by working through the mesh in an orderly fashion until insignificant or zero changes are made. Then the mesh is "relaxed." Values of ψ to the left of the diagonal are recorded equal to the image opposite, in applying Eq. (6.23) to values on the diagonal.

In Fig. 6-28 there is some uncertainty about the boundary conditions. The mesh is worked out so that the values along the diagonal are all equal, since it is the only potential line whose position is known. Then values are

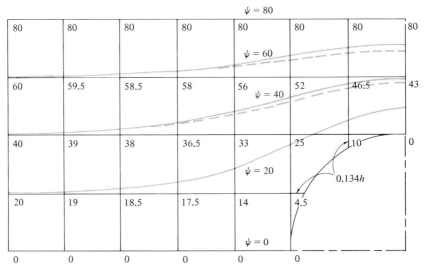

Fig. 6-29. Relaxation network for stream function for two-dimensional flow around a cylinder midway between two parallel walls.

changed by a factor such that far from the corner a change of 20 in ϕ has the same linear dimension as a change of 20 in ψ in Fig. 6-27. Figures 6-27 and 6-28 may be compared with Fig. 6-6a. A grid finer than that used would be more accurate, of course, since the solution to a finite-difference equation gets closer to that of the differential equation as the difference intervals get smaller.

Figure 6-29 shows streamlines for two-dimensional flow around a circular midway between parallel boundaries. Because of symmetry, only one-fourth the flow need be shown. Values of ψ for the coarse grid are given, and the solid streamlines are drawn from them. A finer grid (one-half the size shown) resulted in the streamlines shown by dashed lines. There is, of course, some personal element in interpolation when drawing these lines.

PROBLEMS

6-1. Which of the following scalar functions could represent the velocity potential for an ideal incompressible fluid flow?

(a) $f = x - 3y$ (e) $f = \sin(x - y)$

(b) $f = x^2 + y^2$ (f) $f = \ln(x + y)$

(c) $f = x^2 - y^2$ (g) $f = \ln(x - y)$

(d) $f = \sin(x + y)$ (h) $f = \arctan(y/x)$

6-2. For the fluid flows of Prob. 6-1, what are

(a) the stream functions and

(b) the velocity vectors?

6-3. Let $u = -2x$ and $v = 2y + 2$.

(a) Show that continuity is satisfied.

(b) Show that the flow is irrotational.

(c) What is the stream function?

(d) What is the potential function?

(e) Plot the flow net and compare with that for Example 3-1.

6-4. The velocity components for a two-dimensional incompressible flow are $u = 2x + y$ and $v = -x - 2y$.

(a) Is continuity satisfied? If so, find the stream function.

(b) Is the flow irrotational? If so, find the potential function.

6-5. Find the potential function from which the following velocity components are derivable, when possible.

(a) $u = 2xy, v = x^2 + 1$

(b) $u = -x^2 + 2x + y^2, v = +2xy - 2y$

(c) $u = x^2 - y^2 + x, v = -2xy - y$

6-6. What is the stream function for a southwest wind blowing at 12 ft/s (or 12 m/s) using east as the $+x$ direction and north as the $+y$ direction?

6-7. Two functions, ϕ_1 and ϕ_2, are both harmonic. Show that

(a) $\phi_1 + \phi_2$,

(b) $C\phi_1$ and $C\phi_2$,

(c) $(\phi_1 + C)$ and $(\phi_2 + C)$ are also harmonic.

6-8. Show that a constant added to a stream function or to a potential function does not change the flow pattern.

6-9. Show that the vorticity for two-dimensional flow is equal to

$$\frac{\partial^2 \psi}{\partial x^2} + \frac{\partial^2 \psi}{\partial y^2}$$

6-10. In percolating flow through a uniformly pervious material, the velocity is so small the total head H may be considered to be pressure head and potential head. The velocity through this material is $\mathbf{V} = k$ grad H, where k is the permeability of the material (not dimensionless). Show that this velocity is derivable from a potential function.

6-11. Laminar flow in the x direction between parallel plates has a velocity profile given by

$$u = -\frac{dp}{dx}\frac{b^2 - y^2}{2\mu}$$

where the spacing between plates is $2b$ and y is the distance from the midplane (see sketch for Prob. 4-56).

(a) What is the average flow velocity?

(b) Show that the velocity u is derivable from a potential function.

(c) Show that the average velocity is also derivable from a potential function.

6-12. A fluid is between two parallel plates which are a distance B apart. One plate is at rest and the other moves at a velocity V so that the velocity profile between the plates is given by $u = Vy/B$, where u is the fluid speed at a distance y from the plate at rest. What is the stream function which describes this flow?

6-13. The general expression in polar form for the potential function for two-dimensional steady flow around a corner of angle α may be written as

$$\phi = r^{\pi/\alpha} \cos \frac{\pi\theta}{\alpha}$$

(a) What is the stream function for this flow?

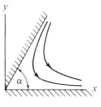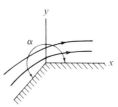

Prob. 6-13

(b) Show that for $\alpha = \pi$, the flow may be represented by $\mathbf{V} = -\mathbf{i}$.

(c) For $\alpha = \pi/2$, show that the flow is that of Example 3-2.

6-14. What is the magnitude of the velocity at the corner in Prob. 6-13

(a) for $0 < \alpha < \pi$ and

(b) for $\pi < \alpha < 2\pi$?

6-15. Plot the streamlines and equipotential lines for $\psi = 0, 2, 4,$ and 6 and $\phi = 0, \pm 2, \pm 4,$ and ± 6, with $\alpha = 3\pi/2$ in Prob. 6-13, to a scale of 2 cm or 1 in. $= 10$ units. Polar graph paper is suggested.

6-16. The pressure is constant and equal to atmospheric along the upper and lower streamlines for the weir flow beyond the weir plate in Fig. 6-4, and along the upper streamline for the jet issuing from the sluice gate in Fig. 6-5. Explain why the spacing between equipotential lines varies along these streamlines.

6-17. Consider two-dimensional incompressible flow through an orifice in a rectangular duct.

(a) Draw a flow net for ideal fluid flow and make a sketch of the pressure variation along the duct center line.

(b) Draw the streamline pattern for a real fluid flow and make a sketch of the pressure variation along the duct centerline. Compare with part (a).

6-18. A line source of strength $2\pi m$ is at the origin, and another source of strength $4\pi m$ is at $(0,a)$.

(a) Where is the stagnation point?

(b) What is the velocity at $(a, 0)$?

6-19. Show that the potential function for a three-dimensional point source of strength Q at the origin is $\Phi = Q/4\pi r$. What is the corresponding stream function?

6-20. What uniform flow will produce a Rankine half-body of maximum width 0.50 m with a line source of strength 3 m²/s?

6-21. For a free-stream velocity $u_s = 60$ cm/s and a line source strength of 1.0 m³/s per meter at the origin, what is the resultant velocity at $r = 30$ cm and $\theta = 120°$ for this two-dimensional half-body flow?

6-22. A line source and a line sink, each of strength 2π m²/s, are placed at $y = 0$ and $x = -6$ m and $x = +6$ m, respectively, in a uniform flow field with $u_s = 6$ m/s.

(a) What is the stream function for this flow?

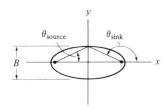

Prob. 6-22

(b) Where are the stagnation points?

(c) Show that the stream function at the stagnation points is $\psi = 0$.

(d) What is the velocity field represented by this flow?

(e) Write the stream function in polar form.

(f) What is the maximum width of the long thin shape around which the fluid flows? (It flows inside as well.)

(g) Where is $v = 0$?

(h) What is the difference between the free-stream pressure and that at (0,5)? Density $= 1000 \text{ kg/m}^3$.

6-23. Repeat Prob. 6-22 for a line source and line sink strength of $1.5\pi \text{ m}^3/\text{s}$ per meter, other parameters being the same.

6-24. A line source of strength 2π and an equal line sink are in a uniform flow at a speed of 4 parallel to a line connecting them. They are 2 units apart. Calculate the thickness ratio of the oval Rankine body represented by the zero streamline.

6-25. A line source of strength $9\pi \text{ m}^2/\text{s}$ is combined with an irrotational vortex of circulation $3\pi \text{ m}^2/\text{s}$, both at the origin. What is the difference in pressure between points at (4,4) and (3,0) for a fluid of density 900 kg/m^3?

6-26. Two equal doublets are placed at the origin with their axes at right angles to each other. What is the velocity at (4,4) in terms of their strength C?

6-27. A line source in two-dimensional flow of strength q is placed at the origin, and a uniform flow of velocity u_s in the $+x$ direction is added to it to produce flow around a two-dimensional half-body (Fig. 6-12).

(a) Show that the stagnation point is at $x = -q/2\pi u_s$.

(b) What is the value of the stream function at the stagnation point?

(c) Show that as $x \to +\infty$, the width of the half body is q/u_s. This is the distance between streamlines for which $\psi = -q/2$.

6-28. Show that the half-body of Fig. 6-12 is one-half its maximum width at $x = 0$.

6-29. What is the maximum flow velocity at the surface of the half-body of Fig. 6-12? At what value of θ (in polar coordinates) is it located? Give values for $q = 3 \text{ m}^2/\text{s}$ and $u_s = 15 \text{ m/s}$.

6-30. A circular cylinder of radius R is in an infinite ideal fluid which flows past the cylinder with a velocity u_s far from the cylinder (Fig. 6-15). The fluid density is ρ and the free-stream pressure is p_s.

(a) Show that the resultant force on the cylinder is zero and

(b) that the force on both the upstream and downstream halves of the cylinder is equal to $2Rp_s - (\rho R u_s^2/3)$ per unit length of cylinder.

6-31. For what values of θ (Fig. 6-15) is the pressure at the surface of the cylinder the same as the free-stream pressure p_s? (See discussion of the pitot cylinder in Sec. 13-1 and Fig. 13-6.)

6-32. (a) What is the stream function for potential flow past a circular cylinder of radius 15 cm normal to a free stream of velocity $u_s = 2.5 \text{ m/s}$?

    (b) What is the difference between the maximum and the minimum pressures on the surface of the cylinder for a fluid density of 1000 kg/m^3?

6-33. Repeat Prob. 6-32 for a circular cylinder of radius 20 cm, other parameters remaining the same.

6-34. (a) What is the strength of the circulation around a cylinder of radius R in a free stream of velocity u_s which will bring the stagnation points together at $\theta = 3\pi/2$? See Fig. 6-16.

    (b) For what values of θ is the pressure on the cylinder equal to the free-stream pressure?

6-35. Obtain an expression for the pressure $(p - p_s)/(\rho u_s^2/2)$ in terms of θ, ρ, u_s, R, and Γ around the surface of a cylinder for steady incompressible flow past the cylinder

    (a) without circulation,

    (b) with stagnation points at $\theta = 5\pi/4$ and $7\pi/4$, and

    (c) with stagnation points coinciding at $\theta = 3\pi/2$. Refer to Fig. 6-16.

6-36. For the three conditions of Prob. 6-35,

    (a) what is the lift force per unit length of cylinder;

    (b) at what values of θ is the pressure on the surface of the cylinder the same as the free-stream pressure p_s?

6-37. Standard air at a free-stream velocity of $u_s = 30 \text{ m/s}$ flows past a long circular cylinder normal to the flow. A circulation of $20\pi \text{ m}^2/\text{s}$ encircles the cylinder, which has a radius of 30 cm. Find on the surface of the cylinder

    (a) the tangential velocity due to the circulation,

    (b) the maximum air speed,

    (c) the stagnation points,

    (d) the difference between the maximum and the minimum pressure, and

    (e) the lift per foot of length of the cylinder.

6-38. A 2-m-diameter cylinder 3 m high is rotated about its axis at 120 r/min in a stream of standard air flowing past the cylinder at 10 m/s. The rotation produces a circulation 50 percent as effective as an irrotational vortex having the same peripheral speed at the surface of the cylinder. What is the lift force on the cylinder?

6-39. In Prob. 6-38, the cylinder is 2.5 m in diameter and 4 m high and is in a stream of standard air moving past at 10 m/s. Calculate the lift force in newtons.

6-40. Show that the lift on a flat plate parallel to a uniform flow with circulation is equal to $u_s\rho\Gamma$ per unit length.

6-41. Calculate the force per unit length of cylinder

    (a) on the right half and

    (b) on the upper half of a cylinder with circulation shown in Fig. 6-16.

6-42. Incompressible two-dimensional flow at a velocity U exists between two parallel plates spaced a distance H apart. It flows past a cylinder of radius R oriented normal to the flow and midway between the plates. What is the potential function for this flow? Use method of images.

6-43. Sketch a two-dimensional flow net for two equal sources at $(0,1)$ and at $(0,-1)$, respectively. What is the potential function for this configuration?

6-44. A line source is located at $(a,0)$, the y axis being a solid impervious boundary. Where on the boundary is the velocity a maximum?

6-45. Estimate the liquid flow rate per square meter of plate area through a perforated plate with 1.5-cm-diameter holes in a square array 5 cm on centers. The average pressure behind the plate is 28 kPa gage, and the issuing jets discharge into the atmosphere.

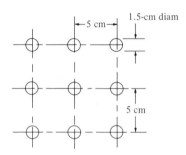

Prob. 6-45

6-46. In Prob. 6-45 the holes are 3.0 cm in diameter and spaced 10 cm on centers. The pressure behind the plate is 40 kPa gage. What is the flow rate in m^3/s per square meter of perforated plate?

6-47. Estimate the flow rate per square meter of total area through slots 1 cm wide on 5 cm centerlines when the average pressure behind the slots is 21 kPa gage and the two-dimensional slots discharge into the atmosphere.

6-48. A plate with a 5-cm-diameter hole is attached to the end of an 10-cm-diameter pipe (this is called an *end-cap orifice*). For a flow rate of $0.030 \ m^3/s$, what is the net force on the end plate? Water flows without frictional effects.

6-49. Repeat Prob. 6-48 for a 10-cm-diameter hole in a 20-cm pipe and a flow rate of $0.10 \ m^3/s$.

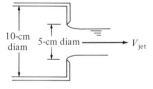

Prob. 6-48

6-50. A flow passage is bounded by the x and y axes (for which $\psi = 0$) and the curve $xy = 12$ in the first quadrant (on which $\psi = 300$). Determine the value of the stream function ψ at the nodes of a grid at integer values of x and y for $x \leq 6$ and $y \leq 6$ by a relaxation method. Sketch in streamlines for $\psi = 200$ and 100.

6-51. In Fig. 6-29, replace the quarter circle with a straight line. This would then represent flow past a square cylinder, one corner pointing into the flow, placed midway between parallel walls. The diagonal of the square cylinder is one-half the distance between the parallel walls. Obtain a streamline pattern from a relaxation method.

6-52. Show that the circulation around a doublet is zero. *Hint:* Use a circular path.

6-53. Show that the circulation around a source is zero.

References

1. L. M. Milne-Thomson, *Theoretical Hydrodynamics*, 4th ed. (New York: The Macmillan Company, 1960).
2. V. L. Streeter, *Fluid Dynamics* (New York: McGraw-Hill Book Company, Inc., 1948).
3. A. H. Shapiro, *The Dynamics and Thermodynamics of Compressible Fluid Flow*, Vol. 1 (New York: The Ronald Press Company, 1953), Pt. III.

Chapter 7
Viscous Flow: The Boundary Layer

In the last chapter we dealt with nonviscous fluids in steady irrotational flows. There were no restrictions on the tangential velocity along a boundary—the fluid moved at whatever speed was indicated by the gradient of the velocity potential function. Although a relative tangential velocity existed between the fluid and a boundary, there was no relative motion between them in a direction normal to the boundary. Adjacent layers of fluid experienced no tangential force (shear) but only normal forces (pressure).

In a viscous fluid, adjacent layers transmit not only normal forces but also tangential forces. Thus, when a viscous fluid is in motion there is an additional condition at a boundary. There is no tangential component of velocity there—no relative motion exists between a fluid and its boundary. This situation exists for a viscous continuum fluid and is known as a no-slip condition. There is a region of retarded flow, then, near a boundary past which a viscous fluid flows and in which the velocity relative to the boundary varies from zero at the boundary to a maximum or free-stream value some distance away. This region of retarded flow is known as the boundary layer, a concept first introduced by Prandtl in 1904. Viscous effects are concentrated within this boundary layer, and beyond it viscous effects are often negligible and the flow there may be treated as a potential flow (Fig. 7-1).

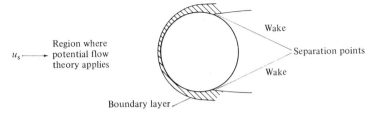

u_s ⟶

Region where
potential flow
theory applies

Wake

Separation points

Wake

Boundary layer

Fig. 7-1.   Flow of a viscous fluid past a circular cylinder (boundary layer thickness not to scale).

Boundary layer theory is extremely important in modern fluid mechanics and is essential to an understanding of convective heat transfer. Definitive treatments of boundary layer theory have been given by Schlichting [1–3] and Prandtl [4]. A brief introduction to thermal boundary layers is given in Chapter 16.

In the present chapter, a discussion of laminar and turbulent flows will first be given, and this will be followed by a study of boundary layers and boundary layer separation in incompressible flow.

7-1. Laminar and Turbulent Flow

The flow of viscous fluids may be divided into two different classes. Smooth, quiet flows without lateral motions which cause mixing are known as *laminar* flows. The motion appears to be like layers or fluid moving alongside one another. Flows with random or irregular fluctuations of mechanical quantities (velocity, pressure, or temperature, for example) are known as *turbulent* flows. Photographs in the frontispiece illustrate these two types of flow.

Smoke rising from a cigaret held at rest in a quiescent environment will have a straight threadlike appearance for a few centimeters. The air flow indicated by this filament is laminar. Above this straight filament the smoke will become wavy, and finally it will show an irregular lateral motion superimposed on the vertical motion. This flow is turbulent. Fluffy snowflakes falling in a crosswind may be observed to have highly random and irregular motions superimposed on their mean motion, and thus their paths indicate a turbulent motion of the air. The gustiness of atmospheric winds is another manifestation of turbulence. Even if the wind seems to be steady, it very likely is turbulent on a scale smaller than gusty winds. Boundary layers along ship hulls and airplane wings are turbulent. The flow of petroleum products and natural gas in long pipelines is turbulent. The wakes of cars, trucks, ships, and aircraft are turbulent. The flow of water in rivers and in the Gulf Stream is turbulent. Combustion processes usually are turbulent; combustion is enhanced thereby. Laminar flows are the exception rather than the rule; turbulent flows are much more common.

Even though turbulent flows are extremely important from a practical engineering point of view, they also are of theoretical interest because they represent examples of nonlinear mechanical systems. The Navier–Stokes equations [Eqs. (4.21), (4.22), or (4.23)] are nonlinear partial differential equations, and exact solutions of them for laminar flows are limited to about a dozen flow situations. Additional numerical solutions of the equations have been made, however. When equations are written for turbulent flows, there are more unknowns than equations. Physical models of turbulent flows generally deal with time-average values of the fluctuating quantities, and much progress has been made in studying the behavior of turbulent flows in this way. A complete understanding of turbulent flows, however, awaits a satisfactory development of statistical theories.

Two important types of forces in the flow of viscous fluids are inertia forces resulting from the motion and viscous forces. When viscous forces are large compared to inertia forces, the flow generally is laminar. Laminar flow is thus associated with slow motions, small flow passages, and fluids of large viscosity. When inertia forces are large compared to viscous forces, the flow generally is turbulent, and this type of flow is associated with high fluid velocities, large flow systems, and relatively low-viscosity fluids. The ratio of inertia to viscous forces in a flow system is called the *Reynolds number* (see Chapter 8) and is expressed as the product of a characteristic flow velocity, a characteristic length describing the flow system and the fluid density all divided by the dynamic viscosity of the fluid, so that the Reynolds number = $VL\rho/\mu$. It is a dimensionless number since it is the ratio of two forces, and the same numerical value of the Reynolds number is obtained for a given flow situation regardless of whether the values for velocity, length, density, and dynamic viscosity are in the cgs system, the SI system, or the ft-slug-sec system of units. Turbulent flows exhibit a number of unique characteristics. The most obvious one is their irregularity. Typical of the variations in the velocity in the mean flow direction is the record shown in Fig. 7-2b. Similar variations of the v and w components would also be measured.

The diffusing action of turbulent flows is another characteristic also readily noticeable. The lateral motions cause a mixing of the fluid and thus

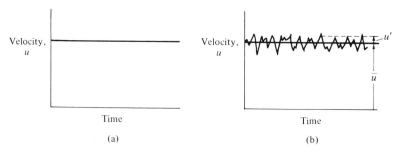

Fig. 7-2. Measured velocity at a point in a viscous flow. (a) Steady laminar flow. (b) Steady turbulent flow.

transport of mass, momentum, and energy (both mechanical and thermal) are enhanced. The exhausts from jet aircraft, smokestacks, and the mouths of smokers show the diffusing nature of turbulent flows.

A third characteristic of turbulent flows is that they always occur at large Reynolds numbers. If the Reynolds number of a particular flow initially is small, instabilities may develop in the laminar flow and then these instabilities, if not damped out, often cause the flow to become turbulent. Such is frequently the case for external flow along a surface for which the Reynolds number length parameter is the distance along the surface in the direction of flow, and thus the Reynolds number increases in the flow direction.

A fourth characteristic is the fluctuating nature of vorticity in turbulent flows. This vorticity is three dimensional: the axes of the vortexes are distributed in all directions. In *isotropic* turbulence they are distributed equally in all directions. Associated with the vorticity of turbulent flows is the dissipative nature of these flows. As a result of work of deformation by shear stresses, the internal energy (thermal) is increased at the expense of a kinetic energy (mechanical) decrease. If there is no source of energy to make up for these mechanical energy losses, the turbulence will decay. An English scientist, L. F. Richardson [5], described the dissipation of energy from larger to smaller eddies and then onto a complete decay of turbulence by a poem:

> Big whorls have little whorls,
> Which feed on their velocity;
> And little whorls have lesser whorls,
> And so on to viscosity (in the molecular sense).

It should be noted that turbulent flows are flows of a continuum; the smallest scales involved in the turbulent process are large compared with the molecular scales of the fluid. Lastly, turbulence is a characteristic of flows, not of fluids.

Osborne Reynolds in 1883 published a description of some classical experiments he conducted on flow through glass pipes. He introduced a dye at the centerline of the pipe and observed the motion of the dye as the liquid flowed along the pipe. He described clearly the two regimes of flow: laminar when the dye maintained a single threadlike filament all along the center of the pipe, and turbulent when the dye quickly dispersed throughout the fluid flowing in the pipe. He discovered that when a dimensionless quantity VD/v (which we now all call the Reynolds number) was below about 2300 the flow was always laminar. Disturbances were damped out. Above that value the flow could be either laminar or turbulent. Reynolds was able to obtain laminar flow up to a value of VD/v of nearly 13,000 by carefully avoiding disturbances in the incoming flow. By the middle of the twentieth century laminar flow had been made to persist at Reynolds numbers in the tens of thousands. In engineering practice, however, the limit generally is considered to be about 2000–2300 since disturbances are almost always present.

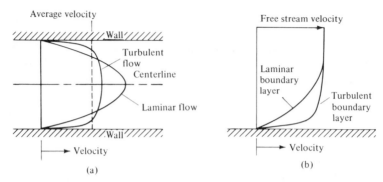

Fig. 7-3. Velocity profiles for viscous flow. (a) Laminar and turbulent flow in a pipe at the same average velocity. (b) External boundary layer flow.

Reynolds was the first to delineate clearly the two flow regimes, laminar and turbulent.

Velocity profiles are different for laminar flows and for turbulent flows in ducts as well as in external flows. Lateral transport of momentum in turbulent flows tends to flatten or to even out the velocity variation with distance from a boundary. Typical velocity profiles are compared in Fig. 7-3.

Shear stresses are also different in laminar and in turbulent flows. Reynolds was the first to express instantaneous values of the fluctuating mechanical parameters in a turbulent flow as an average value plus a fluctuating component. Thus $u = \bar{u} + u'$ (Fig. 7-2b), $v = \bar{v} + v'$, $w = \bar{w} + w'$, and $p = \bar{p} + p'$. He inserted these values into the Navier–Stokes equations of motion for a viscous fluid and obtained, among other things, a term known as the Reynolds stress, describing a shear stress in a turbulent flow which is in addition to the shear stress expressed by $\tau = \mu(du/dy)$ for laminar flow of a Newtonian fluid. It is given by $\tau_{\text{turbulent}} = -\rho \overline{u'v'}$, where $\overline{u'v'}$ is the time average value of instantaneous products of the fluctuating velocity components u' and v',

One important characteristic of turbulent flows, then, is that shear stresses are much greater than in laminar flows for the same velocity gradients. Boussinesq [6] in the 1870s introduced a mixing coefficient A_t for the Reynolds stress in a turbulent flow in order that it could be expressed in a manner similar to that in a laminar flow. Thus,

$$\tau_{\text{turbulent}} = -\rho \overline{u'v'} = A_t \frac{d\bar{u}}{dy}$$

Modern usage defines a turbulent eddy diffusivity or turbulent eddy viscosity $\epsilon = A_t/\rho$ such that shear stresses due to laminar contributions are

$$\tau_{\text{laminar}} = \rho v \frac{du}{dy}$$

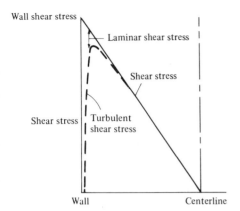

Fig. 7-4.   Shear stress distribution in a channel.

and those due to turbulent contributions are

$$\tau_{\text{turbulent}} = \rho\epsilon \, \frac{d\bar{u}}{dy}$$

The kinematic viscosity v is a fluid property, whereas the turbulent eddy viscosity ϵ is a flow parameter. Values of ϵ many orders of magnitude greater than v have been measured.

It can be shown (Sec. 10-2) that shear stresses in flow inside a circular tube or between parallel plane surfaces vary linearly from the wall to zero at the tube centerline or the centerplane between parallel walls. In a turbulent flow these are shown in Fig. 7-4, where it is shown that nearly all the shear stress away from the wall is due to turbulence. Near the wall turbulence fluctuations die out, and the shear there is laminar shear.

Surface roughness has no effect on laminar flow, but it slows down the flow near a surface and increases wall shear stress for turbulent flow.

The pressure gradient for flow in pipes (pressure drop per unit length of pipe) is different for laminar flow than for turbulent flow. The pressure gradient is directly proportional to the average flow velocity for laminar flow and varies with the velocity to the 1.75–1.85 power of velocity for smooth pipes and with V^2 for rough pipes for turbulent flow.

Wall shear stresses for external flow along a flat plate in a uniform free stream vary with the free-stream velocity to the 1.5 power for laminar flow, but to the 1.8 power for turbulent flow.

Thus we see that laminar flows are different from turbulent flows not only in the detailed structure of the flow but also they differ in measurable ways which are important in engineering practice.

7-2. Description of the Boundary Layer

The development of a boundary layer can perhaps be best described by a study of flow along a flat plate. Consider a uniform flow of an incompressible

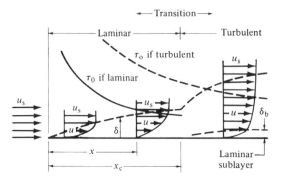

Fig. 7-5.   Flow of a viscous fluid along a flat plate.

fluid at a free-stream velocity u_s approaching the plate (Fig. 7-5). When the fluid reaches the leading edge, large shear stresses are set up near the plate surface, and the fluid particles at the plate surface are brought to rest and those for a short distance normal to the plate are retarded because of viscous shear. The region of retarded flow is called the *boundary layer*, and its thickness is designated as δ. For some longitudinal distance x_c, the flow within the boundary layer is laminar. Downstream from this point the boundary layer flow becomes unstable and eventually becomes turbulent. If the velocity u_s for a given fluid is increased, x_c is decreased such that the product $u_s x_c$ remains essentially constant. The value of this constant varies directly with the kinematic viscosity of the fluid; and if different fluids are used, the ratio $u_s x_c / \nu$ is approximately constant. This ratio is one form of the Reynolds number, which is discussed in more detail in Chapter 8.

The boundary layer thickens in the direction of flow, and thus the velocity change from zero at the plate surface to u_s at a distance δ takes place over an increasingly greater distance normal to the plate. The rate of change of velocity determines the velocity gradient at the plate surface and thus the shear stresses as well. This shear stress for the laminar boundary layer is

$$\tau_0 = \mu \left(\frac{du}{dy} \right)_{y=0} \tag{1.28}$$

over the entire plate surface. As the laminar boundary layer thickens, instabilities set in, and under certain conditions these are not damped out, so that a turbulent boundary layer forms. These instabilities were originally predicted by Tollmein and Schlichting in Germany in the early 1930s and verified experimentally by Dryden, Schubauer, and Skramstad in the United States in 1940. The cross components of velocity (normal to the main flow) produce a mixing of the fluid with an accompanying momentum exchange which is absent in laminar flow. For the turbulent boundary layer, the resulting velocity profile is more rounded with a higher velocity gradient at the plate surface, and the boundary layer is thicker and the shear stresses greater in this instance. Actually, a viscous sublayer exists between the plate surface and the turbulent portion of the boundary layer. This is reasonable to expect, since the inertia forces are very small because of the very low velocities near the plate surface; also viscous forces are relatively large and complete mixing is inhibited by the presence of the plate.

The transition from a laminar to a turbulent boundary layer depends on the roughness of the plate and the turbulence level in the free stream, in addition to the ratio $u_s x_c / v$. Either plate roughness; a high turbulence level in the free stream; or, if the free stream is not uniform, a decelerating free stream will cause transition to take place nearer the leading edge of the plate (at a smaller value of x_c). A surface is considered to be hydrodynamically smooth if its roughness elements have no influence on the flow outside the viscous sublayer but are drowned out by it. Conversely, a surface is considered rough if its roughness elements do have an effect on the flow outside the viscous sublayer. An excellent review of studies on the transition from laminar to turbulent flow was given by Dryden [7].

The drag on the plate due to viscous shear is indicated by the area under the curves of shear stress in Fig. 7-5. The total drag is obviously dependent upon the location of the transition from laminar to turbulent flow within the boundary layer because of the higher shear in the turbulent region. If transition takes place near the leading edge, the total drag is greater than if transition takes place farther downstream. Methods have been developed to inhibit transition and thus reduce drag. These methods include boundary layer suction through a porous surface and a careful design of the shape of the surface (for other than a flat plate).

The longer the flat plate, the greater is the total drag, but the *average* shear stress decreases with length. If, however, the boundary layer changes from laminar to turbulent, the average shear stress will increase, but will then decrease as the plate length increases. For long plates on which the laminar portion of the boundary layer extends for only a relatively short distance, it is customary to assume a turbulent boundary layer over the entire plate.

Up to this point in our discussion, flow past a flat plate has involved no pressure variations either parallel or normal to the plate. In both the laminar and turbulent portions of the boundary layer, the velocity at a fixed distance y_1 from the plate decreases with x, but when transition occurs this velocity will increase. Only at the plate surface itself is the velocity zero (Fig. 7-6).

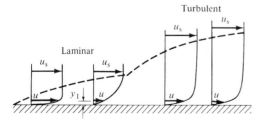

Fig. 7-6. Velocity variation within boundary layer at a fixed distance from the boundary surface.

The pressure gradient is not always zero; it is determined for given boundaries by a solution of the potential flow problem for those boundaries (see Chapter 6). For example, for flow in converging sections the pressure decreases, and for flow in diverging sections the pressure increases in the direction of flow. Flow in a direction of increasing pressure is called flow with a positive, or adverse, pressure gradient and occurs in flow around curved boundaries and towards a stagnation point as well as in diverging ducts. In these instances, the pressure is increased at the expense of a decrease in kinetic energy or velocity, and velocities at a fixed distance from the surface may decrease to zero. When this happens, the boundary layer separates from the surface, and the phenomenon is known as *boundary-layer separation*. The separation point is followed by a region called the *wake*, in which intense eddies exist with an accompanying increase in drag or dissipation of energy.

For flow past a curved surface, the velocity profile has an inflection (change in curvature) beginning where the pressure gradient becomes positive (Fig. 7-7). The particle at y_1 has a decreasing velocity as it flows along the surface, and it finally comes to rest momentarily where the boundary layer separation occurs. If the boundary layer were laminar in the region of the adverse pressure gradient, the velocity at y_1 would be less than if the boundary layer were turbulent at y_1 (Fig. 7-6). Hence the particle in laminar flow would have less kinetic energy and would come to rest farther upstream than if the flow were turbulent in the boundary layer. Separation, then, is retarded if the boundary layer can be made to become turbulent. If drag is due to viscous shear, it is desirable to retard the formation of turbulence in the boundary layer. If drag is largely due to low pressures in the wake, it is desirable to enhance formation of turbulence in the boundary layer so that the separation point is moved downstream. This retarded separation will create a smaller wake.

Flow in a pipe is similar to flow along a flat plate, except that a negative pressure gradient (pressure drop in the flow direction) exists for incompressible flow in a pipe. If fluid is assumed to enter a pipe in such a way that separation of streamlines from the pipe entrance is avoided, the velocity profile at entrance is quite flat. Thus a boundary layer will grow along the

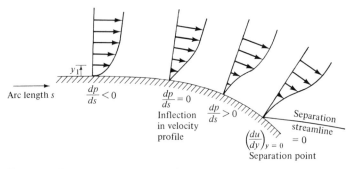

Fig. 7-7. Flow past a curved surface.

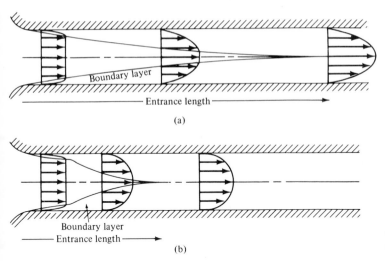

Fig. 7-8. Growth of boundary layer in a pipe (not to scale). (a) Laminar flow. (b) Turbulent flow.

pipe walls. At some point the boundary layers from the walls will meet at the center of the pipe. Beyond this point the velocity profile will not change form, and the flow is called *fully developed flow*. Again, the velocity gradient and the wall shear are greatest at entrance, and decrease to a steady value at and beyond the fully developed region (see Sec. 9-2). In the entrance region or length the centerline velocity must increase, since if some fluid is retarded within the boundary layer, continuity requires that the velocity outside the boundary layer be increased. The growth of the boundary layer in pipes is illustrated in Fig. 7-8.

When water with a free surface flows over a very steep spillway, the boundary layer builds up in thickness as the water flows down the spillway and rapidly becomes turbulent. When the boundary layer thickness equals the water depth, air becomes entrained in the water because of the turbulence, and the water becomes frothy in appearance.

Various boundary layer parameters are of interest. The boundary layer thickness, the local wall shear stress or local friction or drag coefficient, and the average wall shear stress or average friction or drag coefficient are among these.

The boundary layer thickness may be expressed in a number of ways.

1. One definition refers to the actual thickness δ of the region of retarded flow, and this definition is useful for analytical calculations.
2. Since the velocity u within the boundary layer approaches the free-stream velocity u_s asymptotically (Fig. 7-5), in experimental measurements of boundary layer velocity profiles the boundary layer thickness δ' is commonly defined as the distance from the boundary to the point where $u = 0.99u_s$.

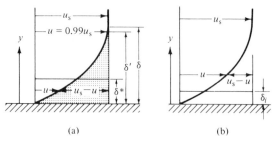

Fig. 7-9. Boundary layer thickness defined: (a) displacement thickness (rectangular area $u_s\delta*$ equals cross-hatched area) and (b) momentum thickness.

3. A *displacement thickness* $\delta*$ is defined as the distance the actual boundary would have to be displaced in order that the actual flow rate be the same as that of an ideal fluid past the displaced boundary. It may be expressed as[†]

$$\delta* = \frac{1}{u_s} \int_0^\delta (u_s - u)\, dy = \int_0^\delta \left(1 - \frac{u}{u_s}\right) dy \tag{7.1}$$

4. A *momentum thickness* δ_i is defined as a distance from the actual boundary such that the momentum flux through this distance δ_i is the same as the deficit of momentum flux through the actual boundary layer. This may be expressed as[†]

$$(\rho\, \delta_i u_s)u_s = \int_0^\delta \rho(u_s - u)u\, dy$$

so that

$$\delta_i = \int_0^\delta \left(1 - \frac{u}{u_s}\right)\frac{u}{u_s}\, dy \tag{7.2}$$

If u/u_s is expressed in terms of y, then $\delta*$ and δ_i can be expressed in terms of δ. The value of δ, in turn, may be found from a solution of the boundary-layer equations. Definition sketches for the displacement and the momentum thickness of the boundary layer are shown in Fig. 7-9.

The local skin-friction drag coefficient is the ratio of local wall shear stress to the dynamic pressure of the free stream, or

$$c_f = \frac{\tau_0}{\rho u_s^2/2} \tag{7.3}$$

The average skin-friction drag coefficient is the ratio of the total friction drag to the product of dynamic pressure of the free stream

---

[†] The upper limit of the integrals should be infinity, but the integrals are zero between δ and infinity, and thus an upper limit of δ gives the same result.

and total area being sheared, or

$$C_f = \frac{\text{total friction drag}}{(\rho u_s^2/2)(\text{area sheared})} \tag{7.4}$$

These will be determined for a flat plate in a uniform free stream (with no pressure gradients) from both the Prandtl boundary-layer equations for a laminar boundary layer and from a momentum analysis for both a laminar and a turbulent boundary layer.

7-3. The Prandtl Boundary-layer Equations

A solution of the Navier–Stokes equations [Eq. (4.22)] in simplified form and the continuity equation [Eq. (3.18b)] for steady, two-dimensional flow in a laminar boundary layer for flow along a flat plate parallel to the flow was made originally by Blasius in 1908 [cf. Ref. 1]. The boundary-layer thickness was assumed to be small as compared with any other characteristic dimension including the radius of curvature (if any) of the boundary. Only those terms in the Navier–Stokes equations of the largest order of magnitude in each equation were retained; those of a lesser order of magnitude were neglected. The simplified Navier–Stokes equations and the continuity equation for two-dimensional flow are known as the *Prandtl boundary-layer equations* (see Appendix IV for their derivation). With x and u taken parallel to the boundary and y and v normal to it, they may be written as

$$u \frac{\partial u}{\partial x} + v \frac{\partial u}{\partial y} = -\frac{1}{\rho} \frac{\partial p}{\partial x} + v \frac{\partial^2 u}{\partial y^2}$$

$$\frac{\partial p}{\partial y} = 0 \tag{7.5}$$

$$\frac{\partial u}{\partial x} + \frac{\partial v}{\partial y} = 0$$

The second condition ($\partial p/\partial y = 0$) implies that the pressure is constant across the boundary layer at any given location. The Bernoulli equation applied to the flow outside the boundary layer determines the pressure within the boundary layer at any location. For flow parallel to a flat plate with a constant free-stream velocity, the pressure is also constant, so that $\partial p/\partial x = 0$ in the absence of gravity. The Prandtl boundary-layer equations then become

$$u \frac{\partial u}{\partial x} + v \frac{\partial u}{\partial y} = v \frac{\partial^2 u}{\partial y^2} \quad \text{and} \quad \frac{\partial u}{\partial x} + \frac{\partial v}{\partial y} = 0 \tag{7.6}$$

with boundary conditions at $y = 0$, $u = v = 0$; at $y = \infty$, $u = u_s$.

SOLUTION OF THE PRANDTL BOUNDARY-LAYER EQUATIONS
FOR A LAMINAR BOUNDARY LAYER

The Prandtl boundary-layer equations are transformed into a single ordinary differential equation by introducing two new variables. The first is a dimensionless y coordinate

$$\eta = y\sqrt{\frac{u_s}{vx}} \tag{7.7}$$

and the second is a stream function

$$\psi = -\sqrt{vxu_s}f(\eta)$$

From these

$$\frac{u}{u_s} = f'(\eta) \tag{7.8}$$

and then Eqs. (7.6) become

$$ff'' + 2f''' = 0 \tag{7.9}$$

with boundary conditions $f = f' = 0$ at $\eta = 0$ and $f' = 1$ at $\eta = \infty$. Here f represents $f(\eta)$ and primes are derivatives with respect to η.

This differential equation was first solved by Blasius in 1908 using a power series expansion about $\eta = 0$ and an asymptotic solution about $\eta = \infty$, with a joining of the solutions at an intermediate point. Numerous solutions have been presented since that time. A simple one by Piercy and Preston [8] will be outlined here.

If $f'' = z$, then Eq. (7.9) becomes

$$fz + 2\frac{dz}{d\eta} = 0$$

Integrating twice, with appropriate boundary conditions, gives

$$\frac{u}{u_s} = \frac{\int_0^\eta \left[\exp\left(-\frac{1}{2}\int_0^\eta f\,d\eta\right)\right]d\eta}{\int_0^\infty \left[\exp\left(-\frac{1}{2}\int_0^\infty f\,d\eta\right)\right]d\eta} = f' \tag{7.10}$$

This is not a direct solution of $ff'' + 2f''' = 0$ since the function $f(\eta)$ is not known, but it may be used to obtain a solution by successive approximations. The method is as follows: Estimate a value for f ($f = 2\eta$, for example), introduce it into Eq. (7.10) and solve for $u/u_s = f'$, integrate f' to get a new approximation for f and repeat until successive approximations agree within acceptable limits. Integration may be done using any numerical method—the trapezoidal rule is satisfactory.

TABLE 7-1. VALUES OF f, f', AND f'', FROM
AN ANALYSIS OF LAMINAR BOUNDARY LAYER
ALONG A FLAT PLATE

	METHOD DESCRIBED	HOWARTH [9]		
η	f'	f'	f	f''
0	0	0	0	0.332
0.6	0.200	0.199	0.060	0.330
1.2	0.394	0.394	0.238	0.317
1.8	0.575	0.575	0.529	0.283
2.4	0.729	0.729	0.922	0.228
3.0	0.846	0.846	1.397	0.161
3.6	0.924	0.923	1.930	0.098
4.2	0.967	0.967	2.498	0.050
4.8	0.988	0.988	3.085	0.022
5.4	0.996	0.996	3.681	0.0079
6.0	0.999	0.999	4.280	0.0024
8.0		1.000	6.297	0.00001

Results are shown in Table 7-1 after five iterations, together with accurate results by Howarth [9].

The boundary layer thickness is at $u/u_s = 0.99$, and from Table 7-1, $\eta = 4.91$ (interpolated from a more detailed table). Thus from Eq. (7.7), $4.91 = \delta\sqrt{u_s/vx}$, or

$$\frac{\delta}{x} = \frac{4.91}{(u_s x/v)^{1/2}} \tag{7.11}$$

The displacement thickness is, from Eq. (7.1),

$$\delta^* = \int_0^\infty \left(1 - \frac{u}{u_s}\right) dy = \int_0^\infty (1 - f')\sqrt{\frac{vx}{u_s}}\, d\eta$$

$$= \sqrt{\frac{vx}{u_s}}\,[\eta_1 - f(\eta_1)]$$

where η_1 is outside the boundary layer. From Table 7-1,

$$[\eta_1 - f(\eta_1)] = 1.72 \text{ (actually closer to 1.73 as } \eta \to \infty)$$

The drag coefficient, from Eq. (7.4), for a plate of width B and length L, is

$$C_f = \frac{B\int_0^L \tau_0\, dx}{(\rho u_s^2/2)(BL)}$$

where

$$\tau_0 = \mu(du/dy)_{y=0} = \mu u_s f''(d\eta/dy)_{y=0} = \mu u_s \sqrt{\frac{u_s}{vx}}\, f''(0)$$

From Howarth's solution, $f''(0) = 0.332$. Then

$$C_f = \frac{2\mu u_s(0.332)}{\rho u_s^2 L}\sqrt{\frac{u_s}{v}}\int_0^L x^{-1/2}\,dx$$

$$C_f = \frac{1.328}{(u_s x/v)^{1/2}} \tag{7.12}$$

7-4. The Momentum Equation for the Boundary Layer

The analytical–numerical solution of the laminar boundary layer along a flat plate with no pressure variation in the outer flow described in the previous section is quite straightforward. Solutions of a like nature for laminar flows have been made for about a dozen flow situations, but the mathematical solutions present difficulties. Simpler approximate methods have been devised which agree well with results for flat plates without pressure gradients and have been extended to other laminar flows such as flat plates and axisymmetric flows with pressure gradients.

One method makes use of the momentum theorem. This will be applied to the flat plate without pressure gradients in the outer flow and compared with the solution of the Prandtl boundary-layer equations given in the preceding section. Then it will be used to obtain results for a turbulent boundary layer for the same flow situation. Results, again, are in very good agreement—not with analytical results, because there are none, but with experiments.

In 1921 von Kármán introduced an approximate and quite accurate method of analyzing a boundary layer, making use of the momentum theorem introduced in Chapter 4. Accuracy of the results depends on the choice of velocity profile in the momentum equation. This determines shear stresses at the boundary surface for laminar flow. An analogy between pipe

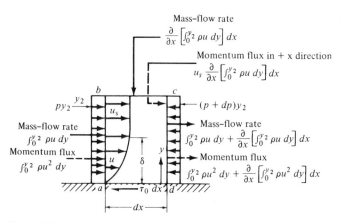

Fig. 7-10.　Infinitesimal length of boundary layer.

flow and external boundary-layer flow suggested by Prandtl makes use of pipe-flow measurements for wall shear and the velocity profile for turbulent flow.

The momentum equation for a boundary layer along a flat surface may be derived by applying the momentum theorem to a region of unit width, of infinitesimal length dx, and for height y_2.

In Fig. 7-10, the mass flow rate and momentum flux into and out of the region $abcd$ are indicated. The mass flow through bc is the difference between that through cd and ab, from continuity. Since streamlines are not exactly parallel to the boundary (Fig. 4-8), some fluid will pass through a line such as bc parallel to the boundary. The x component of velocity of this mass flow is u_s, and the momentum flux is then as shown. The external forces py_2, $(p + dp)y_2$, and $\tau_0\, dx$ are as shown. The pressure is commonly considered to the invariant with y at a given section. Equation (4.3), the momentum equation, becomes

$$\Sigma F_x = \text{net momentum flux out of the region } abcd$$

and thus

$$py_2 - \tau_0\, dx - (p + dp)y_2 = \frac{\partial}{\partial x}\left(\int_0^{y_2} \rho u^2\, dy\right) dx - u_s \frac{\partial}{\partial x}\left(\int_0^{y_2} \rho u\, dy\right) dx$$

which, when simplified, becomes[†]

$$\tau_0 + y_2 \frac{dp}{dx} = \rho \frac{d}{dx} \int_0^{y_2} (u_s - u)u\, dy - \rho \frac{du_s}{dx} \int_0^{y_2} u\, dy \qquad (7.13)$$

The second term on the left-hand side of this equation may be expressed in terms of the free-stream velocity u_s by applying the Bernoulli equation to the region outside the boundary layer. This equation is valid for ideal incompressible fluid flow and is assumed valid even for real fluids beyond the boundary layer:

$$\rho \frac{u_s^2}{2} + p = \text{constant}$$

or, when differentiated with respect to x,

$$\rho u_s \frac{du_s}{dx} + \frac{dp}{dx} = 0$$

so that

$$y_2 \frac{dp}{dx} = -y_2 \rho u_s \frac{du_s}{dx} = -\rho \frac{du_s}{dx} \int_0^{y_2} u_s\, dy$$

[†] Total derivatives are now used since the quantities vary only with x.

Then Eq. (7.13) becomes

$$\tau_0 = \rho \frac{d}{dx} \int_0^\delta (u_s - u)u\,dy + \rho \frac{du_s}{dx} \int_0^\delta (u_s - u)\,dy \qquad (7.14a)$$

The limits of integration have been changed from y_2 to δ, since both integrands vanish beyond δ because of the $(u_s - u)$ term. Equation (7.14a) may be written in terms of the displacement thickness δ^* and the momentum thickness δ_i:

$$\tau_0 = \rho \frac{d}{dx} (u_s^2 \delta_i) + \rho u_s \frac{du_s}{dx} \delta^* \qquad (7.14b)$$

For flow past a flat plate with no pressure gradient (u_s is constant), this becomes

$$\tau_0 = \rho \frac{d}{dx} \int_0^b (u_s - u)u\,dy \qquad (7.15)$$

and this corresponds with the result of Example 4-10, which can be obtained directly and may be written as

$$F = \int_0^x \tau_0\,dx = \rho \int_0^\delta (u_s - u)u\,dy$$

7-5. The Flat Plate in a Uniform Free Stream with No Pressure Gradients: Approximate Momentum Analysis

LAMINAR BOUNDARY LAYER

The boundary-layer thickness, the local shear or local friction coefficient, and the average shear or average friction coefficient over any length of plate may be obtained for any prescribed shape of velocity profile within the boundary layer. The profiles may be assumed to be similar at successive x positions, so that

$$\frac{u}{u_s} = f\left(\frac{y}{\delta}\right)$$

This assumption has been verified experimentally. The function f is a judiciously chosen expression which satisfies sufficient boundary conditions. Three equations for f derived from $u = a + by + cy^2 + dy^3$ are given in the accompanying tabulation, together with the boundary conditions imposed in each instance,

BOUNDARY CONDITIONS			
AT $y = 0$	AT $y = \delta$	EQUATION	
$u = 0$	$u = u_s$	$u/u_s = y/\delta$	(7.16)
$u = 0$	$u = u_s,\ du/dy = 0$	$u/u_s = 2(y/\delta) - (y/\delta)^2$	(7.17)
$u = 0,\ d^2u/dy^2 = 0$	$u = u_s,\ du/dy = 0$	$u/u_s = (3/2)(y/\delta) - (1/2)(y/\delta)^3$	(7.18)

The condition $d^2u/dy^2 = 0$ at $y = 0$ is obtained from Prandtl's simplification of the Navier–Stokes equations relating inertia, pressure, and viscous forces, and implies that since the velocity in the immediate neighborhood of the wall is very small, only shear forces act on the fluid. This assumption in turn requires that the shear stress is constant for all values of y very near the wall and thus du/dy is constant and $d^2u/dy^2 = 0$, when $dp/dx = 0$.

The boundary-layer thickness δ is obtained by equating the value of boundary shear stress in Eq. (7.15) to its definition, $\tau_0 = \mu(du/dy)_{y=0}$. Then,

$$\tau_0 = \rho \frac{d}{dx} \int_0^\delta (u_s - u)u \, dy = \mu \left(\frac{du}{dy} \right)_{y=0}$$

Using the velocity profile given by Eq. (7.18), this is

$$u_s^2 \rho \frac{d}{dx} \int_0^\delta \left(1 - \frac{u}{u_s} \right) \frac{u}{u_s} \, dy = \mu \left(\frac{3}{2} \frac{u_s}{\delta} \right)$$

or

$$\frac{39}{280} \rho u_s^2 \frac{d\delta}{dx} = \frac{3}{2} \mu \frac{u_s}{\delta}$$

A separation of variables gives

$$\delta \, d\delta = \frac{140}{13} \frac{\mu}{\rho u_s} \, dx$$

and integration gives

$$\frac{\delta^2}{2} = \frac{140}{13} \frac{\mu}{\rho u_s} x + C$$

where $C = 0$, since when $x = 0$, $\delta = 0$. In dimensionless form,

$$\frac{\delta}{x} = \frac{4.64}{\sqrt{u_s \rho x / \mu}} = \frac{4.64}{\sqrt{u_s x / \nu}} \tag{7.19}$$

The displacement thickness, from Eq. (7.1), is

$$\delta^* = \int_0^\delta \left(1 - \frac{u}{u_s} \right) dy$$

and for the profile given by Eq. (7.18),

$$\delta^* = 0.375 \, \delta$$

The local skin-friction coefficient c_f is defined as the ratio of the local wall shear stress τ_0 to the dynamic pressure of the free stream:

$$c_f = \frac{\tau_0}{\rho u_s^2 / 2} \tag{7.3}$$

where

$$\tau_0 = \frac{3}{2}\mu\frac{u_s}{\delta} = \frac{0.323\rho u_s^2}{\sqrt{u_s x/v}}$$

so that

$$c_f = \frac{0.646}{\sqrt{u_s x/v}} \tag{7.20}$$

The average skin-friction coefficient over any length of plate x is found by its definition, which is the ratio of the total shear force on the plate to the product of the dynamic pressure of the free stream times the area of the plate (which is x for a unit width),

$$C_f = \frac{\text{drag per unit width}}{(\rho u_s^2/2)x} = \frac{\int_0^x \tau_0\,dx}{(\rho u_s^2/2)x}$$

and when $\tau_0 = 0.323\rho u_s^2/\sqrt{u_s x/v}$ is substituted and the integration performed,

$$C_f = \frac{1.292}{\sqrt{u_s x/v}} \tag{7.21}$$

which is exactly twice the local value at any point x.

Table 7-2. RESULTS OF CALCULATIONS FOR LAMINAR BOUNDARY LAYER

VELOCITY PROFILE	$\dfrac{\delta}{x}$	C_f	$\dfrac{\delta^*}{x}$
Momentum analysis			
$\dfrac{u}{u_s} = \dfrac{y}{\delta}$	$\dfrac{3.46}{\mathrm{Re}_x^{1/2}}$	$\dfrac{1.156}{\mathrm{Re}_x^{1/2}}$	$\dfrac{1.73}{\mathrm{Re}_x^{1/2}}$
$\dfrac{u}{u_s} = 2\left(\dfrac{y}{\delta}\right) - \left(\dfrac{y}{\delta}\right)^2$	$\dfrac{5.48}{\mathrm{Re}_x^{1/2}}$	$\dfrac{1.462}{\mathrm{Re}_x^{1/2}}$	$\dfrac{1.83}{\mathrm{Re}_x^{1/2}}$
$\dfrac{u}{u_s} = \dfrac{3}{2}\left(\dfrac{y}{\delta}\right) - \dfrac{1}{2}\left(\dfrac{y}{\delta}\right)^3$	$\dfrac{4.64}{\mathrm{Re}_x^{1/2}}$	$\dfrac{1.292}{\mathrm{Re}_x^{1/2}}$	$\dfrac{1.74}{\mathrm{Re}_x^{1/2}}$
$\dfrac{u}{u_s} = \sin\dfrac{\pi y}{2\delta}$	$\dfrac{4.80}{\mathrm{Re}_x^{1/2}}$	$\dfrac{1.310}{\mathrm{Re}_x^{1/2}}$	$\dfrac{1.74}{\mathrm{Re}_x^{1/2}}$
Exact solution[a]			
(Blasius)	$\dfrac{4.91}{\mathrm{Re}_x^{1/2}}$	$\dfrac{1.328}{\mathrm{Re}_x^{1/2}}$	$\dfrac{1.73}{\mathrm{Re}_x^{1/2}}$

[a] To be used for engineering calculations.

These results are obtained when the velocity profile given by Eq. (7.18) is used. Results for all three profiles given above [Eqs. (7.16), (7.17), and 7.18)], for a sine-function velocity profile and for the results obtained by solving the Prandtl boundary-layer equations in Sec. 7-3 (called the *Blasius results* since he was the first to solve them) are tabulated in Table 7-2. As mentioned before, and as will be shown in detail in Chapter 8, the ratio $u_s x/\nu$ is a form of the Reynolds number, and is designated as Re_x since it is based on a characteristic length x.

A comparison of Eqs. (7.17), (7.18), and the velocity profile obtained from a solution of the Prandtl boundary-layer equations is shown in Fig. 7-11.

There is remarkable agreement between the Blasius solutions and those based on assumed velocity profiles applied to the momentum equation of the boundary layer.

The qualitative results of Sec. 7-1 concerning the growth of the boundary layer and the decrease in shear along the plate are indicated quantitatively in Table 7-2. This table shows the following:

1. The laminar boundary-layer thickness, however defined, increases as the square root of the distance x from the leading edge and inversely as the square root of the free-stream velocity u_s.
2. The local and average skin-drag coefficients vary inversely with the square root of both x and u_s.
3. The total drag ($F = C_f \rho u_s^2 x/2$ per unit width) varies as the 1.5 power of the free-stream velocity and the square root of the length x.

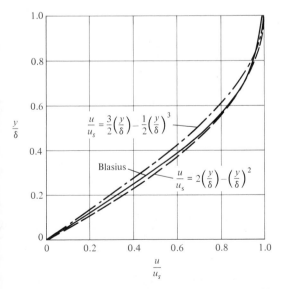

Fig. 7-11.   Laminar boundary layer velocity profiles.

The boundary layer is laminar for values of $u_s x/v$ (the Reynolds number based on x) up to about 300,000–500,000, depending on the plate roughness and level of turbulence in the free stream, as mentioned in Sec. 7-1.

Example 7-1

Standard air flows past a flat plate at a free-stream velocity of 10 m/s. What is the thickness of the boundary layer and the displacement thickness 30 cm from the leading edge?

SOLUTION

The Reynolds number is $\text{Re}_x = (10)(0.3)/1.46 \times 10^{-5} = 2.05 \times 10^5$, and thus the boundary layer may be assumed to be laminar:

$$\frac{\delta}{x} = \frac{4.91}{\text{Re}_x^{1/2}} = \frac{4.91}{453} = 0.0108$$

The boundary layer thickness is then

$$\delta = (0.0108)(300) = 3.2 \text{ mm}$$

and the displacement thickness is $\delta^* = \frac{3}{8}\delta = 1.2$ mm.

TURBULENT BOUNDARY LAYER

The thickness of the turbulent boundary layer may be obtained from Eq. (7.15) in a manner similar to that used to obtain the thickness of the laminar boundary layer. The analysis, however, is not so rigorous as that for the laminar case. A different shape of the velocity profile within the boundary layer will have to be used, of course, and the wall shear will be expressed in a form obtained from measurements on turbulent flow in pipes. We will assume that the boundary layer is turbulent from the leading edge of the plate. This is not strictly correct, but for long plates where the laminar boundary layer exists over a relatively small percentage of the total length, the error in making this assumption is very small. Also, a turbulent boundary layer is often artificially generated in many model tests by mounting a wire near the leading edge of the plate or by roughening the leading portion of the plate surface. This reduces the laminar portion of the boundary layer to a negligibly small region. In instances where the laminar portion of the boundary layer is not negligible, the results for the laminar and the turbulent portions will be combined.

Prandtl suggested that the shape of an external boundary layer velocity profile is like that for turbulent flow in a pipe. The free-stream velocity u_s for the external boundary layer is comparable to the maximum or centerline

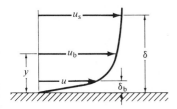

Fig. 7-12. Turbulent boundary layer with laminar or viscous sublayer.

velocity u_m in the pipe, and the boundary-layer thickness is equivalent to the pipe radius. Note the similarity of velocity profiles in Fig. 7-3.

Prandtl suggested that the velocity varies as the seventh root of the distance from the wall for a smooth boundary:

$$\frac{u}{u_s} = \left(\frac{y}{\delta}\right)^{1/7} \tag{7.22}$$

This equation cannot apply at the wall surface, since the velocity gradient at that point is infinite $(du/dy = u_s/7\ \delta^{1/7}y^{6/7})$, and this would indicate an infinite shear, which is not physically possible. The laminar sublayer adjacent to the wall is assumed to have a linear velocity profile, and this profile becomes tangent to the seventh-root profile at the outer edge of the laminar sublayer (at $y = \delta_b$) in Fig. 7-12.

The empirical expression of Blasius for the boundary shear is

$$\tau_0 = 0.0233\ \rho u_s^2 \left(\frac{\nu}{u_s\delta}\right)^{1/4} \tag{7.23}$$

which was obtained from pipe-flow measurements by assuming that if a pipe were unrolled to form a flat surface, the boundary layer thickness δ would correspond to one-half the pipe diameter.

Equation (7.15) becomes

$$0.0233\ \rho u_s^2 \left(\frac{\nu}{u_s\delta}\right)^{1/4} = \rho u_s^2 \frac{d}{dx}\int_0^\delta \left[1 - \left(\frac{y}{\delta}\right)^{1/7}\right]\left(\frac{y}{\delta}\right)^{1/7} dy$$

Integration and separation of variables gives

$$\delta^{1/4}\,d\delta = 0.240\left(\frac{\nu}{u_s}\right)^{1/4} dx$$

so that a completely turbulent boundary layer has a thickness given by

$$\frac{\delta}{x} = \frac{0.382}{(u_s x/\nu)^{1/5}} \tag{7.24}$$

The displacement thickness is $\delta/8$, or

$$\frac{\delta^*}{x} = \frac{0.048}{(u_s x/\nu)^{1/5}} \tag{7.25}$$

and the momentum thickness is $7\delta/72$, or

$$\frac{\delta_i}{x} = \frac{0.037}{(u_s x/v)^{1/5}} \qquad (7.26)$$

As before, the local skin-friction coefficient c_f is

$$c_f = \frac{\tau_0}{\rho u_s^2/2} \qquad (7.27a)$$

where, for the turbulent boundary layer,

$$\tau_0 = 0.0233 \, \rho u_s^2 \left(\frac{v}{u_s \delta}\right)^{1/4} = 0.030 \, \rho u_s^2 \left(\frac{v}{u_s x}\right)^{1/5}$$

and thus

$$c_f = \frac{0.060}{(u_s x/v)^{1/5}} \qquad (7.27b)$$

The average skin-friction coefficient over a finite length of plate x is

$$C_f = \frac{\text{drag per unit width}}{(\rho u_s^2/2)x} = \frac{\int_0^x \tau_0 \, dx}{(\rho u_s^2/2)x}$$

so that for $5 \times 10^5 < u_s x/v < 10^7$,

$$C_f = \frac{0.074}{(u_s x/v)^{1/5}} \qquad (7.28)$$

Below the lower limit of this Reynolds number ($\mathrm{Re}_x = u_s x/v$) the boundary layer normally is laminar, and the shear stress at the boundary expressed by Eq. (7.23) is valid only between the limits given.

For $10^7 < u_s x/v < 10^9$ (Reynolds number between 10^7 and 10^9) an empirical equation obtained by Schlichting [1] agrees quite well with measured data.

Schlichting's equation is

$$C_f = \frac{0.455}{(\log_{10} \mathrm{Re}_x)^{2.58}} \qquad (7.29)$$

Other semiempirical equations relating the skin-friction coefficient C_f with Reynolds number $u_s x/v$ have been given in the literature.

The method of this section for calculations of turbulent boundary layers on smooth flat plates may be used to extend the results to x-Reynolds numbers larger than 10^7 by using additional pipe-flow results. Equation (7.23) was obtained from a straight-line approximation to the smooth pipe curve

Table 7-3. RESULTS OF CALCULATIONS FOR MOMENTUM ANALYSIS OF TURBULENT BOUNDARY LAYER

RANGE OF Re_D	FRICTION FACTOR, f	$\dfrac{u}{u_m}$	$\dfrac{\bar{u}}{u_m}$	WALL SHEAR STRESS, τ_0	C_f	RANGE OF Re_x
$<10^5$	$\dfrac{0.316}{Re_D^{1/4}}$	$\left(\dfrac{y}{R}\right)^{1/7}$	$\dfrac{49}{60}$	$0.0233\rho u_s^2(v/u_s\delta)^{1/4}$	$\dfrac{0.074}{Re_x^{1/5}}$	5×10^5 to 10^7
10^4–10^6	$\dfrac{0.180}{Re_D^{1/5}}$	$\left(\dfrac{y}{R}\right)^{1/8}$	$\dfrac{128}{153}$	$0.0142\rho u_s^2(v/u_s\delta)^{1/5}$	$\dfrac{0.045}{Re_x^{1/6}}$	1.8×10^5 to 4.5×10^7
10^5–10^7	$\dfrac{0.117}{Re_D^{1/6}}$	$\left(\dfrac{y}{R}\right)^{1/10}$	$\dfrac{200}{231}$	$0.0100\rho u_s^2(v/u_s\delta)^{1/6}$	$\dfrac{0.0305}{Re_x^{1/7}}$	2.9×10^7 to 5×10^8

of Fig. 9-6 for Reynolds numbers based on the pipe diameter Re_D less than about 10^5. Blasius used the relation $f = 0.316/Re_D^{1/4}$. Additional relations are listed in Table 7-3. All are straight-line approximations to the slightly curved relation between f and Re_D from Eq. (9.19) and the smooth pipe curve shown in Fig. 9-6. Corresponding shapes of pipe-flow velocity profiles according to Schlichting [1] are listed in the third column of Table 7-3, and calculated results are given in the remaining columns.

There is excellent agreement between these listed drag coefficients and experimental results. This illustrates the usefulness of the momentum analysis and of the original suggestion of Prandtl that an external boundary layer has much in common with turbulent flow in a pipe.

It is seen that for a turbulent boundary layer:

1. The boundary-layer thickness increases as the $\frac{4}{5}$ power of the distance from the leading edge, as compared with $x^{1/2}$ for a laminar boundary layer.
2. The local and average skin-friction coefficients vary inversely as the fifth root of both x and u_s, as compared with the square root for a laminar boundary layer.
3. The total drag varies as the $\frac{9}{5}$ power of the free-stream velocity ($u_s^{9/5}$) and the $\frac{4}{5}$ power of the plate length ($x^{4/5}$), as compared with $u_s^{1.5}$ and $x^{0.5}$ for a laminar boundary layer.

Measurements of the turbulent boundary layer velocity profiles on smooth flat plates and in smooth circular tubes show that the profile consists of three regions: a viscous sublayer next to the boundary surface, a completely turbulent zone farther away, and an intermediate buffer region connecting them. The profile is generally plotted on semilogarithmic scales with a dimensionless linear ordinate $u^+ = u/u_*$ and a dimensionless logarithmic abscissa $y^+ = yu_*/v$, where u_* is defined as a shear velocity since it has dimensions of a velocity, $u_* = \sqrt{\tau_0/\rho}$. The relationship between u^+ and y^+ is known as the law of the wall. It is a universal plot for smooth surfaces and is shown in Fig. 7-13.

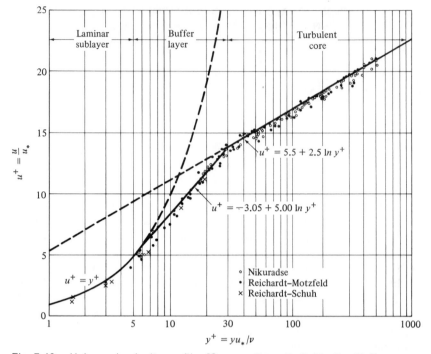

Fig. 7-13.   Universal velocity profile. [Source: From R. C. Martinelli, *Trans. ASME*, Vol. 69 (1947), pp. 947–959. Used with permission of the publishers, The American Society of Mechanical Engineers.]

THE VISCOUS SUBLAYER

The thickness of the viscous sublayer for smooth surfaces corresponds to the distance from the wall where $y^+ = 5$. Thus $5 = u_* \delta_b / \nu$ or $\delta_b = 5\nu/u_*$. The ratio of this viscous sublayer thickness to the boundary layer thickness given by Eq. (7.24) is, when Eq. (7.23) is used to express the wall shear stress τ_0,

$$\frac{\delta_b}{\delta} = \frac{80}{(\mathrm{Re}_x)^{7/10}} \tag{7.30}$$

At $\mathrm{Re}_x = 5 \times 10^5$, $\delta_b/\delta = 0.0082$ and when $\mathrm{Re}_x = 10^7$, $\delta_b/\delta = 0.0010$ which indicates that the viscous sublayer is very thin.

BOTH LAMINAR AND TURBULENT BOUNDARY LAYERS

A boundary layer is generally laminar to begin with and changes to a turbulent condition at some downstream point. The total drag, as well as the drag coefficient, may be calculated by adding the drag owing to the laminar boundary layer to the drag owing to the turbulent portion of the boundary layer, assuming that the turbulent portion beyond the transition point acts as though it were turbulent from the leading edge of the plate.

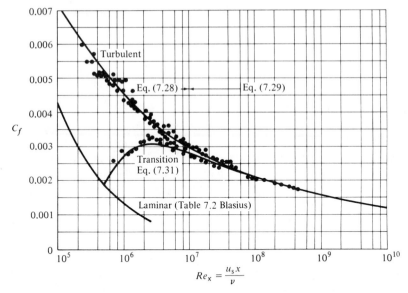

Fig. 7-14.   Drag coefficients for a smooth flat plate. Measurements by C. Wieselsberger, F. Gebers, W. Froude, G. Kempf, and K. E. Schoenherr.

The Prandtl–Schlichting skin-friction equation for a smooth plate is

$$C_f = \frac{0.455}{(\log_{10} Re_x)^{2.58}} - \frac{A}{Re_x} \qquad (7.31)$$

where the value A depends on the critical value of the Reynolds number $(u_s x_c/\nu)$ at which the laminar boundary layer becomes turbulent. Values of A for various values of $u_s x_c/\nu$ are

$u_s x_c/\nu$	10^5	5×10^5	10^6
A	360	1700	3300

Equations (7.28), (7.29), (7.31), and the results in Table 7-3, and also the Blasius equation from Table 7-2 are shown in Fig. 7-14. Measurements by a number of experimenters are in good agreement with the curves shown.

For a laminar boundary layer, the roughness of the plate surface does not affect the aforementioned results, although the boundary layer will become turbulent more readily. For a turbulent boundary layer, the results apply for smooth plates and for plates whose roughness elements are such that the plate behaves as a smooth plate. Schlichting gives an expression for the maximum height of roughness elements in order that a surface may be still considered hydrodynamically smooth. This is

$$k_{\text{adm}} \le 100 \frac{\nu}{u_s} \qquad (7.32)$$

In all instances of flow along flat plates parallel to the flow, the drag force D is

$$D = C_f \left(\frac{\rho u_s^2}{2} \right) A \qquad (7.33)$$

where C_f is the skin-friction drag coefficient, ρ is the fluid density, u_s is the free-stream velocity, and A is the area of the plate subjected to the shearing action of the fluid. The skin-friction drag for many surfaces may be approximated by calculating the skin-friction drag on an equivalent flat plate.

The curves of Fig. 7-14 generally are used to obtain drag coefficients in making numerical calculations; equations from which the curves are plotted need not be used.

PROBLEMS

7-1. Derive Prandtl's equation for a laminar boundary layer by applying the momentum theorem to an element of incompressible fluid of length dx and height dy in the $x-y$ plane in a direction parallel to a flat surface. With a pressure gradient, this is

$$u \frac{\partial u}{\partial x} + v \frac{\partial u}{\partial y} = -\frac{1}{\rho} \frac{dp}{dx} + v \frac{\partial^2 u}{\partial y^2}$$

Forces on fluid element

Momentum flux into and out of fluid element

Prob. 7-1

7-2. Show from the Prandtl boundary-layer equations that at a boundary surface, the velocity profile for a laminar boundary layer has an infinite radius of curvature. That is, at $y = 0$, $\partial^2 u / \partial y^2 = 0$.

7-3. Carry out in detail the steps leading from Eq. (7.13) to Eq. (7.14).

7-4. The shape factor for a boundary layer is defined as the ratio of the displacement thickness to the momentum thickness. It is 2.59 for the Blasius profile. Compare this with the shape factor for
  (a) the laminar boundary layer velocity profile of Eq. (7.18),
  (b) the sine function in Table 7-2, and
  (c) the turbulent boundary layer velocity profile of Eq. (7.22).

7-5. Assume that the shear stress varies linearly in a laminar boundary layer such that $\tau = \tau_0(1 - y/\delta)$. Calculate the displacement and the momentum thickness in terms of δ.

7-6. In Prob. 7-5, calculate the displacement and the momentum thickness in terms of Re_x.

7-7. In Prob. 7-5, calculate the average drag coefficient in terms of Re_x.

7-8. The free-stream pressure in a two-dimensional converging duct decreases linearly such that $p = p_1 - Cx$. What is the form of the boundary layer equation [Eq. (7.14)] for this situation?

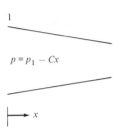

$p = p_1 - Cx$

Prob. 7-8

7-9. Assume $u/u_s = (y/\delta)^{1/8}$ in a turbulent boundary layer. Calculate the boundary layer thickness δ/x and the drag coefficient C_f using Eq. (7.23) for the wall shear stress. Compare results with Eqs. (7.24) and (7.28).

7-10. How does the ratio of displacement thickness to boundary layer thickness (δ^*/δ) for a laminar boundary layer compare with that for a turbulent boundary layer? Determine a qualitative answer from an inspection of the velocity profiles within the boundary layers, and then check the ratio of Eq. (7.25) to Eq. (7.24) with similar ratios obtained from Table 7-2.

7-11. From the definition of displacement thickness and momentum thickness of a boundary layer [Eqs. (7.1) and (7.2)] and from the turbulent velocity profile of Eq. (7.22), derive Eqs. (7.25) and (7.26).

7-12. Carry out in detail the steps leading to Eq. (7.30), which gives the relative thickness of the laminar sublayer.

7-13. Estimate the ratio of the turbulent boundary-layer thickness just after transition at $Re_x = 5 \times 10^5$ to the laminar boundary layer thickness just before transition.

7-14. The velocity profile in a laminar boundary layer is given by

$$\frac{u}{u_s} = \frac{5}{3}\left(\frac{y}{\delta}\right) - \frac{2}{3}\left(\frac{y}{\delta}\right)^3$$

What is the boundary-layer thickness δ/x in terms of $u_s x/\nu$?

7-15. Water at 20°C flows along a smooth flat plate at a free-stream velocity $u_s = 6.0$ m/s. Estimate the water velocity 1.0 cm from the plate surface (a) 3 m and

(b) 9 m from the leading edge of the plate.

Assume a turbulent velocity profile given by Eq. (7.22). *Hint*: First calculate the boundary layer thickness at each location.

7-16. Repeat Prob. 7-15 for water at $20°C$, a free-stream velocity $u_s = 10.0 \, \text{m/s}$ at a distance 1.5 cm from the plate at distances

(a) 5 m and

(b) 15 m from the leading edge of the plate.

7-17. Standard air (kinematic viscosity is $1.46 \times 10^{-5} \, \text{m}^2/\text{s}$) flows along a flat plate at a velocity $u_s = 7.5 \, \text{m/s}$. Estimate the boundary layer thickness 15 cm from the leading edge.

7-18. Repeat Prob. 7-17 for $u_s = 10 \, \text{m/s}$ and 30 cm from the leading edge.

7-19. A smooth flat plate is in a parallel flow stream. What is the ratio of the drag over the upstream half of the plate to that over the entire plate

(a) for a laminar boundary layer over its entire length and

(b) for a turbulent boundary layer over its entire length?

7-20. A smooth flat rectangular plate is four times as long as it is wide and is held parallel to a uniform flow stream. For (a) a completely laminar boundary layer and (b) for a completely turbulent boundary layer, is the total drag greater when the short side is parallel to the flow direction or when the long side is parallel to the flow direction? Get a qualitative answer from physical reasoning, then calculate the ratio numerically.

7-21. In Prob. 7-19, over what percentage of the plate length will the drag be one-half the total drag for parts (a) and (b)?

7-22. (a) Estimate the skin-friction drag on an airship hull with a surface shape equivalent to a cylinder 12 m in diameter and 90 m long, at an air speed of 25 m/s in standard air. Kinematic viscosity is $1.46 \times 10^{-5} \, \text{m}^2/\text{s}$.

(b) What power is needed to overcome drag?

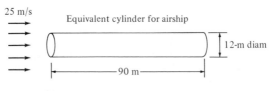

25 m/s

Equivalent cylinder for airship

12-m diam

90 m

Prob. 7-22

7-23. A 6-m smooth wax model of a cargo ship is towed through fresh water at $15°C$ at a speed of 5 knots (2.58 m/s). The wetted hull area is $6.50 \, \text{m}^2$. What is the skin friction drag, assuming a completely turbulent boundary layer equivalent to that on a smooth flat plate?

7-24. (a) Estimate the skin-friction drag on an airship 15 m in diameter with an effective length of 110 m at an air speed of 25 m/s in standard air.

(b) What power would be required to propel the craft?

7-25. A 6.0-m, smooth wax model of a ship is towed through fresh water at 15°C at a speed of 2.60 m/s. The wetted hull area is 6.40 m². What is the skin-friction drag for a completely turbulent boundary layer equivalent to that on a flat plate?

7-26. A ship is 155 m long, has a wetted hull area of 4460 m² and travels at 10.3 m/s in sea water at 20°C.
(a) Estimate the skin-friction drag on the hull, assuming it to be equivalent to a smooth flat plate.
(b) Plating roughness, antifouling paint roughness, and natural fouling roughness add 0.00026 to the drag coefficient; and hull curvature effects add another 0.00010. Estimate the frictional drag of the ship hull when all these are considered. See Fig. 11-1 for a comparison of smooth flat plates with a typical clean ship hull.

7-27. Repeat Prob. 7-26 for a ship speed of 6 m/s.

7-28. The inlet ducting to a fan test setup is 60 cm in diameter and 3.0 m long. Standard air (kinematic viscosity is 1.46×10^{-5} m²/s) enters the duct through a well-rounded entrance, and thus the entrance velocity profile may be assumed flat, with no boundary layer, at 1.25 m/s.
(a) What is the boundary layer thickness 3 m from the entrance?
(b) What is the displacement thickness 3 m from the entrance?
(c) What is the air velocity over the central core of the duct 3 m from the entrance?

7-29. Repeat Prob. 7-28 for an entrance velocity of 2.5 m/s.

7-30. Water at 20°C and at an axial velocity of 15 m/s enters the 30.00-cm-diameter test section of a cavitation-testing water tunnel with a boundary layer thickness equivalent to that from a starting point 45 cm upstream. The velocity profile is flat over the entire cross section except for the boundary layer. Estimate the increase in axial velocity at the end of a 60-cm test section length due to the growth of the boundary layer. Assume flat plate theory; it is valid here.

7-31. Water enters a round pipe with an essentially flat velocity profile. The boundary layer thickens in the direction of flow (Fig. 7-8), which results in an acceleration of the fluid in the core with an accompanying pressure drop given by the Bernoulli equation. If the walls are made slightly diverging to accommodate the boundary layer growth, the

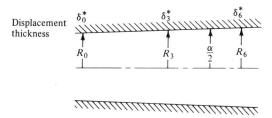

Prob. 7-31

axial pressure may be essentially constant. Assume water at 20°C enters a 90.0-cm-diameter smooth duct at $u_s = 18.0$ m/s with a turbulent boundary layer equivalent to that 22 cm from the leading edge of a flat plate.

(a) Estimate the duct radius at 90 and 180 cm downstream from the duct entrance for a constant pressure core flow, assuming flat plate boundary-layer growth. The boundary must flare out to accommodate the displacement thickness of the boundary layer.

(b) Why is the Bernoulli equation valid in the core flow?

7-32. In order to straighten the flow and reduce the scale of turbulence in a water tunnel, an egg crate type of flow straightener is formed from thin plates. A number of 10-cm square ducts 60 cm long result. For a flow velocity of 2.4 m/s approaching the straightener and a water temperature of 20°C, calculate

(a) the displacement thickness of the boundary layer at the downstream end of the straightener (assume no pressure gradient, although a very small one exists);

(b) the velocity in the core flow at the downstream end of the flow straightener (the velocity increases because of the retardation of flow within the boundary layer);

(c) the pressure drop through the straightener by applying the Bernoulli equation to the core flow; and

(d) the pressure drop through the straightener, by determining the drag per duct and writing the momentum equation between entrance and exit of the duct.

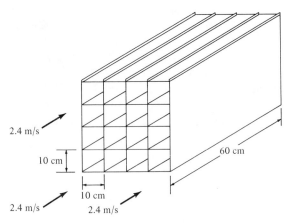

2.4 m/s

10 cm

60 cm

10 cm

2.4 m/s   2.4 m/s

Prob. 7-32

7-33. An incompressible fluid flows past one side of a porous flat surface at a free-stream velocity of u_s and at constant pressure. Fluid is drawn from the boundary layer through the porous surface at a uniform velocity of $v \ll u_s$. The boundary-layer velocity profile is given by

$$\frac{u}{u_s} = f\left(\frac{y}{\delta}\right) = f(\eta)$$

Apply the momentum theorem and show that the local skin-friction coefficient may be expressed as

$$c_f = \frac{\tau_0}{\rho u_s^2/2} = 2\frac{d\delta}{dx}\int_0^1 f(\eta)[1 - f(\eta)]\, d\eta + \frac{2v}{u_s}$$

Refer to Example 4-10. In this problem the fluid removed at the porous surface has an initial velocity of u_s in the $+x$ direction, but after removal from the flow its velocity in this direction is zero.

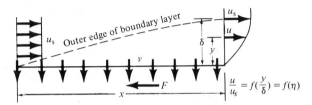

Prob. 7-33

7-34. How would the results of Prob. 7-33 be affected if fluid is injected into the boundary layer uniformly instead of being removed?

7-35. By integrating over a circular area, verify the values of $\bar{u}/u_m$ in Table 7-3.

7-36. Verify the expressions for the wall shear stresses in Table 7-3 from the given expressions for the friction factor f and the values of $\bar{u}/u_m$.

7-37. Verify the expressions for the drag coefficients C_f in Table 7-3.

References

1. H. Schlichting, *Boundary Layer Theory* (translated by J. Kestin), 6th ed. (New York: McGraw-Hill Book Company, Inc., 1968).
2. H. Schlichting, "Boundary-Layer Theory," Sec. 9 of *Handbook of Fluid Dynamics*, edited by V. L. Streeter (New York: McGraw-Hill Book Company, Inc., 1961).
3. H. Schlichting, "Three-Dimensional Boundary Layer Flow," *Proc. Ninth Convention, International Association for Hydraulic Research, Dubrovnic, Yugoslavia, 1961*, pp. 1262–1290.
4. L. Prandtl, *Essentials of Fluid Dynamics* [New York: Macmillan, Inc. (Hafner Press), 1952].
5. L. F. Richardson, *Weather Prediction by Numerical Process* (New York: Cambridge University Press, 1922), p. 66.
6. J. Boussinesq, "Théorie de l'écoulement tourbillant," *Mém. Prés. Acad. Sci., Paris*, Vol. 23 (1877), p. 46.
7. H. L. Dryden, "Transition from Laminar to Turbulent Flow," Sec. A of *Turbulent Flows and Heat Transfer*, edited by C. C. Lin (Princeton, N.J.: Princeton University Press, 1959).

8. J. Piercy and N. A. V. Preston, "A Simple Solution of the Flat Plate Problem of Skin Friction and Heat Transfer," *Phil. Mag.*, Vol. 21 (1936), pp. 995–1005.

9. L. Howarth, "On the Solution of the Laminar Boundary Layer Equations," *Proc. Roy. Soc., Ser. A*, Vol. 164 (1938), p. 547. See also Ref. 1 above.

See also:

A. S. Monin and A. M. Yaglum, *Statistical Fluid Mechanics: Mechanics of Turbulence*, edited by J. L. Lumley (Cambridge, Mass.; The MIT Press, 1971).

O. G. Sutton, *Atmospheric Turbulence* (London: Methuen and Co. Ltd., 1955 and New York: John Wiley and Sons, Inc., 1955).

O. G. Sutton, *The Challenge of the Atmosphere* (New York: Harper & Row, Publishers, 1961).

H. Tennekes and J. L. Lumley, *A First Course in Turbulence* (Cambridge, Mass.: The MIT Press, 1972).

P. K. Chang, *Separation of Flow* (Elmsford, N.Y.: Pergamon Press, Inc., 1970).

Part IV
APPLICATIONS

The remaining chapters all contain items not previously discussed, although they all apply the concepts already presented.

Included are the basics of model studies, viscous flow in ducts, compressible gas flows, dynamic drag and lift, open-channel flow, flow measurements, and principles of turbomachinery.

The final two chapters show similarities between open-channel flow and gas flow, and also present a short introduction to thermal boundary layers superimposed on hydrodynamic boundary layers.

Chapter 8
Dimensionless Numbers and Dynamic Similarity

In fluid mechanics, as well as in other physical systems, experimental data are often presented in dimensionless form. These dimensionless numbers may be obtained in various ways. For example, a Reynolds number may be obtained by a formal dimensional analysis of a system in which viscosity plays a predominate role by taking the ratio of the inertia to viscous forces in a flow system or by writing the differential equations of motion (including viscous terms) in dimensionless form. Dynamic similarity between two or more geometrically similar flow systems is said to exist when the appropriate dimensionless numbers are the same for the systems.

8-1. Introduction

In physical systems, certain quantities, such as mass, length, and time (or force, length, and time), are considered to be fundamental quantities since they cannot be expressed in simpler terms. All other physical quantities may be expressed in terms of these fundamental quantities. Velocity is a length divided by time, dynamic viscosity is a mass divided by length and time, and density is a mass per unit volume (length cubed). The fundamental quantities are said to have dimensions. We measure each fundamental *dimension* in various systems of *units*, such as kilograms and slugs for mass, meters and feet for length, and seconds and years for time.

In any physical system in which two or more quantities are interrelated, it is often convenient to set up dimensionless quantities which in turn are interrelated. The number of dimensionless quantities is always less than the total number of physical quantities, and this fact enables us to correlate experimental data more easily. We may say, for example, that the friction factor for the flow of any incompressible fluid in a smooth pipe depends on the Reynolds number. Such a statement is much simpler than saying that the pressure drop per unit length of pipe for incompressible flow in a smooth pipe depends on the viscosity, density, and average velocity of the fluid and on the pipe diameter. The actual relationship between the dimensionless numbers usually has to be determined from experiment.

We are all familiar with the Mach number, since it appears frequently in the daily press. The Mach number is a dimensionless number, being the ratio of the flow, or flight, velocity to the speed of sound; the inertial to the elastic, or compression, forces in the fluid system; or the kinetic energy of the mean flow to the mean kinetic energy of the gas molecules. Many other dimensionless numbers, though not all, are named in honor of individuals who have done pioneering work in the field associated with the number. Among the dimensionless numbers in the field of fluid mechanics are, in addition to the Mach number, (a) the Reynolds number, Re, in connection with effects of viscosity; (b) the Froude number, Fr, in connection with gravity effects; (c) the Weber number, We, in connection with surface tension effects; (d) the Knudsen number, Kn, in conjunction with slip flow in rarefied gases; (e) the pressure coefficient, C_p, indicative of pressure variations (the cavitation number or index σ is a special form of a pressure coefficient used in conjunction with cavitation phenomena); and (f) the drag and lift coefficients, C_D and C_L, in connection with drag and lift.

In heat transfer, often closely related with fluid mechanics, many other dimensionless numbers are used, noteworthy ones being named after Nusselt, Prandtl, Stanton, Peclet, Schmidt, Eckert, Graetz, and Grashof. This list is by no means complete.

The main advantage of using dimensionless parameters is that similarity of two or more flows generally results when the flows are geometrically similar and when the governing dimensionless parameters for the flows are the same for the two or more flows. Hence, results of studies on one flow can be applied to similar flows. We have already seen that the skin-friction drag coefficients in Fig. 7-14 are valid for all incompressible flows along smooth flat plates in a uniform flow field. It is not important what the fluid is, what the flow velocity is, or how long the plate is. So long as the Reynolds number, $u_s x/v$, is between 10^5 and 10^9, Fig. 7-14 may be used. All flows at $Re_x = 1.5 \times 10^8$, for example, will be similar, and they all have a drag coefficient of $C_f = 0.0020$.

It may be stated that without dimensionless numbers, experimental progress in fluid mechanics (heat transfer as well) would have been almost nil; it would have been swamped by masses of accumulated data.

8-2. Dimensional Analysis of Fluid Systems

The dimensions of a number of quantities used in fluid mechanics are listed in Table 8-1. The MLT and the FLT systems of dimensions are related by the expressions

$$F = Ma = \frac{ML}{T^2} \quad \text{and} \quad M = \frac{FT^2}{L}$$

Table 8-1 DIMENSIONS OF VARIOUS QUANTITIES

	FLT	MLT
Geometrical characteristics		
Length (diameter, height, breadth, chord, span, etc.)	L	L
Angle	None	None
Area	L^2	L^2
Volume	L^3	L^3
Fluid properties[a]		
Mass	FT^2/L	M
Density (ρ)	FT^2/L^4	M/L^3
Specific weight (γ)	F/L^3	M/L^2T^2
Kinematic viscosity (ν)	L^2/T	L^2/T
Dynamic viscosity (μ)	FT/L^2	M/LT
Elastic modulus (K)	F/L^2	M/LT^2
Surface tension (σ)	F/L	M/T^2
Flow characteristics		
Velocity (V)	L/T	L/T
Angular velocity (ω)	$1/T$	$1/T$
Acceleration (a)	L/T^2	L/T^2
Pressure (Δp)	F/L^2	M/LT^2
Force (drag, lift, shear)	F	ML/T^2
Shear stress (τ)	F/L^2	M/LT^2
Pressure gradient ($\Delta p/L$)	F/L^3	M/L^2T^2
Flow rate (Q)	L^3/T	L^3/T
Mass flow rate ($\dot{m}$)	FT/L	M/T
Work or energy	FL	ML^2/T^2
Work or energy per unit weight	L	L
Torque and moment	FL	ML^2/T^2
Work or energy per unit mass	L^2/T^2	L^2/T^2

[a] Density, viscosity, elastic modulus, and surface tension depend upon temperature, and therefore temperature will not be considered a property in the sense used here.

Either system of dimensions may be used (exclusively) in a dimensional analysis.

We are interested in obtaining the most significant and independent dimensionless groups, or parameters, for the particular physical system being analyzed. Actually, the resulting groups will indicate only how to organize a set of experiments and how to plot the resulting experimental data. We cannot determine how one dimensionless variable will vary with another except by experiment (in a few instances of viscous flow this variation may be calculated analytically).

The so-called π (pi) theorem—commonly called the Buckingham π theorem, but first stated by Vaschy and proved in increasing generality by Buckingham, Riabouchinsky, and Martinot-Lagarge, and by Birkhoff[†]—will be explained by means of an example.

Suppose we want to analyze the flow of an incompressible fluid in a round pipe. We judge that the pressure drop per unit length of pipe (this is the pressure gradient $\Delta p/L$) depends on the pipe diameter D, the pipe roughness k (the effective height of the roughness elements), the average flow velocity V, the fluid density ρ, and the fluid viscosity μ. This may be written as

$$\frac{\Delta p}{L} = f(D,k,V,\rho,\mu)$$

and in terms of dimensions in the ELT system as

$$\frac{F}{L^3} = f\left[(L), (L), \left(\frac{L}{T}\right), \left(\frac{FT^2}{L^4}\right), \left(\frac{FT}{L^2}\right) \right]$$

There are $n = 6$ quantities involved and $m = 3$ fundamental dimensions (F, L, and T). The π theorem states that there will be at least one set of $n - m$ (equal to 3 here) independent dimensionless groups in a dimensional analysis. The total number of dimensionless groups is

$$\frac{n!}{(m + 1)!(n - m - 1)!}$$

which is 15 in this example. The independent groups will be designated as $\pi_1, \pi_2, \ldots$, and π_{n-m}.

Each of the three groups ($n - m = 3$) will consist of m (equal to 3 in this example) quantities in common, and these are called *repeating variables*. Three general rules guide the selection of repeating variables:

    a. In a dimensional analysis of any physical system the repeating variables must include among them all of the m fundamental dimensions.

[†] G. Birkhoff, *Hydrodynamics* (Princeton, N.J.: Princeton University Press, 1950).

b. For a fluid system, the most significant groups will result if the repeating variables are chosen so that one is a geometrical characteristic, one is a fluid property, and one is flow characteristic (Table 8-1).

c. The dependent variable should not be used as a repeating variable.

After the repeating variables are chosen, each one of the remaining original quantities is included with each one of the π groups. Let D be the geometrical characteristic, ρ the fluid property, and V the flow characteristic ($\Delta p/L$ also results from flow, but a dependent variable should not be used as a repeating variable). Then

π_1 will contain D, ρ, V, and $\Delta p/L$;
π_2 will contain D, ρ, V, and k; and
π_3 *will contain* D, ρ, V, *and* μ.

In order that these πs be dimensionless, *any one* quantity in each may appear to the first power and the others will appear to some unknown power which can be found. Thus let

$$\pi_1 = D^x \rho^y V^z \frac{\Delta p}{L}$$

$$= (L)^x \left(\frac{FT^2}{L^4}\right)^y \left(\frac{L}{T}\right)^z \frac{F}{L^3} = L^0 F^0 T^0$$

in order to be dimensionless, The values of x, y, and z may be obtained by equating exponents of F, L, and T to zero. For this example, these are

$$y + 1 = 0 \quad \text{for} \quad F$$

$$x - 4y + z - 3 = 0 \quad \text{for} \quad L$$

$$2y - z = 0 \quad \text{for} \quad T$$

from which $x = 1$, $y = -1$, and $z = -2$. Thus,

$$\pi_1 = \frac{\Delta p}{L} \frac{D}{\rho V^2}$$

which is commonly known as one-half the friction factor f for a pipe. Similarly, $\pi_2 = k/D$, which is known as the relative roughness of a pipe, and $\pi_3 = \mu/VD\rho$, which is the reciprocal of the Reynolds number for a pipe. Note that the reciprocal, square, square root, and so forth, or a constant times a dimensionless number, are also dimensionless, and the useful form of the result is determined largely by convention or experience. The result of the example may be expressed as

$$\pi_1 = f(\pi_2, \pi_3)$$

or that the friction factor for pipe flow depends on the relative roughness of the pipe and on the Reynolds number of the flow. The actual relationship is determined experimentally (see Figs. 9-4 and 9-6). A different choice of repeating variables (μ and ρ are both fluid properties, and D and k are both geometrical characteristics) would result in a number of different πs, some of the total of 15 possible. The dimensional analysis is equally as valid for any three independent dimensionless groups, although experience in this instance shows that the most significant ones are those just obtained.

As long as the six chosen quantities are the only ones included in a study of fluid flow in pipes, the experimental results will apply to *any* incompressible fluid flowing at *any* velocity in *any* round straight pipe. High velocity gas flow, for example, cannot be correlated with the resulting πs because the compressibility of the fluid, indicated by the elastic modulus K, was omitted. The dimensionless numbers are independent of the system of units used in making the experiments, so the results are applicable to the foot-pound-second system, the centimeter-gram-second system, the kilogram-meter-second system, or any other system. The advantages of using dimensionless parameters are clearly apparent. We have succeeded in expressing six related quantities in terms of three; this enables us to plot the friction factor as a function of the Reynolds number for a family of curves, each representing a given relative roughness, usable in any consistent system of units.

The success of a dimensional analysis depends on selecting all the pertinent variables involved in any flow situation. If any are omitted, the results will be incomplete. If variables are introduced which are not important, later analysis or experiment will indicate that the dependent variable is not affected by the dimensionless group which contains the extra variable.

8-3. Force Ratios

A second method of obtaining dimensionless parameters is to take the ratio of significant forces that exist in fluid flows.

Among the forces encountered in flowing fluids are those due to inertia, viscosity, gravity, pressure, surface tension, and compressibility. These may be expressed in a semidimensional way in terms of length L, velocity V, density ρ, viscosity μ, gravity g, pressure change Δp, surface tension σ, and compressibility K, rather than simply a force F. Quantities in brackets are retained in order that the type of force may be recognized:

$$\text{Inertia force} = Ma = \frac{\rho L^3 V^2}{L} = [\rho]L^2 V^2$$

$$\text{Viscous force} = \tau A = \mu\left(\frac{du}{dy}\right)L^2 = \mu\left(\frac{V}{L}\right)L^2 = [\mu]VL$$

$$\text{Gravity force} = Mg = \rho L^3[g]$$

$$\text{Pressure force} = (\Delta p)A = [\Delta p]L^2$$

$$\text{Surface tension force} = [\sigma]L$$

$$\text{Compressibility force} = [K]L^2$$

Inertia forces are usually important in any fluid in motion. The ratio of the inertia force to each of the others is then expressed as

$$\frac{\text{inertia force}}{\text{viscous force}} = \frac{\rho L^2 V^2}{\mu V L} = \frac{\rho L V}{\mu}, \quad \text{the Reynolds number Re}$$

$$\frac{\text{inertia force}}{\text{gravity force}} = \frac{\rho L^2 V^2}{\rho L^3 g} = \frac{V^2}{Lg} \text{ or } \frac{V}{\sqrt{Lg}}, \quad \text{the Froude number Fr}$$

$$\frac{\text{pressure force}}{\text{inertia force}} = \frac{\Delta p L^2}{\rho L^2 V^2} = \frac{\Delta p}{\rho V^2} \text{ or } \frac{\Delta p}{\rho V^2/2}, \quad \text{the pressure coefficient } C_p$$

$$\frac{\text{inertia force}}{\text{surface tension force}} = \frac{\rho L^2 V^2}{\sigma L} = \frac{V^2}{\sigma/\rho L} \text{ or } \frac{V}{\sqrt{\sigma/\rho L}}, \quad \begin{array}{l}\text{the Weber}\\\text{number We}\end{array}$$

$$\frac{\text{inertia force}}{\text{compressibility force}} = \frac{\rho L^2 V^2}{K L^2} = \frac{V^2}{K/\rho} \text{ or } \frac{V}{\sqrt{K/\rho}}, \quad \text{the Mach number M}$$

8-4. Normalized Equations of Motion and Energy

A third method of deriving dimensionless parameters is to normalize the differential equations of motion and energy. The Reynolds and Froude numbers may be obtained from the equations of motion, and the Reynolds, Mach, and Prandtl numbers from the energy equation.

It is sufficient to treat only one of the scalar Navier–Stokes equations [Eq. (4.21)] in order to obtain results. If the body force X is due to gravity alone, it may be expressed as

$$X = g_x = g\left(\frac{\partial h}{\partial x}\right)$$

where h is positive upward in the gravitational field. The equations are made dimensionless by dividing each length by a characteristic reference length L and each velocity by a characteristic reference velocity U. The various terms in Eq. (4.21) are replaced by the following:

$$x' = x/L, \qquad y' = y/L, \qquad z' = z/L, \qquad t' = t/(L/U),$$
$$h' = h/L, \qquad u' = u/U, \qquad v' = v/U, \qquad w' = w/U,$$
$$p' = p/\rho U^2$$

Substitution of these into the x component Navier–Stokes equation gives, after division by $\rho U^2/L$,

$$u' \frac{\partial u'}{\partial x'} + v' \frac{\partial u'}{\partial y'} + w' \frac{\partial u'}{\partial z'} + \frac{\partial u'}{\partial t'}$$

$$= \frac{gL}{U^2} - \frac{\partial p'}{\partial x'} + \frac{\mu}{\rho UL} \left(\frac{\partial^2 u'}{\partial x'^2} + \frac{\partial^2 u'}{\partial y'^2} + \frac{\partial^2 u'}{\partial z'^2} \right)$$

$$= \frac{1}{Fr^2} - \frac{\partial p'}{\partial x'} + \frac{1}{Re} \left(\frac{\partial^2 u'}{\partial x'^2} + \frac{\partial^2 u'}{\partial y'^2} + \frac{\partial^2 u'}{\partial z'^2} \right)$$

This has significant physical meaning. It indicates, for example, that if viscous effects govern the flow and if gravity effects are negligible, two geometrically similar flow fields will be kinematically similar and dynamic effects will be similar if the Reynolds number is the same in both flows. Also, if gravity effects predominate and viscous effects are negligible, two flow fields as well as dynamic effects will be similar if the Froude number is the same for both. Thus, for incompressible flow, the normalized equations of motion show that for similarity of flow in geometrically similar situations the Reynolds number and the Froude number are governing parameters.

For compressible flows, the approximate form of the energy equation for boundary layers (not used in this text, but see Schlichting [1], for example) is normalized by the additional parameters $T' = T/T_{ref}$ and $\rho' = \rho/\rho_{ref}$. For only moderate temperature variations, the viscosity μ and thermal conductivity k may be taken as constant. Then the energy equation

$$\rho c_p \frac{\partial T}{\partial t} = u \frac{\partial p}{\partial x} + k \frac{\partial^2 T}{\partial y^2} + \mu \left(\frac{\partial u}{\partial y} \right)^2$$

becomes, after dividing through by $UT'c_p/L$,

$$\rho' \frac{\partial T'}{\partial t'} = \left(\frac{c_p}{c_v - 1} \right) M^2 u' \frac{\partial p'}{\partial x'} + \frac{1}{Re\ Pr} \frac{\partial^2 T'}{\partial y'^2}$$
$$+ \frac{[(c_p/c_v) - 1]M^2}{Re} \left(\frac{\partial u'}{\partial y'} \right)^2$$

where $M^2 = U^2/kRT$ is the square of the Mach number, $Pr = c_p\mu/k$ is the Prandtl number (a fluid property), and $Re = \rho'UL/\mu$. Thus similarity of compressible viscous flows implies the same values of Mach, Prandtl, and Reynolds numbers.

8-5. Dynamic Similitude

Flow systems are considered to be dynamically similar (a) if they are geometrically similar and (b) if the forces acting in one system are in the same ratio to each other as similar forces in the second system. We might also state that flow systems are dynamically similar if the dimensionless parameters obtained in a dimensional analysis of the systems are the same for both. This is essentially the same as conditions (a) and (b). Thus, in the example worked out in Sec. 8-2, any two pipe systems for which the Reynolds number and relative roughness are the same will have dynamically similar flows, and the friction factor will be the same for both. This fact indicates that the pressure drop in a large oil pipeline could be determined by making measurements of the pressure drop in a small pipe carrying water.

All types of model tests made of airplanes, missiles, rivers, harbors, breakwaters, pumps, turbines, and so forth are based on the criterion of dynamic similarity. It should be pointed out, however, that many practical problems confront the experimenter, and it is not always possible to have complete dynamic similarity in making model studies in fluid-flow systems.

When inertia and viscous forces govern the flow in a fluid system, dynamic similarity requires that the Reynolds number be the same for each. Large Reynolds numbers indicate large inertia forces compared to viscous forces, and this condition indicates turbulent flow. Small Reynolds numbers indicate relatively small inertia forces compared to viscous forces and are associated with laminar or viscous flow. The length in the Reynolds number is some characteristic length of the system; it may be a pipe diameter, the length from the leading edge of an airfoil, the diameter of a settling particle, and so forth. The velocity is a characteristic velocity; it may be the free-stream velocity or the so-called shear velocity $\sqrt{\tau_0/\rho}$.

When gravity forces govern the flow, the Froude number should be the same in two dynamically similar systems. The length is a characteristic length; it may be the length of a ship if gravity waves are involved or the water depth in open-channel flow. In the latter instance, Froude numbers less than 1 indicate subcritical flow and Froude numbers greater than 1 indicate supercritical flow. Whether the flow is below or above critical depends on the flow velocity as compared to the velocity of an elementary surface wave (see Sec. 12-2).

If surface tension forces are significant, the Weber number should be the same for dynamic similarity. In most large liquid flow systems, surface tension forces rarely affect the flow and thus models of rivers, for example, must be large enough so that surface tension forces do not affect the model flow either. The role of surface tension forces in cavitation is not yet fully understood.

For compressible flow of gases, the Mach number should be the same in two systems in which there is dynamic similarity. Mach numbers less than, the same as, or greater than the speed of a weak pressure wave (an acoustic wave) are associated with subsonic, sonic, or supersonic flow, respectively. The velocity in the Mach number may be the free-stream velocity or the local velocity at some prescribed point other than the free stream. The reference velocity $\sqrt{K/\rho}$ may be the acoustic velocity in the free stream, the acoustic velocity at a stagnation point where the stream velocity is zero, or the acoustic velocity at a point where the stream velocity is sonic (see Chapter 10).

The pressure coefficient will automatically be the same for two flow systems if the other dimensionless numbers are the same for the particular type of modeling being carried out.

Example 8-1

Oil of kinematic viscosity 1.4×10^{-5} m^2/s flows in a 75-cm pipe at an average velocity of 2.5 m/s. At what velocity should water at 20°C flow in a 7.5-cm pipe for dynamically similar flow?

SOLUTION

The Reynolds number for the two flows must be equal:

$$Re_0 = Re_w$$

$$\frac{V_0 D_0}{\nu_0} = \frac{V_w D_w}{\nu_w} \quad \text{or} \quad \frac{(2.5)(0.75)}{1.4 \times 10^{-5}} = \frac{V_w(0.075)}{1 \times 10^{-6}}$$

$$V_w = 1.79 \text{ m/s}$$

Example 8-2

A river model is built to a scale of $\frac{1}{80}$. What surface velocity in the prototype river is represented by a corresponding surface velocity of 18 cm/s in the river model?

SOLUTION

Subscripts p and m denote prototype and model, respectively:

$$Fr_p = Fr_m$$

$$\frac{V_p}{\sqrt{L_p g_p}} = \frac{V_m}{\sqrt{L_m g_m}}$$

and

$$V_p = V_m \left(\frac{L_p}{L_m}\right)^{1/2} = (0.18)(80)^{1/2} = 1.61 \text{ m/s}$$

Example 8-3

A missile flying at Mach 3 in standard air (15°C) is studied by means of a $\frac{1}{10}$-scale model in a wind tunnel at $-40°C$. What is the wind-tunnel speed, and at what speed does the prototype missile fly?

SOLUTION

$$M_m = M_p = \frac{V_m}{\sqrt{kRT_m}} = \frac{V_p}{\sqrt{kRT_p}} = 3$$

$$V_m = 3\sqrt{(1.40)(287)(233)} = 918 \text{ m/s}$$

$$V_p = 3\sqrt{(1.40)(287)(288)} = 1020 \text{ m/s}$$

▬

Example 8-4

Show that the Reynolds number for a velocity of 40 ft/s and a length of 300 ft in water at 70°F is the same when calculated from these given quantities expressed in SI units.

SOLUTION

From Table A-8 (Appendix II) $v = 1.059 \times 10^{-5}$ ft^2/s. Then Re $= VL/v =$ (40)(300)/1.059 $\times 10^{-5} = 1.133 \times 10^9$.

In the SI system $v = 0.984 \times 10^{-6}$ from Table A-8 (Appendix II); $V = (40)(0.3048) = 12.192$ m/s and $L = (300)(0.3048) = 91.44$ m. Then Re $= VL/v = (12.192)(91.44)/0.984 \times 10^{-6} = 1.133 \times 10^9$.

▬

In addition to the five dimensionless force ratios just given, many other dimensionless numbers exist and should be equal in dynamically similar flows. All are obtainable by dimensional analysis of various systems. Among those in fluid systems are the drag and lift coefficients and the cavitation number, or index.

DRAG COEFFICIENT

Drag may be due to viscous shear (skin-friction drag on an airfoil or ship's hull), to pressure (flow normal to a flat surface), to gravity effects (wave drag of an ocean vessel), or to compressibility effects (high-speed missile). In any case, a dimensional analysis will indicate that the drag coefficient is a function of the Reynolds number, the Froude number, or the Mach number, respectively. For dynamically similar flow, the appropriate one of these three

Table 8-2   MODELING RATIOS[a]

		MODELING PARAMETER			
RATIO	REYNOLDS NUMBER	FROUDE NUMBER, UNDISTORTED MODEL[b]	FROUDE NUMBER, DISTORTED MODEL[b]	MACH NUMBER, SAME GAS[d]	MACH NUMBER, DIFFERENT GAS[d]
Velocity $\dfrac{V_m}{V_p}$	$\dfrac{L_p}{L_m}\dfrac{\rho_p}{\rho_m}\dfrac{\mu_m}{\mu_p}$	$\left(\dfrac{L_m}{L_p}\right)^{1/2}$	$\left(\dfrac{L_m}{L_p}\right)^{1/2}_V$	$\left(\dfrac{\theta_m}{\theta_p}\right)^{1/2}$	$\left(\dfrac{k_m R_m \theta_m}{k_p R_p \theta_p}\right)^{1/2}$
Angular velocity $\dfrac{\omega_m}{\omega_p}$	$\left(\dfrac{L_p}{L_m}\right)^2\dfrac{\rho_p}{\rho_m}\dfrac{\mu_m}{\mu_p}$	$\left(\dfrac{L_p}{L_m}\right)^{1/2}$	c	$\left(\dfrac{\theta_m}{\theta_p}\right)^{1/2}\dfrac{L_p}{L_m}$	$\left(\dfrac{k_m R_m \theta_m}{k_p R_p \theta_p}\right)^{1/2}\dfrac{L_p}{L_m}$
Volumetric flow rate $\dfrac{Q_m}{Q_p}$	$\dfrac{L_m}{L_p}\dfrac{\rho_p}{\rho_m}\dfrac{\mu_m}{\mu_p}$	$\left(\dfrac{L_m}{L_p}\right)^{5/2}$	$\left(\dfrac{L_m}{L_p}\right)^{3/2}_V\left(\dfrac{L_m}{L_p}\right)_H$	c	c
Time $\dfrac{t_m}{t_p}$	$\left(\dfrac{L_m}{L_p}\right)^2\dfrac{\rho_m}{\rho_p}\dfrac{\mu_p}{\mu_m}$	$\left(\dfrac{L_m}{L_p}\right)^{1/2}\left(\dfrac{g_m}{g_p}\right)^{1/2}$	$\left(\dfrac{L_m}{L_p}\right)_H\left(\dfrac{L_p}{L_m}\right)^{1/2}_V\left(\dfrac{g_m}{g_p}\right)^{1/2}$	$\left(\dfrac{\theta_p}{\theta_m}\right)^{1/2}\dfrac{L_m}{L_p}$	$\left(\dfrac{k_p R_p \theta_p}{k_m R_m \theta_m}\right)^{1/2}\dfrac{L_m}{L_p}$
Force $\dfrac{F_m}{F_p}$	$\left(\dfrac{\mu_m}{\mu_p}\right)^2\dfrac{\rho_p}{\rho_m}$	$\left(\dfrac{L_m}{L_p}\right)^3\dfrac{\rho_m}{\rho_p}$	$\dfrac{\rho_m}{\rho_p}\left(\dfrac{L_m}{L_p}\right)_H\left(\dfrac{L_m}{L_p}\right)^2_V$	$\dfrac{\rho_m}{\rho_p}\dfrac{\theta_m}{\theta_p}\left(\dfrac{L_m}{L_p}\right)^2$	$\dfrac{K_m}{K_p}\left(\dfrac{L_m}{L_p}\right)^2$

[a] Subscript m indicates model, subscript p indicates prototype.
[b] For the same value of gravitational acceleration for model and prototype.
[c] Of little importance.
[d] Here θ refers to temperature.

numbers will be the same for model and prototype, and thus the drag coefficient will also be the same for both. This may be expressed as

$$C_{D_m} = C_{D_p} = \frac{D_m}{(\rho_m V_m^2/2)A_m} = \frac{D_p}{(\rho_p V_p^2/2)A_p}$$

and the drag of a prototype may be estimated by measurement of model drag according to this equation. Appropriate velocity ratios are determined from Table 8-2 by the type of modeling.

LIFT COEFFICIENT

The lift coefficient is defined as

$$C_L = \frac{L}{(\rho V^2/2)A}$$

where L is the lift and A is the chord area of the lifting surface. The coefficient is obtainable (except for the constant 2) in a dimensional analysis along with a shape parameter and the angle of attack. It is also a function of the Reynolds number, and if these are equal for a given airfoil shape, for example, the lift coefficients will also be the same for two foils.

CAVITATION NUMBER

The cavitation number is a form of pressure coefficient with a datum, or reference, pressure. This reference pressure may be the vapor pressure, a cavity pressure, or some other reference cavitation pressure. It may be expressed as

$$\sigma = \frac{p - p_{\text{ref}}}{\rho V^2/2}$$

If vapor pressure is the datum,

$$\sigma_v = \frac{p - p_v}{\rho V^2/2}$$

or if a cavity pressure is the datum,

$$\sigma_c = \frac{p - p_c}{\rho V^2/2}$$

It is often assumed that cavitation similarity exists if the cavitation number is the same for two cavitating flow conditions, but this has not been established completely. The cavitation number is used as a modeling parameter for rotating machines (pumps and turbines).

8-6. Modeling Ratios

It is necessary to determine relationships between velocities, angular velocities, discharges, and times for a model and a prototype. The method used to obtain these will be illustrated and the results tabulated.

Velocity ratios may be obtained directly from the appropriate modeling parameter (Re, Fr, We, or M).

Angular velocity is expressed as a velocity divided by a length, and thus

$$\frac{\omega_m}{\omega_p} = \frac{V_m L_p}{L_m V_p} = \left(\frac{V_m}{V_p}\right)\left(\frac{L_p}{L_m}\right)$$

A discharge is a velocity times an area, and thus

$$\frac{Q_m}{Q_p} = \frac{V_m A_m}{V_p A_p} = \left(\frac{V_m}{V_p}\right)\left(\frac{L_m}{L_p}\right)^2$$

Time is the ratio of a length to a velocity, and thus

$$\frac{t_m}{t_p} = \left(\frac{L_m}{V_m}\right)\left(\frac{V_p}{L_p}\right) = \left(\frac{L_m}{L_p}\right)\left(\frac{V_p}{V_m}\right)$$

Example 8-5

The wave drag on a $\frac{1}{16}$-scale model of a merchant ship is determined to be 1.1 N when tested for a prototype speed of 6 m/s. What is the wave drag on the prototype? The model is tested in fresh water; the prototype sails in seawater ($s = 1.025$).

SOLUTION

$Fr_m = Fr_p$, since wave drag is a gravity phenomenon. Thus

$$C_{D_m} = C_{D_p} = \frac{D_m}{(\rho_m V_m^2/2)A_m} = \frac{D_p}{(\rho_p V_p^2/2)A_p}$$

where $V_p/V_m = (L_p/L_m)^{1/2}$ from the Froude law. Then

$$D_p = D_m \frac{\rho_p}{\rho_m} \left(\frac{V_p}{V_m}\right)^2 \left(\frac{L_p}{L_m}\right)^2 = (1.1)(1.025)(4)^2(16)^2 = 4.62 \text{ kN}$$

Distorted models are sometimes used. The size of a river model, for example, is determined by the space available, the reach of the prototype river which is to be modeled, and the discharge required and available. In some instances the model might become quite small, with a scale of $\frac{1}{100}$, for example. Then depths less than 30 cm in the prototype become less than 3 mm in the model, and surface tension or viscosity may affect the flow in the model. But if surface tension or viscosity do not affect the flow in the prototype, they should not affect the flow in the model. To avoid this, a distorted model may be used wherein the vertical scale is larger than the horizontal scale, say, $\frac{1}{40}$ or so. Velocity is determined by gravity forces in the vertical direction. Thus for a distorted model the velocity in the model is related to that in the prototype by the equation

$$\frac{V_m}{V_p} = \left(\frac{L_m}{L_p}\right)_V^{1/2}$$

which is the square root of the vertical scale ratio. The discharge is the flow velocity times an area in a vertical plane made up of a vertical length L_V and a horizontal length L_H. Thus

$$\frac{Q_m}{Q_p} = \frac{V_m}{V_p}\frac{A_m}{A_p} = \left(\frac{L_m}{L_p}\right)_V^{1/2}\left(\frac{L_m}{L_p}\right)_H\left(\frac{L_m}{L_p}\right)_V = \left(\frac{L_m}{L_p}\right)_H\left(\frac{L_m}{L_p}\right)_V^{3/2}$$

The time is a horizontal length divided by a velocity, and thus

$$\frac{t_m}{t_p} = \frac{(L_m)_H}{V_m}\frac{V_p}{(L_p)_H} = \left(\frac{L_m}{L_p}\right)_H\left(\frac{L_p}{L_m}\right)_V^{1/2}$$

Some velocity, rotational speed, discharge, and time ratios are given in Table 8-2.

8-7. Incomplete Similarity

In many instances more than one force ratio is involved in a flow system. In general, for these instances complete dynamic similarity is possible only for full-size models, and since this is usually impractical, incomplete similarity results. An example is the study of the drag of a surface ship, for which the drag is due both to viscous shear along the hull and to waves, a gravity phenomenon. In order to keep both the Reynolds number and the Froude number the same for the model and the prototype, the corresponding velocity ratios are equated to obtain

$$\frac{v_m}{v_p} = \left(\frac{L_m}{L_p}\right)^{3/2}$$

If the model is $\frac{1}{15}$ the size of the prototype, the model fluid viscosity must be $\frac{1}{58}$ the viscosity of water. There is no such liquid. And if the model is tested in water at the same temperature as in the prototype (a practical thing to do), $L_m = L_p$ and a full-size model is then necessary. Actually the model is tested according to the Froude law, and estimates are made of the skin-friction (viscous shear) drag, which amounts to approximately 50 to 80 percent of the total. (Chapters 7 and 11 illustrate the method of calculating skin-friction drag.) Precise accuracy is not possible. The sequence of calculations is as follows:

Total drag$_{model}$ = skin drag$_{model}$ + wave drag$_{model}$
Measured ⟶ *Calculated* ⟶ *By difference*
Total drag$_{prototype}$ = skin drag$_{prototype}$ + wave drag$_{prototype}$
By addition ⟵ *Calculated* ⟵ *Calculated by equating wave drag coefficient for model and prototype*

Some limitations of model tests will be described briefly.

REYNOLDS MODELING

Reynolds modeling is used for studies of pipe flow, lift and drag of airfoils, and drag on almost any shape in incompressible flow; and in boundary layer studies in both incompressible and compressible flow. Viscous effects in gases may be modeled at Mach numbers below about 0.3 without interference from compressible effects. Above that, compressible effects enter in, and these must be modeled properly. Surface roughness should be similar in order that the onset of turbulence in the boundary layer occurs similarly in the model and prototype. Also, the level of turbulence in the free streams should be the same. Thus, flying an airplane through still air is not the same as blowing air past the same airplane in a large wind tunnel at the same velocity, although the Reynolds number may be the same for both. Boundary layer transition will not be the same for model and prototype. A larger portion of the wing surfaces (and the fuselage as well) will most likely have a laminar boundary layer when flying in still air than when in the wind tunnel, and thus a smaller portion will have a turbulent boundary layer in still air than in the wind tunnel. Since the drag is greater for a turbulent boundary layer, the total drag on the airplane flying in still air would be less than the total drag on the airplane in the wind tunnel.

FROUDE MODELING

Froude modeling is used for the measurement of the wave resistance of ships, tidal models of harbors, wave phenomena (beach erosion and breakwaters), river models, and water entry phenomena. The difficulty of ship modeling has already been mentioned and is not yet completely solved. In wave-study models, capillarity may be important and the viscous damping of waves is also not taken into account exactly. Air entrainment in flow over large dams is not modeled precisely in small models. Exact modeling of bed movement and sediment transport for a movable-bed river is difficult if not impossible to achieve, and this type of modeling is not only a science but an art.

MACH MODELING

Mach modeling is done for gas flows at Mach numbers above about 0.3. The viscous effects (Reynolds phenomena) are not entirely absent, even at supersonic flow, because shock interactions with the boundary layer occur and the thickness of a shock is influenced by the Reynolds number. Condensation shock is by no means governed solely by the Mach number.

8-8. Some General Expressions for Dynamic Similitude

Dimensional analysis applied to a rectilinear flow system and to a rotating system result in a number of dimensionless groups commonly used in practice, some of which have already been discussed.

RECTILINEAR FLOW SYSTEMS

Suppose that a force F (drag or lift, for example) is believed to depend on several linear dimensions of the system, a, b, c, and d; the flow velocity V; the fluid density ρ; the fluid viscosity μ; the acceleration of gravity g; the pressure variations in the system Δp; the fluid surface tension σ; and the fluid compressibility K. Then we may write

$$F = f(a, b, c, d, V, \rho, \mu, g, \Delta p, \sigma, K)$$

There are $12 - 3 = 9$ independent dimensionless groups to be obtained. If a, V, and ρ are chosen as repeating variables, the results of a dimensional analysis are

$$2\pi_1 = \frac{F}{\rho V^2 a^2/2}, \quad \text{the drag or lift (or any force) coefficient}$$

$$\pi_2 = \frac{a}{b}$$

$$\pi_3 = \frac{a}{c}$$

$$\pi_4 = \frac{a}{d}$$

$$\pi_5 = \frac{\rho V a}{\mu}, \quad \text{the Reynolds number Re}$$

$$\pi_6 = \frac{V^2}{ag} \text{ or } \frac{V}{\sqrt{ag}}, \quad \text{the Froude number Fr}$$

$$2\pi_7 = \frac{\Delta p}{\rho V^2/2}, \quad \text{the pressure coefficient } C_p$$

$$\pi_8 = \frac{\rho V^2 a}{\sigma} \text{ or } \frac{V}{\sqrt{\sigma/\rho a}}, \quad \text{the Weber number We}$$

$$\pi_9 = \frac{\rho V^2}{K} \text{ or } \frac{V}{\sqrt{K/\rho}}, \quad \text{the Mach number M}$$

Thus we may state that

$$\pi_1 = f(\pi_2, \pi_3, \pi_4, \pi_5, \pi_6, \pi_7, \pi_8, \pi_9)$$

or that when constants such as 2 are included,

$$\frac{F}{\rho V^2 a^2/2} = f\left(\frac{a}{b}, \frac{a}{c}, \frac{a}{d}, \frac{\rho V a}{\mu}, \frac{V}{\sqrt{ag}}, \frac{\Delta p}{\rho V^2/2}, \frac{V}{\sqrt{\sigma/\rho a}}, \frac{V}{\sqrt{K/\rho}}\right)$$

or that

$$C_D \text{ or } C_L = f\left(\frac{a}{b}, \frac{a}{c}, \frac{a}{d}, \text{Re, Fr, } C_p, \text{We, M}\right)$$

As previously mentioned, it may not be possible to eliminate all but one or two effects in a flow situation, and in these instances compromises must be made. For example, the drag or lift coefficient for an airfoil (or hydrofoil) section may depend on some geometrical parameters and only the Reynolds number in a low-velocity air stream (compressible and gravity effects are absent). They may depend on both Re and M in a high subsonic gas flow. They may depend on both Re and Fr for a hydrofoil at shallow submergence where surface waves are set up. If the hydrofoil cavitates, the Weber number We and the pressure coefficient in the form of the cavitation number may also be involved.

ROTATING SYSTEMS

Rotating systems include pumps, compressors, turbines, propellers, fluid couplings, torque converters, and so forth. Suppose the volumetric flow rate Q through a rotating machine or system is believed to depend on the efficiency η, the energy per unit mass of fluid e ($H = e/g$ is often called the head on a pump or turbine for incompressible flow), the power supplied P, the rotational speed N, the diameter of the rotor D, the fluid density ρ, the fluid viscosity μ, the fluid compressibility K, the torque T, and the thrust F. Then

$$Q = f(\eta, H, P, N, D, \rho, \mu, K, T, F)$$

We will obtain at least $11 - 3 = 8$ independent dimensionless groups, one of which is already dimensionless (η). Thus there are seven groups to be found from a dimensional analysis. If D, ρ, and N are chosen as repeating variables, the results are

$$\pi_1 = \frac{Q}{ND^3}$$

$$\pi_2 = \frac{e}{N^2D^2} \quad \left(\text{generally used in the form } \frac{H}{N^2D^2}\right)$$

$$\pi_3 = \frac{P}{\rho N^3 D^5}$$

$$\pi_4 = \frac{\rho ND^2}{\mu}, \quad \text{a form of the Reynolds number Re}$$

$$\pi_5 = \frac{K}{\rho N^2 D^2} \text{ or } \frac{ND}{\sqrt{K/\rho}}, \quad \text{a form of the Mach number M}$$

$$\pi_6 = \frac{T}{\rho N^2 D^5}, \quad \text{the torque coefficient } C_{\text{torque}}$$

$$\pi_7 = \frac{F}{\rho N^2 D^4}, \quad \text{the thrust coefficient } C_{\text{thrust}}$$

$$\pi_8 = \eta, \quad \text{the efficiency}$$

Thus we may state that

$$\frac{Q}{ND^3} = f\left(\eta, \frac{H}{N^2D^2}, \frac{P}{\rho N^3 D^5}, \text{Re, M}, C_{\text{torque}}, C_{\text{thrust}}\right)$$

The ratio of $\pi_1^{1/2}/\pi_2^{3/4}$ is called the specific speed of a pump:

$$N_{s(P)} = \frac{N\sqrt{Q}}{e^{3/4}} \tag{8.1}$$

This equation is truly dimensionless, but in North American pump practice N is commonly expressed in revolutions per minute, Q in gallons per minute, and e in terms of the head H in feet of fluid. Thus Eq. (8.1) is generally stated as

$$N_{s(P)} = \frac{N\sqrt{Q}}{H^{3/4}} \tag{8.2}$$

The specific speed of a pump as used in Eq. (8.2) is not truly dimensionless but is dimensional owing to the use of H rather than gH and to the units used for N and Q. Since $e = gH$, Eq. (8.1) should be used for rating pumps for use on the moon, for example, where the value of g is different from the value of g on the earth. A pump producing a high head at a relatively low discharge has a low specific speed; this is characteristic of a centrifugal pump. One producing a low head at a relatively large flow rate has a relatively high specific speed; this is characteristic of an axial-flow or propeller-type pump. The intermediate range of moderate heads at moderate discharge is characteristic of mixed-flow pumps. Typical values of specific speed for centrifugal pumps as defined in North America range from about 500 to 5000; for mixed-flow pumps, from 4000 to 10,000; and for axial-flow or propeller pumps, from 10,000 to 15,000 per stage. One stage consists of one rotor or impeller, and the specific speed generally applies to a single stage of a multistage pump.

The ratio $\pi_3^{1/2}/\pi_2^{5/4}$ is called the *specific speed of a turbine:*

$$N_{s(T)} = \frac{N\sqrt{P}}{\rho^{1/2}e^{5/4}} \tag{8.3}$$

Hydraulic turbines pass water, and the value of the water density is generally omitted. The power P is commonly expressed as brake horsepower (bhp), and the speed and head are expressed the same way as in pump practice. The American turbine practice is to designate the specific speed of a turbine in a dimensional form as

$$N_{s(T)} = \frac{N\sqrt{\text{bhp}}}{H^{5/4}} \tag{8.4}$$

A turbine operating under a high head with a relatively low flow rate has a relatively low specific speed; this is characteristic of a Pelton wheel, or impulse turbine. A turbine operating under a relatively low head at a large flow rate has a relatively high specific speed; this is characteristic of a propeller-type, or Kaplan, turbine. The intermediate range of moderate heads and moderate flow rates is characteristic of a mixed-flow, or Francis, type of turbine. Typical values of specific speed for impulse turbines are about 5; for Francis turbines, from 20 to 100; and for propeller, or Kaplan, turbines, about 100 to 200.

The dimensionless groups listed earlier for rotating systems or machines are applicable to dynamically similar operating conditions for geometrically similar systems. They are widely used by pump and turbine manufacturers in predicting the performance of large units from laboratory tests on small units. They may be used to predict the performance of a given machine at a speed other than the rated speed, assuming equal efficiencies at the two speeds. Some examples will illustrate the principles involved.

Example 8-6

A centifugal pump is rated at 63 L/s (liters per second) at a head of 41 m at a speed of 1750 r/min. For the same efficiency of operation, what would be the flow rate and head when the speed of the pump is reduced to 1450 r/min?

SOLUTION

From π_1, $Q_1/N_1 = Q_2/N_2$ since the diameter is the same. Thus $Q_2 = 63(1450/1750) = 52.2$ L/s.

From π_2, $H_1/N_1^2 = H_2/N_2^2$ since the diameter is the same. Thus $H_2 = 41(1450/1750)^2 = 28.1$ m.

Example 8-7

A model turbine with a 42-cm runner (rotor) is tested under a head of 5.64 m at a speed of 374 r/min. The measured power output is 16.5 kW at an efficiency of 89.3 percent. The prototype runner is 409 cm in diameter. What are the head, speed, flow rate, and power output for the prototype turbine for dynamically similar flow? (Actually, the efficiency of the larger unit would be about 3 percent greater than for the model in this instance.) What type of turbine is this?

SOLUTION

For geometrical similarity, all linear dimensions are in the same ratio. Thus,

$$H_p = H_m \frac{D_p}{D_m} (5.64)(409/42) = 54.9 \text{ m}$$

From π_2,

$$N_p = N_m \left(\frac{D_m}{D_p}\right)^{1/2} = 374(42/409)^{1/2} = 120 \text{ r/min}$$

The flow rate may be obtained from the expression for π_1, or from the fact that the model test is based on the Froude law, since gravity flow is involved. From π_1, as well as from the Froude law, $Q_p/Q_m = (L_p/L_m)^{5/2}$. The flow in the model is found from Sec. 5-3 to be $Q_m = P_m/\gamma_m H_m \eta_m$, where the w term in the energy equation is equivalent to the H term in the dimensionless groups in this section. Substituting,

$$Q_m = (16.52)/(9810)(5.64)(0.893) = 0.334 \text{ m}^3/\text{s}$$

and

$$Q_p = 0.334(409/42)^{5/2} = 98.8 \text{ m}^3/\text{s}.$$

The power output of the prototype turbine at the same efficiency as the model may be found from Sec. 5-3 or from π_3. From π_3,

$$P_p = P_m \left(\frac{N_p}{N_m}\right)^3 \left(\frac{D_p}{D_m}\right)^5 = 16.52 \left(\frac{120}{374}\right)^3 \left(\frac{409}{42}\right)^5 = 47{,}800 \text{ kW}$$

The specific speed of the model (which is the same as that for the prototype) is

$$N_{s(T)} = \frac{374\sqrt{22.15}}{18.5^{5/4}} = 46$$

indicating a Francis turbine.

The model power of 16.52 kW = 22.15 horsepower (hp), and the head of 5.64 m = 18.5 ft; these values are used to compute the specific speed, in conformance with North American practice.

—————

PROBLEMS

8-1. Show the dimensional equivalence of the following:
(a) Energy per unit mass and velocity squared
(b) Time rate of change of energy and force times velocity
(c) Head and energy per unit weight
(d) Energy per unit volume and dynamic pressure

8-2. A fluid property known as the Prandtl number is used in heat transfer studies. It is given by $\text{Pr} = \mu c_p/k$ (see Chapter 16) and is a dimensionless number. What are the dimensions of k, the thermal conductivity, in the MLT and in the FLT system of dimensions? Include θ for temperature.

8-3. The resistance F of a surface ship depends on the ship speed V, the ship length L, the hull surface roughness k, the fluid density ρ, the fluid viscosity μ, and the acceleration of gravity g in connection with wave resistance. Express these variables in dimensionless form.

8-4. The power P required to drive a fan or blower depends on the fluid density ρ, the fluid viscosity μ, the impeller diameter D, the volumetric flow rate Q, and the rotational speed N. Express these variables in dimensionless form. Show that $\rho N D^2 / \mu$ is a form of the Reynolds number.

8-5. The size d of spray or drops formed when liquid flows from a nozzle depends on a characteristic velocity in the nozzle V, the nozzle tip diameter D, the fluid density ρ, the fluid viscosity μ, the surface tension between the liquid and air σ, and the acceleration of gravity g. Express these variables in dimensionless form.

8-6. The depth y_2 downstream of a hydraulic jump depends on the upstream depth y_1, the unit discharge q (volumetric flow rate per unit width of channel), and the acceleration of gravity g. Express these variables in dimensionless form. Compare with Eq. (12.14).

8-7. The torque T needed to rotate a disk in a viscous fluid depends on the disk diameter D, the angular speed N, and the density ρ and the viscosity μ of the fluid. Express these in dimensionless form.

8-8. The pressure gradient $\Delta p / L$ for fully developed laminar flow in a circular tube depends on the flow rate Q, the tube diameter D, the fluid viscosity μ, and the fluid density ρ. Express these variables in dimensionless form.

8-9. The boundary layer thickness δ on a smooth flat plate in an incompressible flow without pressure gradients depends on the free-stream velocity u_s, the fluid density ρ, the fluid viscosity μ, and the distance from the leading edge of the plate x. Express these variables in dimensionless form. Compare with Eqs. (7.19) and (7.24).

8-10. The fluid velocity u within a boundary layer depends on the free-stream velocity u_s, the fluid density ρ, the fluid viscosity μ, the wall shear stress τ_0, the normal distance from the boundary y, and the distance from the leading edge of the boundary surface x. Express these variables in dimensionless form. Discuss the term y/x in connection with the results of Chapter 7 which showed that u/u_s depends on y/δ.

8-11. Suppose the transition point x_t between a laminar and a turbulent boundary layer on an airfoil depends on the chord length C, the free-stream velocity u_s, the fluid viscosity v, the turbulence intensity u' (the root-mean-square values of the turbulent velocity fluctuations about u_s), and the scale of the turbulence L. Show that this may be expressed as

$$\frac{x_t}{C} = f\left(\frac{u_s C}{v}, \frac{u'}{u_s}, \frac{L}{C}\right)$$

8-12. The resistance to flow per unit length of pipe R/L for a circular tube of diameter D depends on the tube roughness k (a length), the fluid density ρ and viscosity μ, and the average flow velocity V. Show that these may be expressed dimensionlessly as

$$\frac{R}{LD\rho V^2} = f\left(\frac{VD\rho}{\mu}, \frac{k}{D}\right)$$

The product LD represents a length squared, i.e., an area.

8-13. Laminar flow occurs along a flat surface, passes over a step, and reattaches itself to a second flat surface parallel to the first at a distance x_R from the step. This reattachment distance x_R depends on the step height s, the free-stream velocity u_s, the fluid density ρ, the fluid viscosity μ, and the boundary-layer thickness δ at the step. Express these variables in dimensionless form.

8-14. The flow rate Q of a liquid through an orifice in a pipeline depends on the orifice diameter d, the pipe diameter D, the pressure drop across the orifice Δp, and the fluid density ρ and viscosity μ. Show that these are related by the dimensionless groups

$$\frac{Q}{d^2\sqrt{\Delta p/\rho}} = f\left(\frac{d}{D}, \frac{\mu}{\rho d\sqrt{\Delta p/\rho}}\right)$$

Compare this with Eq. (13.14).

8-15. The flow rate Q over a rectangular weir depends on the head on the weir crest H, the weir crest height Z (Fig. 13-15), the breadth of the weir crest b, the fluid density ρ, the fluid viscosity μ, the fluid surface tension σ, and the acceleration of gravity g, which may be considered a flow parameter. Express these variables in dimensionless form, and identify the Froude, Reynolds, and Weber numbers in your results.

8-16. Repeat Prob. 8-15 for a V-notch weir of angle θ (see Fig. 13-18), the variable b being replaced by θ.

8-17. The thrust of a ship propeller F depends on the propeller diameter D, the rotational speed N, the fluid density ρ, the fluid viscosity μ, and the speed of the propeller through the fluid V called the advance velocity, and the acceleration of gravity g. Express these variables in dimensionless form.

8-18. The lift of a given airfoil shape in a subsonic flow depends on the angle of attack α, the chord length C, and span S, the free-stream velocity u_s, the fluid viscosity μ, and the fluid density ρ. Express these variables in dimensionless form to show that

$$\frac{L}{\rho u_s^2 CS} = f\left(\alpha, \frac{S}{C}, \frac{u_s \rho C}{\mu}\right)$$

Note: The application of the π theorem will give C^2, representing an area, instead of CS, the actual wing area.

8-19. The longitudinal velocity u in a circular tube for turbulent flow depends on the distance from the wall y, the tube radius R, the wall shear stress τ_0, the fluid density ρ and viscosity μ, and the pipe roughness k (a length). Arrange these in dimensionless form.

8-20. Water at 15°C (see Appendix II for properties) flows in a 25-cm-diameter pipe at an average velocity of 3.00 m/s. The pressure drop in 30 m of pipe length is 1.15 kPa.

    (a) For dynamically similar flow, at what average velocity should air at 40°C ($\mu = 1.8 \times 10^{-5}$ kg/m s) and 350 kPa abs flow in a 15-cm pipe?

    (b) What will be the pressure drop for this air in 100 m of 15-cm pipe of similar roughness? Refer to π_1 in the analysis of Sec. 8-2.

8-21. Water at 20°C flows in a 30-cm pipe at an average velocity of 3.45 m/s. For dynamically similar flow, calculate the average flow velocity for

    (a) air at 35°C and 3.50×10^5 N/m² in a 13-cm pipe and

    (b) gasoline at 15°C in a 20-cm pipe.

8-22. An aircraft is to fly in air at 0°C and 75 percent of sea level standard pressure at a speed of 100 m/s. A $\frac{1}{15}$-scale model is to be tested in a wind tunnel at a maximum speed of 100 m/s in order to avoid compressibility effects. The air in the wind tunnel is at 25°C. What should be the pressure in the wind tunnel for dynamically similar flow as for the prototype? Give answer in terms of sea level atmospheres.

8-23. Repeat Prob. 8-22 for a $\frac{1}{20}$-scale model.

8-24. Tests in a wind tunnel are to be used to determine the lift and drag of hydrofoils with a 30-cm chord for a boat designed to travel at 18 m/s in water at 20°C. In order to avoid compressibility effects in the wind tunnel, assume air at 15°C is to be used at a maximum velocity of 60 m/s. What wind tunnel pressure should be used with the 30-cm hydrofoils in the wind tunnel? The hydrofoils are to be run deep so that only viscous effects need to be considered.

8-25. The drag of an airship in standard air at 15 m/s is to be studied by means of a $\frac{1}{15}$-scale model.

    (a) What is the required pressure in a wind tunnel for an air speed of 60 m/s at 15°C?

    (b) At what speed should the model be towed in water at 20°C for dynamically similar flow? Towing must be deep to avoid gravity waves, or at high pressure in a water tunnel in order to avoid cavitation.

8-26. Repeat Prob. 8-25 for a $\frac{1}{20}$-scale model.

8-27. Repeat Prob. 8-24 for 40-cm hydrofoils.

8-28. The drag and bending moment on a structure in a 30-m/s wind is to be studied on a $\frac{1}{20}$-scale model in a pressurized wind tunnel where the air is eight times the density of atmospheric air, but at the same temperature.

    (a) What should be the wind speed in the wind tunnel?

    (b) What prototype bending moment is indicated by a measured bending moment of 35 N m on the model?

8-29. A maximum flow rate of $0.170 \text{ m}^3/\text{s}$ is available for a river model. This corresponds to $850 \text{ m}^3/\text{s}$ for the prototype. What is the maximum size of the model in terms of prototype size?

8-30. A spillway model is built to a scale of $\frac{1}{45}$. When the depth of water over the crest is 5 cm, the flow rate is 42 L/s (liters per second). To what head and flow rate in the prototype does this correspond?

8-31. The flow rate in a river is $1400 \text{ m}^3/\text{s}$. A distorted model with a horizontal scale of $\frac{1}{70}$ and a vertical scale of $\frac{1}{20}$ is built for laboratory tests. What is the flow rate in the model?

8-32. A tank containing water drains through a 2-cm-diameter orifice in 6 min, 20 s. How long would it take for kerosene to drain from a geometrically similar tank through a 8-cm-diameter orifice? Neglect viscous effects, and consider only the effect of gravity.

8-33. The tank containing water in Prob. 8-32 is on the moon, where the acceleration of gravity is $\frac{1}{6}$ that on earth. How long will it take for the tank to drain on the moon?

8-34. A maximum flow rate of $0.20 \text{ m}^3/\text{s}$ is available for a river model. This corresponds to $945 \text{ m}^3/\text{s}$ for the prototype river. What is the largest size model which may be used?

8-35. The flow in a river is $1500 \text{ m}^3/\text{s}$. A distorted model with a horizontal scale of $\frac{1}{60}$ and a vertical scale of $\frac{1}{16}$ is built for laboratory testing. What is the flow rate in the model?

8-36. A model of the Mississippi River watershed at the Waterways Experiment Station at Vicksburg, Miss. is built to a scale of $\frac{1}{100}$ vertically and $\frac{1}{2000}$ horizontally. What flow rate in the river is represented by a flow rate of $0.040 \text{ m}^3/\text{s}$ in the model?

8-37. A $\frac{1}{120}$-scale model of the entrance channel to Depoe Bay, Oregon, was built to study wave energy entering the harbor area. For prototype wave periods ranging from 6 to 31 s, what wave periods are appropriate in the model?

8-38. The wave resistance of a $\frac{1}{40}$-scale model of a ship is studied in fresh water. The prototype ship travels at 28 knots (1 knot $= 0.515$ m/s).
(a) What is the model speed?
(b) What wave drag in the prototype corresponds to a wave drag of 9.0 N for the model?

8-39. The wave resistance of a $\frac{1}{30}$-scale model of a ship is studied in fresh water. The prototype ship travels at 10.4 m/s.
(a) What should be the model speed?
(b) What wave drag in the prototype ship corresponds to a wave drag of 12.4 N for the model?

8-40. A 6.00-m model of a 130-m ship is towed through fresh water at 2.60 m/s with a total drag of 109 N. The wetted hull area of the model is 6.50 m^2.
(a) What is the skin-friction drag on the model?
(b) What is the wave drag for the model?

(c) What is the corresponding speed for the prototype?

(d) Estimate the wave drag for the prototype ship.

(e) What is the skin-friction drag for the prototype ship? Use curve for a typical ship hull as shown in Fig. 11-1.

(f) What power is required to propel the prototype and to tow the model at their appropriate speed?

8-41. Repeat Prob. 8-40 for a model speed of 3.00 m/s and a total drag of 138 N.

8-42. A $\frac{1}{1800}$-scale model of a tidal estuary is operated to satisfy the Froude law. What length of time in the model represents 12.4 h in the prototype? This is approximately a tidal period.

8-43. A tidal model is built to a scale of $\frac{1}{3600}$ horizontally and $\frac{1}{81}$ vertically. For a tidal period of 12.4 h in nature, what is the tidal period in the model?

8-44. A tidal model built to a scale of $\frac{1}{3000}$ horizontally and $\frac{1}{70}$ vertically is tested in a laboratory. For a tidal period of 12.4 h in nature, what is the tidal period in the model?

8-45. A $\frac{1}{40}$-scale model of a river is used to study a hydraulic structure. A force on one part of the model is 2.65 N. What is the comparable force on the prototype?

8-46. A model of a dam for a hydroelectric project is to be built for flow studies. The maximum flow in the river is 3400 m³/s, and the maximum supply available in the laboratory for model studies is 112 L/s. What is the largest distorted model that may be built, in relation to the river size, in order that dynamically similar flows may be achieved? The horizontal scale is to be reduced to one-third that of the vertical scale; that is, $(L_m/L_p)_V = 3(L_m/L_p)_H$.

8-47. A $\frac{1}{50}$-scale model of a river contains some bridge piers. The force on one pier in the model is 1.75 N. What force would be expected on a full-size pier in the prototype, assuming similarity for inertia and gravity forces?

8-48. An airplane is to fly in standard air at 75 m/s. A $\frac{1}{6}$-scale model is made and tested in a wind tunnel with standard air at 420 m/s. Considering inertia, viscous, and compressibility forces, is the flow past the model dynamically similar to that past the prototype? Explain.

8-49. Fundamental studies on boundary layers for compressible flow are made in wind tunnels with air flowing past a model at rest and in firing ranges in which a model is fired through a gas which is at rest. In which type of test would free-flight conditions be better simulated? Discuss your answer.

8-50. One of the advantages of using hydrogen rather than air for cooling electrical generators is the reduction in windage losses. Estimate these losses for hydrogen as a percentage of those for air in a given machine. Assume hydrogen and air at essentially the same pressure and temperature. Power loss is proportional to the product of a drag force and a velocity.

8-51. A supersonic firing range consists of a closed chamber containing xenon gas at 20°C. At what velocity should a small missile be fired through the chamber in order that the steady flow pattern past it be dynamically similar to a free flight at 1200 m/s in standard air at sea level?

8-52. Repeat Prob. 8-51 for a free-flight speed of 1500 m/s.

8-53. Write expressions for the ratio of prototype power to model power for dynamically similar flows based on the
(a) Reynolds number,
(b) Froude number, and
(c) Mach number.

8-54. A pump discharges 250 L/s under a head of 3.70 m at an efficiency of 82 percent at 1150 r/min. Estimate the flow rate, head, and brake power for a speed of 1450 r/min, assuming the same efficiency. Water is being pumped.

8-55. A centrifugal pump is rated at 30 L/s at a head of 37 m when pumping water at 1750 r/min. The pump efficiency is 74 percent. Estimate the flow rate, head, and brake power at 1450 r/min, assuming the same efficiency.

8-56. A pump delivers 0.250 m^3/s under a head of 4.00 m at an efficiency of 85 percent at 1150 r/min. Estimate the flow rate, head, and input power at a higher speed of 1450 r/min at the same efficiency. Water is pumped.

8-57. A 40-cm-diameter model of a 245-cm-diameter propeller is tested in accordance with the results of Prob. 8-17, with the exception that viscous forces are considered negligible. A torque of 20 N m at 450 r/min develops a thrust of 245 N at a speed of advance of 2.60 m/s.
(a) To what speed of advance and rotational speed for the prototype propeller do these test conditions correspond?
(b) Estimate the thrust and torque for the prototype propeller, assuming the same efficiency and fluid for both model and prototype.
(c) What is the propeller efficiency? This is defined as the ratio of power output to power input.

8-58. A 50-cm-diameter turbine wheel is tested at 1008 r/min under a head of 122 m and delivers 1.10 kW. It is a $\frac{1}{7}$-scale model of a large turbine which is to operate under a net head of 1475 m.
(a) What is the speed of the prototype turbine?
(b) What is the power output of the prototype turbine at the same efficiency as the model?
(c) What type of turbine is this?

8-59. An aircraft is designed to fly at 200 m/s at 3000 m ($T = -4.5°C$ and $p = 70$ kPa abs). Discuss the possibility of testing a $\frac{1}{20}$-scale model in water.

8-60. A laboratory model of a hydroelectric project contains a hydraulic jump which dissipates 10 W. The model is built to a scale of $\frac{1}{30}$. What power would be dissipated in the prototype jump?

8-61. In making model tests of the following situations, state for each whether the Reynolds number, the Froude number, the Mach number, or a combination of them are significant.

(a) Liquid flow in a pipe
(b) High-velocity gas flow in a pipe
(c) The drag of a parachute
(d) Flow over a spillway in a river model
(e) The drag of a submarine at a 100-ft depth
(f) The total drag on a cargo ship
(g) Boundary-layer growth on a supersonic missile
(h) The movement of sediment in a river
(i) Silt settling in a reservoir
(j) Beach erosion

References

1. H. Schlichting, *Boundary Layer Theory* (translated by J. Kestin), 6th ed. (New York: McGraw-Hill Book Company, Inc., 1968).

The reader is referred to the following sources for additional information on dimensional analysis and dynamic similarity:

P. W. Bridgman, *Dimensional Analysis* (New Haven, Conn.: Yale University Press, 1931; Paperback Y-82, 1963).

W. J. Duncan, *Physical Similarity and Dimensional Analysis* (London: Edward Arnold and Co., 1953).

M. Holt, "Dimensional Analysis," Sec. 15 of *Handbook of Fluid Dynamics*, edited by V. L. Streeter (New York: McGraw-Hill Book Company, Inc., 1961).

H. L. Langhaar, *Dimensional Analysis and Theory of Models* (New York: John Wiley and Sons, Inc., 1951).

I. S. Pearsall, *Cavitation* (London: Mills & Boon, Ltd., 1972).

D. J. Schuring, *Scale Models in Engineering* (Elmsford, N.Y.: Pergamon Press, Inc., 1977).

M. S. Yalin, *Theory of Hydraulic Models* (New York: Macmillan, Inc., 1971).

H. A. Becker, *Dimensionless Parameters: Theory and Methodology* (New York: John Wiley & Sons, Inc., 1976).

Chapter 9
The Flow of Viscous Fluids in Ducts

9-1. Basic Considerations

The effects of fluid viscosity on the flow of liquids and gases in ducts are of both theoretical and practical interest. Laminar flow in many instances can be analyzed theoretically, but turbulent flow has thus far defied rigorous analysis. Both theory and experiment have been necessary to reach the present state of understanding of turbulent flow. In either case, viscosity is important, and the Reynolds number with the hydraulic diameter[†] as the characteristic length ($Re = VD_h/v$) will be a significant parameter. Flow in ducts generally is laminar up to a Reynolds number of about 2300; then it passes through a transition regime (intermittently laminar and turbulent) before becoming entirely turbulent. For engineering calculations, it is generally assumed laminar below a Reynolds number of 2000 and turbulent above 2000.

Duct flow is important because:

1. Studies of turbulent flow in pipes have led to a better understanding of turbulent flow in general. Recall that experimental results from

---

[†] The hydraulic diameter D_h = 4 (area of the duct cross section) divided by the duct perimeter ($D_h = 4A/P$). Thus for a circular tube the hydraulic diameter equals the tube diameter.

studies of turbulent flow in pipes were used in the analysis of turbulent boundary layers [Eq. (7.23)].

2. We need to be able to estimate the head-loss term h_L in the energy equation for incompressible flow in order to determine the power given up by a liquid to a turbine, the power requirements for pumping liquids through pipelines, or the flow rate for a given situation. The term h_L represents the mechanical energy loss only in available energy.

3. Flow of real gases is never reversible, and the wall shear stress (commonly referred to as wall friction) affects the flow of gases even in short nozzles, though not appreciably in certain respects. In long pipes, wall friction has a pronounced effect on gas flow.

4. Many systems of engineering interest involve both fluid flow and heat transfer in ducts, and an understanding of the flow process is a necessary prerequisite to an understanding of the heat-transfer process.

5. Flow through cascades (in turbines, compressors, or vaned elbows, for example) may be studied as flow *around* a single blade or as duct flow *between* blades.

Incompressible laminar and turbulent flow in both the entrance and the fully developed flow regions in ducts; incompressible flow through contractions, expansions, and pipe fittings; and flow of mixtures in pipes will be treated in this chapter.

The development of the velocity profile in the entrance region of a pipe was discussed in Sec. 7-2, and the growth of the boundary layer was indicated in Fig. 7-8. The flow was assumed to be one dimensional at entrance. For

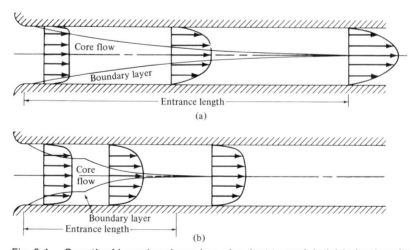

Fig. 9-1. Growth of boundary layer in a pipe (not to scale): (a) in laminar flow; (b) in turbulent flow.

both laminar and turbulent flow (shown again in Fig. 9-1), the wall shear stress is very large at entrance and generally decreases in the direction of flow to a fixed value. The magnitude of the pressure gradient dp/dx also generally decreases in the flow direction to a fixed value, but at a rate slightly less than that for the wall shear stress. Finally, the velocity profile also changes, and eventually it becomes adjusted to a fixed profile. The wall shear stress, the pressure gradient, and the velocity profile all approach their fixed values asymptotically, and thus it is difficult to set a precise length for the entrance region. It could be defined as the region or length required for any one of these three quantities to reach a fixed value.

When the wall shear stress, the pressure gradient, and the velocity profile have reached constant conditions, the flow is called *fully developed flow*.

Fluid flowing with an average velocity V through a cross section in a circular tube has a rate of flow of momentum of $\beta \rho V^2 (\pi D^2/4)$, from Eq. (4.6). The momentum theorem applied to a control volume of length dx in the tube, assumed to be of constant diameter, is (see Fig. 9-2)

$$p\frac{\pi D^2}{4} - (p + dp)\frac{\pi D^2}{4} - \tau_0 \pi D \, dx = \frac{\pi D^2}{4}\frac{d}{dx}(\beta \rho V^2)\, dx$$

so that

$$-\frac{dp}{dx} = \frac{4}{D}\tau_0 + \rho V^2 \frac{d\beta}{dx} + \beta \rho V \frac{dV}{dx} \tag{9.1}$$

since $\beta V \, d(\rho V)/dx = 0$ from continuity.

For fully developed, incompressible flow, $d\beta/dx = 0$, $dV/dx = 0$, and the wall shear stress τ_0 which depends on the shape of the velocity profile near the tube wall does not change with x. Thus for this situation

$$-\frac{dp}{dx} = \frac{\Delta p}{L} = \frac{4\tau_0}{D} \tag{9.2}$$

The pressure drops in the direction of flow, and thus dp is negative for a positive dx. The *magnitude* of the pressure drop over a length of tube L is generally designated as $\Delta p = p_1 - p_2$, where section 1 is upstream of section 2. All of the pressure drop for fully developed incompressible flow goes into

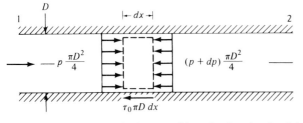

Fig. 9-2. Flow through elemental length of a circular tube.

overcoming wall shear. This wall shear stress may be conveniently determined by measuring the pressure drop over a given length of pipe.

For developing flow in the entrance of a tube, $\beta = 1$ initially and approaches a fully developed value at some distance downstream of the entrance. For incompressible, fully developed flow $\beta = \frac{4}{3}$ for laminar flow in a circular tube and about 1.03 for turbulent flow. Similar values apply for compressible flow. In the entrance region the overall pressure drop overcomes the wall shear and increases the flow momentum. For incompressible flow, Eq. (9.1) becomes

$$-\frac{dp}{dx} = \frac{4\tau_0}{D} + \rho V^2 \frac{d\beta}{dx} \tag{9.3}$$

Flow becomes truly fully developed only after $d\beta/dx = 0$, for only then is there no further change in the shape of the velocity profile. However, the velocity profile near the tube wall becomes fixed before that in the interior of the flow, so that the wall shear becomes fixed while there is still some variation in β and dp/dx. This is why measurements of wall shear stress indicate a shorter entrance length than do measurements of dp/dx or of the velocity near the tube axis.

For compressible flow of a gas beyond the entrance of a circular tube, the normalized shape of the velocity profile is essentially constant and thus $d\beta/dx \approx 0$. However, V varies with x because of thermodynamic changes which affect the density ρ. Hence the wall shear stress τ_0 and the pressure gradient dp/dx vary with x, and Eq. (9.1) takes the form

$$-\frac{dp}{dx} = \frac{4\tau_0}{D} + \beta \rho V \frac{dV}{dx} \tag{9.4}$$

The pressure drop thus goes into overcoming wall shear and increasing the momentum of the flow. Equation (9.4) will be integrated in Sec. 10-10.

9-2. Fully Developed Incompressible Flow in Ducts

In engineering practice it is customary to express the pressure gradient (pressure drop per unit length of pipe) in the form of the so-called Darcy–Weisbach equation, which was also developed by a dimensional analysis in Sec. 8-2. This equation is

$$\frac{\Delta p}{L} = \frac{f}{D} \frac{\rho V^2}{2} \tag{9.5}$$

where f is the friction factor, $\rho V^2/2$ is the dynamic pressure of the mean flow, and D is the pipe diameter. An alternative form in terms of the head loss due to friction h_f is

$$h_f = \frac{\Delta p}{\gamma} = f \left(\frac{L}{D}\right) \frac{V^2}{2g} \tag{9.6}$$

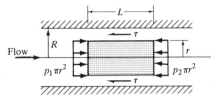

Fig. 9-3.   Cylindrical element in a round pipe.

The value of the friction factor has to be known so that these equations may be used in making calculations. The dimensional analysis of Sec. 8-2 indicated that the friction factor f was a function of the relative roughness (k/D) of the pipe surface and of the Reynolds number of the mean flow $(VD\rho/\mu)$. The friction factor can be determined analytically for laminar flow and semiempirically for turbulent flow. These two methods will be discussed separately, and then the final results will be given in graphic form.

For fully developed incompressible flow in a round pipe, the momentum theorem applied to a cylindrical element of length L and radius r (Fig. 9-3) gives

$$p_1 \pi r^2 - p_2 \pi r^2 - \tau 2\pi r L = 0$$

or

$$\tau = \frac{p_1 - p_2}{L}\left(\frac{r}{2}\right) \tag{9.7}$$

which is valid for both laminar and turbulent flow. Equation (9.7) shows that the shear stress varies linearly with the pipe radius, being zero at the center and a maximum at the pipe walls, where the wall shear stress τ_0 becomes

$$\tau_0 = \frac{p_1 - p_2}{L}\left(\frac{R}{2}\right) = \frac{\Delta p}{L}\left(\frac{D}{4}\right) \tag{9.2}$$

A comparison of Eq. (9.7) with Eq. (9.5) shows that the wall shear stress is related to the friction factor by the equation

$$\tau_0 = f \frac{\rho V^2}{8} \tag{9.8}†$$

It will be convenient to introduce a shear velocity v_* defined as

$$v_* = \sqrt{\frac{\tau_0}{\rho}} = V\sqrt{\frac{f}{8}} \tag{9.9}$$

For circular tubes the tube diameter D is the characteristic length parameter used in the various equations. For noncircular ducts the hydraulic

---

† The friction factor may be defined as the ratio of the local wall shear stress to the dynamic pressure of the mean flow, or $f' = \tau_0/(\rho V^2/2)$. The Darcy–Weisbach friction factor is four times this value or $f = 4f'$.

diameter D_h is used. This is defined as

$$D_h = \frac{4A}{P}$$
(9.10)

where A is the cross-sectional flow area and P is the duct perimeter.

9-3. Fully Developed Laminar Incompressible Flow in Ducts

CIRCULAR TUBES

For laminar flow, the shear stress may be expressed not only as in Eq. (9.7) but also in terms of the velocity gradient:

$$\tau = \mu \frac{du}{dy} = -\mu \frac{du}{dr}$$

where u is the varying velocity throughout the cross section and $du/dy = -du/dr$ (since $y = R - r$, $dy = -dr$) is the velocity gradient. Then, from Eq. (9.7),

$$\frac{du}{dr} = -\frac{\tau}{\mu} = -\frac{\Delta p}{L} \frac{r}{2\mu}$$

Integration with the boundary condition $u = 0$ at $r = R$ gives

$$u = \frac{1}{4\mu} \left(\frac{\Delta p}{L} \right) (R^2 - r^2)$$
(9.11a)

which indicates a parabolic velocity profile. This equation could also be obtained by integrating the appropriate Navier–Stokes equation in cylindrical coordinates, namely

$$\frac{dp}{dx} = \mu \left(\frac{d^2u}{dr^2} + \frac{1}{r} \frac{du}{dr} \right) = \frac{\mu}{r} \frac{d}{dr} \left(r \frac{du}{dr} \right)$$

with the boundary condition that $u = 0$ at $r = R$, and $du/dr = 0$ at $r = 0$.

The maximum velocity u_{max} is at the center of the tube (at $r = 0$) and is

$$u_{max} = \frac{\Delta p}{L} \left(\frac{R^2}{4\mu} \right)$$

Thus the velocity profile of Eq. (9.11a) could also be written as

$$\frac{u}{u_{max}} = 1 - \left(\frac{r}{R} \right)^2$$
(9.11b)

The average velocity V is

$$V = \frac{Q}{A} = \frac{\int u \, dA}{A} = \frac{\int_0^R 2\pi r u \, dr}{\pi R^2} = \frac{\Delta p}{L} \left(\frac{R^2}{8\mu} \right) = \frac{u_{max}}{2}$$
(9.12)

or one-half the maximum. The pressure gradient, in terms of the flow rate, is

$$\frac{\Delta p}{L} = \frac{128 \mu Q}{\pi D^4} \tag{9.13}$$

The friction factor f for laminar flow is obtained by combining Eqs. (9.5) and (9.13) to get

$$f = \frac{64}{VD\rho/\mu} = \frac{64}{Re_D} \tag{9.14}$$

This expression has been verified experimentally and is valid for engineering calculations of both *smooth* and *rough* circular pipes for Reynolds numbers up to about 2000. It is possible to achieve laminar flow for Reynolds numbers as high as 40,000 or more in carefully controlled laboratory experiments, but instabilities are usually present in piping systems, limiting laminar flow to Reynolds numbers no greater than about 2000.

Similar relations for fully developed laminar flow of a power-law non-Newtonian fluid in circular tubes are developed in Appendix III.

Example 9-1

What is the pressure drop in 15 m of 6-mm tubing for a flow of Univis J-43 hydraulic fluid ($\mu = 0.014$ kg/ms and $s = 0.848$ from Fig. A-1 of Appendix II) at 50°C at a velocity of 2.00 m/s? What is the wall shear stress?

SOLUTION
The density is $\rho = (0.848)(1000) = 848$ kg/m^3. Then

$$Re_D = \frac{VD\rho}{\mu} = \frac{(2.00)(0.006)(848)}{0.014} = 727$$

$$f = \frac{64}{Re_D} = \frac{64}{727} = 0.0880$$

$$\Delta p = \frac{f(L/D)\rho V^2}{2} = \frac{(0.0880)(15/0.006)(848)(2.00)^2}{2} = 373 \text{ kPa}$$

From Eq. (9.2)

$$\tau_0 = \frac{\Delta p D}{4L} = \frac{(373,000)(0.006)}{(4)(15)} = 37.3 \text{ Pa}$$

NONCIRCULAR DUCTS
Various methods of analysis have been used to determine the friction factor $[f = (\Delta p/L)(2D_h/\rho V^2)]$ and Reynolds number ($Re = VD_h\rho/\mu$) relation for noncircular ducts. Results of these analyses for various cross sections are

Table 9-1
FRICTION
FACTORS FOR
CONCENTRIC
ANNULUS, LAMINAR
FLOW[1]

Table 9-2
FRICTION
FACTORS FOR
RECTANGLE
LAMINAR FLOW[1]

Table 9-3
FRICTION
FACTORS FOR
CIRCULAR
SEGMENT
LAMINAR FLOW[2]

r_1/r_2	f Re	a/b	f Re	α	f Re
0.0001	71.78	0	96.00	0	62.2
0.001	74.68	1/20	89.91	10	62.2
0.01	80.11	1/10	84.68	20	62.3
0.05	86.27	1/8	82.34	30	62.4
0.10	89.37	1/6	78.81	40	62.5
0.20	92.35	1/4	72.93	60	62.8
0.40	94.71	2/5	65.47	90	63.1
0.60	95.59	1/2	62.19	120	63.3
0.80	95.92	3/4	57.89	150	63.7
1.00	96.00	1	56.91	180	64.0

Table 9-4 LAMINAR FLOW FRICTION FACTORS[3]

	CIRCULAR SECTOR	ISOSCELES TRIANGLE	RIGHT TRIANGLE
α	f Re	f Re	f Re
0	48.0	48.0	48.0
10	51.8	51.6	49.9
20	54.5	52.9	51.2
30	56.7	53.3	52.0
40	58.4	52.9	52.4
50	59.7	52.0	52.4
60	60.8	51.1	52.0
70	61.7	49.5	51.2
80	62.5	48.3	49.9
90	63.1	48.0	48.0

given in Tables 9-1 through 9-4. Note that flow in a rectangle of zero aspect ratio represents flow between parallel plates and corresponds to flow in a concentric annulus of radius ratio approaching unity. Also note that a circular sector with $\alpha = 90°$ is identical to a circular segment with $\alpha = 90°$.

Example 9-2

The same oil as in Example 9-1 flows in an equilateral triangle duct 2 cm on each side. What is the pressure drop in 15 m of this duct for a flow velocity of 2 m/s?

SOLUTION
From Table 9-4, $f = 53.3/Re$. The duct area is $A = \sqrt{3}$ cm^2, the perimeter is $P = 6$ cm, and the hydraulic diameter is $D_h = 4A/P = (4)(\sqrt{3})/6 = 1.155$ cm $= 0.01155$ m. Then

$$Re = \frac{VD_h\rho}{\mu} = \frac{(2)(0.01155)(848)}{0.014} = 1399$$

$$f = \frac{53.3}{1399} = 0.0381$$

$$\Delta p = \left(\frac{fL}{D_h}\right)\left(\frac{\rho V^2}{2}\right) = \frac{(0.0381)(15/0.01155)(848)(2)^2}{2} = 83.9 \text{ kPa}$$

9-4. Fully Developed Incompressible Turbulent Flow in Ducts

CIRCULAR TUBES

In turbulent flow in a pipe, the radial components of velocity cause an interchange of momentum between adjacent layers of fluid and as a result the velocity profile is flatter than that for laminar flow. The higher the Reynolds number, the flatter the velocity profile. Since the velocity is zero at the pipe wall, this results in a very large velocity gradient at the pipe wall with a resulting higher wall shear stress than for laminar flow at the same Reynolds number.

The analysis of turbulent flow is not as simple as that for laminar flow. The Navier–Stokes equations may be written for turbulent flow to include velocity and pressure fluctuations. These equations involving statistical correlations of fluctuating quantities are known as the Reynolds equations. The number of unknowns, unfortunately, is greater than the number of equations. G. I. Taylor in 1935 suggested that, in the absence of mean velocity gradients, turbulence was not only homogeneous but also isotropic—independent of the orientation or location of the coordinate axes. The Reynolds equations then became simplified, and theoretical analyses and experimental studies have been enhanced by the assumption of isotropy. A number of statistical theories have received attention, notably those by Taylor, Burgers, von Kármán, Howarth, Dryden, Kolmogoroff, and Lin [see Ref. 4], but none has yet received universal acceptance.

A number of phenomenological theories have been advanced [5]. These are based on observation of transverse transport of fluid owing to a mixing process in turbulent flow. Saint-Venant in 1843 introduced a mixing co-efficient into fluid stress equations. Prandtl in 1925 made use of a mixing length which represents the transverse distance that fluid elements or lumps travel in exchanging momentum. This mixing length may be illustrated by comparing it with the mean distance molecules in a gas move transverse to the mean flow direction between molecular collisions. In 1930, von Kármán made use of a similar length parameter. Taylor in 1915 suggested that vorticity was conserved in this lateral transport of fluid elements or lumps, rather than momentum.

Reynolds in about 1880 showed that the turbulent shear stress in two-dimensional flow could be expressed as

$$\tau = -\rho \overline{u'v'}$$

where u' and v' are the instantaneous fluctuations of velocity from the mean in a direction parallel to the mean flow and normal to it, respectively. In a velocity gradient, Prandtl suggested that

$$u' \propto l \frac{du}{dy}$$

where l is the mixing length just mentioned, and v' is proportional to u'. Thus the turbulent shear stress could be written as

$$\tau = \rho l^2 \left| \frac{du}{dy} \right| \frac{du}{dy}$$

where the proportionality factor is included in l. Near a boundary the mixing length was considered to vary with y, the normal distance from the boundary, such that $l = \kappa y$. Thus, at the boundary,

$$\tau_0 = \rho \kappa^2 y^2 \left(\frac{du}{dy} \right)^2$$

This equation may be integrated to obtain

$$\frac{u}{\sqrt{\tau_0/\rho}} = \frac{u}{v_*} = \frac{1}{\kappa} \ln y + C$$

where v_* is the shear velocity. Experiments show that $\kappa = 0.4$. Thus, since $u = u_{max}$ at the pipe centerline (at $y = R$),

$$\frac{u_{max} - u}{\sqrt{\tau_0/\rho}} = \frac{u_{max} - u}{v_*} = 2.5 \ln \frac{R}{y} \tag{9.15}$$

It was assumed by von Kármán that the turbulent velocity fluctuations vary with a length l, which he considered to vary with du/dy divided by d^2u/dy^2, and with the mean velocity gradient du/dy. In addition, he assumed the turbulent shear to vary with $\rho l^2 (du/dy)^2$ as did Prandtl. Thus von

Kármán's analysis shows that

$$\sqrt{\frac{\tau}{\rho}} = -\kappa \frac{(du/dy)^2}{d^2u/dy^2}$$

and this gives, upon integration,

$$\frac{u_{max} - u}{\sqrt{\tau_0/\rho}} = \frac{u_{max} - u}{v_*}$$

$$= -\frac{1}{\kappa}[\ln(1 - \sqrt{1 - y/R}) + \sqrt{1 - y/R}] \tag{9.16}$$

where $\kappa = 0.36$, which fits experimental data quite well.

Neither Prandtl's nor von Kármán's concept is valid at the pipe centerline where the transverse momentum transport is zero or at the pipe wall where the flow is laminar. This laminar region is called the viscous sublayer and was discussed in Sec. 7-5. The results given by Eq. (9.16) apply quite well to very high Reynolds numbers. Turbulent flow at lower Reynolds numbers may be described by a power-law equation of the form of Eq. (7.22).

Three flow regimes are generally considered: (1) the hydraulically smooth regime, in which the roughness elements are submerged within the laminar sublayer (see Sec. 7-5; for this regime, $0 \leq v_*k/v \leq 5$); (2) the transition regime, in which the roughness elements protrude partly beyond the laminar sublayer (for this regime, $5 \leq v_*k/v \leq 70$); (3) the hydraulically rough regime, in which all roughness elements protrude beyond the laminar sublayer (for this regime, $v_*k/v > 70$).[†]

An empirical expression by Blasius for the wall shear stress for turbulent flow in a *smooth* pipe is

$$\tau_0 = 0.0395\,\rho V^2 \left(\frac{v}{VD}\right)^{1/4} = \frac{0.0395\,\rho V^2}{Re_D^{1/4}} \tag{9.17}$$

where V is the average velocity and D is the pipe diameter. [This was transformed in Eq. (7.23) to apply to a flat plate.] The friction factor is

$$f = \frac{8\tau_0}{\rho V^2} = \frac{0.316}{Re_D^{1/4}} \tag{9.18}$$

and this equation is known as *Blasius' law of pipe friction for smooth pipes.* It applies for Reynolds numbers no greater than about 100,000 and was obtained by assuming a velocity profile of the form

$$\frac{u}{u_{max}} = \left(\frac{y}{R}\right)^{1/n} \tag{7.22}$$

[†] Note that v_*k/v is a Reynolds number based on the shear velocity and roughness height k.

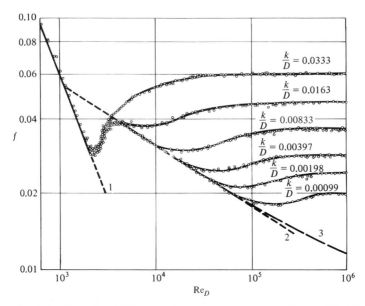

Fig. 9-4. Results of Nikuradse's measurements on pipes artificially roughened with sand grains. Curve 1: Eq. (9.14), $f = 64/\text{Re}_D$. Curve 2: Eq. (9.18), $f = 0.316/\text{Re}_D^{0.25}$. Curve 3: Eq. (9.19), $1/\sqrt{f} = 2 \log(\text{Re}_D\sqrt{f}) - 0.8$.

outside or beyond the laminar sublayer (but extending not quite to the pipe centerline) with $n \approx 7$, where u is the velocity at a distance y from the pipe wall. Actually, values of n experimentally determined by J. Nikuradse (see Schlichting [5]) vary from 6 at $\text{Re}_D = 4 \times 10^3$ to 10 at $\text{Re}_D = 3.2 \times 10^6$. This variation in n suggested a logarithmic form for the velocity profile and for the friction formula at higher Reynolds numbers. The result is

$$\frac{1}{\sqrt{f}} = 2 \log(\text{Re}_D \sqrt{f}) - 0.8 \tag{9.19}$$

and this equation is known as Prandtl's law of pipe friction for smooth pipes. Experiments by J. Nikuradse have verified this equation up to a Reynolds number of 3.4×10^6.

For *rough* pipes, the classical experiments are those by J. Nikuradse in which uniform sizes and sand grains were cemented to the inside surfaces of smooth pipes so that a range of relative roughness (k/D, the ratio of sand-gain diameter to pipe diameter) from about 0.001 to 0.033 was obtained. Friction factors and Reynolds numbers were determined for all test runs, and the results are shown in Fig. 9-4.

The analysis of flow in *rough* pipes (generalized results also apply to smooth pipes) involves the shear velocity v_*, defined as

$$v_* = \sqrt{\frac{\tau_0}{\rho}} = \sqrt{\frac{f}{8}}\, V \tag{9.20}$$

In this flow regime, Prandtl's semiempirical analysis gives a universal velocity distribution law as

$$\frac{u_{max} - u}{v_*} = 5.75 \log \frac{R}{y} = 2.5 \ln \frac{R}{y} \qquad [9.15]$$

where u_{max} is the centerline velocity, u is the variable velocity at a distance y from the pipe walls, and R is the pipe radius. From Eq. (9.15) the average velocity V may be calculated as

$$V = u_{max} - 3.75v_* \qquad (9.21)$$

Combining Eqs. (9.20) and (9.21) gives an expression for the ratio of the centerline velocity to the average velocity as

$$\frac{u_{max}}{V} = 1 + 1.33\sqrt{f} \qquad (9.22)$$

This equation may be used as a means of estimating the average velocity for fully developed turbulent flow in a round pipe from a single measurement of the centerline velocity, although a more accurate determination results from measurements which give the actual velocity profile.

A typical measured velocity profile for fully developed turbulent flow in a circular tube is shown in Fig. 9-5.

In the completely rough regime of flow, a semiempirical analysis also indicates that the velocity profile may be given in terms of the equivalent sand-grain roughness k as

$$\frac{u}{v_*} = 5.75 \log \frac{y}{k} + 8.5 = 2.5 \ln \frac{y}{k} + 8.5 \qquad (9.23)$$

When Eq. (9.23) is combined with Eq. (9.21) and the resulting constant adjusted for experimental results of Nikuradse, an expression for the friction factor becomes

$$f = \frac{1}{\left(2 \log \dfrac{D}{2k} + 1.74\right)^2} \qquad (9.24)$$

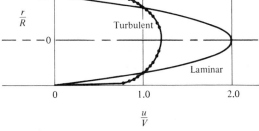

Fig. 9-5. Measured turbulent velocity profile at Re = 80,000 in a smooth circular tube compared with laminar velocity profile at Re < 2000.

An equation by Colebrook and White encompasses all flow regimes. This equation is

$$\frac{1}{\sqrt{f}} = 1.74 - 2 \log\left(\frac{2k}{D} + \frac{18.7}{\mathrm{Re}_D \sqrt{f}}\right) \tag{9.25}$$

which becomes Eq. (9.24) as Re_D gets very large, and Eq. (9.19) for smooth pipes for which $k \to 0$.

The application of theoretical analyses and test results for turbulent flow in commercial pipes is difficult because the absolute roughness k of these pipes is not well known. In addition, it is known that roughness spacing and shape, as well as size, affects pipe friction. Experiments on commercial pipes have led to the use of an *equivalent sand-grain roughness* for these pipes, and the results useful for engineering calculations are shown in Fig. 9-6. Calculations are only as accurate as the knowledge of pipe roughness, and estimates of pressure drops or head losses owing to pipe friction may be considered good if they are within 5 or 10 percent of the actual values.

A study of Figs. 9-4 and 9-6 indicates that (1) for laminar flow, the friction factor depends only on the Reynolds number and is not affected by pipe roughness; (2) for hydraulically smooth pipes, the friction factor for turbulent flow also depends only on the Reynolds number; (3) for the hydraulically rough regime, the friction depends only on the relative roughness of the pipe surface and is not dependent on the Reynolds number; (4) for the smooth-to-rough transition regime in rough pipes, the friction factor depends on both the relative roughness and the Reynolds number.

Three major types of problems may arise in connection with friction losses in pipes. These problems involve the determination of the following:

1. The pressure drop, or head loss, for a given flow of a given fluid in a given pipe. The relative roughness k/D and the Reynolds number are determined and the friction factor obtained from Fig. 9-6. The pressure drop, or head loss, is calculated from Eq. (9.5) or (9.6).
2. The discharge for a given pipe and fluid with a specified pressure drop, or head loss. Two methods may be used:
   a. A friction factor may be assumed and the velocity calculated from Eqs. (9.5) or (9.6). If the Reynolds number is high, the friction factor will depend only on the relative roughness of the pipe, and this is a good initial assumption. The Reynolds number based on the first estimate of velocity is then calculated and a better friction factor obtained from Fig. 9-6 if the Reynolds number is low enough so that the hydraulically rough regime is not reached. Successive trials will give the desired flow rate.
   b. A discharge may be assumed, and the corresponding velocity used to calculate an estimated Reynolds number. The pressure drop, or head loss, may then be determined as in the type problem

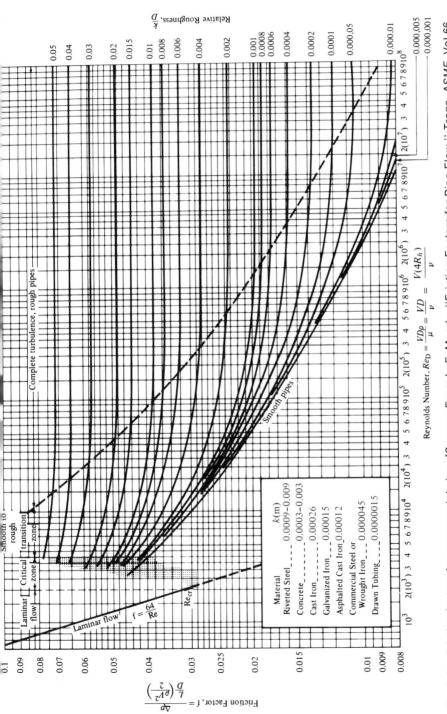

Fig. 9-6. Friction factors for commercial pipe. [Source: From L. F. Moody, "Friction Factors for Pipe Flow," *Trans. ASME*, Vol.66 (1944), pp. 671-684. Used with permission of the publishers, The American Society of Mechanical Engineers.]

referred to in (a) and compared with the specified value. Too large a calculated head loss indicates too large an assumed flow rate, and vice versa. Successive trials may be made until the calculated head loss agrees with the given value, or three trials may be plotted (Q vs. h_f) and the desired flow rate corresponding to the given head loss interpolated graphically.

3. The pipe size required to carry a given flow rate of a given fluid with a specified pressure drop or head loss. A pipe diameter may be assumed and the pressure drop, or head loss, calculated as in the type problem referred to in 2(a) and compared with the given value. If the calculated head loss is too large, the assumed pipe diameter was too small, and vice versa. Successive trials may be made until the calculated head loss agrees with the given value, or three trial solutions may be plotted (D vs. h_f) and the desired pipe diameter obtained by graphic interpolation. The next larger commercial pipe size should be chosen for an engineering design.

Example 9-3

What is the pressure drop in 200 m of smooth pipe 100 mm in diameter when oil with a viscosity 0.050 kg/m s and density 900 kg/m³ flows at (a) 0.50 m/s and (b) 3.0 m/s?

SOLUTION

(a) $Re = VD\rho/\mu = (0.5)(0.1)(900)/0.05 = 900$, indicating laminar flow. Then $f = 64/900 = 0.0711$ and

$$\Delta p = (f L/D)(\rho V^2/2)$$
$$= (0.0711)(200/0.1)(900)(0.5)^2/2 = 16.0 \quad \text{kPa}$$

(b) $Re = 5400$ and $f = 0.0360$ from Fig. 9-6. Then

$$\Delta p = (0.0360)(200/0.1)(900)(3.0)^2/2 = 291.6 \quad \text{kPa}$$

Example 9-4

What is the flow rate for water at 15°C in a 25-cm cast iron pipe when the head loss is 5.0 m in 300 m of pipe?

SOLUTION

Method 1. $k/D = 0.00026/0.25 = 0.00104$ and, for a high Re_D, $f = 0.020$. Then from Eq. (9.6)

$$\frac{V^2}{2g} = \frac{h_f D}{f L} = \frac{(5)(0.25)}{(0.020)(300)} = 0.208 \quad \text{and} \quad V = 2.02 \text{ m/s}$$

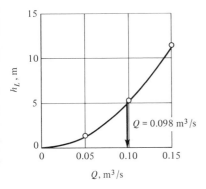

Fig. 9-7.   See Example 9-4.

For this first estimate of velocity, $Re_D = VD/v = (2.02)(0.25)/1.14 \times 10^{-6} = 4.4 \times 10^5$, and for this Re_D and given relative roughness, $f = 0.021$. Then

$$\frac{V^2}{2g} = \frac{(5)(0.25)}{(0.021)(300)} = 0.198 \quad \text{and} \quad V = 1.97 \text{ m/s}$$

The Reynolds number for this velocity is essentially the same as above, and thus $f = 0.021$. The flow rate is

$$Q = VA = (1.97)\left(\frac{\pi}{4}\right)(0.25)^2 = 0.097 \text{ m}^3/\text{s}$$

Method 2. If flow rates of 0.05, 0.10, and 0.15 m³/s are assumed, the corresponding head losses are calculated to be 1.33, 5.13, and 11.42 m, respectively. A graphic interpolation gives $Q = 0.098$ m³/s (Fig. 9-7).

Example 9-5

What is the flow rate for water at 15°C in a commercial steel pipe, 250-mm diameter, when the head loss in 300 m of pipe is 5.00 m?

SOLUTION

$k = 0.000045$, so the relative roughness is $k/D = 0.000045/0.25 = 0.00018$, and for high Re, $f = 0.0133$. Then

$$V^2/2g = h_f D/f L = (5.00)(0.25)/(0.0133)(300) = 0.313 \quad \text{m}$$

and

$$V = \sqrt{(2)(9.807)(0.313)} = 2.48 \quad \text{m/s}$$

For this velocity, $Re = VD/v = (2.48)(0.25)/1.140 \times 10^{-6} = 5.4 \times 10^5$. (The viscosity is found in Table A-8 in Appendix II.) For this Re, $f = 0.0152$.

Then $V^2/2g = h_f D/fL = 0.274$ and $V = 2.32$ m/s. For this velocity, Re = 5.1×10^5 and f is again 0.0152. Thus the flow rate is $Q = VA = (2.32)(\pi/64) = 0.114$ m^3/s = 114 L/s.

Example 9-6

Water flows at a rate of 91 L/s over a distance of 500 m in a commercial steel pipe with a pressure drop not to exceed 825 kPa. What is the minimum size pipe which can be used? Dynamic viscosity is 10^{-6} kg/m s.

SOLUTION

Method 1. From Eq. (9.5), $\Delta p = (fL/D)(\rho V^2/2) = (fL/D)(\rho Q^2/2A^2)$, where $A = \pi D^2/4$. Then

$$D = \left[\frac{8f L\rho Q^2}{\pi^2 \, \Delta p}\right]^{1/5}$$

A reasonable value of the friction factor may be assumed to obtain a first approximation for D. If this assumed value of f is in error by a factor of 2, the value of D will be within 15 percent ($2^{1/5} = 1.15$) of its correct value. Let $f_1 = 0.020$. Then

$$D = \left[\frac{(8)(0.020)(500)(1000)(0.091)^2}{(\pi^2)(825,000)}\right]^{1/5} = 0.152 \text{ m}$$

Then $V = Q/A = 0.091/(\pi/4)(0.152)^2 = 5.01$ m/s and the Reynolds number is

$$\text{Re}_D = \frac{VD}{\nu} = \frac{(5.01)(0.152)}{10^{-6}} = 7.6 \times 10^5$$

and

$$\frac{k}{D} = \frac{0.000045}{0.152} = 0.00030$$

so that from Fig. 9-6, $f = 0.0158$. Then $D = 0.145$ m from the above expression, and a recalculation for Re_D gives about the same value (7.6×10^5), but $k/D = 0.000045/0.145 = 0.00031$ and $f = 0.0158$ as before. Thus $D = 145$ mm.

Method 2. If diameters of 120, 150, and 200 mm are assumed, the corresponding pressure drops are 2190, 700, and 163 kPa, respectively. A graphic interpolation gives $D = 144$ mm.

NONCIRCULAR DUCTS, TURBULENT FLOW

The results for pressure drop or head loss for fully developed turbulent flow in circular tubes may be generally applied to noncircular ducts if the diameter

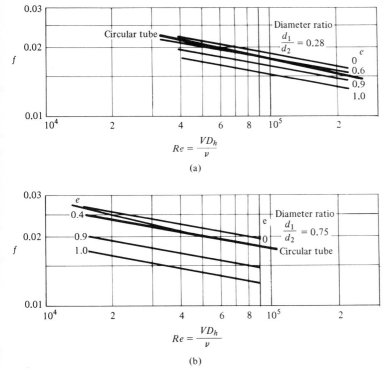

Fig. 9-8.   Friction factors for smooth eccentric annulus compared with friction factors for smooth circular tubes; after V. K. Jonsson [6].

D is replaced by the hydraulic diameter D_h [Eq. (9.10)] in the expressions for relative roughness, Reynolds number, and pressure drop or head loss. They may be used for square ducts, for rectangular ducts where the ratio of the two sides does not exceed about 8, for equilateral triangular ducts, for hexagonal ducts, and for concentric annular ducts with a ratio of inner to outer diameter up to at least 0.75 [6]. For these concentric annular ducts, the friction factor is about 6–8 percent greater than that for a smooth circular tube at Reynolds numbers from 15,000 to 150,000.

The effect of eccentricity on the friction factor in smooth annular ducts is shown in Fig. 9-8. The eccentricity is defined as $e = s/(r_2 - r_1)$, where s is the distance between the axis of the outer and the inner tube, and r_2 and r_1 are the radii of these tubes, respectively. When $e = 0$, the tubes are concentric; and when $e = 1$, the inner tube is in contact with the outer tube.

In ducts with tall, narrow, triangular cross sections, both laminar and turbulent flow may coexist at a section [7], and the analysis of this type of flow is not so straightforward as in ducts more nearly circular.

9-5. Steady Incompressible Flow in the Entrance Region of Ducts

In many instances of engineering interest, ducts are short and a knowledge of entrance-flow phenomena is necessary.

Equation (9.3) indicates that, for developing flow in the entrance region of a duct, the overall pressure drop overcomes wall shear and increases the flow momentum. In calculating this pressure drop, it is customary to use fully developed friction factors and to add a correction term k_L to account for entrance effects. Thus

$$\frac{p_1 - p}{\rho V^2 / 2} = \frac{fL}{D_h} + k_L \tag{9.26}$$

where p_1 is the pressure at the duct inlet, and L is the distance from the duct entrance to a location where the pressure is p. The correction term k_L accounts for the development of the velocity profile, incremental viscous dissipation in the entrance region relative to that for fully developed flow, and separation losses, if any. The value of k_L is a function of position along the duct, but becomes a constant in the fully developed flow region. For practical purposes, it is the fully developed value of k_L that is of greatest use and which is referred to in the hydraulic literature as an entrance loss coefficient. It depends on the Reynolds number of the flow and on the shape of the duct entrance as well as on the shape and surface of the duct itself. Values of k_L have been calculated and measured for laminar flow in ducts and have been measured for turbulent flow in some ducts.

The length of duct required to attain essentially fully developed conditions is called the *hydrodynamic entrance length* L_e. For engineering calculations, it is generally sufficient to associate the entrance length with the distance from the duct entrance that is needed for the pressure gradient to become within a specified percentage of the fully developed pressure gradient.

Values of k_L and of L_e for both laminar and turbulent flow in various ducts will be listed.

LAMINAR FLOW

The pressure drop in the entrance length L_e may be obtained from the Bernoulli equation written along the duct axis. This is valid since there is no shear in the core flow in the entrance region (Fig. 9-1a). Thus

$$p_1 - p_e = (\rho u_{max}^2 / 2) - \rho V^2 / 2 = \left[\left(\frac{u_{max}}{V} \right)^2 - 1 \right] (\rho V^2 / 2) \tag{9.27}$$

For any duct in which the ratio of maximum velocity to average velocity for fully developed flow is known, the pressure drop in the entrance region is given by Eq. (9.27). If values of both f and k_L are known, the entrance length may be obtained by combining Eqs. (9.26) and (9.27). Thus

$$\frac{L_e}{D_h} = \frac{1}{f} \left[\left(\frac{u_{max}}{V} \right)^2 - 1 - k_L \right] \tag{9.28}$$

Friction factors are known as functions of Re (Tables 9-1 through 9-4) and thus values of L_e/D_h Re may be calculated from Eq. (9.28) for ducts in which k_L values are known.

For a *circular tube*, $u_{max}/V = 2$. The value of k_L has been determined both analytically as well as experimentally with an average value of about 1.30 [1]. Thus since $f\,\text{Re} = 64$, the entrance length for laminar flow in a circular tube is

$$\frac{L_e}{D} = \frac{\text{Re}_D}{64}(2^2 - 1 - 1.30) = 0.0265\,\text{Re}_D \tag{9.29}$$

The pressure drop for fully developed flow in this length of pipe is

$$\Delta p = \frac{64}{\text{Re}_D}\left(\frac{L_e}{D}\right)(\rho V^2/2) = 1.70(\rho V^2/2)$$

and thus the pressure drop in the entrance region is greater than that for fully developed flow in an equal length of pipe by the factor

$$\frac{\Delta p_e}{\Delta p} = \frac{3}{1.70} = 1.76 \tag{9.30}$$

The average wall shear stress in the entrance region is greater than that for fully developed flow in an equal length of pipe by the factor

$$\frac{(\bar{\tau}_0)_e}{\tau_0} = 1.76 - \frac{2(\beta - 1)}{1.70} = 1.76 - \frac{2(1.33 - 1)}{1.70} = 1.37 \tag{9.31}$$

These values (1.76 and 1.37) are approximate and apply to the entrance region defined as the region where the pressure gradients and the wall shear stresses, respectively, are developing toward a fixed value. If the entrance region were defined as the region in which the velocity profiles are developing, both values would be lower, although the shear–stress ratio would always be less than the pressure-gradient ratio.

Some values of k_L and L_e/D_h Re for laminar flow in various ducts are given in Tables 9-5 through 9-8.

TURBULENT FLOW

Both analysis and experiment indicate the same general results for turbulent flow as for laminar flow, namely, that the wall shear stress becomes fixed in a shorter entrance region than the entrance region required for the pressure gradient to become fixed. This region, in turn, is shorter than the region required for the velocity profile to become fixed. The entrance region is shorter, however, for turbulent flow than for laminar flow on any similar basis of comparison.

Values of the entrance loss coefficient k_L in Eq. (9.26) as well as the entrance length L_e are determined experimentally. For a square entrance, flow separation results in a high k_L value of about 0.50–0.55 for circular tubes and concentric annuli [8]. For a rounded entrance, the initial boundary layer flow is generally laminar (Fig. 9-1b) and the entrance region pressure drop does not substantially exceed (it may even be less) the corresponding fully developed pressure drop. Thus k_L values for smooth circular tubes and concentric annuli are about 0.08 or less under these conditions. If an obstruc-

Table 9-5   ENTRANCE EFFECTS,
CONCENTRIC ANNULUS,
LAMINAR FLOW (see Table 9-1)

r_1/r_2	k_L
0.0001	1.13
0.001	1.07
0.01	0.97
0.05	0.86
0.10	0.81
0.20	0.75
0.40	0.71
0.60	0.69
0.80	0.69
1.00	0.69

Table 9-6   ENTRANCE EFFECTS,
RECTANGLE, LAMINAR FLOW
(see Table 9-2)

a/b	k_L	L_c/D_h Re
0	0.69	0.0059
1/8	0.88	0.0094
1/5	1.00	0.0123
1/4	1.08	0.0146
1/2	1.38	0.0254
3/4	1.52	0.0311
1	1.55	0.0324

Table 9-7   ENTRANCE
EFFECTS, CIRCULAR SEGMENT,
LAMINAR FLOW (see Table 9-3)

α	k_L
0	1.74
10	1.73
20	1.72
30	1.69
40	1.65
60	1.57
90	1.46
120	1.39
150	1.34
180	1.33

Table 9-8   ENTRANCE EFFECTS, LAMINAR FLOW
(see Table 9-4)

α	CIRCULAR SECTOR k_L	ISOSCELES TRIANGLE k_L	RIGHT TRIANGLE k_L
0	2.97	2.97	2.97
10	2.06	2.14	2.40
20	1.71	1.85	2.09
30	1.58	1.79	1.94
40	1.53	1.83	1.88
50	1.50	1.95	1.88
60	1.49	2.14	1.94
70	1.48	2.38	2.09
80	1.47	2.72	2.40
90	1.46	2.97	2.97

tion causes the boundary layer to be turbulent at the tube or annulus entrance, k_L values are larger, depending upon the size of the boundary layer trip.

Entrance lengths for turbulent flow in smooth tubes and concentric annuli are based on the length required for the pressure gradient to become within a specified percentage of the fully developed pressure gradient (5 percent, for example). On this basis L_e for these ducts (annulus diameter ratio up to 0.75) with square or rounded entrances are about 30 hydraulic diameters or less [6, 8]. The effect of increasing the eccentricity and the diameter ratio for the eccentric annulus is to increase the entrance length; $L_e = 91$ hydraulic diameters for an eccentricity $e = 1.0$ and a diameter ratio of 0.75. The range of Reynolds number for these annulus results is about 15,000 to 150,000.

Figure 9-9 shows the development of the pressure gradient in the entrance region of a smooth circular tube with a square entrance and with a rounded entrance [8]. The entrance lengths are seen to be about 30 hydraulic diameters or less based on the 5-percent pressure-gradient criterion. For the square entrance, the pressure-gradient ratio is less than unity at small values of x/D_h because of the effects of pressure recovery downstream of the vena

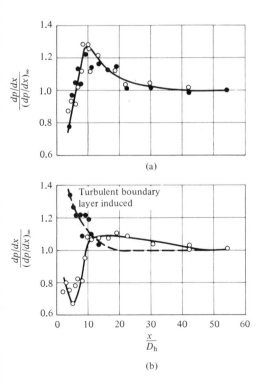

(a)

(b)

Fig. 9-9.   Local pressure gradients in a smooth tube [8]. (a) Square entrance Re = 60,000–70,000. (b) Rounded entrance, Re = 49,000.

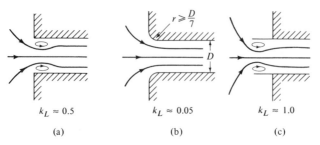

$k_L \approx 0.5$ $k_L \approx 0.05$ $k_L \approx 1.0$

(a)   (b)   (c)

Fig. 9-10. Pipe entrances with pipes flowing full. (a) Square. (b) Round. (c) Reentrant.

contracta (Fig. 9-10a). It is even negative at x/D_h values nearer zero where the pressure increases in the direction of flow in the expansion following the vena contracta. For the rounded entrance the solid curve in Fig. 9-9b indicates an initial laminar boundary layer which undergoes a transition to turbulence (Fig. 9-1) at the minimum point on the curve. With a boundary layer trip to induce a turbulent boundary layer at the tube entrance, the expected monotonically decreasing ratio of pressure gradients is obtained.

For both laminar and turbulent flow, local as well as average wall shear stresses and local as well as average pressure gradients generally are greater in the entrance region than in the fully developed flow region of a duct. The increase in local or average pressure gradients is greater than the increase in local or average wall shear stresses, since additional pressure drop must be provided to increase the momentum in the entrance region.

Example 9-7

A fluid enters a 20-mm-diameter smooth tube at a Reynolds number of 1500 with a flat velocity profile. What is the dimensionless pressure drop $\Delta p/(\rho V^2/2)$ in (a) 1.0 m of pipe and (b) 10.0 m of pipe?

SOLUTION
From Eq. (9.29) the entrance length is about $(0.0265)(1500) = 40$ diameters, or 0.8 m. Thus it is a large portion of the short pipe and a smaller part of the long pipe:

$$\frac{\Delta p}{\rho V^2/2} = f\,\frac{L}{D_h} + k_L, \quad \text{where} \quad f = \frac{64}{Re_D} \quad \text{and} \quad k_L = 1.30$$

(a) $\dfrac{\Delta p}{\rho V^2/2} = \dfrac{64}{1500}\,\dfrac{1.000}{0.020} + 1.30 = 3.42$

(b) $\dfrac{\Delta p}{\rho V^2/2} = \dfrac{64}{1500}\,\dfrac{10.000}{0.020} + 1.30 = 22.6$

From these results it is obvious that the entrance effects decrease for longer pipes and are important only for relatively short pipes.

▬▬

Example 9-8

Water drains from an open tank through a 100-mm-diameter smooth pipe with a square entrance from the tank (Fig. 9-10a). What head is required to produce a flow velocity of 3 m/s in the pipe for a pipe length of (a) 2 m, (b) 20 m, and (c) 200 m?

SOLUTION

$Re_D = VD/v = (3)(0.100)/10^{-6} = 3 \times 10^5$ and, from Fig. 9-6, $f = 0.0143$. The energy equation from the free surface in the tank to the discharge end of the pipe shows that the head must impart a velocity head to the flow, overcome the fully developed friction head loss, and the entrance head loss. Thus

$$h = \frac{V^2}{2g} + \frac{\left(\frac{fL}{D}\right)V^2}{2g} + \frac{k_L V^2}{2g}, \qquad \text{where} \quad k_L = 0.5$$

(a) $h = \left(\frac{V^2}{2g}\right)\left(1 + \frac{fL}{D} + h_L\right) = \left[\frac{(3)^2}{2g}\right](1 + 0.3 + 0.5) = 0.82 \text{ m}$

(b) $h = \left[\frac{(3)^2}{2g}\right](1 + 3.0 + 0.5) = 2.06 \text{ m}$

(c) $h = \left[\frac{(3)^2}{2g}\right](1 + 30 + 0.5) = 14.2 \text{ m}$

The numerical terms in the parentheses indicate that the entrance head loss was 28 percent (0.5/1.8), 11 percent, and 1.6 percent, respectively, of the total head loss in the three instances. Results indicate that entrance effects decrease for longer pipes and are important only for relatively short pipes.

▬▬

9-6. Contractions, Expansions, and Pipe Fittings; Turbulent Flow

Losses in contractions, expansions, and pipe fittings are largely due to separation effects and are known as *form* losses, as contrasted to pipe friction losses due to wall shear. Form losses generally are a result of vortex shedding with a subsequent vortex decay from viscous dissipation. Wall shear also accounts for a large part of the losses in a well-designed contraction, but in any case the losses are due to the viscosity of the fluid. The losses are generally expressed in terms of the velocity head $V^2/2g$ at the downstream end of the duct element. The effects actually extend for some distance downstream of the loss-producing element, because of the distance required for the velocity profile to reach its fully developed shape. It is customary in engineering calculations, however, to consider the loss as a localized loss and to be concentrated in the immediate neighborhood of the loss-producing section. In all instances except for expansions, the losses are expressed as

$$h_L = k_L \frac{V_2^2}{2g} \tag{9.32}$$

where k_L is an experimentally determined loss coefficient. For expansions, the loss may be written as

$$h_L = k_L \frac{(V_1 - V_2)^2}{2g} \tag{9.33}$$

section 1 being upstream from section 2.

Losses of this type are often referred to as *minor* losses, but they are minor only for relatively long pipelines; for short pipe systems, they often make up the major part of the total losses.

CONTRACTIONS

When a real fluid passes around a corner, separation occurs (Fig. 9-10), and the eddies in the separation zone give rise to head losses. These losses may be minimized by rounding the corners; the optimum is attained when the radius of curvature is one-seventh of the pipe diameter. A projecting entrance as shown in Fig. 9-10c is known as a Borda mouthpiece, for which the contracted flow section is one-half the pipe cross section if the pipe does not flow full (see Problem 4-57).

Abrupt changes in pipe size produce separation, and the loss coefficient is a function of the diameter ratio D_2/D_1 (Fig. 9-11). Typical values of k_L are given in Fig. 9-12.

A gradual contraction may be designed with a minimum head loss. Typical examples are contractions for subsonic wind tunnels, water tunnels, and fire nozzles. Separation is to be avoided near the downstream end of wind and water tunnel contractions, especially in a water tunnel, in order that low pressures, which enhance cavitation, may be avoided. In addition, a very flat velocity profile is desired in the test section immediately following the contraction, and this type of profile is typical of a contracting flow stream. Most of the total head at the contraction entrance is pressure head, which is

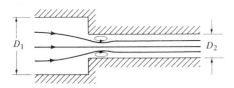

Fig. 9-11. Abrupt decrease in pipe size (contraction).

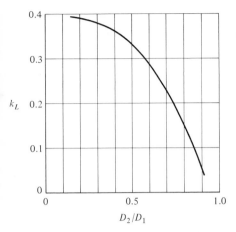

Fig. 9-12. Typical loss coefficients for an abrupt contraction.

uniform across the section. Most of the total head downstream is velocity head, and since very little of it originates from the nonuniform velocity head upstream, it is very uniform across the downstream section. A typical value of the kinetic energy correction factor α, which is a measure of the flatness of the velocity profile, at the downstream end of a well-designed contraction with $D_2/D_1 = \frac{1}{3}$ is 1.002, indicating a very flat profile. The loss coefficient in this contraction may well be as low as 0.03, based on the downstream velocity head.

EXPANSIONS

The head loss in a sudden expansion may be determined by writing the continuity, momentum, and energy equations for the isolated region indicated by dashed lines in Fig. 9-13. These equations are, respectively,

$$V_1 A_1 = V_2 A_2$$

$$p_1 A_1 + p_1(A_2 - A_1) - p_2 A_2 = V_1 A_1 \rho(V_2 - V_1)$$

neglecting wall shear, and

$$\frac{V_1^2}{2g} + \frac{p_1}{\gamma} = \frac{V_2^2}{2g} + \frac{p_2}{\gamma} + h_L$$

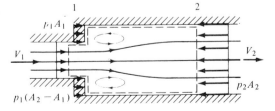

Fig. 9-13. Flow through a sudden enlargement in a pipe.

The pressure at the point of enlargement is approximately p_1, since the streamline spacing is essentially the same as at section 1. Solving for the head loss gives

$$h_L = \frac{(V_1 - V_2)^2}{2g} \tag{9.34}$$

A submerged pipe discharging into a large reservoir or tank represents a sudden enlargement for which $V_2 \to 0$, and for this the head loss is $h_L = V_1^2/2g$. Thus, one pipe velocity head is lost because of viscous dissipation as the jet leaving the pipe eventually is brought to rest.

The high losses owing to separation in a sudden enlargement may be reduced considerably if the enlargement is made gradual in order that separation be avoided. An enlargement of this type is known as a subsonic diffuser. If the total angle of divergence is made too small, both the wall friction and the fabrication costs will be large. An optimum divergence angle is about $7°$ or $8°$, for which the loss coefficient of Eq. (9-33) is about 0.14; it may be as low as 0.06 for a diffuser with a carefully designed parabolic curvature at the inlet with a length equal to one-half or more of the inlet diameter.

Flow in a diffuser is very complex. Diffusers are designed to convert velocity head into pressure head in subsonic wind tunnels, water tunnels, draft tubes for hydraulic turbines, and pump volutes, for example. Their ability to make this conversion is often used as a measure of efficiency and is known as the *pressure efficiency*, as contrasted to energy efficiency. The pressure efficiency may be written as

$$\eta_p = \frac{p_2 - p_1}{\alpha_1 \dfrac{\rho V_1^2}{2} - \alpha_2 \dfrac{\rho V_2^2}{2}} \tag{9.35a}$$

and the energy efficiency as

$$\eta_e = \frac{\alpha_1 \dfrac{V_1^2}{2g} - h_L}{\alpha_1 \dfrac{V_1^2}{2g}} \tag{9.35b}$$

Subscripts 1 and 2 refer to upstream and downstream sections, respectively, where the pressure is p, the velocity V, the kinetic energy correction factor α, the fluid density ρ, and the head loss in the diffuser h_L. In Eq. (9.35b), the efficiency is the ratio of energy at exit to the energy at inlet, the only energy at inlet being kinetic energy since pressure itself does not represent an ability to do work. Also, the pressure recovery for an incompressible fluid is independent of the pressure level at inlet, and thus the pressure intensity at inlet should not be included. Loss coefficients for diffusers are defined in various ways, and care should be exercised in their use.

Studies by Robertson and Ross [9] indicate that the pressure efficiency is lower for a thicker boundary layer at the entrance to a diffuser and decreases with divergence angle from 5° to 10°. The energy efficiency, however, is quite independent of these factors as long as separation is avoided. Typical values range from 83 to 93 percent for pressure efficiency and from 91.5 to 96.5 percent for energy efficiency for similar geometries and flows.

PIPE FITTINGS

Losses through pipe fittings may be given in the form of Eq. (9.32) or in terms of an equivalent pipe length. In either instance, calculations are approximate at best, and accurate information would require that direct measurements be made. Typical loss coefficients for various fittings are given in Table 9-9.

The pressure drop, or head loss, owing to energy dissipation in an elbow is largely due to a secondary flow superimposed on the main flow. As a result of centrifugal effects, the fluid in the faster core is directed outward and, in order to satisfy continuity (the outward flowing fluid must be replaced), the slower fluid near the boundaries is directed inward. This condition is shown in Fig. 9-14.

Example 9-9

Water flows in a 20-cm pipe which enlarges abruptly to a 30-cm pipe diameter. For a flow rate of 110 L/s (a) what is the head loss, (b) what is the pressure

Table 9-9 RESISTANCE COEFFICIENTS FOR VALVES AND FITTINGS, k_L

VALVE OR FITTING	NOMINAL DIAMETER, CM					
	2.5	5	10	15	20	25
Globe valve, wide open:						
Screwed	9	7	5.5			
Flanged	12	9	6	6	5.5	5.5
Gate valve, wide open:						
Screwed	0.24	0.18	0.13			
Flanged		0.35	0.16	0.11	0.08	0.06
Foot valve, wide open			0.80 for all sizes			
Swing check valve, wide open						
Screwed	3.0	2.3	2.1			
Flanged			2.0 for all sizes			
Angle valve, wide open:						
Screwed	4.5	2.1	1.0			
Flanged		2.4	2.1	2.1	2.1	2.1
Regular elbow, 90°						
Screwed	1.5	1.0	0.65			
Flanged	0.42	0.37	0.31	0.28	0.26	0.25
Long-radius elbow, 90°						
Screwed	0.75	0.4	0.25			
Flanged		0.3	0.22	0.18	0.15	0.14

SOURCE: Reproduced, with permission, from Engineering Data Book: Pipe Friction Manual (Cleveland: Hydraulic Institute, 1979).
Note: The k_L values listed may be expressed in terms of an equivalent pipe length for a given installation and flow by equating $k_L = fL_e/D$ so that $L_e = k_L D/f$.

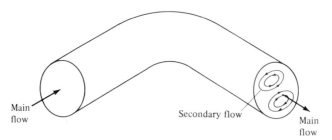

Fig. 9-14. Secondary flow in an elbow.

rise, (c) what is the pressure rise for a gradual enlargement, and (d) what is the pressure rise for a well-designed gradual enlargement?

SOLUTION

Let subscript 1 refer to the 20-cm-diameter duct and subscript 2 refer to the 30-cm-diameter duct.

(a) From Eq. (9.34), $h_L = (V_1 - V_2)^2/2g$, where

$$V_1 = \frac{Q}{A_1} = \frac{(0.110)}{(\pi/4)(0.20)^2} = 3.50 \text{ m/s}$$

and

$$V_2 = \frac{Q}{A_2} = \frac{(0.110)}{(\pi/4)(0.30)^2} = 1.56 \text{ m/s}$$

Thus $h_L = (3.50 - 1.56)^2/2g = 0.192$ m.

(b) The energy equation between sections 1 and 2 is

$$\frac{V_1^2}{2g} + \frac{p_1}{\gamma} = \frac{V_2^2}{2g} + \frac{p_2}{\gamma} + h_L$$

Thus

$$p_2 - p_1 = \gamma\left(\frac{V_1^2}{2g} - \frac{V_2^2}{2g} - h_L\right)$$

$$= 9810(0.624 - 0.124 - 0.192)$$

$$= 3020 \text{ Pa}$$

(c) For a gradual enlargement, $k_L = 0.14$ in Eq. (9.33) and

$$p_2 - p_1 = \gamma\left[\frac{V_1^2}{2g} - \frac{V_2^2}{2g} - \frac{k_L(V_1 - V_2)^2}{2g}\right]$$

$$= 9810[0.624 - 0.124 - (0.14)(0.192)]$$

$$= 4640 \text{ Pa}$$

(d) For a well-designed gradual enlargement (a diffuser), $k_L = 0.06$ and

$$p_2 - p_1 = 9810[0.624 - 0.124 - (0.06)(0.192)]$$
$$= 4790 \text{ Pa}$$

Results indicate a greater pressure recovery when a gradual enlargement is used and even better pressure recovery when a well-designed diffuser is used.

9-7. Applications

The form of the energy equation as applied to liquids for one-dimensional flow

$$\frac{V_1^2}{2g} + \frac{p_1}{\gamma} + z_1 - w = \frac{V_2^2}{2g} + \frac{p_2}{\gamma} + z_2 + h_L$$

indicates that, in the absence of external work, a decrease in total head occurs in the direction of flow. The total head at a section in a pipe system is the sum of the kinetic head and the piezometric head. A line connecting the values of total head at successive points along a piping system is known as the *total head line*, and a line connecting values of piezometric head at successive points along a piping system is known as the *piezometric head line*. The vertical distance between these lines at any section is the velocity head at that section. The energy or total head line *drops* in the direction of flow by an amount equal to the head loss h_L, which occurs in the direction of flow. For a constant-area pipe, this drop is due to friction and the slope of the total head line is equal to h_L/L. This slope is defined as the sine rather than the tangent of the angle with the horizontal. For entrances, contractions, expansions, pipe fittings, and exits the loss is generally considered to be more or less concentrated at the location of these items, although the effects of the loss-producing elements extend for some distance downstream from each of them. An example of the total head line and the piezometric head line for a piping system is shown in Fig. 9-15. If a pump is in the system, the total head line *rises* an amount equal to the work term w in the energy equation.

The height of the piezometric head line above the pipe center line represents the pressure head p/γ in the pipe. If the piezometric head line is below the pipe centerline, the pressure in the pipe is subatmospheric and the liquid in the pipe is under a partial vacuum. If the piezometric head line is below the pipe by a height equal to more than the height h_b of a barometer containing the liquid at the same temperature, vapor pressure will exist in the pipe and flow may cease. This condition, then, limits the height to which a liquid may be siphoned. If the piezometric head line is below the pipe centerline by an amount h_b or more in a water supply system, for example, either the pressure in the mains must be increased or the particular pipe must be laid at a lower level.

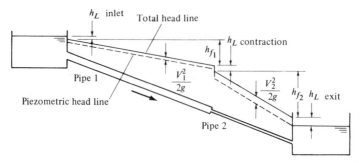

Fig. 9-15. Total head line and piezometric head line for flow between two open reservoirs.

PIPES IN SERIES

If two or more pipes are in *series*, the same flow passes through each pipe. If the pipes are designated with integer subscripts $(1, 2, 3,$ and so forth), the total head loss through the entire system is the sum of the losses through each individual pipe and fitting. These statements are expressed as

$$Q_0 = Q_1 = Q_2 = Q_3 = \cdots \tag{9.36a}$$

or

$$Q_0 = V_1 A_1 = V_2 A_2 = V_3 A_3 = \cdots \tag{9.36b}$$

and if h_L is the head loss for fittings and valves,

$$\Sigma h_L = h_{f1} + h_{f2} + h_{f3} + \cdots + h_L \tag{9.37}$$

PIPES IN PARALLEL

If two or more pipes are connected in *parallel*, the total flow rate is the sum of that through each individual branch and the head loss through one branch is the same as for all the others. These statements may be expressed as

$$Q_0 = Q_1 + Q_2 + Q_3 + \cdots \tag{9.38a}$$

or

$$Q_0 = V_1 A_1 + V_2 A_2 + V_3 A_3 + \cdots \tag{9.38b}$$

and

$$h_{L_1} = h_{L_2} = h_{L_3} = \cdots \tag{9.39}$$

The head loss through any one branch may be considered as being due purely to friction, or the valve and fitting losses may be expressed either in terms of an equivalent pipe length or as a loss coefficient times the velocity head in the pipe. Then Eq. (9.39) may be written as

$$\left(f_1\frac{L_1}{D_1} + \Sigma k_{L_1}\right)\frac{V_1^2}{2g} = \left(f_2\frac{L_2}{D_2} + \Sigma k_{L_2}\right)\frac{V_2^2}{2g}$$

$$= \left(f_3\frac{L_3}{D_3} + \Sigma k_{L_3}\right)\frac{V_3^2}{2g} = \cdots$$

so that

$$\frac{V_2}{V_1} = \sqrt{\frac{(f_1L_1/D_1) + \Sigma k_{L_1}}{(f_2L_2/D_2) + \Sigma k_{L_2}}}$$

and so on. The lengths, diameters, and loss coefficients are presumably known, and values of friction factors are estimated as in the type problem referred to in item 2 of Sec. 9-4. Then, in order to determine the flow distribution or the head loss through the system, Eqs. (9.38) may be written and V_1 (and thus Q_1), for example, estimated from

$$Q_0 = V_1A_1 + \frac{V_2}{V_1}V_1A_2 + \frac{V_3}{V_1}V_1A_3 + \cdots$$

from which V_2, V_3, and so forth (and thus Q_2, Q_3, and so forth) may also be estimated. A check of the assumed friction factors by calculating the respective Reynolds numbers based on the estimated velocities and the relative roughness of the pipes should be made and the new values of the friction factors used for an improved set of calculations. However, this check is often unnecessary.

Example 9-10

A new commercial steel pipe 200 mm in diameter and 1000 m long is in parallel with a similar pipe 300 mm in diameter and 3000 m long. The total flow in both pipes is 0.20 m³/s. What is the head loss through the system? Use water at 20°C (kinematic viscosity is 10^{-6} m²/s) and consider only frictional head loss.

SOLUTION
The relative roughness of the pipes are 0.000225 and 0.00015, respectively. At large Reynolds number the corresponding friction factors are 0.014 and 0.013. These are estimates, and a trial solution for the velocity in each pipe is made from these data. Then Reynolds numbers and more accurate friction factors are obtained iteratively. Using subscripts 1 and 2 for the smaller and larger pipes, respectively,

$$\frac{V_2}{V_1} = \sqrt{\frac{f_1}{f_2}\frac{L_1}{L_2}\frac{D_2}{D_1}} = \sqrt{\left(\frac{0.014}{0.013}\right)\left(\frac{1000}{3000}\right)\left(\frac{300}{200}\right)} = 0.734$$

The pipe areas are 0.0314 and 0.0707 m^2. Then, from continuity, $Q = V_1 A_1 + V_2 A_2$ or $0.20 = 0.0314 V_1 + (0.734 V_1)(0.0707)$ and $V_1 = 2.40$ m/s and $V_2 = 1.76$ m/s. Corresponding Reynolds numbers are

$$\text{Re}_1 = \frac{(2.40)(0.20)}{10^{-6}} = 4.8 \times 10^5 \quad \text{and} \quad f_1 = 0.0156$$

$$\text{Re}_2 = \frac{(1.76)(0.30)}{10^{-6}} = 5.3 \times 10^5 \quad \text{and} \quad f_2 = 0.0150$$

Then a recalculation of V_2/V_1 gives a value of 0.721, from which $V_1 = 2.43$ m/s. The head loss is the same for each pipe, and for pipe 1

$$h_f = \left(\frac{f_1 L_1}{D_1}\right)\left(\frac{V_1^2}{2g}\right) = \frac{(0.0156)(1000/0.20)(2.43)^2}{2g}$$

$$= 23.5 \text{ m}$$

Example 9-11

In Fig. 9-15 the open reservoirs are 20 m apart in elevation. Pipe 1 is 60 cm in diameter and 200 m long. Pipe 2 is 30 cm in diameter and 100 m long. Both are made of commercial steel. The entrance to pipe 1 is square ($k_L = 0.5$) and the junction between pipes 1 and 2 is an abrupt contraction. Assume the water viscosity to be 10^{-6} m^2/s. What is the flow rate?

SOLUTION

$k_1/D_1 = 0.000045/0.6 = 0.000075$ and $f_1 = 0.0115$ if Re_1 is large; $k_2/D_2 = 0.000045/0.3 = 0.00015$ and $f_2 = 0.0128$ if Re_2 is large. The energy equation written between the free surfaces of the reservoirs, with all losses shown in Fig. 9-15 included, is

$$0 + 0 + 20 = 0 + 0 + 0 + h_{L\,\text{inlet}} + h_{f-1} + h_{L\,\text{contraction}} + h_{f-2} + h_{L\,\text{exit}}$$

$$20 = \frac{(0.5)V_1^2}{2g} + \frac{(f_1 L_1/D_1)V_1^2}{2g} + \frac{k_L V_2^2}{2g} + \frac{(f_2 L_2/D_2)V_2^2}{2g} + \frac{V_2^2}{2g}$$

From continuity $V_2 = 4V_1$, and thus the energy equation may be written as

$$20 = \frac{V_1^2}{2g}\left\{0.5 + \frac{(0.0115)(200)}{0.6} + (0.35)(16) + \left[\frac{(0.0128)(100)}{0.3}\right]16 + 16\right\}$$

so that $V_1 = 2.04$ m/s and $V_2 = 8.16$ m/s. Better values of the friction factors are obtained after calculating the Reynolds numbers for the first approximations to the velocities.

$$\text{Re}_1 = \frac{(2.04)(0.60)}{10^{-6}} = 1.22 \times 10^6 \quad \text{and then} \quad f_1 = 0.013$$

$$\text{Re}_2 = \frac{(8.16)(0.30)}{10^{-6}} = 2.45 \times 10^6 \quad \text{and then} \quad f_2 = 0.0135$$

A recalculation of the value of V_1 with these new f values is

$$20 = \frac{V_1^2}{2g}\left[0.5 + \frac{(0.013)(200)}{0.6} + (0.35)(16) + \frac{(0.0135)(100)(16)}{0.3} + 16\right]$$

so that $V_1 = 2.00$ m/s and $V_2 = 8.00$ m/s. Both new Reynolds numbers are essentially the same as above. Then the flow rate is

$$Q = V_1 A_1 = (2.00)\left(\frac{\pi}{4}\right)(0.60)^2 = 0.565 \text{ m}^3/\text{s}$$

9-8. Empirical Pipe-Flow Equations

A number of empirical pipe-flow equations for water in pipes have been used, and the equation of Hazen and Williams is perhaps the most widely used. This equation is

$$V = 1.318C(R_h)^{0.63}S^{0.54} \qquad \text{ft/s}$$

$$Q = 1.318C(R_h)^{0.63}S^{0.54}A \quad \text{ft}^3/\text{s}$$

where

R_h = hydraulic radius of the pipe, A/P ($R_h = D/4$ for a round pipe), in feet;

S = slope of the total head line h_f/L;

A = pipe cross-sectional area;

C = roughness coefficient.

In SI units the Hazen–Williams equations are

$$V = 0.850CR_h^{0.63}S^{0.54} \qquad \text{m/s} \tag{9.40a}$$

and

$$Q = 0.850CR_h^{0.63}S^{0.54}A \quad \text{m}^3/\text{s} \tag{9.40b}$$

where the hydraulic radius R_h is in meters. The roughness coefficient C is the same whether the equations in the technical English system or in the SI system are used. The different coefficients (1.318 and 0.850) indicate that the Hazen–Williams equations are not dimensionally consistent.

Roughness values C for the Hazen–Williams equation are given in Table 9-10. The Hazen–Williams equation is based on the premise that the Reynolds numbers are large and the pipes are reasonably rough so that the flow regime is in the range labeled complete turbulence, rough pipes in Fig. 9-6. In this range the friction factors or roughness coefficients are independent of the Reynolds number. The equation is relatively easy to use with electronic hand calculators, but special graphs, alignment charts, and

Table 9-10 HAZEN-WILLIAMS ROUGHNESS VALUE

TYPE OF PIPE	C
Extremely smooth pipes	140
New steel or cast iron	130
Wood, average concrete	120
New riveted steel, clay	110
Old cast iron, brick	100
Old riveted steel	95
Badly corroded cast iron	80
Very badly corroded iron or steel	60

slide rules are also available for its use. Equations of this type are very useful in the analysis of pipe problems.

Flow in parallel pipe systems may readily be solved. Since $R_h = D/4$ for a round pipe, we may write Eq. (9.40b) as

$$Q = \frac{0.850\pi CD^{2.63}}{4^{1.63}} \left(\frac{h_L}{L}\right)^{0.54}$$

Thus from Eq. (9.38a),

$$Q_0 = h_L^{0.54}(C_1' + C_2' + C_3' + \cdots + C_n')$$

where

$$C' = \frac{0.850\pi CD^{2.63}}{4^{1.63}L^{0.54}}$$

which has a fixed value for each pipe. Therefore, any assumed head loss h_L through the parallel system will give flows in each pipe in the correct proportion, though the total may not be correct. The flow in each branch may be corrected by the same factor needed to correct the total flow to the given Q_0, and the head loss may be determined directly from Eq. (9.40).

Example 9-12

Determine the flows in each of the two parallel pipes in Example 9-10 by using the Hazen–Williams equation.

SOLUTION

$C = 130$ from Table 9-10. Assume a head loss of $h_L = 20$ m. Then for the 200-mm pipe, $h_L/L = 20/1000$ and

$$Q_{200} = (0.850)(130)\left(\frac{0.200}{4}\right)^{0.63}\left(\frac{20}{1000}\right)^{0.54}\left(\frac{\pi}{4}\right)(0.200)^2$$

$$= 0.0636 \text{ m}^3/\text{s}$$

For the 300-mm pipe $h_L/L = 20/3000$ and

$$Q_{300} = (0.850)(130)\left(\frac{0.300}{4}\right)^{0.63}\left(\frac{20}{3000}\right)^{0.54}\left(\frac{\pi}{4}\right)(0.300)^2$$

$$= 0.1021 \text{ m}^3/\text{s}$$

The total flow for the assumed 20-m head loss would be 0.1657, whereas the actual flow is 0.200 m^3/s. Thus a factor of 1.207 applied to each branch will result in a flow of 0.200 m^3/s total:

$$Q_{200} = (0.0636)(1.207) = 0.0768 \text{ m}^3/\text{s}$$

$$Q_{300} = (0.1021)(1.207) = 0.1232 \text{ m}^3/\text{s}$$

for a total of 0.200 m^3/s. These results compare with 0.0763 and 0.1237 m^3/s, respectively, from Example 9-10.

9-9. Pipe Networks

Complex pipe networks conveying water may be analyzed quite readily using the Hazen-Williams equation. The flow distribution for a given network may be desired, and this is generally an indeterminate problem which must be solved by successive trials, or iterations. In the design of a network, the flow and pressures at various points may be specified and the pipe sizes determined. This is also an indeterminate type of problem which must be solved by successive trials, or iterations.

A network consists of a finite number of loops containing any number of individual pipes, some of which may be common to two loops. In Fig. 9-16 the simple network has two loops, pipe 2 being common to both loops. Two

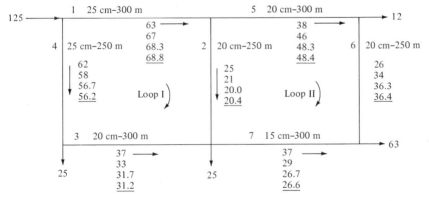

Fig. 9-16. See Example 9-13. Flow distribution by Hardy–Cross method. Pipe numbers are labeled. Flow rates are in liters per second (L/s). See Table 9-11.

conditions must be met for a balanced flow in the network:

1. The net flow into any junction must be zero. This means that the flow rate into the junction must equal the flow rate out of the junction.
2. The net head loss (or pressure drop) around a loop must be zero. If a loop is traversed in either direction, a balanced flow must result in a return to the original condition (head or pressure) at the starting point.

The procedure for determining the flow distribution in a given network involves assigning flows in each pipe so that continuity at each junction is satisfied (condition 1). Then the head loss around each loop is calculated and, if not zero, adjustments to the assumed flows are made either by pure estimate or by a method of iteration known as the Hardy–Cross method. The correction for each loop is given by

$$\Delta Q = -\frac{\text{net head loss for the assumed flows}}{1.85\,\Sigma h_L/Q_0 \text{ for the assumed flows}}$$

This equation is derived as follows:

In a given loop in a network let Q be the actual, or balanced, flow rate and Q_0 the assumed flow rate, so that $Q = Q_0 + \Delta Q$. Then, since the Hazen–Williams equation (as well as others) may be expressed as $h_L = nQ^x$, we may write

$$nQ^x = n(Q_0 + \Delta Q)^x$$

$$= n\left[Q_0^x + xQ_0^{x-1}\,\Delta Q + \frac{x(x-1)}{2}\,Q_0^{x-2}(\Delta Q)^2 + \cdots\right]$$

If ΔQ is truly small compared to Q_0, terms beyond the second may be neglected. For a balanced loop or network,

$$\Sigma h_L = \Sigma nQ^x = \Sigma nQ_0^x + \Delta Q\,\Sigma xnQ_0^{x-1} = 0$$

Solving for ΔQ we get

$$\Delta Q = -\frac{\Sigma nQ_0^x}{\Sigma xnQ_0^{x-1}} = -\frac{\Sigma h_L}{1.85\,\Sigma h_L/Q_0} \tag{9.41}$$

since $x = 1.85$ (the reciprocal of 0.54) in the Hazan-Williams equation.

The procedure is as follows:

1. Assume any reasonable flow distribution, in both magnitude and direction, in all pipes so that the total flow into each junction is algebraically zero. This should be indicated on a diagram of the pipe network.
2. Set up a table to analyze each closed loop in the network semi-independently.

3. Compute the head loss h_L in each pipe.
4. For each loop, consider the flow rate Q_0 and the head loss h_L to be positive for clockwise flow in the loop and negative for counterclockwise flow.
5. Compute the algebraic head loss Σh_L in each loop.
6. Compute the total head loss per unit discharge h_L/Q_0 for each pipe. Determine the sum of the quantities $\Sigma h_L/Q_0 = \Sigma n Q_0^{0.85}$ for each loop. From the definitions of head loss and flow direction, each term in this sum is necessarily positive.
7. Determine the flow correction for each loop from

$$\Delta Q = -\frac{\Sigma h_L}{1.85 \, \Sigma h_L/Q_0} \qquad [9.41]$$

This correction is to be applied algebraically to each pipe in the loop. For a pipe which is in common with another loop, the flow correction for that pipe is the net effect of the corrections for both loops.
8. Indicate corrected flows on the diagram of the pipe network as in step 1. A check on the corrections of step 7 will be shown by a continuity check at each pipe junction.
9. Repeat steps 1 thru 8 until either the head loss for a loop is balanced within desired limits or the flow corrections are made as small as desirable.

Flow corrections may be made by either of two methods:

1. Corrections for all loops may be made before any corrections are applied. The head loss and the value of h_L/Q_0 for a pipe in common with two loops need be calculated but once and the results used in both loops.
2. Correction for a loop may be applied to each pipe in that loop before calculating correction in the next or successive loops.

Example 9-13

Given the pipe network shown in Fig. 9-16, determine the flow rate through each pipe. The pipe sizes and lengths are shown alongside each pipe. Assume a C value of 100.

SOLUTION
Assumed flow and direction are indicated near each pipe, together with corrected flows for each trial as indicated in Table 9-11. Corrections were made to the nearest 1 L/s for the first and second trials, and to the nearest 0.1 L/s for the third trial. Final results are shown underlined in Fig. 9-16. This type of problem may be solved readily on programmable computers.

Table 9-11   DATA FOR EXAMPLE 9-13

LOOP	PIPE	DIAMETER (cm)	L (m)	Q_0 (L/s)	h_L (m)	h_L/Q_0	Q_0 (L/s)	h_L (m)	h_L/Q_0	Q_0 (L/s)	h_L (m)	h_L/Q_0
					FIRST TRIAL			SECOND TRIAL			THIRD TRIAL	
I	1	25	300	+63	+3.23	0.0513	+67	+3.62	0.0540	+68.3	+3.75	0.0549
	2	20	250	+25	+1.44	0.0576	+21	+1.04	0.0495	+20.0	+0.95	0.0475
	3	20	300	−37	−3.58	0.0968	−33	−2.89	0.0875	−31.7	−2.69	0.0849
	4	25	250	−62	−2.61	0.0421	−58	−2.31	0.0398	−56.7	−2.22	0.0391
					−1.52	0.2478		−0.54	0.2308		−0.21	0.2264

$$\Delta Q = -\frac{(-1.52)}{(1.85)(0.2478)} \qquad \Delta Q = -\frac{(-0.54)}{(1.85)(0.2308)} \qquad \Delta Q = -\frac{(-0.21)}{(1.85)(0.2264)}$$

$$= +3.3 \text{ L/s} \qquad\qquad = +1.3 \text{ L/s} \qquad\qquad = +0.5 \text{ L/s}$$

LOOP	PIPE	DIAMETER (cm)	L (m)	Q_0 (L/s)	h_L (m)	h_L/Q_0	Q_0 (L/s)	h_L (m)	h_L/Q_0	Q_0 (L/s)	h_L (m)	h_L/Q_0
II	5	20	300	+38	+3.76	0.0989	+46	+5.35	0.1163	+48.3	+5.86	0.1213
	6	20	250	+26	+1.55	0.0596	+34	+2.55	0.0750	+36.3	+2.88	0.0793
	7	15	300	−37	−12.09	0.3268	−29	−9.25	0.3190	−26.7	−7.94	0.2973
	2	20	250	−25	−1.44	0.0576	−21	−1.04	0.0495	−20.0	−0.95	0.0477
					−8.22	0.5429		−2.39	0.5598		−0.15	0.5456

$$\Delta Q = -\frac{(-8.22)}{(1.85)(0.5429)} \qquad \Delta Q = -\frac{(-2.39)}{(1.85)(0.5598)} \qquad \Delta Q = -\frac{(-0.15)}{(1.85)(0.5456)}$$

$$= +8.2 \text{ L/s} \qquad\qquad = +2.3 \text{ L/s} \qquad\qquad = +0.1 \text{ L/s}$$

9-10. Flow of Mixtures in Pipes

Two-phase flow in pipes (gas–liquid, solids in liquids, and solids in gases) is quite complicated and in most instances is not well understood. There are some situations, however, in which either empirical results or experiments based on dimensional analysis which result in semiempirical equations have been found useful. The hydraulic conveying of paper pulp and solids will be discussed briefly.

PAPER STOCK

Wood fibers for making paper are suspended in water as a means of handling and are conveyed in pipelines at consistencies (ratio of weight of air-dry pulp to the weight of water in a given total volume of mixture) up to about 6 percent. The effect of the presence of the wood fibers is to increase the wall shear of the more or less homogeneous mixture as compared to that for water flowing alone. The pressure drop, or head loss, due to friction is not noticeably affected for consistencies below about 1.3 percent.

Experiments have been made on a practical basis with consistencies as one of the major parameters. An analysis and check [10] of experiments conducted in Germany [11] resulted in correlations that can be expressed in terms of the Darcy–Weisbach equation. The friction factor is a function of the type of pulp and a pseudo-Reynolds number. The effective friction factor to be used in the equation

$$h_f = f \frac{L}{D} \left(\frac{V^2}{2g} \right)$$

is given by

$$f = \frac{250 \, K'}{\text{Re}'^{1.63}} \tag{9.42}$$

In this expression K' depends on the type of paper pulp, and is

1.0 for unbleached sulfite, southern kraft, and cooked groundwood;
0.9 for soda, sulfate, bleached sulfate, and reclaimed paper;
1.2 for Canadian kraft and groundwood.

The value of the pseudo-Reynolds number for $1.3 < C < 6$ is

$$\text{Re}' = 0.26 \frac{D^{0.205} V \rho}{C^{1.157}} \tag{9.43}$$

where D is the pipe diameter in meters, V the average flow velocity in meters per second, ρ the density of water ($1000 \, \text{kg/m}^3$), and C is the stock consistency in percent.

SOLIDS

In the hydraulic conveying of solids in pipes (sand, gravel, and coal, for example) two flow regimes are generally considered.

1. *Homogeneous* transport applies to fine particles which are maintained in suspension, and the mixture flows like a homogeneous fluid. In laminar flow, however, the particles may settle out. For turbulent flow, the usual methods for calculating head loss apply if the density and apparent viscosity of the mixture are used. The concentration of fine particles should be low so that the mixture does not become non-Newtonian.
2. In *heterogeneous* flow, particles may tend to slide at the surface of a more or less stationary bed formed by the settled particles along the bottom of a horizontal pipe, or they may move as a sliding bed. For this regime it is possible to estimate head losses, but the actual physical situation is not well understood.

An empirical equation from Worster and Durand (see Smith [12]) for this type of flow gives the relative increase in pressure gradient for a mixture compared to that for water alone. This equation is

$$\frac{(\Delta p/L)_m - (\Delta p/L)_w}{(\Delta p/L)_w} = 121 C \left[\frac{gD(s_s - 1)}{V^2} \frac{V_s}{\sqrt{gd(s_s - 1)}} \right]^{1.5} \tag{9.44}$$

where subscripts m, w, and s refer to mixture, water, and solids, respectively; C is particle concentration by volume; D is pipe diameter; s_s is the specific gravity of the solids; V the mean flow velocity of the water; V_s the settling velocity of the particles in still water; and d is the particle diameter (or equivalent) for which 85 percent of the particles are smaller (15 percent are larger than d).

This equation has been verified in tests with sand, gravel, and manganese dioxide up to $\frac{3}{16}$ in. in mean diameter in pipes from 1 to 3 in. in diameter, with $\frac{1}{12}$-in. sand and $\frac{1}{2}$-in. coal in 3-in. pipes, and with $\frac{1}{12}$-in. sand and 1-in. coal in 6-in. pipes [12, 13]. Specific gravities of coal, sand, gravel, and manganese dioxide are 1.4, 2.6, 2.6, and 4.1, respectively.

Equation (9.44) indicates that the increase in pressure drop for a slurry of solids in water varies inversely as the cube of the flow velocity. The equation also involves the ratio of two forms of the Froude number: one form is based on the flow velocity and the pipe diameter; the other on the settling velocity of the particles and the particle diameter. The expression has not been verified for slurries with a wide variation of particle sizes and thus should be used with caution in those instances.

In order to use Eq. (9.44), the settling velocity of the particles must be determined by direct measurement. Natural particles are rarely spheres, and the drag on nonspherical particles, such as sand and gravel, is greater than that on spherical particles of the same material and mass.

Over a wide range of Reynolds numbers the drag coefficient for spheres (Fig. 11-4) and nonspherical particles is essentially constant, and the settling velocity for a given shape of material varies as the square root of its size. Then $V_s/\sqrt{d}$ in Eq. (9.44) is a constant, regardless of particle size, and the equation may be applied to any size distribution of particles so long as they are not so small that the drag coefficient increases. An average value of $V_s/\sqrt{d}$ measured for a number of particles would be used in estimating the pressure drop for a heterogeneous mixture of solids in water.

PROBLEMS

9-1. Crude oil with a viscosity of 4.5×10^{-6} m²/s flows in a 25-mm-diameter tube.
(a) What is the maximum average flow velocity for which the flow may be considered laminar?
(b) What is the pressure drop in 50 m of tubing at that flow?
(c) What is the wall shear stress?

9-2. Water at 20°C ($v = 1.00 \times 10^{-6}$ m²/s) flows in a 1.00-cm circular tube.
(a) What is the maximum average velocity for laminar flow?
(b) What is the pressure drop in 50 m of tubing?

9-3. An oil flows from a large open cup through a vertical tube 45 cm long with an inside diameter of 1 mm at a rate of 20 cm³/min. The oil surface in the cup is 60 cm above the end of the tube. What is the kinematic viscosity of the oil? Assume fully developed laminar flow throughout the entire length of the tube, and neglect the exit velocity head. Verify these assumptions as being valid.

9-4. (a) Compare the cost of pumping fuel oil through 150 m of 10-cm commercial steel pipe at 6.5 L/s at 10°C (kinematic viscosity 0.002 m²/s) with that at 40°C (kinematic viscosity 2.6×10^{-4} m²/s).

(b) Would it be economical to heat the oil from 10°C to 40°C in order to reduce pumping costs? For fuel oil $c_p = 2350$ J/kg °C.

9-5. Olive oil at 15°C ($\mu = 8.2 \times 10^{-2}$ kg/m s and $\rho = 918$ kg/m³) is pumped at a rate of 10 L/s. What size pipe will convey this oil with a pressure drop of 80 kPa in 31 m of pipe?

9-6. At what radial distance from the pipe axis will the velocity be equal to the mean velocity for fully developed laminar flow?

9-7. Laminar flow exists in a round pipe. If the flow rate is reduced, what effect does this have on

(a) the wall shear stress and

(b) the friction factor?

9-8. Laminar flow occurs in the annulus between two coaxial cylinders of radius R_1 and R_2 ($R_2 > R_1$), respectively. The pressure drop over a length L is Δp. Show that the average velocity of flow is

$$V = \frac{\Delta p}{8\mu L}\left[(R_1^2 + R_2^2) - \frac{(R_2^2 - R_1^2)}{\ln R_2/R_1}\right]$$

where μ is the dynamic viscosity of the fluid. *Hint*: Let τ be the shear stress at a radius r. Apply the momentum theorem to a cylindrical shell in the annular region to get

$$\frac{dp}{dL} + \frac{1}{r}\frac{d(\tau r)}{dr} = 0$$

Since τ is a function of r only and p of L only, integrate first with respect to r, then with respect to r again substituting $\tau = -\mu(du/dr)$ and multiplying through by dr/r with boundary conditions $u = 0$ at $r = R_1$ and at R_2. This gives an expression for the varying velocity $u = u(r)$.

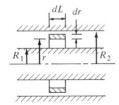

Prob. 9-8

9-9. In Prob. 9-8, show that for $R_2/R_1 = 2$, $f = 95.2/Re$ for fully developed laminar flow in an annulus. In general

$$f\,Re = \frac{64\,[(R_2/R_1) - 1]^2}{\left(\dfrac{R_2}{R_1}\right)^2 + 1 - \dfrac{(R_2/R_1)^2 - 1}{\ln(R_2/R_1)}}$$

9-10. Laminar flow occurs in a circular tube. A thin diametrical plate divides the flow area into two equal semicircular segments. By what factor is the pressure gradient increased for the same flow rate?

9-11. Laminar flow occurs in a circular tube. By what factor is the pressure gradient increased when
(a) a wire 1/100 the tube diameter, and
(b) 1/1000 the tube diameter is placed along the tube axis?

9-12. Oil with a density of 900 kg/m^3 and a kinematic viscosity of 9×10^{-5} m^2/s flows in a 5- $\times$ 5-cm square duct at an average velocity of 4.00 m/s. What is the pressure drop in 30 m of duct length?

9-13. What is the pressure drop in 30 m of 2.5- $\times$ 10-cm rectangular duct for the flow of Prob. 9-12?

9-14. Univis J-43 hydraulic fluid at 65°C flows through a 25-mm smooth tube. The pressure drop in a 15-m length is 3.5 kPa. What is the flow rate in L/s?

9-15. An oil with a kinematic viscosity of 3.7×10^{-5} m^2/s and density 930 kg/m^3 flows at an average velocity of 3.0 m/s in an equilateral triangle duct 3 cm on each side. What is the pressure drop in 15 m of this duct?

9-16. Oil ($\rho = 900$ kg/m^3 and $v = 2 \times 10^{-4}$ m^2/s) flows in a 5- $\times$ 5-cm square duct at an average velocity of 4.0 m/s. What is the pressure drop in 20 m of duct length?

9-17. By what factor is the pressure drop in a 2.5 $\times$ 10-cm duct in Prob. 9-16 greater or less than that for the square duct of the same area?

9-18. Given a circular duct and a noncircular duct, both having the same perimeter P. A given fluid of viscosity v flows at a rate Q through both ducts.
(a) Show that the Reynolds number is the same for both flow situations and is $Re = 4Q/Pv$.
(b) Show that the pressure gradient for the flow in the noncircular duct is related to that for the flow in the circular duct by the expression

$$\frac{(\Delta p/L)_n}{(\Delta p/L)_c} = \frac{f_n}{f_c}\left(\frac{A_c}{A_n}\right)^3$$

where subscripts c and n indicate circular and noncircular ducts, respectively. In general the friction factors are not the same for laminar flow [Eq. (9.14) and Tables 9-1 through 9-4), but are the same for turbulent flow in smooth ducts.

9-19. Explain why the friction factor decreases as the velocity of a given fluid increases in a circular tube for fully developed laminar flow.

9-20. Use Fig. 9-6 to obtain answers to the following questions.

(a) For what flow regime does the pressure drop in a pipe vary as the square of the flow rate?

(b) What is the friction factor at $Re_D = 10^5$ for a smooth pipe? For $k/D = 0.0001$? For $k/D = 0.001$?

(c) Over what range of Reynolds numbers is the friction factor constant for a 15-cm cast iron pipe?

(d) Suppose the absolute roughness of a given pipe were to increase over a period of years to three times its initial value. Would this have greater effect on the pressure drop for a given turbulent flow at high Reynolds numbers or at low Reynolds numbers?

(e) For what flow regime does f depend only on Re_D?

(f) For what regime does f depend only on k/D?

(g) For what flow regime does f depend on both Re_D and k/D?

(h) The friction factor is 0.06 for a smooth pipe. What is the friction for a pipe relative roughness $k/D = 0.001$ at the same Reynolds number?

(i) Repeat part (h) for $f = 0.015$.

9-21. When the flow rate through a given smooth pipe is 4 L/s the friction factor is 0.06. What friction factor can be expected if the flow rate is increased to 24 L/s?

9-22. What is the pressure drop per kilometer of pipe when water at 20°C flows in a 40-cm commercial steel pipe at a flow rate of 0.50 m³/s?

9-23. What is the pressure drop per kilometer of pipe when water at 15°C flows in a 25-cm cast iron pipe at flow rate of 225 L/s?

9-24. What is the head loss and pressure drop per kilometer of 60-cm commercial steel pipe through which gasoline at 15°C flows at an average velocity of 4.5 m/s?

9-25. Water flows in a new horizontal 30-cm cast iron pipe. In checking for the magnitude of a leak in the pipeline, two pressure gages 600 m apart upstream of the leak indicate a pressure difference of 140 kPa. Two gages 600 m apart downstream of the leak indicate a pressure difference of 133 kPa. Estimate the magnitude of the leak.

9-26. What is the frictional head loss and the power to overcome it for the flow of gasoline at 15°C in a 60-cm commercial steel pipe 100 km in length? The average flow velocity is 4.00 m/s.

9-27. Crude oil (specific gravity $s = 0.87$ and kinematic viscosity is 4.6×10^{-6} m²/s) is to be pumped through a 30-cm-diameter class H cast iron pipe. This pipe will safely withstand 2400 kPa internal pressure. How far apart should pumping stations be placed for a flow of 100 L/s?

9-28. A 10-cm-diameter smooth pipe carries olive oil at an average velocity of 1.40 m/s. What is the pressure gradient when the oil temperature is

(a) 20°C and

(b) 40°C?

(c) Explain why the pressure gradient is reduced so little for the warmer, less viscous oil.

9-29. What is the equivalent sand-grain roughness for a circular duct when a head loss of 1.14 m is measured over a 3.65-m length of 25-mm-diameter pipe for water flowing at 2.00 m/s at a temperature of 20°C?

9-30. Water flows in a 15-cm pipe at a rate of 90 L/s. The total head drops 3.80 m between two sections 30 m apart along the pipe.

(a) What is the friction factor?

(b) What is the relative roughness of the pipe?

(c) Identify the pipe material, assuming it is in new condition.

9-31. Repeat Prob. 9-30 for water at room temperature in a 15-cm pipe at a flow rate of 90 L/s. The head drops 10.0 m in 80 m of pipe length.

9-32. Glycerin (density is 1250 kg/m³ and dynamic viscosity is 1.0 N s/m²) flows through a 10-cm-diameter pipe 60 m long and at an upward slope of 10° at a flow rate of 25 L/s. At the lower inlet the pressure is 800 kPa gage.

(a) What is the Reynolds number of the flow?

(b) What is the pressure drop from viscous shear?

(c) What is the average wall shear stress?

(d) What is the pressure at the higher, downstream end of the pipe?

9-33. Part of the Jordan River water system in Israel consists of a 275-cm-diameter prestressed concrete pipe (roughness $k = 0.00030$ m) which conveys 17 m³/s. What is the pressure drop per kilometer of pipeline?

9-34. A proposal for a trans-Mediterranean aqueduct involves water at a flow rate of 566 m³/s flowing in a 45.7-m-diameter duct over a distance of 800 km. Estimate the pressure drop. Assume that for such a large diameter duct, it is essentially hydraulically smooth.

9-35. Water at 20°C flows in a 15-cm-diameter pipe. What flow rate will produce a pressure drop of 90 kPa in 300 m of horizontal pipe?

9-36. Water flows in a 5-cm-diameter smooth pipe at 20°C. A piezometer tube 15.0 m upstream from the free discharge at the end of the pipe shows water to rise 3.65 m above the pipe centerline at the discharge end. What is the flow rate?

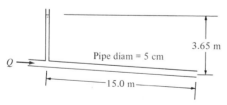

Prob. 9-36

9-37. Oil with a viscosity 5×10^{-4} m²/s and a density of 800 kg/m³ is pumped through a cast iron pipe 10 cm in diameter at a rate of 25 L/s.

(a) What is the pressure drop in 100 m of pipe?

(b) What size pipe would reduce the pressure drop to one-third this value for the same flow rate?

9-38. The total head drops 5.50 m in a length of 450 m of 30-cm cast iron pipe. What is the flow rate for water at 20°C?

9-39. What is the flow rate for water at 10°C which results in a pressure drop of 17.2 kPa in 300 m of cast iron pipe 60 cm in diameter?

9-40. Fully developed flow at a flow rate Q exists in a pipe of diameter D_1 and length L. The pressure drop is Δp_1. What would be the pressure drop in terms of Δp_1 for the same Q and L if the pipe diameter were doubled for

(a) laminar flow, and

(b) turbulent flow?

9-41. It is desired to convey a hydraulic fluid (kinematic viscosity is 9×10^{-6} m²/s and specific gravity is 0.848) at a rate of 2.50 L/s through a smooth pipe with a pressure drop not to exceed 14 kPa per 30 m of pipe. What size pipe should be used?

9-42. Diesel oil ($s = 0.85$ and kinematic viscosity is 4.6×10^{-6} m²/s) is to be pumped from a tanker to a storage tank at atmospheric pressure through 180 m of cast iron pipe at a rate of 32 L/s. The oil level in the storage tank is 22 m higher than that in the tanker. The pump in the tanker can develop a discharge pressure of 415 kPa gage. Consider losses due only to pipe friction. What size pipe will be necessary?

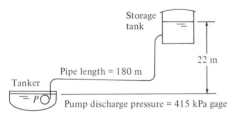

Prob. 9-42

9-43. Plot a curve of the ratio of the average to the centerline velocity ($V/u_{\max}$) for fully developed turbulent flow in a smooth circular tube as a function of the Reynolds number of the flow. Use a linear scale for $V/u_{\max}$ as ordinate and a log scale for Re as abscissa. Refer to Eq. (9.22).

9-44. Repeat Prob. 9-43 for a pipe of relative roughness 0.002.

9-45. Fully developed flow of standard air occurs in a 5-cm smooth circular tube at an average velocity of 40 m/s. Plot a velocity profile based on the universal velocity distribution shown in Fig. 7-13.

9-46. Fully developed flow in a smooth tube is at an average velocity of 60 cm/s. What is the maximum (centerline) velocity when

(a) Re = 1600 and

(b) Re = 10⁶?

9-47. Fully developed flow in a 30-cm cast iron pipe is at an average velocity of 1.80 m/s. What is the centerline velocity for
(a) fuel oil at $10°C$ ($v = 0.002$ m^2/s) and
(b) crude oil at $20°C$ ($v = 5.6 \times 10^{-6}$ m^2/s)?

9-48. The wall shear stress and hence the pressure gradient may be estimated for fully developed turbulent flow from velocity measurements. Water flows at $V = 0.86 \, u_{max} = 2.10$ m/s in a 15-cm-diameter pipe.
(a) What is the wall shear stress?
(b) What is the pressure drop in 30 m of pipe?

9-49. From Eqs (9.15) and (9.21) show that

$$\frac{V - u}{v_*} = 2.5 \ln \frac{R}{y} - 3.75$$

9-50. Measurements in fully developed turbulent flow of a liquid in a pipe indicate that the velocity midway between the pipe wall and the pipe axis is 0.9 times the centerline velocity.
(a) What is the average velocity of flow in terms of the centerline velocity?
(b) What is the relative roughness of the pipe?

9-51. Direct measurements indicate that for fully developed turbulent flow in a circular tube of relative roughness 0.0018 and a smooth tube at Re $= 82,000$ the average velocity corresponds to that at $y/R = 0.25$. At what value of y/R is the velocity equal to the average velocity for
(a) the power-law velocity profile $u/u_m = (y/R)^{1/7}$ and
(b) the universal velocity profile given by Eq. (9.15)?

9-52. Turbulent flow occurs in a circular tube. A thin diametrical plate divides the flow area into two equal semicircular segments. By what factor is the pressure gradient increased for the same flow rate
(a) for the completely rough flow regime and
(b) for smooth tubes or the smooth-to-rough flow regime? Compare with laminar flow in Prob. 9-10.

9-53. Compare the pressure gradient in a rectangular duct with that for a circular duct, the flow rate and duct perimeter being the same in each instance. Consider aspect ratios of 1:1, 2:1, 3:1, and 4:1 for the rectangular duct.

9-54. Show that the hydraulic diameter of a hexagon is equal to the diameter of the inscribed circle.

9-55. Show that the hydraulic diameter of an isosceles triangle is equal to the diameter of the inscribed circle.

9-56. Standard air flows through an 0.5- by 1.0-m rectangular duct made of smooth aluminum sheet at a rate of 4.50 m^3/s. What diameter of round duct of the same material would convey this flow with the same pressure gradient?

9-57. Air at 40°C and 100 kPa abs pressure flows at 3.0 m/s through the 8.75- by 35-cm smooth duct in the space between studs in the wall of a house. What diameter round duct of the same material would convey the same flow with the same pressure gradient?

9-58. Standard air flows at a velocity of 30 m/s in the annular space between smooth circular tubes 25 and 100 mm in diameter, respectively. What is the pressure drop per 100 m of length for
(a) a concentric annulus and
(b) an eccentricity of 0.9?

9-59. Water at 60°C flows through the annular space between smooth circular tubes 75 and 100 mm in diameter, respectively, at an average velocity of 1.50 m/s. What is the pressure drop per 30 m of length for
(a) a concentric annulus and
(b) an eccentricity of 0.9?

9-60. A 50- by 100-cm rectangular air duct is made of smooth aluminum sheet. It conveys 10.0 m³/s of standard air. What is the pressure drop in 100 m of this duct?

9-61. Water at 15°C flows at an average velocity of 3.5 m/s in the annular space between smooth circular tubes 10 and 2.5 cm in diameter, respectively. What is the pressure drop in 20 m of length for
(a) a concentric annulus, and
(b) an eccentricity of 1.0?

9-62. Explain qualitatively why the total force resulting from the pressure drop in the entrance region for laminar flow is greater than the total shear force along the pipe wall.

9-63. A fluid of constant density ρ enters a pipe of radius R with a uniform velocity V. At a downstream section the velocity varies with the radius r according to the equation

$$u = 2V\left(1 - \frac{r^2}{R^2}\right)$$

Let the pressures at sections 1 (inlet) and 2 (downstream section) be p_1 and p_2, respectively. Show that the frictional force of the pipe walls on the fluid between sections 1 and 2 is

$$F = \pi R^2(p_1 - p_2 - \tfrac{1}{3}\rho V^2)$$

9-64. What is the average wall shear stress in the entrance region for laminar flow in a circular tube of diameter D, an average flow velocity V, and a fluid density ρ? Compare the result with the wall shear stress for fully developed flow, $\tau_0 = (16/\text{Re})(\rho V^2/2)$, and Eq. (9.31).

9-65. Fuel oil at 30°C drains from a tank through 2 m of 25-mm diameter pipe. The oil level in the tank is 3 m above the exit end of the tube.

(a) What is the flow rate through the tube if fully developed flow is assumed to exist throughout the entire pipe? Assume exit velocity head is negligible, then confirm it. The entrance is well rounded.

(b) What is the entrance length for conditions assumed in part (a)?

(c) Will the flow rate calculated in part (a) be changed significantly if the additional pressure drop in the entrance region is taken into account?

9-66. Suppose a 50-mm pipe is used instead of a 25-mm pipe in Prob. 9-65.

(a) What is the flow rate if fully developed flow is assumed to exist throughout the entire pipe and the exit velocity is neglected?

(b) What is the flow rate if fully developed flow is assumed to exist but the exit velocity head is considered?

(c) What is the flow rate if the additional pressure drop in the entrance region and the exit velocity head are both included?

(d) Which flow condition (a, b, or c) most closely resembles the true physical situation? Note that the entrance length is about five diameters or one-sixth of the pipe length.

9-67. Some ducts are to be provided with pressure taps to measure the friction factor for fully developed laminar flow. Each duct is attached to a supply reservoir with a short rounded entrance. Estimate the minimum distance from the tank to the location of the first pressure tap to ensure fully developed flow based on pressure gradient for flow at Re = 2000 for

(a) a circular tube,

(b) a square duct, and

(c) a 1:5 rectangle.

9-68. Fuel oil ($v = 5 \times 10^{-4}$ m^2/s) drains from a tank through 200 cm of 3-cm-diameter pipe. The level of oil in the open tank is 250 cm above the exit end of the pipe. Answer parts (a), (b), and (c) of Prob. 9-65.

9-69. Suppose a 6-cm pipe is used instead of the 3-cm pipe in Prob. 9-68. Answer parts (a), (b), (c), and (d) of Prob. 9-66.

9-70. Water flows at a rate of 200 L/s through a contraction from a 30-cm pipe to a 13-cm pipe. Compare the

(a) head loss and

(b) pressure drop through this sudden contraction with that for a well-designed gradual contraction.

9-71. Water flows through a 20-cm pipe which enlarges abruptly to 40 cm in diameter. A differential manometer containing mercury shows a deflection of 12 cm when connected across the enlargement. What is the flow rate?

9-72. A pipe whose area is 0.24 m^2 enlarges abruptly to a pipe of 0.48 m^2 area. A few diameters downstream the second pipe enlarges abruptly to 0.96 m^2 in area. The measured pressure rise from the first to the third pipe (a few diameters downstream of the last enlargement) is

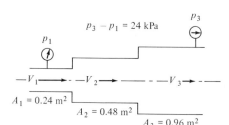

Prob. 9-72

24 kPa. What is the flow rate when the pipes are horizontal and convey water? Assume one-dimensional flow and neglect pipe friction.

9-73. Pipes of diameters D_1, D_2, and D_3 (in order of increasing size) are in series and connected by sudden enlargements. For given D_1 and D_3, what is D_2 in order that the total loss due to the enlargements is a minimum?

9-74. A pipe of diameter D_1 enlarges suddenly to $D_2 = 2D_1$. A distance L downstream the pressure is the same as that immediately upstream of the enlargement. For a friction factor f in the larger pipe, what is L in terms of D_2?

9-75. Water flows at a rate of 170 L/s in a horizontal 15-cm pipe which is enlarged to a 30-cm pipe. Estimate the head loss and pressure rise between the two pipes for
(a) a sudden enlargement,
(b) a 7° diffuser with an abrupt change from the 15-cm pipe to the diffuser cone, and
(c) a 7° diffuser with a parabolic transition between the 15-cm pipe and the diffuser cone.

9-76. The measured pressure rise through a 15- to 30-cm-diameter diffuser in a water tunnel is 0.82 $\rho V_1^2/2$ where V_1 is the average velocity at the upstream end. Kinetic energy correction factors at the upstream and downstream ends are $\alpha_1 = 1.004$ and $\alpha_2 = 1.56$, respectively.
(a) What is the pressure efficiency?
(b) What is the energy efficiency?

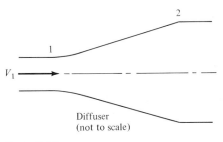

Prob. 9-76

9-77. A 15-cm pipe is joined to a 30-cm pipe by a reducing flange. For water flowing at a rate of 115 L/s, what is the head loss

(a) when the water flows from the smaller to the larger pipe and

(b) when the water flows from the larger to the smaller pipe?

9-78. A liquid flows in a pipe with $k/D = 0.002$ at a Reynolds number of 10^5. This pipe is replaced by a similar pipe of two-thirds the diameter of the original pipe. What is the flow velocity in the smaller pipe in terms of that in the original larger pipe if the pressure drop is the same in each instance? Both pipes are of the same length.

9-79. A pipe is replaced by one twice its diameter. How will the flow rate for the larger pipe compare with that for the smaller pipe, assuming the head loss across each pipe to be the same? Consider both laminar and turbulent flow in both smooth and rough pipes.

9-80. Consider two identical open tanks, one with a vertical discharge pipe of length L_1 and the other with a vertical discharge pipe of length $L_2 = L_1 + \Delta L$. Both pipes flow full. Let the water depth in each tank be h feet, the pipe diameters be D, the entrance loss coefficient be k_L, and the friction factor f be the same in each pipe. Show that the flow velocity V_1 in the short pipe is related to the flow V_2 in the longer pipe by

(a) $V_1 > V_2$ if $hf/D > 1 + k_L$,

(b) $V_1 = V_2$ if $hf/D = 1 + k_L$, and

(c) $V_1 < V_2$ if $hf/D < 1 + k_L$.

  Note: One might ask whether an open tank will drain more quickly through a short drainpipe or through a long drainpipe. Results for a vertical pipe indicate that whether the flow rate from one tank is greater than the other is independent of the pipe length.

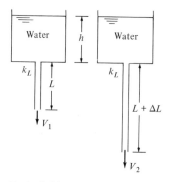

Prob. 9-80

9-81. Water drains from a large tank through a 10-cm galvanized iron pipe 2.50 m long under a head of 4.0 m. The friction factor may be assumed to be 0.022. The pipe flows full at exit. What is the flow velocity for

(a) a rounded entrance and

(b) a reentrant type of entrance (Fig. 9-10)?

9-82. Water for an impulse turbine is supplied from a reservoir 250 m higher in elevation than the nozzle exit. The commercial steel pipeline is 40 cm in diameter and 2700 m long. The nozzle contracts the flow to a 10-cm jet. What is the power represented by the jet? Assume losses in the nozzle to be 2 precent of the jet velocity head. *Hint:* Assume an f; solve for the velocity in the pipe; compute an Re; obtain a more accurate f value.

9-83. Which 2-m-long cast iron pipe would drain a large water tank in less time:
(a) a 75-mm pipe with a rounded entrance or
(b) a 100-mm pipe with a projecting inlet (see Fig. 9-10c)?

9-84. A 60-cm-diameter welded commercial steel pipeline conveys 0.54 m³/s of oil ($s = 0.83$ and $\mu = 5 \times 10^{-3}$ kg/m s) with pumping stations every 75 km. What is the power required to pump this oil?

9-85. A 60-cm riveted steel pipe ($k = 0.0009$ m) 9000 m long connects two open reservoirs whose levels differ by 45 m.
(a) What is the flow rate from the higher to the lower reservoir?
(b) What power would be required to pump 625 L/s from the upper to the lower reservoir?
(c) What power would be required to pump 625 L/s from the lower to the upper reservoir?

9-86. In Prob. 9-85c, what percent saving in power would result if a 40-cm riveted steel pipe 6000 m long is connected in parallel to one-half the existing 60-cm pipe for a total flow of 625 L/s?

9-87. For turbulent flow in the hydraulically rough regime, does a change in the total flow through a group of pipes in parallel affect the relative distribution of flow through the individual pipes? Explain your answer.

9-88. A 60-cm riveted steel pipe ($k = 0.0009$ m) 4500 m long in series with a 40-cm riveted steel pipe 3000 m long connects two reservoirs whose levels differ by 30 m. What is the flow rate from the higher to the lower reservoir?

9-89. Piping consisting of 150 m of 20-cm commercial steel pipe with three 90° elbows ($k_L = 0.3$) is connected to an open reservoir with a projecting entrance. The free discharge is 10 m below the free surface in the reservoir. What is the flow rate of water?

9-90. What diameter commercial steel pipe is required to convey 550 L/s of water from one open reservoir to another 15 m below the first through 300 m of pipe?

9-91. A total flow of 850 L/s of water at 20°C passes through two asphalt-dipped cast iron pipes connected in parallel. One is 30 cm in diameter and 300 m long, and the other is 40 cm in diameter and 600 m long. What is the flow rate through the 40-cm pipe? Consider only frictional losses from wall shear.

9-92. A total flow of 1.00 m³/s of water at 15°C passes through two cast iron pipes in parallel. One is 25 cm in diameter and 300 m long, and the other is 30 cm in diameter and 500 m long. What is the flow rate through the 25-cm pipe? Consider only frictional losses.

9-93. Determine the flow through each pipe in the network. Assume $C = 100$ for all pipes. Will the distribution be different if C is other than 100 but the same for all pipes?

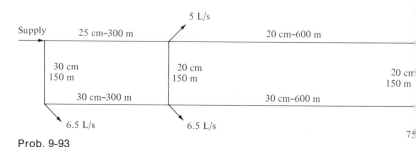

Prob. 9-93

9-94. Determine the flow through all pipes in the network. Assume $C = 110$ for all pipes.

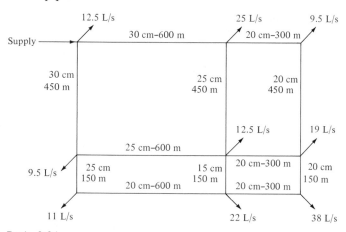

Prob. 9-94

References

1. T. S. Lundgren, E. M. Sparrow, and J. B. Starr, "Pressure Drop due to the Entrance Region in Ducts of Arbitrary Cross Section," *Trans. Am. Soc. Mech. Engrs., Basic Eng.,* Vol. 86, Ser. D, No. 3 (1964), pp. 620–626.

2. E. M. Sparrow and A. Haji-Sheikh, "Flow and Heat Transfer in Ducts of Arbitrary Shape with Arbitrary Thermal Boundary Conditions," *J. Heat Transfer, Trans. Am. Soc. Mech. Engrs.,* Vol. 88, Ser. C, No. 4 (1966), pp. 351–358.

3. E. M. Sparrow and A. Haji-Sheikh, "Laminar Heat Transfer and Pressure Drop in Isosceles Triangular, Right Triangular, and Circular Sector Ducts," *Ibid.,* Vol. 87, Ser. C, No. 3 (1965), pp. 426–427.

4. S. K. Friedlander and L. Topper, Editors, *Turbulence: Classical Papers on Statistical Theory* [New York: John Wiley and Sons, Inc. (Interscience), 1961].
5. H. Schlichting, *Boundary Layer Theory*, translated by J. Kestin (New York: McGraw-Hill Book Company, Inc., 1968).
6. V. K. Jonsson, "Experimental Studies of Turbulent Flow Phenomena in Eccentric Annuli," Ph.D. Thesis, University of Minnesota (1965).
7. E. R. G. Eckert and R. M. Drake, *Heat and Mass Transfer* (New York: McGraw-Hill Book Company, Inc., 1959), pp. 159–160.
8. R. M. Olson and E. M. Sparrow, "Measurements of Turbulent Flow Development in Tubes and Annuli with Square or Rounded Entrances," *A.I.Ch.E. Journal*, Vol. 9 (1963), pp. 766–770.
9. J. M. Robertson and D. Ross, "Effect of Entrance Conditions on Diffuser Flow," *Trans. Am. Soc. Civil Engrs.*, Vol. 118 (1953), pp. 1068–1097.
10. R. E. Durst, A. J. Chase, and L. C. Jenness, "An Analysis of Data on Stock Flow in Pipes," *J. Tech. Assn. Pulp and Paper Ind.*, Vol. 35, No. 12 (1952).
11. W. Brecht and H. Heller, "A Study of the Pipe Friction Losses of Paper Stock Suspensions," *Ibid.*, Vol. 33, No. 9 (1950).
12. R. A. Smith, "Experiments on the Flow of Sand-Water Slurries in Horizontal Pipes," *J. Inst. Chem. Engrs.*, Vol. 33 (1955), pp. 85–92.
13. *Ibid.*, and D. M. Newitt, J. F. Richardson, M. Abbott, and R. B. Turtle, "Hydraulic Conveying of Solids in Horizontal Pipes," and Discussion by R. C. Worster, *Ibid.*, pp. 93–113.

The reader is referred to the following additional information:
W. H. Graf, *Hydraulics of Sediment Transport* (New York: McGraw-Hill Book Company, Inc., 1971), pp. 421–502.
G. B. Wallis, *One-Dimensional Two-Phase Flow* (New York: McGraw-Hill Book Company, Inc., 1969).

Chapter 10
The Flow of Compressible Gases

Situations in which relatively large variations in fluid density occur with associated large variations in velocity exist in many gas flows and involve thermodynamic effects. The study of this type of flow is often referred to as *gas dynamics*. The behavior of the gas depends to a large extent on the speed of the gas flow in relation to that of a weak pressure wave, measured in terms of the Mach number M. At low subsonic speeds (M < 0.2), the density variations are so small that the flow may be considered incompressible. At higher subsonic speed, the density variations increase and their effects become more and more pronounced. At supersonic speeds (M > 1), the effects are very pronounced and abrupt changes in velocity and pressure, which increase in magnitude as the Mach number increases, occur across a shock, for example.

The following assumptions will be made in the treatment given in this chapter:

1. Gases will be considered perfect ($p = \rho R T$ and $c_v^{'} = $ constant). This simplifies the equations used and gives quite accurate results for moderate pressures, temperatures, or Mach numbers. For pressures below 50 atm (5×10^6 Pa) or stagnation temperatures below $550°$K,

accuracy is good for any Mach number; and for higher pressures and temperatures, errors increase with Mach number, up to a maximum which depends on the pressure and temperature.[†]

2. Flow will generally be considered adiabatic (without heat transfer). In addition it will be considered reversible, and thus isentropic, with the exception of flow across a shock. Diabatic flow (with heat transfer) is discussed in Sec. 10-4. Frictional flow of compressible gases in pipes is presented in Sec. 10-5.
3. The flow will be considered one dimensional.
4. Changes in potential energy (elevation) will not be considered, since they are trivial (if not zero) compared to changes in kinetic energy and enthalpy.
5. No external work is done on or by the gas.

With conditions, 2, 3, 4, and 5, the steady-flow energy equation [Eq. (5.3a)] will have the forms

$$\frac{V_1^2}{2} + h_1 = \frac{V_2^2}{2} + h_2 = h_0 = \text{constant} \tag{10.1a}$$

and

$$V\,dV + dh = 0 \tag{10.1b}$$

Thus there will be a mutual interchange of kinetic energy with enthalpy, resulting in opposite changes in velocity and temperature.

Straightforward application of the continuity, momentum, and energy equations together with thermodynamic relationships for perfect gases, the isentropic relation $p/\rho^k = \text{constant}$, and the second law of thermodynamics will enable us to make a quantitative study of compressible gas flow. Results will apply with remarkable accuracy to high-speed flight and flow through gas-turbine, steam-turbine, and rocket nozzles. Flow with friction through pipes will be taken up in Sec. 10-5.

10-1. The Velocity of Sound

The velocity of a plane, weak, pressure pulse in a gas may be determined by applying the continuity and momentum principles across a wave front traveling in a duct of area A. The momentum theorem applied to the dashed region of Fig. 10-1 (wall shear is negligible) is

$$(p + dp)A - pA = cA\rho[-(c - dV) - (-c)]$$

so that

$$dp = \rho c\,dV \tag{10.2}$$

[†] C. DuP. Donaldson, "Note on the Importance of Imperfect-Gas Effects and Variation of Heat Capacities on the Isentropic Flow of Gases," *NACA R.M.* No. L8J14 (1948).

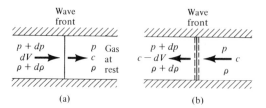

Fig. 10-1. Weak pressure wave in a duct of area *A*. (a) Observer at rest. (b) Observer moving along with wave front.

From continuity,

$$cA\rho = (c - dV)A(\rho + d\rho)$$

and if the second-order differentials are neglected,

$$\frac{d\rho}{\rho} = \frac{dV}{c} \tag{10.3}$$

Combining Eqs. (10-2) and (10-3), and using Eq. (1-20c),

$$c^2 = \frac{dp}{d\rho} = \frac{K}{\rho}$$

For a gas,

$$c^2 = \left(\frac{\partial p}{\partial \rho}\right)_{\text{isentropic}}$$

Since the changes in pressure and temperature are extremely small, they are considered to be reversible. In addition, the temperature gradients are also small and the process is very rapid, consequently no heat is transferred. Thus the process is essentially isentropic, for which $p/\rho^k = $ constant. The logarithmic form is

$$\ln p - k \ln \rho = \ln(\text{constant})$$

from which $dp/d\rho = kp/\rho = kRT$. Thus the velocity of sound in a perfect gas is

$$c = \sqrt{(\partial p/\partial \rho)_{\text{isentropic}}} = \sqrt{kp/\rho} = \sqrt{kRT} = \sqrt{K/\rho} \tag{10.4}^{\dagger}$$

since the isentropic elastic modulus for a perfect gas is kp. For air, $c = 20.04\sqrt{T}$ m/s, where T is in degrees Kelvin.

---

† The acoustic velocity in any medium is $c = \sqrt{K/\rho}$. For gas–liquid mixtures the acoustic velocity becomes less than that for either the liquid or gas alone. For a liquid with a small concentration of gas nuclei, the elastic modulus of the mixture is reduced, with no appreciable reduction in density, and thus the acoustic velocity is reduced. For a gas with minute liquid droplets, the density of the mixture is increased, with no appreciable change in elastic modulus, and again the acoustic velocity for the mixture is reduced. The velocity of sound in a 14 percent water and 86 percent air mixture is about 30 m/s (depending on the pressure, temperature, and impressed frequency) as compared with about 1460 m/s for water and about 335 m/s for air.

10-2. The Mach Number

The Mach number M has been defined (Chapter 8) as the ratio of (a) the flow velocity to the sound velocity, (b) the inertia to the elastic forces in a flow system, or (c) the kinetic energy of the mean flow to the mean kinetic energy of the gas molecules. In any case,

$$M = \frac{V}{c} \tag{10.5}$$

The velocity V may be either the local velocity or the relative velocity between the free stream and a body immersed in the stream. For example, a local velocity may be the velocity at the throat or at the exit of a nozzle. The velocity of either an aircraft or a missile in free flight through still air and the free-stream velocity past a test body in a wind tunnel are examples of relative velocity between a stream and a body immersed in the stream.

The reference sonic velocity may be (a) the local sonic velocity c determined from the local temperature, (b) the sonic velocity c_0 at the stagnation condition, or (c) the velocity c^* where the flow is, or would be, sonic. These are interrelated by the steady-flow energy equation [Eq. (10.1a), with $h = c_p T$, $c_p = Rk/(k - 1)$, and $c = \sqrt{kRT}$] applied to each of the three states:

$$\frac{V^2}{2} + \frac{1}{k-1}c^2 = \frac{1}{k-1}c_0^2 = \frac{k+1}{2(k-1)}c^{*2} \tag{10.6}$$

We will use the local sound velocity c as a reference throughout this chapter.

For steady flow, the energy equation for adiabatic flow with the assumptions listed at the beginning of this chapter is

$$\frac{V^2}{2} + h = \text{constant}$$

There are two extreme situations for a given gas condition:

1. If all kinetic energy is converted to enthalpy, the velocity is zero and the temperature is a maximum.
2. If all the enthalpy could be converted to kinetic energy (a hypothetical condition), the temperature would drop to absolute zero and the velocity would be a maximum.

In the first instance, the velocity is zero and the speed of sound is a maximum; in the second instance, the speed of sound is zero and the velocity is a maximum. This is indicated in three expressions for the constant in the energy equation: (1) for an arbitrary given state where the velocity is V and the temperature T, (2) for the stagnation state, and (3) for the zero-temperature state. With $\Delta h = c_p \Delta T$ and $T = c^2/kR$, the energy equation may be written for these three states as

$$V^2 + \frac{2}{k-1}c^2 = \frac{2}{k-1}c_0^2 = V_{max}^2 \tag{10.7}$$

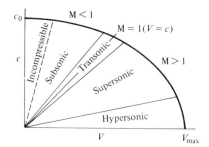

Fig. 10-2. Ellipse showing ranges of compressible gas flow.

A plot of this equation is shown in Fig. 10-2 and is known as the steady-flow adiabatic ellipse.

10-3. Isentropic Flow of a Perfect Gas

Between any two points or sections in an isentropic flow the following energy, continuity, isentropic, and gas equations apply:

$$\frac{V_1^2}{2} + h_1 = \frac{V_2^2}{2} + h_2 \qquad\qquad [10.1a]$$

$$V_1 A_1 \rho_1 = V_2 A_2 \rho_2 \qquad\qquad [3.17]$$

$$\frac{p_1}{\rho_1^k} = \frac{p_2}{\rho_2^k} \qquad\qquad [1.14]$$

$$h_2 - h_1 = c_p(T_2 - T_1) = \frac{Rk}{k-1}(T_2 - T_1) \qquad\qquad [1.9 \text{ and } 1.13]$$

STAGNATION TEMPERATURE
The stagnation temperature exists at a point of zero velocity. Thus if the fluid proceeds to this point adiabatically

$$\frac{V^2}{2} + h = h_0$$

from which

$$T_0 = T + \frac{V^2}{2c_p} \qquad\qquad (10.8a)$$

and the stagnation temperature rise varies as the square of the gas velocity. Recall that $V^2 = M^2 c^2 = M^2 kRT$ and that $c_p = Rk/(k-1)$. Thus

$$\frac{T_0}{T} = 1 + \frac{k-1}{2} M^2 \qquad\qquad (10.8b)$$

Actually, Eqs. (10.8a) and (10.8b) are valid for any adiabatic flow of a perfect gas at all Mach numbers, whether reversible or not. Thus they may be applied across a shock wave, which is not reversible, as in Sec. 10-6.

STAGNATION PRESSURE

The stagnation pressure p_0 is, by definition, the pressure reached isentropically, and is often called the *isentropic stagnation pressure*. It is also called the reservoir pressure, since for any flow condition a reservoir pressure p_0 may be imagined from which flow proceeds isentropically to a pressure p and Mach number M. If the flow is entirely isentropic, p_0 is constant throughout the flow; if nonisentropic, p_0 changes from section to section. In an irreversible adiabatic flow p_0 decreases in the direction of flow, and the decrease in p_0 is a measure of the irreversibility of the flow or the increase in entropy of the gas. From Eqs. (1.14b) and (10.8b)

$$\frac{p_0}{p} = \left(\frac{T_0}{T}\right)^{k/(k-1)} = \left(1 + \frac{k-1}{2}M^2\right)^{k/(k-1)} \tag{10.9a}$$

and this is the defining equation for stagnation pressure for *both* subsonic and supersonic flow. Expanding this expression[†] gives, after simplifying,

$$p_0 = p + \frac{\rho V^2}{2}\left[1 + \frac{1}{4}M^2 + \frac{2-k}{24}M^4 + \frac{(2-k)(3-2k)}{192}M^6 + \cdots\right] \tag{10.9b}$$

This may be compared with $p_0 = p + \rho V^2/2$ for incompressible flow.

The term in brackets in Eq. (10.9b) indicates the effect of the increase in gas density due to compressibility and is called the *compressibility factor*. Values of this factor for $k = 1.4$ (air, for example) range from 1 as the Mach number approaches zero (no compressibility effects) to 1.276 as the Mach number approaches unity. This means that the dynamic pressure increases 27.6 percent over the dynamic pressure if incompressible flow is incorrectly assumed because of the compressibility of the gas as the Mach number approaches unity.

DUCT FLOW

For incompressible flow, the continuity equation $VA = $ constant indicates that an increase in flow area is associated with a decrease in velocity and a

---

[†] $(1 + x)^a = 1 + ax + a(a-1)\dfrac{x^2}{2!} + a(a-1)(a-2)\dfrac{x^3}{3!} + \cdots$ which is convergent for $x^2 < 1$.

Mathematically, the resulting expansion given by Eq. (10.9b) is valid for $M < \sqrt{5}$ for a gas with $k = 1.4$. Physically, Eqs. (10.9a) and (10.9b) are valid only for isentropic flow and may be applied to subsonic initial flow when flow is *toward* a stagnation point. If flow is *from* a reservoir (a stagnation condition), Eq. (10.9a) is valid for any subsequent Mach number and Eq. (10.9b) is valid for subsequent Mach numbers less than $\sqrt{5} = 2.236$ (for $k = 1.4$), provided the flow is isentropic.

decrease in flow area is associated with an increase in velocity. For compressible flow, this is not always the case, because of the changes in fluid density. The relationship between area and velocity changes is a function of the local Mach number and may be found by combining the continuity, energy, and second-law equations in differential form. These, for isentropic flow, are

$$\frac{dV}{d} + \frac{dA}{A} + \frac{d\rho}{\rho} = 0 \qquad \text{from [3.24]}$$

$$V\,dV + dh = 0 \qquad [10.1b]$$

and

$$dh = \frac{dp}{\rho} \qquad (1.6)$$

From the second and third of these equations, $\rho = -dp/V\,dV$, and if this is substituted in the first equation,

$$\frac{dA}{A} = -\frac{dV}{V}\left(1 - \frac{V^2}{dp/d\rho}\right) = -\frac{dV}{V}\left(1 - \frac{V^2}{c^2}\right)$$

so that[†]

$$\frac{dA}{dV} = \frac{A}{V}(\mathrm{M}^2 - 1) \qquad (10.10)$$

Whether the area and the velocity decrease or increase is determined by the sign of dA/dV, which in turn depends on the magnitude of the local Mach number. The various possibilities are listed in Table 10-1.

Subsonic gas flow is similar to incompressible flow insofar as the velocity increases if the flow area decreases and, conversely, the velocity decreases if the flow area increases. The velocity may be sonic ($\mathrm{M} = 1$) *only* where the area is constant and is not changing, such as in the throat of a nozzle. (Con-

---

[†] An alternative derivation is as follows: For a nonviscous fluid, the equation of motion may be written as

$$\frac{dp}{\rho} + V\,dV = 0$$

and since $c^2 = dp/d\rho$ we get

$$c^2\frac{d\rho}{\rho} + V\,dV = 0$$

Combining this equation with the continuity equation [Eq. (3.22)] gives

$$\frac{dV}{V}\left(1 - \frac{V^2}{c^2}\right) + \frac{dA}{A} = 0$$

which is equivalent to Eq. (10.10).

Table 10-1   AREA AND VELOCITY CHANGES FOR GAS
FLOW IN DUCTS [EQ(10.10)]

MACH NUMBER M	$\dfrac{dA}{dV}$	dA	dV	$\dfrac{dV}{dx}$		AS A FUNCTION OF $\dfrac{dA}{dx}$
<1	$-$	$+$	$-$		Velocity decreases	
		$-$	$+$		Velocity increases	
>1	$+$	$+$	$+$		Velocity increases	
		$-$	$-$		Velocity decreases	
$=1$	0	0			Velocity can be sonic only in the throat of a nozzle or in a pipe (see Sec. 10-10)	

stant-area flow—pipe flow—is discussed in Sec. 9-5.) For supersonic gas flow, an increase in flow area produces an *increase* in velocity and a decrease in flow area produces a *decrease* in velocity. Supersonic gas flow is analogous to the flow of traffic in a multilane roadway. If the roadway narrows to fewer lanes, traffic speed is reduced, and if the roadway widens to more lanes, traffic speed is increased.

Additional relationships between changes in velocity, Mach number, temperature, pressure, and area may be obtained by combining the differential forms of the continuity, energy, isentropic, perfect gas, and Mach number equations.

The isentropic relationship $p/\rho^k =$ constant may be written as

$$\frac{dp}{p} - k \frac{d\rho}{\rho} = 0 \tag{10.11}$$

The perfect gas equation $p = \rho R T$ may be written as

$$\frac{dp}{p} = \frac{d\rho}{\rho} + \frac{dT}{T} \tag{10.12}$$

The Mach number is defined as $\mathrm{M} = V/\sqrt{kRT}$ and may be written as

$$\frac{d\mathrm{M}}{\mathrm{M}} = \frac{dV}{V} - \frac{dT}{2T} \tag{10.13}$$

The results of various combinations are:

$$\frac{dV}{V} = \left[\frac{1}{1 + \frac{k-1}{2}M^2} \right] \frac{dM}{M} \qquad (10.14)$$

which, since the term in brackets is always positive, indicates that both velocity and Mach number either increase or decrease together;

$$\frac{dT}{T} = \left[\frac{-(k-1)M^2}{1 + \frac{k-1}{2}M^2} \right] \frac{dM}{M} \qquad (10.15)$$

which, since the terms in brackets is always negative, indicates that temperature changes are opposite to Mach number changes, that is, temperatures decrease with an increase in Mach number and temperatures increase with a decrease in Mach number;

$$\frac{dp}{p} = \left[\frac{-kM^2}{1 + \frac{k-1}{2}M^2} \right] \frac{dM}{M} \qquad (10.16)$$

which, since the term in brackets is always negative, indicates that pressure changes are also opposite to Mach number changes;

$$\frac{dA}{A} = \left[\frac{-(1-M^2)}{1 + \frac{k-1}{2}M^2} \right] \frac{dM}{M} \qquad (10.17)$$

which indicates that area changes and Mach number changes depend on the magnitude of the Mach number, since the term in brackets may be either positive or negative. Results are the same as those for Eq. (10.10) listed in Table 10-1 which relate area and velocity changes. For subsonic flow (M < 1), area and Mach number changes are opposite. For sonic flow (M = 1), $dA/A = 0$, and this flow can occur only in a throat where the flow cross section is not changing. For supersonic flow (M > 1), area and Mach number changes are in the same direction (both increase or decrease together).

FLOW THROUGH NOZZLES

Gas flowing through a converging or a converging–diverging nozzle is usually supplied from a pressure tank or reservoir in which the velocity is zero or essentially so. Thus the supply reservoir is in a known stagnation condition, and the velocity, temperature, and pressure at any other section in the flow are given by Eqs. (10.1a), (10.8b), and (10.9a), respectively. These may be written as

$$V = \sqrt{2c_p T_0 \left(1 - \frac{T}{T_0}\right)} \qquad (10.18a)$$

in terms of the temperature T at any arbitrary section for *any* adiabatic flow. For isentropic flow,

$$V = \sqrt{2c_p T_0 \left[1 - \left(\frac{p}{p_0} \right)^{(k-1)/k} \right]} \qquad (10.18b)$$

in terms of the pressure p at any arbitrary section.

The temperature at any section where the Mach number M is known is given by

$$\frac{T_0}{T} = 1 + \frac{k-1}{2} M^2 \qquad (10.19)$$

and the pressure at any section where the Mach number M is known is given by

$$\frac{p_0}{p} = \left(1 + \frac{k-1}{2} M^2 \right)^{k/(k-1)} \qquad (10.20)$$

Densities may be calculated from the gas equation $p = \rho R T$ or from

$$\frac{\rho_0}{\rho} = \left(1 + \frac{k-1}{2} M^2 \right)^{1/(k-1)} \qquad (10.21)$$

At the section where the velocity is sonic (the throat), the Mach number is unity and the flow is called *critical flow*. If conditions at this section are designated by an asterisk, the critical temperature T^* from Eq. (10.19) is

$$\frac{T^*}{T_0} = \frac{2}{k+1} \qquad (10.22)$$

which is valid for *any* adiabatic flow (both isentropic flow and flow with friction). The critical pressure p^* from Eq. (10.20) or from the isentropic relation and Eq. (10.22) is

$$\frac{p^*}{p_0} = \left(\frac{2}{k+1} \right)^{k/(k-1)} \qquad (10.23)$$

which is valid *only* for isentropic flow. Both T^*/T_0 and p^*/p_0 depend only on the specific heat ratio of the gas. For air ($k = 1.4$), $T^*/T_0 = \frac{5}{6}$ and $p^*/p_0 = 0.528$. Thus air flowing from a supply tank (reservoir) will have its temperature reduced to $\frac{5}{6}$ the tank temperature and its pressure reduced to 52.8 percent of the tank pressure at the location where the velocity becomes sonic (M $= 1$).

If the gas expands (pressure drops), Eq. (10.16) indicates an increase in the Mach number. This is true for both subsonic and supersonic flow. Equations (10.10) and (10.17) indicate that an area reduction in subsonic flow and an area increase in supersonic flow are necessary to accelerate the gas. The relationship between the area A^* where the Mach number is unity (the throat) and the area A at any other section where M $\gtrless 1$ may be obtained by integrating Eq. (10.17) to obtain

$$\frac{A}{A^*} = \frac{1}{M}\left(\frac{1 + \dfrac{k-1}{2}M^2}{(k+1)/2}\right)^{(k+1)/2(k-1)}$$

(10.24)

which depends only on the Mach number for a given gas.

From continuity,

$$\frac{V}{V^*} = \frac{A^*\rho^*}{A\rho} = M\sqrt{\frac{\dfrac{k+1}{2}}{1 + \dfrac{k-1}{2}M^2}}$$

(10.25)

which relates the velocity V at a section where $M \gtrless 1$ to the velocity V^* where the velocity is sonic ($M = 1$).

Note that for isentropic flow, values of T/T_0, p/p_0, ρ/ρ_0, A/A^*, and V/V^* depend only on the Mach number for a given gas, and these functions may be tabulated. An example is shown in Table A-9 (Appendix V) for isentropic flow of a gas for which $k = 1.4$. Gas tables of this type may be used for numerical calculations of flow problems.

The results may easily be applied to the analysis of flow through a given nozzle to the design of a nozzle for given flow conditions.

For a *converging nozzle* (Fig. 10-3), if the receiver and supply pressures are equal, there will be no flow [Eq. (10.18b)]. As p_3 is reduced, V_1 will increase to a sonic value. This occurs when p_1 becomes critical [Eq. (10.23)]. For values of receiver pressure above critical, the nozzle exit pressure and receiver pressure are equal. If the receiver pressure is reduced below the critical pressure, the flow through the nozzle is not affected, since it is sonic at exit and cannot exceed that value. [Suppose V_1 is supersonic. Then the velocity would have to be sonic between the supply tank and nozzle exit. But this is impossible from Eq. (10.10) and Table 10-1, since the area is decreasing and sonic flow cannot occur if the area is changing.] The flow rate is given by

$$\text{mass flow rate} = V_1 A_1 \rho_1$$

The exit velocity may be found from Eq. (10.18a) or (10.18b). The area A_1 is assumed given. The density ρ_1 may be obtained from Eq. (10.21) or from the gas equation, since if either p_1 or T_1 is given, the other may be found from the isentropic relation. If V_1 is sonic, the mass flow rate for air flow becomes

$$\dot{m} = c_1 A_1 \frac{p_1}{RT_1} = \sqrt{kRT_1}(A_1)\frac{0.528p_0}{R(5T_0/6)}$$

(10.26a)

In SI units,

$$\dot{m} = 0.0404 A_1 p_0/\sqrt{T_0} \quad \text{kg/s}$$

(10.26b)

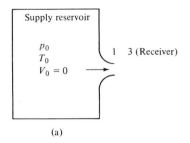

(a)

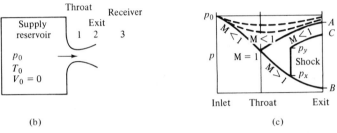

(b)   (c)

Fig. 10-3.   Gas flow through nozzles. (a) Converging nozzle. (b) Converging–diverging nozzle. (c) Axial pressures in nozzles: from inlet to throat for converging nozzle; from inlet to throat to exit for converging–diverging nozzle.

If the exit flow is sonic and the receiver pressure less than critical, the flow becomes supersonic beyond the exit and dissipates itself through a series of successive shocks outside the nozzle.

The axial pressure distribution between the inlet and throat of a converging–diverging nozzle shown in Fig. 10-3c applies to a converging nozzle.

For a *converging–diverging nozzle* (Fig. 10-3b), there is no flow if $p_3 = p_0 (= p_1, = p_2$, and so forth). As the receiver pressure p_3 is lowered, flow exists throughout the nozzle, with a minimum pressure and maximum velocity at the throat. The pressures vary along the axis according to the dashed lines of Fig. 10-3c. As the receiver pressure is lowered to A, the flow in the throat will become sonic and the pressure there will be critical. Flow will be subsonic both upstream and downstream from the throat. There is only one other exit pressure (at B) for which isentropic flow with sonic velocity in the throat may occur. The flow rate is the same whether the exit pressure is at A or at B, since the throat conditions are the same in each instance. The flow is supersonic, however, beyond the throat if the exit pressure is at B. As for a converging nozzle, if the receiver pressure is below B, a series of successive shocks occur beyond the end of the nozzle exit and full expansion does not occur within the nozzle. If the exit pressure is below A but somewhat above B, a shock will be set up in the nozzle at a point depending on the exit pressure and the nozzle shape. In this instance the flow may be considered isentropic only up to the shock and beyond the shock, but not through the shock. It should be noted that for flow which is nonisentropic owing to viscous or frictional effects rather than to shocks, the flow in the nozzle throat is subsonic.

If the throat velocity is sonic, the flow rate is given by Eq. (10.26a). In general, the mass flow rate is always given by the product $VA\rho$. Values of velocity and density may be obtained for any section where the pressure or temperature is specified.

Example 10-1

Air at 700 kPa abs and 40°C flows from a tank through a converging nozzle whose exit area is 5×10^{-4} m². What is the exit pressure, exit temperature, and flow rate when the receiver pressure is (a) 500 kPa abs and (b) atmospheric (101 kPa abs)? See Fig. 10-3.

SOLUTION
(a) Since the receiver pressure is greater than critical (370 kPa abs), flow will be subsonic at the nozzle exit, and $p_1 = 500$ kPa abs:

$$T_1 = T_0(p_1/p_0)^{(k-1)/k} = 313(500/700)^{0.286} = (313)(0.9081)$$
$$= 284.2°\text{K}$$

$$\dot{m} = V_1 A_1 \rho_1$$

where

$$V_1 = \sqrt{2c_p(T_0 - T_1)} = \sqrt{(2)(1005)(28.8)} = 240 \text{ m/s}$$
$$\rho_1 = p_1/RT_1 = 5 \times 10^5/(287.1)(284.2) = 6.13 \text{ kg/m}^3$$

then $m = V_1 A_1 \rho_1 = (240)(5 \times 10^{-4})(6.13) = 0.736$ kg/s.
(b) The receiver pressure is below critical, so the exit pressure is critical at $p_1 = 370$ kPa abs. The exit temperature is, from Eq. (10.22), $T_1 = 5T_0/6 = 261°$K; V_1 is sonic, so

$$V_1 = \sqrt{kRT_1} = 20.0\sqrt{261} = 323 \text{ m/s}$$
$$\rho_1 = p_1/RT_1 = (370 \times 10^3)/(287.1)(261) = 4.94 \text{ kg/m}^3$$

then $\dot{m} = V_1 A_1 \rho_1 = (323)(5 \times 10^{-4})(4.94) = 0.798$ kg/s.

━━━

Example 10-2

If a converging–diverging nozzle with a throat area of 5×10^{-4} m² and an exit area of 10×10^{-4} m² is attached to the supply tank of Example 10-1, what is the flow rate, exit pressure, temperature, and Mach number for complete expansion in the nozzle? See Fig. 10-3.

SOLUTION

For complete expansion the exit pressure is at B in Fig. 10-3c. The flow is critical in the nozzle throat, and thus the flow rate is 0.798 kg/s as in Example 10-lb. The area ratio is $A_2/A^* = 2$, and from Table A-9 (Appendix V) $M_2 = 2.20$ and corresponding values of $p_2/p_0 = 0.0935$ and $T_2/T_0 = 0.508$ indicate values for $p_2 = 65.45$ kPa abs and $T_2 = 159°$K $= -114°$C.

———

10-4. Diabatic Flow of a Perfect Gas Without Friction

Adiabatic flow implies flow without heat transfer to or from the fluid. Diabatic flow is flow in which heat *is* added to or removed from the fluid. In this section we will assume the flow is frictionless, implying a nonviscous (ideal) gas. This condition is approximated in short ducts with large heat transfer. In addition, the gas is assumed to be perfect, and the perfect gas relation $pv = RT$ applies. We will write some differential equations expressing the continuity equation, the Euler equation of motion (applicable only to a nonviscous fluid), the energy equation (including heat transfer), the perfect gas equation, and the definition of the Mach number. These will be combined to give differential equations from which variations in velocity, pressure, density, temperature, and Mach number in the direction of flow may be obtained as a function of Mach number for both heat addition to the gas and heat removal from the gas. Only qualitative results will be illustrated to show the comparison with similar variations in adiabatic flow without friction, as indicated by Eqs. (10.14) through (10-17).

The pertinent equations are

Continuity

$$\frac{dV}{V} + \frac{dA}{A} + \frac{d\rho}{\rho} = 0$$

Euler's equation of motion

$$\frac{dp}{\rho} = -V\,dV$$

Energy

$$dh + V\,dV = dq$$

Perfect gas

$$\frac{dp}{p} = \frac{d\rho}{\rho} + \frac{dT}{T}$$

Mach number

$$\frac{dM}{M} = \frac{dV}{V} - \frac{dT}{2T}$$

These may be combined (after considerable manipulation) to give

$$\frac{dV}{V} = \frac{1}{1-M^2}\left(\frac{dq}{h} - \frac{dA}{A}\right)$$

$$\frac{dp}{p} = \frac{-kM^2}{1-M^2}\left(\frac{dq}{h} - \frac{dA}{A}\right)$$

$$\frac{d\rho}{\rho} = \frac{-1}{1-M^2}\left(\frac{dq}{h} - \frac{dA}{A}\right) - \frac{dA}{A}$$

$$\frac{dT}{T} = \frac{1-kM^2}{1-M^2}\left(\frac{dq}{h} - \frac{dA}{A}\right) + \frac{dA}{A}$$

$$\frac{dM}{M} = \frac{1+kM^2}{2(1-M^2)}\left(\frac{dq}{h} - \frac{dA}{A}\right) - \frac{dA}{A}$$

where

$$h = c_p T = \frac{Rk}{k-1} T = \frac{pk}{\rho(k-1)}$$

Just how the velocity, pressure, temperature, and Mach number will change in the direction of flow depends on whether the flow is subsonic ($M < 1$) or supersonic ($M > 1$) and on whether $(dq/h) - (dA/A)$ is positive, zero, or negative. In the case of density, temperature, and Mach number variations the results will also depend on whether the area increases, decreases, or remains constant (the dA/A terms affect the results).

For a duct of constant cross section ($dA = 0$) without friction but with heat added (dq is positive), the following results are obtained from a determination of the sign of each expression:

1. For the subsonic case, the velocity and Mach number increase and the pressure and density decrease in the direction of flow (the term $1 - M^2$ is positive). The temperature increases if the Mach number is less than $1/\sqrt{k}$ and *decreases* if the Mach number is greater than $1/\sqrt{k}$ but less than unity. Heat addition has the effect of *cooling* the gas in this range of Mach numbers. (See the discussion of the Rayleigh line in Sec. 15-2.)

2. For the supersonic case, the velocity and Mach number decrease, and the pressure, density, and temperature increase in the direction of flow. In all instances the term $1 - M^2$ is negative, and the sign of dV/V and so forth may be determined easily.

In all instances for diabatic flow, all variables change at an infinite rate at a Mach number $M = 1$, and the results are nonvalid at this point. Thus both subsonic flow and supersonic flow approach the sonic condition as heat is added, but cannot go beyond this condition (subsonic flow cannot become supersonic, and supersonic flow cannot become subsonic). Results for heat removal should be determined by the student.

For a duct of constant cross section without friction but with heat transfer, the continuity equation, momentum theorem, and energy equations in finite form may be combined with the perfect gas law and the definition of the Mach number to give equations which may be used in making calculations. These are

Continuity

$$V_1\rho_1 = V_2\rho_2 = \text{constant mass flow intensity } G$$

Momentum theorem

$$p_1 - p_2 = V_1\rho_1(V_2 - V_1) = V_2\rho_2(V_2 - V_1)$$

from which

$$p_1 + V_1^2\rho_1 = p_2 + V_2^2\rho_2$$

This equation indicates that the thrust function is constant. In terms of mass flow intensity G,

$$p_1 + GV_1 = p_2 + GV_2$$

or

$$p_1 + \frac{G^2}{\rho_1} = p_2 + \frac{G^2}{\rho_2} \tag{10.27}$$

which indicates that the pressure and velocity as well as pressure and density are linearly related. Equation (10.27) is known as the equation of the Rayleigh line. This is discussed in detail in Chapter 15.

Energy equation

$$h_1 + \frac{V_1^2}{2} + q = h_2 + \frac{V_2^2}{2}$$

and since $h + V^2/2$ is the stagnation enthalpy h_0,

$$h_{01} + q = h_{02}$$

from which

$$q = c_p(T_{02} - T_{01}) \tag{10.28}$$

and thus the heat transfer directly affects the change in stagnation enthalpy and the stagnation temperature.

Mach number

$$\frac{M_2}{M_1} = \frac{V_1 c_2}{V_2 c_1} = \frac{V_1}{V_2}\sqrt{\frac{T_2}{T_1}}$$

Perfect gas

$$\frac{p_1}{\rho_1 T_1} = \frac{p_2}{\rho_2 T_2}$$

These equations may be combined to give the ratio of stagnation temperatures and static temperatures in terms of Mach number at two sections between which heat is transferred in a frictionless duct.

$$\frac{T_{02}}{T_{01}} = \frac{T_2}{T_1} \frac{\left(1 + \dfrac{k-1}{2} M_2^2\right)}{\left(1 + \dfrac{k-1}{2} M_1^2\right)} \tag{10.29}$$

and

$$\frac{T_2}{T_1} = \frac{M_2^2 (1 + kM_1^2)^2}{M_1^2 (1 + kM_2^2)^2} \tag{10.30}$$

10-5. Flow of Compressible Gases in Pipes with Friction

Subsonic gas flow in pipes differs from liquid flow in pipes primarily in that the gas density decreases and hence the velocity increases in the direction of flow. The general flow equation for compressible flow in a duct of constant cross section was derived in Sec. 9-1 as Eq. (9.4). The wall shear stress may be expressed in terms of the friction factor from Eq. (9.8), so that the pressure gradient is

$$-\frac{dp}{dx} = \frac{f}{D} \frac{\rho V^2}{2} + \beta \rho V \frac{dV}{dx} \tag{10.31}$$

This equation is similar to the Darcy–Weisbach equation [Eq. (9.5)] except for the last term on the right, which represents the pressure drop required to increase the flow momentum. The value of β is about 1.03 for fully developed incompressible turbulent flow. Thus a one-dimensional analysis, with $\beta = 1.0$, is justified.

A dimensional analysis would indicate that the friction factor for compressible flow could depend upon the Mach number as well as the relative roughness of the pipe and the Reynolds number. Experiments have shown, however, that the dependence on Mach number for subsonic flow is negligible and that the friction factor may be obtained in the same manner as for incompressible flow.

Equation (10.31) with $\beta = 1$ may be written in dimensionless form as

$$\frac{dp}{\rho V^2/2} + f \frac{dx}{D} + 2 \frac{dV}{V} = 0 \tag{10.32}$$

This general equation will be integrated directly for isothermal flow and used in the analysis for adiabatic flow in pipes with friction.

ISOTHERMAL FLOW

In order to integrate Eq. (10.32), the variable density and velocity will have to be expressed in terms of the variable pressure, and the variation in the friction factor f will have to be investigated.

If all conditions are known at some upstream section (section 1), those at any arbitrary section downstream can be expressed in terms of known values at section 1.

From the perfect gas equation of state,

$$\frac{p}{\rho} = \frac{p_1}{\rho_1} = RT$$

a constant, from which $\rho = p\rho_1/p_1$.

From continuity,

$$V\rho = V_1\rho_1$$

then $V = V_1 p_1/p$. The differential form of this last equation (for later use) is

$$\frac{dV}{V} + \frac{dp}{p} = 0 \qquad \text{or} \qquad \frac{dV}{V} = -\frac{dp}{p}$$

The first term in Eq. (10.32) can then be written as

$$\frac{2}{\rho_1 V_1^2 p_1} \, p\, dp$$

The friction factor depends on the relative roughness of the pipe, which is assumed to be constant, and on the Reynolds number $\text{Re} = VD\rho/\mu$, which is also constant, since $V\rho$ is constant, from continuity, and the dynamic viscosity depends only on the temperature, which is constant. Hence the friction factor is constant and can be found in the usual manner for known conditions at any section.

If $L = x_2 - x_1$, Eq. (10.32) can now be integrated to obtain

$$p_1^2 - p_2^2 = \rho_1 V_1^2 p_1 \left(f\frac{L}{D} - 2\ln\frac{p_2}{p_1} \right) \qquad (10.33a)$$

or

$$p_1^2 - p_2^2 = k\text{M}_1^2 p_1^2 \left(f\frac{L}{D} - 2\ln\frac{p_2}{p_1} \right) \qquad (10.33b)$$

in terms of the initial Mach number. These equations give the pressure at some distance L downstream of any initial section 1 where conditions are known. The logarithmic term is often small compared to fL/D and may be neglected for a first approximation. Then this first approximation for p_2 should be used to calculate the magnitude of the logarithmic term, and an iterative process used to calculate p_2 precisely.

Equation (10.33b) may be solved for fL/D to give

$$f\frac{L}{D} = \frac{1}{k\mathrm{M}_1^2}\left[1 - \left(\frac{p_2}{p_1}\right)^2\right] - 2\ln\frac{p_1}{p_2} \qquad (10.34)$$

This equation gives the distance L from section 1, where conditions are known, to some downstream section where p_2 is specified. It is a dimensionless equation and indicates that the dimensionless length fL/D is related to the pressure ratio p_2/p_1 by a family of curves, one for each initial Mach number, for a given gas (given k). These relationships are shown in Fig. 10-4.

A dimensional plot of pressure vs. length would have the same general appearance as the curves in Fig. 10-4. Equation (10.32), by letting $dV/V = -dp/p$, contains only two differentials, dp and dx. Thus the pressure gradient, which is the slope of the curves of Figs. 10-4 and 10-5, can be shown to be

$$\frac{dp}{dx} = \frac{\dfrac{pf}{2D}}{1 - \dfrac{p}{\rho V^2}} = \frac{\dfrac{f}{D}\dfrac{\rho V^2}{2}}{k\mathrm{M}^2 - 1} \qquad (10.35)$$

As $\mathrm{M} \to 0$, this is essentially the Darcy equation for liquid pipe flow; thus at low Mach numbers, isothermal flow may be considered as incompressible. The pressure gradients will be within 5 percent at Mach numbers up to 0.184 for air. A thermodynamic analysis will indicate that the entropy increases as the fluid goes from 1 to 0 in Fig. 10-5, and decreases from 0 to 3. This latter decrease is impossible since it contradicts the second law of

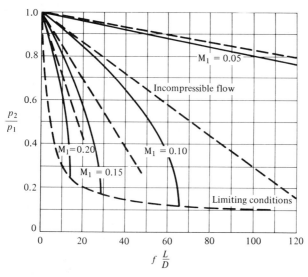

Fig. 10-4. Conditions along pipe for isothermal flow of a gas at various initial Mach numbers ($k = 1.4$).

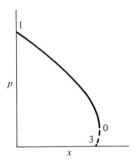

Fig. 10-5. Pressure along pipe for gas flow.

thermodynamics. Thus flow can exist only up to 0, which is called the limiting point. Associated with this limiting point are a limiting Mach number, a limiting or minimum pressure, and a limiting or maximum length of pipe. These will be indicated by asterisks (M^*). Flow from 3 to 0 represents supersonic flow.

• *Limiting or Maximum Mach Number.* At the limiting point the pressure gradient is infinite. Thus for dp/dx to be infinite, the denominator of Eq. (10.35) must be zero. Then

$$M^* = \frac{1}{\sqrt{k}}$$

(10.36)

and the velocity can increase along the pipe only until the Mach number equals $1/\sqrt{k} = 0.845$ for air, for example.

• *Limiting or Minimum Pressure.* Recall that from continuity

$$V_1 p_1 = V p = V^* p^*$$

so that

$$\frac{p^*}{p_1} = \frac{V_1}{V^*} = \frac{M_1 c_1}{M^* c^*} = \frac{M_1}{M^*} = M_1 \sqrt{k}$$

since the acoustic velocity remains constant for isothermal conditions. Therefore,

$$\frac{p^*}{p_1} = M_1 \sqrt{k}$$

(10.37)

and the pressure drops along the pipe to a minimum which depends on the initial Mach number.

• *Limiting or Maximum Length.* The limiting or maximum length of pipe can be found by inserting the expression for the minimum pressure, Eq. 10.37, into the general expression for $f L/D$, Eq. (10.34), to get

$$\frac{fL^*}{D} = \frac{1}{k M_1^2} - 1 - \ln \frac{1}{k M_1^2} \tag{10.38}$$

A number of conclusions may be drawn concerning subsonic isothermal flow:

1. The pressure drops at an increasing rate along the pipe, in contrast with a constant pressure gradient for fully developed liquid flow.
2. The velocity and Mach number increase up to a maximum Mach number of $1/\sqrt{k}$. This increase ceases at a limiting or maximum length, which must be at the end of the pipe. If assumed initial conditions at some upstream section result in a maximum length of pipe which is less than a given pipe length, then the initial conditions will have to be adjusted (lower M_1) so that the given pipe length becomes at least equal to or less than the maximum length for the adjusted initial conditions.
3. The limiting pressure ratio and the limiting or maximum length depend only on the gas (k value) and the initial Mach number. Thus tables may be prepared and used for making pipe calculations. Then

$$\frac{p_2}{p_1} = \frac{M_1}{M_2} \tag{10.39}$$

and from Eq. (10.38)

$$\frac{fL}{D} = \left(\frac{fL^*}{D}\right)_{M_1} - \left(\frac{fL^*}{D}\right)_{M_2} \tag{10.40}$$

These equations relate pressures and lengths between two sections for which the Mach numbers are given.

It is interesting to speculate whether heat is added or removed from the gas as it flows through a pipe. The general energy equation becomes

$$\frac{V_1^2}{2} + q = \frac{V_2^2}{2}$$

and since $V_2 > V_1$ for subsonic flow, q must be positive and heat is added to the gas. The tendency to cool as a result of expansion is greater than the tendency to heat as a result of wall friction, and therefore heat must be added to maintain a constant temperature. For supersonic flow, $V_2 < V_1$, q is negative, and heat is removed from the gas.

ADIABATIC FLOW

Integration of Eq. (10.32) again requires that the variable density and velocity be expressed in terms of the variable pressure, and the variation of the friction factor must again be investigated.

The momentum equation [Eq. (10.32)] when multiplied by $k\text{M}^2/2$ may be written as

$$\frac{dp}{p} + \frac{k\text{M}^2}{2} f \frac{dx}{D} + k\text{M}^2 \frac{dV}{V} = 0$$

The continuity equation in differential form is

$$\frac{dV}{V} + \frac{d\rho}{\rho} = 0$$

The perfect gas equation in differential form is

$$\frac{dp}{p} = \frac{d\rho}{\rho} + \frac{dT}{T}$$

From the definition of the Mach number $\text{M} = V/c = V/\sqrt{kRT}$,

$$\frac{d\text{M}}{\text{M}} = \frac{dV}{V} - \frac{dT}{2T}$$

The energy equation $h + V^2/2 = \text{constant}$ may be written in differential form as

$$\frac{dT}{T} + (k - 1)\text{M}^2 \frac{dV}{V} = 0$$

We have five simultaneous equations in six differential variables: dp/p, dV/V, $d\rho/\rho$, dT/T, $d\text{M}/\text{M}$, and $f\,dx/D$. The first five (dependent) variables may be expressed in terms of $f\,dx/D$, the independent variable. The results are

$$\frac{dp}{p} = -\frac{1 + (k - 1)\text{M}^2}{1 - \text{M}^2} \frac{k\text{M}^2}{2} f \frac{dx}{D}$$

$$\frac{dV}{V} = \frac{1}{1 - \text{M}^2} \frac{k\text{M}^2}{2} f \frac{dx}{D}$$

$$\frac{d\rho}{\rho} = -\frac{1}{1 - \text{M}^2} \frac{k\text{M}^2}{2} f \frac{dx}{D}$$

$$\frac{dT}{T} = -\frac{\text{M}^2(k - 1)}{1 - \text{M}^2} \frac{k\text{M}^2}{2} f \frac{dx}{D}$$

and

$$\frac{d\text{M}}{\text{M}} = \frac{1 + \frac{1}{2}(k - 1)\text{M}^2}{1 - \text{M}^2} \frac{k\text{M}^2}{2} f \frac{dx}{D}$$

A thermodynamic analysis shows that since dx is positive in the direction of flow, f must be positive. Therefore the preceding equations indicate that for subsonic flow (M < 1) the velocity and Mach number increase, while the pressure, density, and temperature decrease in the direction of flow. The opposite is true in each instance for supersonic flow (M > 1).

• *Limiting or Maximum Mach Number.* The equation for dp/p can be solved for the pressure gradient:

$$\frac{dp}{dx} = -\frac{f\,kp}{2D}M^2\left[\frac{1 + (k-1)M^2}{1 - M^2}\right] = -\frac{f}{D}\frac{\rho V^2}{2}\left[\frac{1 + (k-1)M^2}{1 - M^2}\right]$$

(10.41)

As M → 0, this is essentially the Darcy equation for liquid pipe flow; thus for low Mach numbers, adiabatic gas flow may also be treated as incompressible. The pressure gradients will be within 5 percent at Mach numbers up to 0.185 for air. Figure 10-5, which is applicable for adiabatic flow as well as isothermal, indicates that the pressure drops to the limiting point 0, and at that point dp/dx is infinite. From Eq. (10.41), the limiting Mach number is 1 for adiabatic flow.

• *Limiting or Minimum Pressure.* To obtain the limiting pressure, we combine the equations for dp/p and dM/M to obtain

$$\frac{dp}{p} = -\frac{1 + (k-1)M^2}{M[1 + \frac{1}{2}(k-1)M^2]}dM$$

which, when integrated between some given section and the point where M = 1 (the limiting point), gives

$$\frac{p^*}{p_1} = M_1\sqrt{\frac{2[1 + \frac{1}{2}(k-1)M_1^2]}{k+1}}$$

(10.42)

The integration is evident if $dM = dM^2/2M$.

• *Limiting or Maximum Length.* The expression for dM/M can be rearranged to give

$$f\frac{dx}{D} = \frac{1 - M^2}{kM^4[1 + \frac{1}{2}(k-1)M^2]}dM^2$$

by letting $dM = dM^2/2M$. Integrating between a given section and the point where M = 1,

$$\frac{\bar{f}L^*}{D} = \frac{1 - M_1^2}{kM_1^2} + \frac{k+1}{2k}\ln\frac{(k+1)M_1^2}{2[1 + \frac{1}{2}(k-1)M_1^2]}$$

(10.43)

where $\bar{f}$ is the average friction factor. At high Reynolds numbers, especially for rough pipes, the friction factor depends only on the pipe roughness, and thus the friction factor could conceivably be constant for adiabatic flow.

If not,

$$\bar{f} = \frac{1}{L^*} \int_0^{L^*} f \, dx$$

The value of f at section 1 may be assumed to equal $\bar{f}$, and after solving for conditions at some downstream point, the value of f throughout the entire pipe can be examined and an appropriate average value used to recalculate downstream conditions.

- *Limiting Temperature and Velocity.* The preceding expressions for dT/T, dV/V, and dM/M can be solved simultaneously by eliminating $f \, dx/D$ to obtain, after integration,

$$\frac{T^*}{T_1} = \frac{2[1 + \frac{1}{2}(k-1)M_1^2]}{k+1} \tag{10.44}$$

and

$$\frac{V^*}{V_1} = \frac{1}{M_1} \sqrt{\frac{2[1 + \frac{1}{2}(k-1)M_1^2]}{k+1}} \tag{10.45}$$

These equations may be used in the following manner: for the flow conditions known at some arbitrary section 1, the limiting conditions can be calculated. These limiting conditions are the same for corresponding flow conditions at any other section. Therefore

$$\frac{p_2}{p_1} = \frac{p^*/p_1}{p^*/p_2}$$

$$\left(\frac{\bar{f}L}{D}\right)_{1-2} = \left(\frac{\bar{f}L^*}{D}\right)_1 - \left(\frac{\bar{f}L^*}{D}\right)_2$$

$$\frac{T_2}{T_1} = \frac{T^*/T_1}{T^*/T_2}$$

$$\frac{V_2}{V_1} = \frac{V^*/V_1}{V^*/V_2}$$

Flow near the limiting condition would be very nearly adiabatic, because the high heat-transfer rates required to maintain isothermal flow would be very difficult to achieve.

If adiabatic flow is assumed for a given initial subsonic condition, the pressure at any downstream point where the Mach number is M_2 is always slightly less than it would be if the flow were assumed to be isothermal. The ratio is

$$\frac{p_{2 \, ad}}{p_{2 \, iso}} = \left[\frac{1 + \frac{1}{2}(k-1)M_1^2}{1 + \frac{1}{2}(k-1)M_2^2}\right]^{1/2} \tag{10.46}$$

which is less than 1 for subsonic flow and depends on the ratio of specific heat capacities k for the gas and on the initial and downstream Mach numbers. For air with $M_2 = 0.854$ (the limiting Mach number for isothermal flow), $p_{2\,ad}/p_{2\,iso} \geq 0.934$. Similarly, if adiabatic flow is assumed for a given initial subsonic condition, the temperature is always slightly less at any prescribed point where the Mach number is M_2 than if isothermal flow is assumed. The ratio is

$$\frac{T_{2\,ad}}{T_{2\,iso}} = \frac{1 + \frac{1}{2}(k-1)M_1^2}{1 + \frac{1}{2}(k-1)M_2^2} \tag{10.47}$$

which is less than 1 for subsonic flow. For air with $M_2 = 0.854$, $T_{2\,ad}/T_{2\,iso} \geq 0.875$; and for $M_2 = 0.5$, this ratio ≥ 0.952.

Therefore, except at high Mach numbers, subsonic isothermal and subsonic adiabatic flow do not differ appreciably.

Example 10-3

It is desired to pump methane ($k = 1.31$ and $R = 518$ J/kg °K) through a 30-cm commercial steel pipe. The discharge from a compressor is at a pressure of 2000 kPa abs, a temperature of 60°C (dynamic viscosity is 1.3×10^{-5} kg/m s), and a velocity of 15 m/s. For both isothermal and adiabatic flow with friction, find (a) the minimum pressure possible; (b) the maximum length of pipe possible; (c) the maximum velocity possible; (d) the pressure, temperature, velocity, and Mach number at a distance equal to one-half the maximum length from the compressor; and (e) the location of a second compressor if the compressor inlet pressure is 500 kPa abs.

SOLUTION

$$p_1 = 2000 \text{ kPa abs} \qquad \rho_1 = \frac{p_1}{RT_1} = 11.59 \text{ kg/m}^3$$

$$T_1 = 333°K$$

$$c_1 = \sqrt{kRT_1} = \sqrt{(1.31)(518)(333)} = 475 \text{ m/s}$$

$$M_1 = \frac{V_1}{c_1} = \frac{15}{475} = 0.0316$$

$$Re_1 = \frac{V_1 D \rho_1}{\mu_1} = \frac{(15)(0.30)(11.59)}{1.3 \times 10^{-5}} = 4.0 \times 10^6$$

$$\frac{k}{D} = \frac{0.000045}{0.30} = 0.00015 \qquad \text{so that} \quad f_1 = 0.0134$$

See the accompanying Table 10-2.

Table 10-2   COMPARISON OF ISOTHERMAL AND ADIABATIC GAS FLOW
IN A LONG PIPE IN EXAMPLE 10-3

ISOTHERMAL FLOW	ADIABATIC FLOW

(a) $\dfrac{p^*}{p_1} = M_1\sqrt{k} = 0.0316\sqrt{1.31}$

$\quad = 0.0362$

$\quad p^* = (0.0362)(2000) = 72.3 \text{ kPa abs}$

(a) $\dfrac{p^*}{p_1} = M_1\sqrt{\dfrac{2[1 + \frac{1}{2}(k-1)M_1^2]}{k+1}}$

$\quad = 0.0316\sqrt{\dfrac{2[1 + (0.155)(0.0316)^2]}{2.31}}$

$\quad = 0.0294$

$\quad p^* = (0.0294)(2000) = 58.8 \text{ kPa abs}$

(b) $\dfrac{fL^*}{D} = \dfrac{1}{kM_1^2} - 1 - \ln\dfrac{1}{kM_1^2}$

$\quad = \dfrac{1}{(1.31)(0.0316)^2} - 1$

$\quad - \ln\dfrac{1}{(1.31)(0.0316)^2}$

$\quad = 764 - 1 - 6.6$

$\quad = 756$

$\quad L^* = \dfrac{756D}{f} = \dfrac{(756)(0.30)}{0.0134}$

$\quad L^* = 16{,}925 \text{ m}$

(b) $\dfrac{\bar{f}L^*}{D} = \dfrac{1 - M_1^2}{kM_1^2}$

$\quad + \dfrac{k+1}{2k}\ln\dfrac{(k+1)M_1^2}{2[1 + \frac{1}{2}(k-1)M_1^2]}$

$\quad = \dfrac{1 - 0.000999}{1/764}$

$\quad + \dfrac{2.31}{2.62}\ln\dfrac{(2.31)(0.000999)}{2(1.00015)}$

$\quad = 763 - 6$

$\quad = 757$

$\quad L^* = \dfrac{757D}{f} = 16{,}950 \text{ m}$

When T^* is found, Re^* will be calculated and f^* determined. A new $\bar{f}$ may then give a slightly different L^*.

(c) $V^* = c^*M^* = \dfrac{475}{\sqrt{1.31}} = 415 \text{ m/s}$

(c) $V^* = c^* = \sqrt{kRT^*}$, where

$\quad T^* = T_1\dfrac{2[1 + \frac{1}{2}(k-1)M_1^2]}{k+1}$

$\quad = 333\dfrac{2(1.00015)}{2.31} = 288°\text{K}$

$\quad V^* = \sqrt{(1.31)(518)(288)} = 442 \text{ m/s}$

Thus

$\quad Re^* = \dfrac{V^*\rho^*D}{\mu^*} = \dfrac{V_1\rho_1 D}{\mu^*}$

$\quad Re^* = \dfrac{(15)(0.30)(11.59)}{1.2 \times 10^{-5}} = 4.35 \times 10^6$

$\quad f^* = 0.0133 \text{ and } \bar{f} = 0.01335$

$\quad L^* = \dfrac{757D}{\bar{f}} = 17{,}010 \text{ m, a truer value}$

Table 10-2 (*continued*)

ISOTHERMAL FLOW	ADIABATIC FLOW

(d) At $L_2 = 8460$ m,

$$\frac{fL_2}{D} = \frac{756}{2} = 378$$

$$1 - \left(\frac{p_2}{p_1}\right)^2 = kM_1^2\left(f\frac{L_2}{D} + 2\ln\frac{p_1}{p_2}\right)$$

For a first approximation,

$$1 - \left(\frac{p_2}{p_1}\right)^2 = \frac{1}{764}(378) = 0.495$$

$$\frac{p_2}{p_1} = 0.711$$

and $+2\ln(p_1/p_2) = +0.68$, and a recalculation gives

$$\frac{p_2}{p_1} = 0.710$$

$$p_2 = (0.710)(2000) = 1420 \text{ kPa abs}$$

$$T_2 = 60°C = 333°K$$

$$V_2 = \frac{V_1\rho_1}{\rho_2} = V_1\frac{p_1}{p_2} = 15\frac{(11.59)}{8.23}$$

$$= 21.1 \text{ m/s}$$

(e) If $p_3 = 500$ kPa abs,

$$\left(\frac{fL}{D}\right)_{1-3} = \frac{1}{kM_1^2}\left[1 - \left(\frac{p_3}{p_1}\right)^2\right] - 2\ln\frac{p_1}{p_3}$$

$$= 764\left[1 - \frac{1}{16}\right] - 2\ln 4$$

$$= 713$$

$$L_3 = \frac{713D}{f} = 15,960 \text{ m}$$

(d) At $L_2 = 8505$ m,

$$\frac{p_2}{p_1} = \frac{p^*/p_1}{p^*/p_2} = \frac{0.0294}{\text{function of } M_2}$$

Eq. 10-43 with $fL_2/D = 378.5$ is solved for $M_2 = 0.0446$. Then

$$\frac{p^*}{p_2} = M_2\sqrt{\frac{2[1 + \frac{1}{2}(k-1)M_2^2]}{k+1}}$$

$$p_2 = 1417 \text{ kPa abs}$$

$$T_2 = \frac{T^*}{2[1 + \frac{1}{2}(k-1)M_2^2]/(k+1)}$$

$$= \frac{288}{0.866} = 333°K$$

and the temperature has not dropped noticeably. Since $c_1 \approx c_2$,

$$V_2 = V_1\frac{M_2}{M_1}$$

$$= 60\frac{0.0446}{0.0316}$$

$$= 21.2 \text{ m/s}$$

(e) If $p_3 = 500$ kPa abs, $(fL/D)_3$ is a function of M_3, which can be found by trial from Eq. 10-42 to be $M_3 = 0.126$:

$$\left(\frac{fL^*}{D}\right)_3 = \frac{1 - M_3^2}{kM_3^2}$$

$$- \frac{k+1}{2k}\ln\frac{2[1 + \frac{1}{2}(k-1)M_3^2]}{(k+1)M_3^2}$$

$$= 47.3 - 3.5 = 43.8$$

$$L_3^* = \frac{(43.8)(0.30)}{(0.01335)} = 984 \text{ m}$$

$$L_3 = (L^*)_1 - (L^*)_3$$

$$= 17,010 - 984$$

$$= 16,020 \text{ m}$$

The results of Example 10-3 indicate that for given initial conditions: (a) the minimum pressure for adiabatic flow is slightly less than for isothermal flow; (b) the maximum length for both types of flow is essentially the same, that for adiabatic flow being less than 1 percent greater than for isothermal flow; (c) the pressure and temperature at a point halfway to the limiting point are essentially the same for either type of flow; (d) the rate of pressure drop is very large near the limiting point, and practical considerations rule out the advisability of having pipes longer than 80–90 percent of the maximum length; and (e) for practical purposes, since there is always some uncertainty in the values of the friction factor, and purely isothermal or purely adiabatic flow in pipes is rarely if every achieved, either isothermal or adiabatic flow may be assumed in making engineering calculations.

Example 10-4

Carbon dioxide ($R = 188.5$ J/kg °K, $k = 1.30$, and $\mu = 2.2 \times 10^{-5}$ kg/m s) enters a 60-mm smooth tube at a pressure of 1200 kPa abs, a velocity of 30 m/s, and a density of 20 kg/m^3. The tube discharges the gas into the atmosphere. Compare the maximum allowable tube length for isothermal flow with that for adiabatic flow.

SOLUTION

The initial Mach number and the friction factor are both needed in Eqs. (10.38) and (10.43).

$$M_1 = V_1\sqrt{kp_1/\rho_1} = 30/\sqrt{(1.30)(1.20 \times 10^6)/(20)} = 0.1074$$

$$Re = VD\rho/\mu = (30)(0.06)(20)/2.2 \times 10^{-5} = 1.64 \times 10^6$$

so $f = 0.0107$

For isothermal flow, Eq. (10.38) gives

$$fL^*/D = \frac{1}{(1.30)(0.1074)^2} - 1 - \ln\frac{1}{(1.30)(0.1074)^2}$$

$$= 66.8 - 1 - 4.20 = 61.6$$

and

$$L^* = (61.6)(0.06)/0.0107$$
$$= 345 \text{ m}$$

For adiabatic flow, Eq. (10.43) gives

$$fL^*/D = 66.03 + \frac{2.3}{2.6}\ln 0.01322 = 66.03 - \frac{2.3}{2.6}\ln 75.6 = 62.2$$

and

$$L^* = (62.2)(0.06)/0.0107 = 349 \cdot \text{ m}$$

As in Example 10-3, the maximum or limiting length for isothermal and adiabatic flow is essentially the same, that for adiabatic flow being slightly greater by about 1 percent. The higher the initial Mach number, the greater is the difference.

10-6. Normal Shock Waves

A normal shock wave is one in which the plane of the shock wave is at right angles to the flow streamlines. Normal shocks may occur in the diverging section of a nozzle, the diffuser throat of a supersonic wind tunnel, in pipes, and forward of a blunt-nosed body. In all instances the flow is supersonic upstream of the shock, and it will be shown that it is always subsonic downstream of the shock.

Equations relating conditions downstream of the shock to conditions upstream may be obtained by writing the continuity, energy, and momentum equations across the shock. As before, a perfect gas will be assumed. The flow through the shock is adiabatic but not reversible, so the isentropic relationship generally will not hold. Actually, the flow *does* approach isentropic flow for very weak shocks (for upstream Mach numbers not greater than about 1.2), but the assumption of irreversible adiabatic flow in all instances makes the results completely general. The shock forms an essential discontinuity, although it has a finite thickness of about two mean-free-path lengths (about 10^{-5} cm) for upstream Mach numbers greater than 2. The thickness increases for weaker shocks, being about 10^{-3} cm at an upstream Mach number of about 1.007. The abrupt change in gas density across the shock may be detected optically, since the index of refraction of the gas depends on the density. Apparatus in common use includes the interferometer, the schlieren, and the spark shadowgraph, which detect density changes, density gradients, and the rate of change of density gradients, respectively.

It is common practice to designate conditions upstream of a shock with a subscript x and those downstream by the subscript y, and these are indicated in Fig. 10-6.

The continuity equation is

$$V_x A \rho_x = V_y A \rho_y$$

or

$$V_x \rho_x = V_y \rho_y \tag{10.48a}$$

The energy equation is

$$\frac{V_x^2}{2} + h_x = \frac{V_y^2}{2} + h_y = h_0 \tag{10.49}$$

which indicates that the stagnation enthalpy is the same on both sides of the shock. For a perfect gas, $h_0 = c_p T_0$, and thus the stagnation temperature for a perfect gas is also the same on both sides of the shock.

The momentum equation applied to the dashed region in Fig. 10-6 is

$$p_x A - p_y A = V_x A \rho_x (V_y - V_x) = V_y A \rho_y (V_y - V_x)$$

Fig. 10-6.   A normal shock.

or

$$p_x + V_x^2\rho_x = p_y + V_y^2\rho_y \qquad (10.50a)$$

which indicates that thrust function is the same on both sides of a normal shock.

In terms of the Mach number $M = V/\sqrt{kRT}$, Eqs. (10.48a), (10.49a), and (10.50a) become

$$\frac{p_x M_x}{\sqrt{T_x}} = \frac{p_y M_y}{\sqrt{T_y}} \qquad (10.48b)$$

$$T_x\left(1 + \frac{k-1}{2} M_x^2\right) = T_y\left(1 + \frac{k-1}{2} M_y^2\right) \qquad (10.49b)$$

and

$$p_x(1 + kM_x^2) = p_y(1 + kM_y^2) \qquad (10.50b)$$

Eliminating pressure p and temperature T from these we get

$$\frac{1 + kM_x^2}{M_x\left(1 + \dfrac{k-1}{2} M_x^2\right)^{1/2}} = \frac{1 + kM_y^2}{M_y\left(1 + \dfrac{k-1}{2} M_y^2\right)^{1/2}} = f(M,k) \qquad (10.51)$$

which indicates that a particular Mach number function is the same on both sides of the shock. This function is plotted in Fig. 10-7.

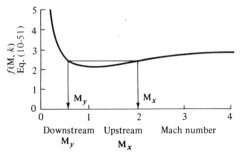

Fig. 10-7.   Mach numbers across a normal shock ($k = 1.4$).

A thermodynamic analysis indicates that the pressure rises across the shock, and from Eq. (10.50b) the Mach number decreases. Thus in Fig. 10-7, points on the right side of the curve minimum at $M = 1$ represent upstream conditions, and those to the left represent downstream conditions. It is seen that the higher the upstream Mach number, the lower the downstream Mach number, and vice versa. Equation (10.51) may be solved for M_y^2 to obtain

$$M_y^2 = \frac{M_x^2 + \dfrac{2}{k-1}}{\dfrac{2k}{k-1} M_x^2 - 1} \tag{10.52}$$

which shows that there is one, and only one, downstream Mach number associated with a given upstream Mach number for a given gas. With this expression for M_y^2, Eqs. (10.49b) and (10.50b) for the temperature and pressure ratios across the shock become

$$\frac{T_y}{T_x} = \frac{2(k-1)}{(k+1)^2 M_x^2}\left(1 + \frac{k-1}{2} M_x^2\right)\left(\frac{2k}{k-1} M_x^2 - 1\right) \tag{10.53}$$

and

$$\frac{p_y}{p_x} = \frac{2k}{k+1} M_x^2 - \frac{k-1}{k+1} \tag{10.54}$$

From these,

$$\frac{p_y}{p_x} = \frac{V_x}{V_y} = \frac{k+1}{2}\,\frac{M_x^2}{1 + \dfrac{k-1}{2} M_x^2} \tag{10.55}$$

The ratio of stagnation to static pressure for flow toward a stagnation point for $M_x > 1$ may be obtained by a combination of Eqs. (10.54) and (10.9a), since the gas must first pass through the shock and then progress to the stagnation pressure from the subsonic condition behind the shock. With the usual subscripts, this ratio may be written as

$$\frac{p_{0y}}{p_x} = \left(\frac{p_y}{p_x}\right)_{\text{Eq. (10.54)}}\left(\frac{p_{0y}}{p_y}\right)_{\text{Eq. (10.9a)}}$$

$$= \left(\frac{k+1}{2} M_x^2\right)^{k/(k-1)}\left(\frac{2k}{k+1} M_x^2 - \frac{k-1}{k+1}\right)^{1/(1-k)} \tag{10.56}$$

Finally, the reduction in stagnation pressure may be obtained from Eqs. (10.9a), (10.52), and (10.54) by considering a gas to flow from a reservoir at pressure p_{0x} through a suitable nozzle to a supersonic condition, then through a normal shock to a stagnation pressure p_{0y}. Because of the entropy increase across the shock, the reservoir pressure is not recovered, and p_{0y} is less than p_{0x}. From Eq. (10.9a),

$$\frac{p_{0y}}{p_{0x}} = \frac{p_y}{p_x} \left(\frac{1 + \dfrac{k-1}{2} M_y^2}{1 + \dfrac{k-1}{2} M_x^2} \right)^{k/(k-1)}$$

If the expression for p_y/p_x in Eq. (10.54) and that for M_y^2 in Eq. (10.52) are substituted, algebraic simplication gives

$$\frac{p_{0y}}{p_{0x}} = \left(\frac{\dfrac{k+1}{2} M_x^2}{1 + \dfrac{k-1}{2} M_x^2} \right)^{k/(k-1)} \left(\frac{2k}{k+1} M_x^2 - \frac{k-1}{k+1} \right)^{1/(1-k)} \qquad (10.57)$$

It should be noted that, although a gas may be considered to flow from a stagnation condition to a supersonic condition isentropically such that Eq. (10.9a) is valid, a gas may not flow from this same supersonic condition to a stagnation condition through a shock wave in a manner such that Eq. (10.9a) applies.

For a normal shock, the values of M_y, T_y/T_x, p_y/p_x, ρ_y/ρ_x, p_{0y}/p_x, and p_{0y}/p_{0x} depend only on the upstream Mach number M_x for a given gas. These values, given in Eqs. (10.52)–(10.57), respectively, may be tabulated. An example is given in Table A-10 (Appendix V) for a normal shock in a gas for which $k = 1.4$.

Example 10-5

If for the converging–diverging nozzle in Example 10-2 a shock were to exist in the diverging portion of the nozzle where the pressure is 280 kPa abs (p_x in Fig. 9-3c), what is the Mach number just upstream and just downstream of the shock, the pressure just downstream of the shock, and the flow rate in the nozzle?

SOLUTION

From Table A-9, the Mach number where the pressure is $280/700 = 0.4$ of the reservoir pressure for isentropic flow is $M_x = 1.223$. From Table A-10, $M_y = 0.828$ and $p_y/p_x = 1.58$ so that $p_y = 442$ kPa abs and the flow rate is the same as in Example 10-2, namely, 0.798 kg/s, since the flow is sonic at the same conditions in the throat with or without the shock. Equations (10.9a), (10.52), and (10.54) could be used instead of the tables to find M_x, M_y, and p_y/p_x, respectively.

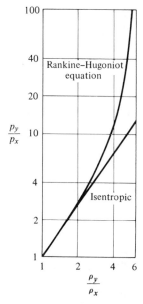

Fig. 10-8. Rankine–Hugoniot and isentropic curves for $k = 1.40$.

The density ratio across a normal shock may be expressed in terms of the pressure ratio across the shock by eliminating M_x^2 in Eqs. (10.54) and (10.55). The result is

$$\frac{\rho_y}{\rho_x} = \frac{\left(\dfrac{k+1}{k-1}\right)\dfrac{p_y}{p_x} + 1}{\dfrac{p_y}{p_x} + \left(\dfrac{k+1}{k-1}\right)} \qquad (10.58)$$

which is known as the *Rankine–Hugoniot equation.* From Eq. (10.54), as the upstream Mach number M_x approaches infinity, the pressure ratio p_y/p_x also approaches infinity. From Eq. (10.58), the density ratio under these circumstances approaches a finite maximum value, which is 6 for air or any other gas for which $k = 1.4$.

A log–log plot of Eq. (10.58) together with the isentropic relation between pressure and density for $k = 1.4$ is shown in Fig. 10-8. The curves show that flow through a normal shock approaches an isentropic flow for weak shocks (flow in which the upstream Mach number M_x approaches unity). This approach to isentropic flow is also indicated by the near-unity values of p_{0y}/p_{0x} in Table A-10 (Appendix V) for M_x values near unity. The strength of a shock may be defined as the ratio of the pressure rise through the shock to the upstream pressure. That is

$$\text{shock strength} = \frac{p_y - p_x}{p_x} = \frac{p_y}{p_x} - 1 \qquad (10.59)$$

The entropy increase across a shock may be obtained by integrating Eq. (1.6) ($T\,ds = dh - v\,dp$). This gives

$$s_y - s_x = c_p \ln\left(\frac{T_y}{T_x}\right) - R \ln\left(\frac{p_y}{p_x}\right)$$

Since $T_{0y} = T_{0x}$, this can be shown to be

$$s_y - s_x = -R \ln\left(\frac{p_{0y}}{p_{0x}}\right) \tag{10.60}$$

Thus flow through shocks is increasingly irreversible as M_x increases (Table A-10 in Appendix V).

Additional treatment of normal shocks is given in Chapter 15, where the Fanno and Rayleigh lines for gas flow are compared with the specific energy and specific thrust diagrams for open channel flow.

10-7. Oblique Shocks

An oblique shock is a general form of discontinuity in a supersonic gas flow which is inclined from a direction normal to the oncoming flow. A normal shock is thus a special form of an oblique shock. Oblique shocks occur in most supersonic flows, although there need not necessarily be shocks simply because supersonic flow exists.

An oblique shock may be obtained or visualized by superimposing a tangential component of velocity to the upstream and downstream velocities for a normal shock. This is equivalent to observing the shock while moving along the normal shock front with a velocity V_t (Fig. 10-9). Subscripts 1 and 2 will be used for the upstream and downstream conditions, respectively, rather than x and y, in order to distinguish between the normal shock results and the oblique shock results. Subscripts t and n refer to the tangential and normal components, respectively.

Oblique shock equations may be obtained from the continuity, momentum (in both tangential and normal directions), and energy equations; or from a transformation of the normal shock equations. The transformation method will be used here. From Fig. 10-9, the upstream Mach number is $M_1 = V_1/c_1$, and $V_{1n} = V_1 \sin\beta$. Thus

$$\frac{V_{1n}}{c_1} = M_1 \sin\beta$$

Also, the downstream Mach number is $M_2 = V_2/c_2$, and $V_{2n} = V_2 \sin(\beta - \theta)$. Thus

$$\frac{V_{2n}}{c_2} = M_2 \sin(\beta - \theta)$$

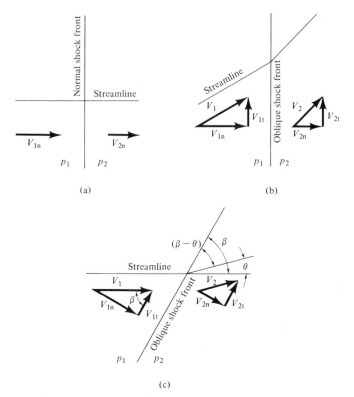

Fig. 10-9. Oblique shock relations obtained from normal shock. (a) Normal shock. (b) Oblique shock. (c) Oblique shock angles.

where β is called the *shock angle* with respect to the upstream flow direction, and θ is the deflection angle of the streamlines as the fluid passes through the oblique shock. Thus in the normal shock equation [Eqs. (10.52)–(10.55)] M_x is replaced by $M_1 \sin \beta$ and M_y is replaced by $M_2 \sin(\beta - \theta)$. This gives for the Mach number downstream of the oblique shock

$$M_2^2 = \frac{1}{\sin^2(\beta - \theta)} \frac{M_1^2 \sin^2 \beta + \dfrac{2}{k - 1}}{\dfrac{2k}{k - 1} M_1^2 \sin^2 \beta - 1} \tag{10.61}$$

The temperature ratio T_2/T_1 is, after rearranging Eq. (10.53),

$$\frac{T_2}{T_1} = 1 + \frac{2(k - 1)}{(k + 1)^2} \frac{M_1^2 \sin^2 \beta - 1}{M_1^2 \sin^2 \beta} (kM_1^2 \sin^2 \beta + 1) \tag{10.62}$$

The pressure ratio p_2/p_1 is, after rearranging Eq. (10.54),

$$\frac{p_2}{p_1} = 1 + \frac{2k}{k + 1} (M_1^2 \sin^2 \beta - 1) \tag{10.63}$$

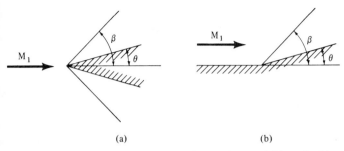

Fig. 10-10.   Supersonic flow past (a) a wedge and (b) an inside corner.

The density ratio is

$$\frac{\rho_2}{\rho_1} = \frac{(k + 1)M_1^2 \sin^2 \beta}{2 + (k - 1)M_1^2 \sin^2 \beta} \tag{10.64}$$

The ratios of thermodynamic properties (temperature, sound speed, pressure, and density) depend only on the normal component of velocity and are not affected by the motion of the observer. The normal component V_{1n} must, of course, be supersonic; that is, $M_1 \sin \beta > 1$. Thus there is a minimum wave angle β for a given upstream Mach number M_1, the maximum wave angle β being that for a normal shock. Hence

$$\frac{\sin^{-1} 1}{M_1} \leqq \beta \leqq \frac{\pi}{2} \tag{10.65}$$

A useful relation between M_1, θ, and β may be obtained by noting that in Fig. (10-9), $\tan \beta = V_{1n}/V_{1t}$ and $\tan(\beta - \theta) = V_{2n}/V_{2t}$. From these

$$\frac{\tan(\beta - \theta)}{\tan \beta} = \frac{V_{2n}}{V_{1n}} = \frac{\rho_1}{\rho_2} = \frac{2 + (k - 1)M_1^2 \sin^2 \beta}{(k + 1)M_1^2 \sin^2 \beta} \tag{10.66}$$

Dividing both numerator and denominator of the last expression by $2M_1^2 \sin^2 \beta$ gives, after simplification,

$$\frac{1}{M_1^2} = \sin^2 \beta - \frac{(k + 1)}{2} \frac{\sin \beta \sin \theta}{\cos(\beta - \theta)} \tag{10.67}$$

In this and preceding equations, θ is the deflection or turning angle of the streamlines as they pass through the oblique shock. It could thus be the half-angle of a wedge placed at a zero angle of attack in a supersonic flow, or the angle of a corner inside which a supersonic flow turns. Both these situations are shown in Fig. 10-10.

Equation (10.67) is shown graphically in Fig. 10-11. The curves indicate that for a given M_1 there are two possible wave angles associated with a given half-wedge or deflection angle. The larger wave angle represents a strong

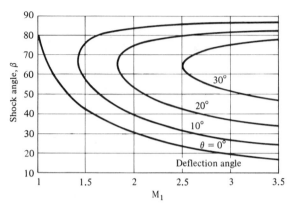

Fig. 10-11. Relation between shock angle, deflection angle, and upstream Mach number for oblique shocks [$k = 1.40$ in Eq. (10.67)].

shock, and the flow downstream is usually subsonic. The smaller wave angle represents a weak shock, and the flow downstream is usually supersonic but less than M_1. Also, for a given M_1 there is a maximum half-wedge or deflection angle associated with this M_1. This represents the maximum angle of the half-wedge for which the shock remains attached.

The Rankine–Hugoniot equation [Eq. (10.58)] is also valid for oblique shocks, since both pressure and density are thermodynamic properties and do not depend on the motion of an observer. Normal shock tables (Table A-10 in Appendix V) may be used for oblique shocks if M_x for a normal shock is taken as $M_1 \sin \beta$ for an oblique shock. Then $M_y = M_2 \sin(\beta - \theta)$ and values of p_y/p_x, ρ_y/ρ_x, T_y/T_x, and p_{0y}/p_{0x} for normal shocks become values of p_2/p_1, ρ_2/ρ_1, T_2/T_1, and p_{02}/p_{01}, respectively.

Example 10-6

Optical measurements of the shock angle for a two-dimensional wedge of half-angle $\theta = 10°$ in a supersonic wind tunnel indicate that $\beta = 35$ deg. (a) What is the Mach number of the air flow? (b) What is the pressure ratio across the oblique shock?

SOLUTION
From Eq. (10.67),

(a) $\dfrac{1}{M_1^2} = \sin^2 35 - \dfrac{2.4}{2} \dfrac{\sin 35 \sin 10}{\cos 25}$

$M_1 = 2.25$

Also, from Fig. 10-11, $M_1 = 2.25$

b) The shock Table A-10 may be used by letting M_x be equivalent to $M_1 \sin \beta = 1.29$, and thus $p_y/p_x = 1.775$ from Table A-10.

PROBLEMS

10-1. Calculate the velocity of sound at $20°C$ for each of the gases listed in Table A-2 of Appendix II, in m/s.

10-2. What is the speed in m/s for a free-flight Mach number of 2.75 in
(a) air,
(b) helium, and
(c) xenon?

10-3. What is the speed of sound in the U.S. standard atmosphere at
(a) sea level,
(b) 11,000 m, and
(c) 47,000 m?
Refer to Table 2-3.

10-4. Small models of missiles are shot through controllable-pressure firing ranges consisting of a gas at rest.
(a) Which of the gases listed in Table A-2 would be most suitable in order to keep the missile speed at a minimum?
(b) What speed is required to achieve a Mach number of 3.0 at $20°C$ in xenon and in helium?
(c) What Mach number corresponds to a missile speed of 450 m/s in xenon and in helium at $10°C$?

10-5. Standard air at rest ($15°C$) is accelerated isentropically. What is the Mach number when the speed becomes 500 m/s?

10-6. Three reference temperatures are used to define three Mach numbers for a given flow. These three reference temperatures result in three reference sonic velocities, one for each temperature. These are
(a) c, the local sonic velocity at the local temperature;
(b) c^*, the sonic velocity where the flow is changed isentropically to a Mach number of 1; and
(c) c_0, the sonic velocity at the stagnation point.
If at a point in a flow stream of air the velocity is 150 m/s and the local temperature is $230°K$, what are c, c^*, and c_0?

10-7. Solve Prob. 10-6 for air at 150 m/s at a local temperature of $230°C$.

10-8. Air at rest at $90°C$ is accelerated isentropically.
(a) What is the air speed in m/s when the Mach number becomes 0.8?
(b) What is the air speed when the flow becomes sonic?
(c) What is the Mach number when the air speed becomes 600 m/s?

10-9. Equation (10.9b) is valid for isentropic flow of air, hydrogen, nitrogen, and oxygen ($k = 1.4$) for M < 2.236. What are the upper limits of M for the other gases listed in Table A-2 (Appendix II) for which this equation may be used?

10-10. Calculate the compressibility factor at $M = \frac{1}{4}, \frac{1}{2}, \frac{3}{4}$, and 1 for air, methane, and xenon, and plot the results for air. What is the effect of the ratio of specific heat capacities k on the compressibility factor?

10-11. What is the stagnation temperature rise $(T_0 - T)$ for a craft traveling at a Mach number of 2.20 in the U.S. standard atmosphere at
(a) sea level,
(b) 2000 m $(T = 2°C)$, and
(c) 5000 m $(T = -17.5°C)$?

10-12. (a) Show that the ratio of stagnation density ρ_0 to density ρ, where the Mach number is M, for isentropic flow is

$$\frac{\rho_0}{\rho} = \left(1 + \frac{k-1}{2} M^2\right)^{1/(k-1)}$$

and by expanding in a power series, show that for very small M

$$\frac{\Delta\rho}{\rho} \approx \frac{1}{2} M^2$$

(b) Suppose gas flow is defined as incompressible if the variation in density is no greater than 2 percent $(\rho_0/\rho \leq 1.02)$. What is the maximum Mach number for which the flow may then be considered incompressible? Show that this has no significant dependency on the type of gas. *Hint:* Calculate the M^4 term in the series expansion for the range of k values in Table A-2 and compare with the M^2 term.

10-13. If air flow is to be considered incompressible when there is a maximum of
(a) 1 percent,
(b) 2 percent, and
(c) 3 percent change in density from the mean value,
What are the corresponding maximum flow Mach numbers?

10-14. In Prob. 10-13, what are the corresponding velocities for standard air at sea level $(T = 15°C)$?

10-15. Air flows at a speed of 250 m/s at a temperature of 0°C and a pressure of 600 kPa abs.
(a) What is the Mach number?
(b) What are the stagnation temperature and pressure?
(c) Compare the pressure rise $(p_0 - p)$ with the dynamic pressure $(\rho V^2/2)$ of the free stream.

10-16. Show that for steady isentropic duct flow, area change is related to fluid acceleration by the equation

$$\frac{dA}{dx} = \left(\frac{1}{c^2} - \frac{1}{V^2}\right) A a_x$$

where A is the duct area, c the local sonic velocity, V the actual velocity, and a_x the convective acceleration.

10-17. Given

$$V\,dV + c^2\,\frac{d\rho}{\rho} = 0$$

Is this valid for any adiabatic flow or for just isentropic flow? Explain.

10-18. Calculate the critical pressure ratio [Eq. (10.23)] for all the gases listed in Table A-2 (Appendix II). Plot a curve relating p^*/p_0 vs k.

10-19. Use Table A-9 (Appendix V) to obtain the answers to the following isentropic flows for air:

(a) At what Mach number is the temperature eight-tenths of the stagnation temperature?

(b) At what Mach number is the stagnation pressure 30 percent greater than the static pressure?

(c) What Mach numbers could exist at sections in a converging–diverging nozzle at which the areas are 1.4 times the throat area?

(d) What would be the ratio of the pressure for subsonic flow to that for supersonic flow in part (c)?

(e) What duct area, in terms of throat area, at the exit of a converging–diverging nozzle fed from a reservoir produces an exit velocity twice the throat velocity?

(f) The density at a section in a nozzle is one-third that at the throat. What is the Mach number at this section?

(g) What is the temperature at a section in a nozzle where the density is one-third that at the throat [see part (f)] in terms of the reservoir temperature?

(h) What ratio of exit area to throat area for a converging–diverging nozzle results in an exit pressure equal to 20 percent of the pressure in the reservoir which feeds the nozzle?

10-20. Solve Prob. 10-15b using gas tables (Table A-9 of Appendix V).

10-21. An impact tube on an aircraft registers a stagnation pressure of 51 kPa gage where the ambient static pressure is 96 kPa abs. The air temperature is 5°C.

(a) What is the free-flight Mach number for the aircraft?

(b) What is its speed?

(c) What is the temperature on the nose of the aircraft?

10-22. An object moves at 220 m/s through standard air at sea level. What is

(a) the gage pressure and

(b) the temperature at the forward stagnation point?

10-23. What reservoir pressure must be used to produce a sonic flow at the exit of a converging nozzle when air exhausts into a standard atmosphere?

10-24. What vacuum must be used to produce sonic flow of air drawn from a standard atmosphere through a converging nozzle?

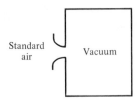

Prob. 10-24

10-25. Compute the mass flow rate for air discharging from a tank at 275 kPa gage and 60°C through a converging nozzle with an exit area of 18.0 cm². The air discharges isentropically into a standard atmosphere.

10-26. Air is in a tank at 700 kPa abs and 40°C and discharges through a converging nozzle of exit area 10.0 cm². Plot a curve relating mass flow rate vs. receiver pressure from 700 to 200 kPa abs, for isentropic flow.

10-27. Air in a tank at 700 kPa abs and 25°C discharges into a standard atmosphere through a converging nozzle. For isentropic flow
(a) what is the exit pressure,
(b) what is the air temperature at the nozzle exit, and
(c) what nozzle exit diameter will discharge 0.80 kg/s?

10-28. A perfect gas ($k = 1.40$ and $R = 500$ J/kg °K) flows from a reservoir at 825 kPa abs and 225°C through a converging nozzle. The flow is isentropic, and the receiver pressure is 140 kPa abs. Find
(a) the gas velocity at the nozzle exit and
(b) the mass flow rate for a nozzle exit diameter of 2.50 cm.

10-29. Air in a reservoir at 825 kPa abs and 50°C discharges through a converging nozzle to the atmosphere. Design a nozzle for a flow rate of 1.50 kg/s. Specify exit diameter.

10-30. In Prob. 10-29 design a converging–diverging nozzle for full expansion at the same flow rate. Specify throat and exit diameters.

10-31. A converging–diverging nozzle fed from a reservoir has an exit area three times the throat area. What is the ratio of exit pressure to reservoir pressure for isentropic flow of
(a) air and
(b) carbon dioxide? $M_{exit} > 1$.

10-32. Air flows isentropically from a reservoir at 1400 kPa abs pressure and at 60°C temperature through a converging–diverging nozzle to a receiver at atmospheric pressure. Full expansion takes place within the nozzle.
(a) What is the exit area in terms of the throat area?
(b) What would be the exit area in terms of the throat area for helium at the same reservoir conditions?

10-33. Air is drawn isentropically from a standard atmosphere at sea level (15°C and 101.3 kPa abs) through a converging–diverging nozzle.

Static pressures at two different locations are 8.00×10^4 and 4.00×10^4 Pa abs, respectively.

(a) What is the Mach number at each of these locations?

(b) What is the velocity at each of these locations?

10-34. Specify the throat dimensions for the nozzle of a two-dimensional supersonic wind tunnel using air, for a test section 25×25 cm and a test section Mach number of 3.50.

10-35. A gas turbine nozzle has an exit Mach number of 1.30 and is fed from a stagnation condition of $870°C$ and 5 atm absolute pressure. What is the exit gas velocity for a gas with $k = 1.30$ and $R = 290$ J/kg°K.

10-36. A gas flows in a duct of constant cross section without friction, but heat removal takes place. How do the velocity, pressure, density, temperature, and Mach number change in the direction of flow for

(a) subsonic flow and

(b) supersonic flow?

Refer to the differential equations in Sec. 10-4.

10-37. Air flows through a short tube without friction. Heat is supplied to increase the initial Mach number of 0.3 at a temperature of $40°C$ to a final number of 0.6. How much heat must be supplied per unit mass of air?

10-38. As heat is added to a subsonic frictionless gas flowing in a constant-area duct, the Mach number increases to a maximum of 1.0 before choking occurs. The gas temperature increases to a maximum at a Mach number of $1/\sqrt{k}$, then decreases to T^*, the value associated with a Mach number of 1.

(a) At what subsonic Mach number is the gas temperature $T = T^*$ for air?

(b) What is T_{max}/T^* for air?

10-39. Air flows in a short tube without appreciable friction. Heat is supplied to increase the Mach number from an inlet value of 0.4 at a temperature of $35°C$ to an exit value of 0.7. What is the heat flow per unit mass of air?

10-40. Air flows in a constant-area duct without friction at an initial temperature of $T_1 = 40°C$ and an initial Mach number of $M_1 = 0.5$. Heat is added to the air at a rate of 56 kJ/kg between sections 1 and 2.

(a) What is the stagnation temperature at each section?

(b) What is the ratio of stagnation temperature at each section to that where the Mach number is 1? (Find T_{01}/T_0^* and T_{02}/T_0^*).

(c) What is the Mach number at section 2?

10-41. Subsonic flow of a gas occurs at constant temperature in a pipe with friction.

(a) Does the magnitude of the pressure gradient dp/dx increase, remain constant, or decrease in the direction of flow?

(b) Does the Reynolds number increase, remain the same, or decrease in the direction of flow?

(c) Does the Mach number increase, remain the same, or decrease in the direction of flow?

(d) Is heat added to the gas or removed from it in order that the temperature be maintained constant?

10-42. Plot curves of p_1/p_2 vs. fL/D similar to those in Fig. 10-4 for $M_1 = 0.1$ for $k = 1.3$ and $k = 1.67$ in order to show the effect of the ratio of specific heat capacities on isothermal flow in a pipe with friction.

10-43. A straight pipe in which gas flows is inclined at an angle θ with the horizontal.

(a) Show that Eq. (10.32) becomes

$$\frac{dp}{\rho V^2/2} + \frac{f\,dx}{D} + \frac{g\sin\theta\,dx}{V^2/2} + 2\frac{dV}{V} = 0$$

(b) Show that the pressure gradient for isothermal flow becomes

$$\frac{dp}{dx} = \frac{\dfrac{pf}{2D} + \dfrac{pg\sin\theta}{V^2}}{1 - \dfrac{p}{\rho V^2}}$$

(c) Show that the limiting pressure for isothermal flow in an inclined pipe is the same as for a horizontal pipe [Eq. (10.37)].

10-44. For given initial conditions, compare the dimensionless limiting length fL^*/D for carbon dioxide with that for xenon at $M_1 = 0.01$ and at $M_1 = 0.1$. Generalize your results in a qualitative statement concerning the effect of the ratio of specific heat capacities k on the limiting length.

10-45. Calculate the ratio of fL^*/D for adiabatic flow to the value of fL^*/D for isothermal flow of air at initial Mach numbers of

(a) 0.05,

(b) 0.1,

(c) 0.2,

(d) 0.3,

(e) 0.4, and

(f) 0.5.

Make a qualitative statement regarding the limiting length for adiabatic flow versus that for isothermal flow of a given gas in a given pipe with given initial conditions.

10-46. The friction factor f is constant for isothermal flow of a gas in a pipe with friction. For subsonic adiabatic flow, is $\bar{f} \le f_1$ or is $\bar{f} \ge f_1$; that is, how does the average friction factor compare with the initial value?

10-47. The Mach number for air flow in a pipe increases from 0.3 to 0.7 because of pipe friction. The initial pressure $p_1 = 140\,\mathrm{kPa}$ abs. What is the pressure at the section where $M_2 = 0.7$

(a) if the flow is isothermal and

(b) if the flow is adiabatic?

10-48. Methane flows through a 20-cm horizontal commercial steel pipe at a rate of 4 kg/s. The inlet pressure is 2000 kPa abs and the temperature is 30°C. Over what length of pipe will the pressure drop to
(a) 50 percent and
(b) 30 percent of its initial value?
Assume isothermal flow.

10-49. What is $\bar{f}L^*/D$ for adiabatic flow in a pipe with friction
(a) for the initial Mach number M_1 approaching zero,
(b) for M_1 approaching 1, and
(c) for very large values of M_1 (supersonic flow)? Here $k = 1.40$.

10-50. Air enters a 30-cm commercial steel pipe from a compressor at 115°C and 1400 kPa abs at an average velocity of 12.0 m/s.
(a) What is the limiting length for isothermal flow?
(b) What is the limiting length for adiabatic flow?

10-51. The limiting Mach number for adiabatic flow in a pipe with friction is 1.0. In Prob. 10-50, what is the Mach number
(a) 30 m and
(b) 300 m from the limiting length for adiabatic flow?

10-52. The limiting Mach number for isothermal flow of air in a pipe with friction is 0.845. In Prob. 10-50, what is the Mach number
(a) 30 m and
(b) 300 m from the limiting length for isothermal flow?

10-53. The Mach number for air flow in a pipe increases from 0.1 to 0.6 because of pipe friction. The pressure at the section where $M_1 = 0.1$ is $p_1 = 550$ kPa abs. What is the pressure at the section where $M_2 = 0.6$
(a) when the air temperature remains constant and
(b) when the flow is adiabatic (but not isentropic)?

10-54. Methane flows through a 20-cm insulated pipe at a rate of 11 kg/s. At inlet the pressure is 700 kPa abs and the temperature is 50°C. Calculate
(a) the initial Mach number,
(b) the Mach number at a section where the temperature is 40°C, and
(c) the pressure at the section where the temperature is 40°C.

10-55. Natural gas is pumped through 50-cm commercial steel pipe at a rate of 150 kg/s. Compressors raise the pressure from 3100 to 5250 kPa abs, and the gas is transmitted to the next compressor station. Assume the gas has properties of methane and is at 40°C. How far apart are the compressor stations?

10-56. Use Table A-10 to obtain answers to the following situations involving normal shocks in air:
(a) The stagnation pressure on the nose of a blunt-nosed projectile is 3.5 times the free-stream pressure of the atmosphere. What is the free-flight Mach number?
(b) What is the Mach number just downstream of the shock in part (a)?

(c) The strength of a shock may be defined as the ratio of the pressure increase across it to the upstream pressure $(p_y - p_x)/p_x$. What Mach number would produce a shock of strength 5?

(d) The ratio of downstream to upstream stagnation pressure across a normal shock is a direct measure of its irreversibility. The entropy increase is $\Delta s = -R \ln(p_{0y}/p_{0x})$. At what Mach number is the entropy increase 218 J/kg °K?

10-57. Use Tables A-9 and A-10 in solving the following situations involving air flow. Consider the flow to be isentropic except across shocks.

(a) A normal shock occurs in a converging–diverging nozzle fed from a reservoir. The shock occurs where $M = 2.2$. What is the ratio of the pressure just downstream from the shock to the reservoir pressure p_y/p_{0x}?

(b) What is the stagnation temperature just downstream from the shock in terms of the reservoir temperature?

(c) What is the area of the nozzle where the shock forms in part (a) in terms of the throat area?

(d) What is the exit area for conditions of part (a) in terms of the throat area for an exit pressure 60 percent of the reservoir pressure ($p_C = 0.60\, p_0$ in Fig. 10-3c)?

10-58. (a) Does the pressure ratio across a normal shock p_y/p_x increase or decrease as the upstream Mach number M_x increases?

(b) Given the pressure p_x, the temperature T_x, the Mach number M_x, and the ratio of specific heat capacities k for a gas, which of these must be known in order that the Mach number M_y downstream of a normal shock may be calculated?

(c) At a Mach number of 0.8, is the stagnation pressure for a given flow greater, equal, or less if compressible effects are considered than if they are ignored?

(d) The Mach number at a given location in a diverging duct is 1.4. Does the Mach number increase, remain the same, or decrease downstream? No shocks are present.

(e) Repeat part (d) for a Mach number of 0.8.

10-59. Derive Eq. 10.60, which gives the entropy increase for flow through a shock.

10-60. Show that the velocities on the two sides of a normal shock may be related by the expression $V_x V_y = (c^*)^2$. *Hint:* Solve for p_y and p_x in terms of k, ρ, V, and $(c^*)^2$ from Eq. (10.6). Equate $p_y - p_x$ from these and from Eq. (10.50a) and simplify, using Eq. (10.48a).

10-61. Show that the downstream Mach number for a normal shock approaches a limiting minimum value as the upstream Mach number increases without limit. What is this minimum Mach number for

(a) air,

(b) carbon dioxide, and

(c) xenon?

10-62. Explain why the presence of normal shocks in a converging–diverging nozzle from a reservoir does not affect the flow rate through the nozzle.

10-63. A converging–diverging nozzle is attached to a pressure tank containing a gas at stagnation conditions. Explain why a shock cannot occur in the converging portion of the nozzle. Assume isentropic flow, except across the shock.

10-64. Air flows from a reservoir at 600 kPa abs and 50°C through a converging–diverging nozzle. A shock occurs at a section where the pressure is 150 kPa abs.

(a) What is the static pressure just beyond the shock?

(b) What is the stagnation pressure downstream of the shock? Note the decrease in stagnation pressure.

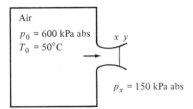

Prob. 10-64

10-65. An air stream at 277°K static temperature has a stagnation temperature of 490°K. What is the Mach number just downstream of a normal shock?

10-66. The density of air is quadrupled in passing through a normal shock. What is the Mach number of the upstream flow?

10-67. A total pressure tube used to measure the air speed in a supersonic wind tunnel measures a pressure 7.0 times the static pressure. What is the Mach number of the flow?

10-68. A normal shock occurs in air flowing in a duct at $M = 2.5$ where the upstream static pressure is 82 kPa abs.

(a) What is the static pressure downstream of the shock?

(b) What is the stagnation pressure across the shock?

(c) What is the decrease in stagnation pressure across the shock?

(d) What is the increase in static pressure across the shock?

10-69. Air from a reservoir at a pressure p_0 flows through a converging–diverging nozzle of throat area A^* and an exit area $2.5A^*$. A normal shock occurs beyond the throat where $A = 1.45A^*$.

(a) What is the exit Mach number?

(b) What is the receiver pressure in terms of p_0?

(c) What receiver pressure would produce isentropic supersonic flow without shocks?

10-70. A supersonic aircraft flies horizontally overhead at 3000 m at a Mach number of 1.6 in still air. What is the time interval between the instant the aircraft is directly overhead and the instant the shock wave is detected by an observer on the ground? Assume $c = 335$ m/s, and that the aircraft is a point source.

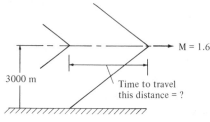

3000 m

Time to travel this distance = ?

M = 1.6

Prob. 10-70

10-71. Air is drawn from a standard atmosphere through a converging–diverging nozzle attached to a vacuum tank. The throat area is 10.0 cm^2, and the exit area at the vacuum tank is 16.0 cm^2. A normal shock occurs where the nozzle area is 13.0 cm^2. What is the pressure in the tank?

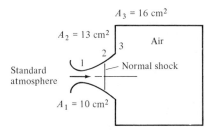

$A_3 = 16$ cm^2

$A_2 = 13$ cm^2

Air

Standard atmosphere

Normal shock

$A_1 = 10$ cm^2

Prob. 10-71

10-72. Air at 600 m/s and $5°C$ passes through a normal shock. At a downstream location the pressure is 4.0 times the pressure just upstream of the shock. What is the Mach number at that location?

10-73. From Eqs. (10.61), (10.63), and (10.64) show that as M_1 becomes very large

$$M_2^2 \rightarrow \frac{k - 1}{2k \sin^2(\beta - \theta)}$$

$$\frac{p_2}{p_1} \rightarrow \frac{2k}{k + 1} M_1^2 \sin^2 \beta$$

$$\frac{\rho_2}{\rho_1} \rightarrow \frac{k + 1}{k - 1} \quad \text{(the same as for a normal shock)}$$

10-74. A two-dimensional wedge is used to measure the Mach number of the flow in a supersonic wind tunnel using air. The total wedge angle is $20°$, and the wave angle β is $50°$.

  (a) What is the Mach number in the wind tunnel?

  (b) What is the smallest Mach number for which this wedge could be used to determine the Mach number?

10-75. What is the Mach number M_2 downstream of the oblique shock of Prob. 10-74?

10-76. Repeat Prob. 10-74 for $\theta = 20°$ and $\beta = 80°$.

10-77. What is the Mach number downstream of the oblique shock of Prob. 10-76?

References

The reader is referred to the following sources for additional information on the flow of compressible gases:

A. B. Cambel, "Compressible Flow," Sec. 8 of *Handbook of Fluid Dynamics*, edited by V. L. Streeter (New York: McGraw-Hill Book Company, Inc., 1961).

N. A. Hall, *Thermodynamics of Fluid Flow* (Englewood Cliffs, N. J.: Prentice-Hall, Inc., 1950).

H. W. Liepmann and A. Roshko, *Elements of Gas Dynamics* (New York: John Wiley and Sons, Inc., 1956).

J. A. Owczarek, *Fundamentals of Gas Dynamics* (Scranton, Pa.: International Textbook Company, 1964).

A. H. Shapiro, *The Dynamics and Thermodynamics of Compressible Fluid Flow*, Vol. 1 (New York: The Ronald Press Company, 1953).

Chapter 11
Dynamic Drag and Lift

In this chapter, the fluid forces acting on a body in a flow stream owing to the relative motion between the fluid and the body are considered. Buoyant and gravity forces on the body are not included since they are static and not dynamic effects and they act regardless of the relative motion between the fluid and the body.

11-1. Fluid Forces on a Body in a Flow Stream

When an extensive, incompressible viscous fluid flows past a body submerged in it or when a body moves through this fluid at rest, two types of forces act over the surface of the body. These surface forces are due to pressure and to viscous shear. Over any infinitesimal area of the body surface, the pressure force is normal to that area and the viscous shear force is parallel or tangent to the area. The components of these forces when taken *in the direction of motion* of the body (or of the approaching fluid with respect to the body) and then summed up over the entire body surface result in what is called *profile drag*. Waves may be set up at a liquid surface as a result of the motion of a body (a ship, boat, or hydrofoil, for example) and drag owing to these waves is called *wave drag*. In compressible gas flow, compression shocks contribute to what is also called wave drag.

If the components of the pressure and viscous shear forces over an infinitesimal area are taken *normal* to the direction of motion of the body (or of the fluid with respect to the body) and summed up over the entire body surface, the resulting force is called *lift*. Associated with this lift force on an airfoil or blade element of finite span is a drag force called *induced drag*.

In a steady flow of an ideal (nonviscous) fluid of infinite extent, only forces from pressure exist and the drag force in all instances except free-cavity flows is zero. Lift, however, may be produced in an ideal fluid by superimposing an irrotational vortex, or circulation, around the body. In fact, vortices, or circulation, are required in order to produce lift in a *viscous* fluid as well as in an ideal (nonviscous) fluid.

11-2. Drag

In the absence of wave or induced drag, the total drag is the profile drag, which in turn is due to pressure and viscous shear. The profile drag may be due entirely to viscous shear, entirely to pressure, or to a combination of both. In the latter instance, viscous shear forces may play an important role, however, in the development of the boundary layer and in influencing the point of separation of the boundary layer from the surface of the body. This in turn affects the size of the wake and the pressure differences within it, both of which partially determine the pressure drag. Positive pressure differences over the leading portion of the body also contribute to pressure drag.

The drag is expressed as the product of a drag coefficient, the dynamic pressure of the free stream, and some characteristic area. The drag coefficient was found in Chapters 7 and 8 to be a function of a number of parameters including the body shape, Reynolds number, Mach number, Froude number, surface roughness, and free-stream turbulence. In any instance, the drag is expressed as

$$\text{drag} = C \frac{\rho u_s^2}{2} A \tag{11.1}$$

where

C = drag coefficient (the drag coefficient was defined in Sec. 8-4 as the ratio of the drag force to the force represented by the product of the dynamic pressure of the free stream and some area);

C_f = skin-friction drag coefficient;

C_D = drag coefficient for other forms of drag;

$\rho u_s^2/2$ = dynamic pressure of the free stream;

A = area being sheared for pure skin-friction drag, the chord area for lifting vanes, or the projected frontal area for other shapes.

SKIN-FRICTION DRAG

Pure, viscous shear drag was discussed in Chapter 7 for parallel flow past a smooth flat plate. The drag coefficient C_f was shown to depend on whether the boundary layer was laminar or turbulent. If the boundary layer is laminar, C_f depends on the Reynolds number of the flow based on the free-stream velocity u_s and the length of the plate x. If the boundary layer is turbulent, C_f depends on the Reynolds number of the flow, the roughness of the plate, and on the location of the transition from a laminar to a turbulent boundary layer, which in turn depends on the plate roughness and on the turbulence level of the free stream.

Resulting curves of drag coefficient as a function of Reynolds number for smooth flat surfaces are given in Fig. 7-14, and again in Fig. 11-1, which also includes an approximate curve for rough surfaces typical of a ship's hull [1].

Additional information on ship resistance may be found in books on naval architecture [e.g., 2].

The skin-friction drag of surface ships, submarines, airships, and so forth, in addition to flat surfaces, may be approximated using the data of Fig. 11-1.

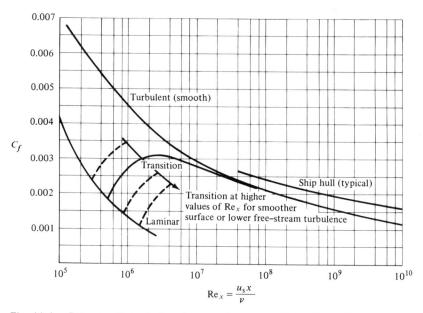

Fig. 11-1. Drag coefficients for plane surfaces parallel to flow. (Source: From H. E. Saunders, *Hydrodynamics in Ship Design*, Vol. 2 (1957), p. 100. Used with permission of the publishers, The Society of Naval Architects and Marine Engineers.)

Example 11-1

A 4.60-m smooth model of an ocean vessel is towed in fresh water at 2.83 m/s with a total measured drag of 75 N. The wetted surface of the hull is 3.50 m². Estimate the skin-friction drag.

SOLUTION

The skin-friction drag may be estimated by considering the wetted surface of the hull as equivalent to a flat plate 4.60 m long with a total area of 3.50 m²:

$$Re_x = \frac{u_s x}{v} = \frac{(2.83)(4.60)}{1.1 \times 10^{-6}} = 1.2 \times 10^7$$

From Fig. 11-1, $C_f = 0.0029$. Then the skin-friction drag is

$$drag_{friction} = C_f \left(\frac{\rho u_s^2}{2} \right) A = \frac{(0.0029)(1000)(2.83)^2(3.50)}{2} = 40.6 \quad N$$

PRESSURE DRAG

Pure pressure drag exists for flow past a flat plate normal to a stream. Any shear forces act normal to the approaching stream and thus do not contribute to the drag force directly. They may, however, affect the growth of the boundary layer along the surface and have a minute effect on the pressure distribution.

The drag coefficient for pure pressure drag depends on the shape of the surface and the Reynolds number of the flow based on some characteristic dimension D. The flow pattern and pressure distribution for two-dimensional flow around a plate of infinite length L is shown in Fig. 11-2. The drag coefficient for a finite plate depends on the ratio of D/L, as well as Re_D,

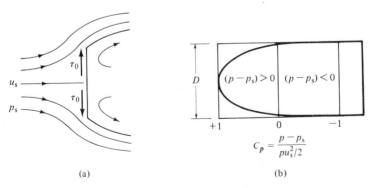

(a)   (b)

Fig. 11-2.   Two-dimensional flow past a flat plate normal to the free stream. (a) Flow pattern, with shear forces normal to the free stream. (b) Pressure distribution according to measurements by A. Fage and F. C. Johansen.

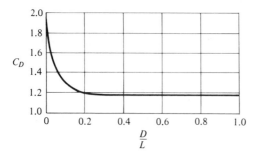

Fig. 11-3.   Effect of aspect ratio on drag coefficient for rectangular plate normal to the flow (from measurements by C. Wieselsberger and O. Flachsbart).

because of end effects. For values of D/L from 0.4 to 1.0 at $\mathrm{Re}_D > 1000$, $C_D \approx 1.16$, which is slightly greater than that for a circular disk ($C_D = 1.12$) for the same range of Re_D. Figure 11-3 indicates the effect of D/L on C_D for a rectangular plate normal to the flow.

Example 11-2

Calculate the drag force on a 3-m billboard 30 m wide at ground level in a 25-m/s wind normal to the billboard. Assume standard air.

SOLUTION
The drag will be one-half that for a 6- by 30-m rectangle, since the flow is essentially the same as that past the top half of a 6- by 30-m rectangle. Thus;

$$\mathrm{Re}_D = \frac{u_s D}{v} = \frac{(25)(6)}{1.46 \times 10^{-5}} = 1.0 \times 10^7 \text{ which is greater than } 10^3$$

From Fig. 11-3, $C_D = 1.2$ for $D/L = 0.2$. Thus the drag is

$$\mathrm{drag} = \frac{(1.2)(1.225)(25)^2(3)(30)}{2} = 41 \quad \mathrm{kN}$$

COMBINED SKIN-FRICTION AND PRESSURE DRAG (PROFILE DRAG)
The flow past a circular cylinder or a sphere may be analyzed analytically with certain approximations for laminar flow. Results of experiments for turbulent flow may be explained on the basis of boundary layer growth and separation. It is informative to look into these flows in some detail. Typical curves relating drag coefficients with Reynolds number are shown in Fig. 11-4. Each curve represents the results of numerous measurements.

   The curve for very slow motion (called *creeping motion*) for the cylinder is from a solution of the Navier–Stokes equations by Lamb [3]. He obtained

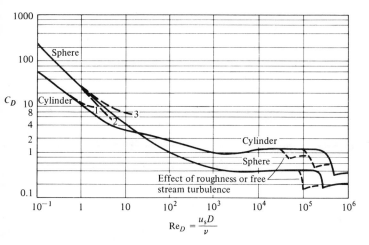

Fig. 11-4. Drag coefficients for spheres and infinite circular cylinders. Curve 1: Lamb's solution for cylinder. Curve 2: Stokes' solution for a sphere [Eq. (11.3)]. Curve 3: Oseen's solution for a sphere [Eq. (11.4)].

a drag coefficient given by

$$C_D = \frac{8\pi}{2\,\text{Re}_D - \text{Re}_D \ln \text{Re}_D}$$

which is valid for Re_D less than about 0.5 [4].

A similar solution for a sphere was made by Stokes in 1850 and gave the drag as

$$\text{drag} = 3\mu u_s \pi D \tag{11.2}$$

and is valid for $\text{Re}_D < 0.1$. Combining Eqs. (11.1) and (11.2), we find that for this range of Reynolds number

$$C_D = \frac{24}{\text{Re}_D} \tag{11.3}$$

Stokes' solution also showed that one-third of the drag was pressure drag and two-thirds was due to viscous shear.

Oseen in 1910 made an improvement in the Stokes solution by including, in part, the inertia terms Stokes omitted. Oseen's solution gave

$$C_D = \frac{24}{\text{Re}_D}\left(1 + \frac{3}{16}\,\text{Re}_D\right) \tag{11.4}$$

which is valid for $\text{Re}_D < 1$.[†]

[†] The Stokes, Oseen, and experimental curves coincide at low Re_D, and it is difficult to place an upper limit to the range of validity of Eqs. (11.3) and (11.4). At $\text{Re}_D = 1$, the Oseen equation gives a C_D 18 percent higher than the Stokes equation; experimental data lie between them.

It is interesting to note that experimental results lie midway between the Stokes and Oseen curves in Fig. 11-4. Thus, since the plot is logarithmic, the experimental data follow the equation

$$C_D = \frac{24}{\text{Re}_D} \left(1 + \frac{3}{16} \text{Re}_D\right)^{1/2} \tag{11.5}$$

quite accurately for Reynolds numbers up to 100.

If a sphere falls through a fluid of infinite extent (fluid dimensions much greater than the sphere diameter), the buoyant and drag forces at terminal or steady velocity are equal to the gravity force on the sphere. Thus for $\text{Re}_D < 0.1$ Stokes' law will apply and

$$\gamma_f \frac{\pi D^3}{6} + 3\mu u_s \pi D = \gamma_s \frac{\pi D^3}{6}$$

Thus, if the fall velocity u_s, the fluid and sphere specific weights γ_f and γ_s, respectively, and the sphere diameter D are known, the fluid viscosity is

$$\mu = \frac{D^2(\gamma_s - \gamma_f)}{18 \, u_s} \tag{11.6}$$

and this equation provides an extremely simple method for measuring dynamic viscosity. If the fluid is of finite extent, the influence of the boundaries of the container is such as to indicate an apparent drag coefficient higher than that in an infinite fluid. If, for example, the sphere falls in the center of a vertical cylinder of diameter D_c, the relative velocity of the fluid adjacent to the the sphere increases, the drag increases, and the sphere will fall slower than in an infinite fluid. The measured velocity u_m should be corrected to its equivalent velocity in an infinite fluid u_s by the equation

$$u_s = \left(1 + 2.4 \frac{D}{D_c}\right) u_m \tag{11.7}$$

For a number of particles uniformly distributed in a fluid, mutual interference will cause them to fall more slowly than if each particle fell alone. The settling velocity of natural particles such as sand and gravel is less than that for equivalent spheres, since the drag coefficient increases with increasing departure from a spherical shape.

Example 11-3

A 1.5-mm steel sphere ($\rho_s = 7800 \text{ kg/m}^3$) falls steadily at a velocity of 3.20 mm/s in an oil of density $\rho_f = 870 \text{ kg/m}^3$ contained in a vertical cylinder 100 mm in diameter. What is the oil viscosity?

SOLUTION
From Eq. (11.7),

$$u_s = \left(1 + 2.4 \frac{1.5}{100}\right)(3.20) = 3.32 \quad \text{mm/s}$$

From Eq. (11.6)

$$\mu = \frac{(0.0015)^2(7800 - 870)(9.81)}{(18)(0.00332)} = 2.56 \quad \text{kg/m s}$$

Equation (11.6) is valid, since $\text{Re}_D = u_s D \rho_f / \mu = (0.00332)(0.0015)(870)/2.56 = 0.0016$, which is less than 0.1 and is well within the range of validity for the Stokes equation.

Example 11-4

Missouri River sand has a mean diameter of about 0.17 mm. What is the settling velocity of this sand in water at 15°C? Assume the particles are spheres with a specific gravity 2.65. Then $v = 1.14 \times 10^{-6}$ m^2/s from Table A-8 (Appendix II).

SOLUTION
Equating drag = weight − buoyancy gives

$$C_D u_s^2 = \frac{4dg}{3}\left(\frac{\rho_s}{\rho_f} - 1\right)$$

$$= \frac{(4)(1.7 \times 10^{-4})(9.807)}{3}(2.65 - 1) = 3.66 \times 10^{-3}$$

To solve for u_s, assume a C_D, then calculate u_s, Re, and C_D from Eq. (11.5). Repeat until C_D values agree:

C_D	10	10.8	10.7	
u_s	0.0191	0.0184	0.0185	Thus $u_s = 1.85$ cm/s
Re	2.85	2.75	2.76	
C_D	10.4	10.7	10.7	

As the Reynolds number increases (this may be produced by an increase in u_s for a given sphere in a given fluid), the laminar boundary layer separates from the aft portion of the sphere and a laminar wake is established. Further increase in Re_D results in a turbulent wake, with a rather constant drag coefficient, such that the drag varies approximately as u_s^2. In this region,

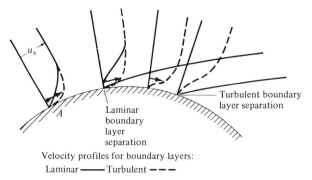

Velocity profiles for boundary layers:
Laminar ——— Turbulent — — —

Fig. 11-5. Separation of laminar and turbulent boundary layers from a cylinder.

vortices are formed and shed in alternate directions behind the sphere or cylinder. This formation and shedding of vortices causes a time-varying asymmetry in the pressure distribution around the aft surface which, in turn, results in alternating transverse forces that may set the body in vibration. Shapes other than spheres and cylinders also exhibit these characteristics. The singing of telephone wires and power lines and the fluttering of venetian blinds are well-known examples. Near $Re_D = 300,000$ the laminar boundary layer suddenly becomes turbulent and the boundary layer separation point moves aft, resulting in a smaller wake and hence smaller drag. This movement of the separation point accounts for the sudden drop in the drag coefficient near a Reynolds number of 300,000 for the sphere and about 500,000 for the cylinder.

The boundary layer may become turbulent at a Reynolds number lower than these values if the sphere or cylinder is rough or if the turbulence level, or intensity, in the flow stream is increased.

The drag curve for rough spheres or for a high turbulence intensity in the free stream is indicated in Fig. 11-4. This reduction in drag coefficient is taken advantage of in the roughening of golf balls.

The shift in separation point for flow past a sphere or circular cylinder is indicated in Fig. 11-5. Fluid particles near the boundary at A are moving faster in a turbulent boundary layer than in a laminar boundary layer. Thus they have more kinetic energy with which to overcome the adverse pressure gradient (increasing pressure) behind the sphere or cylinder. Hence, the particles in the turbulent boundary layer are able to move further aft before coming to rest and being separated from the boundary. As a result, there is a smaller wake and less drag.

This same general situation exists for any body shape for which the drag is significantly affected by a wake.

In the 1920s, the turbulence level, or intensity, in subsonic wind tunnels was measured by determining the critical value of Re_D at which the drag decreased suddenly (or for which the drag coefficient became 0.3, for example) for a smooth sphere mounted on a force-measuring system in the wind tunnel. The higher the intensity of turbulence, the lower was this critical Reynolds number. This was surely an ingenious device. Recent developments in hot-wire anemometry have made possible more detailed studies of turbulence fluctuations and intensities.

WAVE DRAG

When a boat or ship travels on the surface of water, bow and stern waves are generated. Energy is required to generate these waves, and this energy originates from the propulsion system of the boat or ship. The propulsion system, then, supplies the energy or force to overcome the skin-friction drag (plus that of the miscellaneous appurtenances such as rudder and propeller shaft support struts) and to set up these surface waves. That portion of the total drag attributed to wave generation is called the *wave drag*. Wave drag also exists for seaplane hulls, and for submarines and hydrofoils submerged but at a depth shallow enough for surface waves to form.

Wave drag is not measured directly, but it is taken as the residual drag remaining after all drag which can be calculated or estimated is subtracted from the total measured drag. There are, of course, sources of uncertainty in these estimates.

Example 11-5

In Example 11-1, the total measured drag was given as 75 N. The skin-friction drag was calculated to be 40.6 N. The wave drag, then, is the difference between the two, namely $75 - 40.6 = 34.4$ N.

In supersonic gas flows, the drag of airfoils is influenced by Mach waves (shock patterns), and this type of drag is also called *wave drag*.

Drag may be measured directly on a model or determined from velocity and pressure measurements in the wake with the application of the momentum equation.

One of the most complete compilations of drag coefficient data is that by Hoerner [5]. An interesting discussion of drag is contained in a short paperback, *Shape and Flow* by Shapiro [6].

11-3. Lift

The phenomenon of lift produced in an ideal (nonviscous) fluid by the addition of a free vortex[†] (circulation) around a cylinder in a rectilinear flow stream described in Sec. 6-3 is known as the Magnus effect. It was mentioned by Newton in 1672 and was investigated experimentally by Magnus in 1853.

[†] A free vortex is an irrotational vortex.

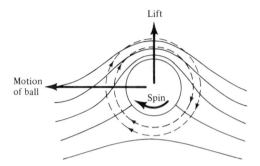

Fig. 11-6. Effect of bottom spin on a ball moving in a viscous fluid.

In a real (viscous) fluid, this effect may be produced by a ping-pong ball, for example, by making it spin as it travels through the air. Because the relative velocity between the air and the ball is zero at the surface of the ball, the spin of the ball produces a circulation approximating a free vortex outside the boundary layer. A top spin produces a downward force, and a bottom spin an upward force. Spin about a vertical axis produces a sideward force, known as a "hook" or "slice" in golf, for example. In the case of both the ideal fluid and the real fluid, a circulation is necessary for the production of a lift force. Figure 11-6 illustrates the effect of bottom spin on a Ping-Pong ball.

Lift is expressed as the product of a lift coefficient, the dynamic pressure of the free stream, and the chord area of the lifting vane. The lift coefficient was found in Chapter 8 to depend on a number of parameters, including the shape and angle of attack of the vane, Reynolds number, Mach number, Froude number for shallow hydrofoils, and aspect ratio.

The lift force is generally defined by the equation

$$\text{lift} = C_L \frac{\rho u_s^2}{2} A \tag{11.8}$$

where C_L is the lift coefficient, $\rho u_s^2/2$ is the dynamic pressure of the free stream, and A is the chord area of the lifting vane.

Lifting vanes are shapes which produce lift and include such things as kites, airfoils, hydrofoils, and propeller blades. Because all lifting vanes are fundamentally the same, it is sufficiently informative to study the behavior of an airfoil.

It was shown in Sec. 6-3 that the lift on a circular cylinder in a nonviscous fluid of density ρ, with a free-stream velocity u_s and with a circulation of strength Γ was

$$\text{lift} = \rho u_s \Gamma \tag{6.21}$$

per unit length of cylinder.

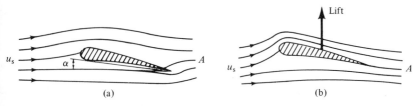

Fig. 11-7.   Nonviscous flow past an airfoil. (a) Without circulation; with no lift or drag. (b) With circulation; lift but no drag. Circulation is added to the flow in (a) to produce the flow pattern in (b).

The theorem of Kutta (1902) and Joukowski (1905) extends this result to include any shape under the same flow conditions. The problem here is to determine the circulation for a given situation or shape of lifting vane. Qualitatively, this may be stated with reference to Fig. 11-7, which shows the flow lines past an airfoil at a given angle of attack α, with and without circulation. The circulation required to swing the trailing stagnation stream-line A tangent to the trailing edge of the airfoil is the desired quantity. The addition of circulation results in a higher velocity and lower pressure over the upper surface and a lower velocity and higher pressure over the lower surface of the airfoil. The result is a lift force upward.

Joukowski discovered a mathematical method for transforming circles into airfoil shapes and was thus able to transform the streamline pattern around a circular cylinder into the streamline pattern around the airfoils. This method enabled him to calculate the theoretical lift for these "transformed" airfoils. Von Karman and von Mises improved Joukowski's mathematical models so that shapes more closely resembling actual wing sections were obtained. Joukowski's theory indicated that the lift coefficient for an airfoil in an ideal fluid when the airfoil thickness and camber approach zero (becoming a flat plate) is

$$C_L = 2\pi \sin \alpha_0 = 2\pi\alpha_0 \qquad (11.9)$$

for small angles of attack. The angle α_0 is the difference between the actual angle of attack and the angle of attack for which the lift is zero. For a flat plate and symmetrical airfoils without camber (curvature), α_0 is the same as α, since the lift is zero for zero angle of attack. Equation (11.9) states that the slope of a line relating C_L (ordinate) to α_0 (abscissa) is 2π, and this is very closely approximated in practice for thin airfoils at small values of α (stall conditions must be avoided).

Thus ideal flow theory predicts actual lift performance amazingly well but gives zero drag in all instances in an infinite fluid for steady flow.

The theory of lift in a real (viscous) fluid is largely credited to Lanchester (in about 1907) and its later improvement to Prandtl. An airfoil in a real fluid must create its own circulation, or vortex field, just as the spinning

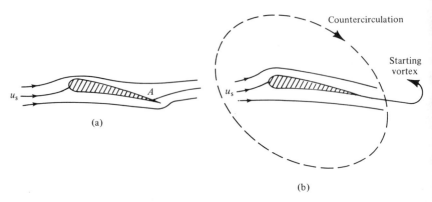

Fig. 11-8. Starting vortex (a) at beginning of motion and (b) after establishment of countercirculation.

Ping-Pong ball, in order to experience lift. The starting vortex is indicated in Fig. 11-8. As the motion begins, it is very slow and approaches that of irrotational flow in an ideal fluid. The fluid particles passing around the trailing edge must move very rapidly and must approach a stagnation condition at A. But because of the viscosity of the fluid, they have less velocity than if the fluid were ideal, and the fluid separates from the trailing edge of the airfoil in the form of a vortex (Fig. 11-8b). As this vortex passes from the airfoil, an opposing reaction starts a countercirculation opposite in direction to that of the trailing vortex. It is this induced countercirculation which produces the lift.

For a finite lifting vane (the preceding discussion applies to a vane of infinite span), additional explanations are necessary since a vortex cannot terminate within the fluid. The lift of a finite wing is zero at its tips, and thus the circulation would appear to be zero there also, so that the vortex system cannot extend out to infinity. A closed vortex loop is necessary, and this loop consists of tip vortices which trail behind the airfoil tips back to the

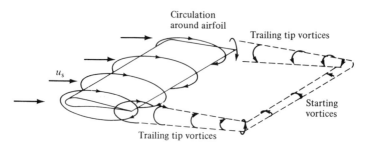

Fig. 11-9. Closed-loop vortex pattern for a finite wing.

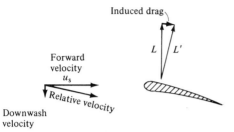

Fig. 11-10.   Downwash (for an airfoil of finite span) which produces induced drag.

starting vortex. This condition is shown schematically in Fig. 11-9. The starting vortex degenerates to zero with time because of viscous dissipation, and the vortex pattern is thus more horseshoe shaped than rectangular or toroidal.

The tip vortices have low-pressure cores, and for ship propellers they may be photographed or seen as threadlike cavities in the shape of helices peeling off the blade tips. Under certain conditions they may be observed as vapor trails behind aircraft flying at high altitudes. Air within the vortex core expands and cools, and vapors may condense out to become visible.

INDUCED DRAG

For an airfoil of finite span, the tip vortices produce a downwash, which is a motion normal to that of the approaching stream. The resulting flow vector is inclined from that representing the motion of the airfoil in the fluid (see Fig. 11-10). Thus the true drag and lift are also inclined from the direction of motion. The component of the force L' in the direction of the vector u_s is called the *induced* drag. The greater the aspect ratio (span/chord) the less the induced drag. For testing purposes, the induced drag may be essentially eliminated by the use of end plates on a finite foil. Work must be expended to produce the trailing tip vortices, and this is at the expense of drag on the foil—much in the same manner as wave drag for a surface ship.

Common methods of plotting experimental lift–drag data include graphs of C_L and C_D vs. α and polar diagrams. Typical data so plotted are shown in Fig. 11-11. One of the most desirable characteristics of an airfoil is a high lift–drag ratio, and the corresponding angle of attack for the maximum value of this ratio is easily obtained by determining the point of tangency for a line passing through the origin. This is point A in Fig. 11-11b. Also, the drag at zero lift is the profile drag, and thus the additional drag at finite lift is the induced drag. At $\alpha = 8°$, for example, the induced drag is as indicated in Fig. 11-11b. In Fig. 11-11c, the slope of the polar diagram and the stall angle are seen to increase slightly with the Reynolds number. Figure 11-11d shows the effect of changes in aspect ratio of the airfoil.

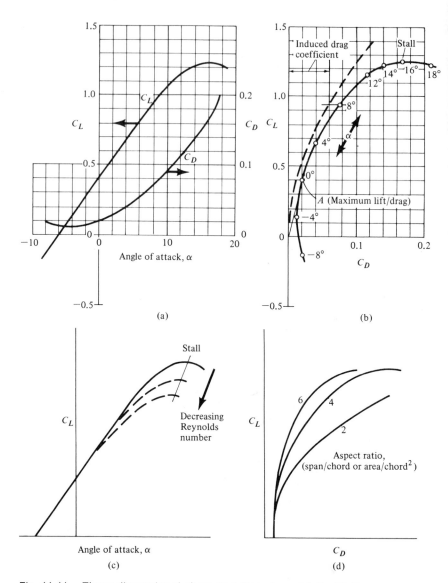

Fig. 11-11. Three-dimensional characteristics of an airfoil of finite span.
(a) Typical lift and drag coefficients for an airfoil. (b) Typical polar diagram.
(c) Effect of Reynolds number on lift coefficient and stall angle. (d) Effect of
aspect ratio.

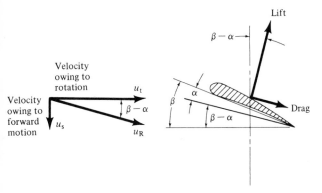

Fig. 11-12.   Section of propeller blade at radius r.

PROPELLER BLADES

A propeller is a lifting vane for which both lift and drag on a blade element contribute towards thrust and torque. A blade element of a typical ship propeller is shown in Fig. 11-12. If the forward motion of the propeller in still water is u_s, the water approaches the blade element parallel to the shaft at this velocity. If the peripheral speed of the propeller is $u_t = \omega r$, the water approaches the blade at a tangential velocity u_t, the resulting relative motion being represented by the vector u_R. The angle of attack is α, and the blade angle at the particular radius shown is β. The resulting lift and drag forces on the blade element are as shown in Fig. 11-12. The thrust, parallel to the shaft for a blade element of length Δr is

$$\text{thrust} = \text{lift}\cos(\beta - \alpha) - \text{drag}\sin(\beta - \alpha) \tag{11.10}$$

and the torque is

$$\text{torque} = [\text{lift}\sin(\beta - \alpha) + \text{drag}\cos(\beta - \alpha)]r \tag{11.11}$$

These equations indicate that both thrust and torque are influenced by both lift and drag. The total thrust and torque, of course, is due to that of all blade elements for all blades constituting the entire propeller.

There is an optimum angle of attack for the particular foil section which makes up the propeller blade. At a given value of u_s, u_t will vary linearly with the radial distance to a blade element, and thus $\beta - \alpha$ will vary. If α is to be constant, the blade angle β must vary, and this accounts for the twist in propeller blades.

If the driving motor or engine runs best at a given fixed speed, then u_t is fixed for each blade element, and variations in forward speed will vary u_s and hence the angle $\beta - \alpha$. In order to keep α constant, the blade angle β should be controlled as a function of forward speed u_s. Thus controllable-pitch propellers for airplanes, propeller pumps, and propeller turbines are often used.

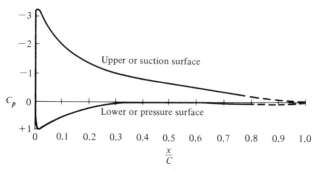

Fig. 11-13. Dimensionless plot of measured pressure distribution over an NACA 0015 airfoil at 8° angle of attack; $Re_x = 6 \times 10^4$.

MEASUREMENT OF LIFT

Lift of airfoils may be calculated or they may be measured directly in a wind tunnel or under noncavitating conditions in a water tunnel or towing basin. For two-dimensional flow, measurements in a tunnel may be made indirectly by an integration of pressures measured over the walls of the tunnel test section or, in general, by an integration of the pressures over the airfoil section.

The algebraic difference between the predominantly negative pressures on the top surface of the foil and the predominantly positive pressures on the bottom surface will result in a net force normal to the chord of the airfoil. As a first approximation, the component of this force normal to the approaching free stream is the net lift on the airfoil.

If p_x is the surface pressure at a distance x from the leading edge of an airfoil of chord length C, p_s is the free-stream pressure, ρ is the free-stream density, and u_s is the free-stream velocity of approach, then the surface pressures may be expressed as a dimensionless pressure coefficient.

$$C_p = \frac{p_x - p_s}{\rho u_s^2/2} = 1 - \left(\frac{u_x}{u_s}\right)^2 \tag{11.12}$$

A typical dimensionless plot of measurements made on an NACA 0015 airfoil (symmetrical foil with maximum thickness 15 percent of the chord length) in a small wind tunnel is shown in Fig. 11-13. The average height of the area between curves, expressed dimensionlessly in the same manner as C_p, is called the normal force coefficient C_N. In terms of the angle of attack α, the lift coefficient is

$$C_L = C_N \cos \alpha - C_C \sin \alpha \tag{11.13}$$

where C_C is the chordwise force coefficient normal to C_N. Since $C_C \sin \alpha$ is usually very small compared to $C_N \cos \alpha$,

$$C_L \approx C_N \cos \alpha \tag{11.14}$$

The average height of the area between the pressure curves for the top and bottom surfaces of the airfoil represents C_N and is found as follows:

$$C_N = \frac{\text{area between curves}}{\text{length equivalent to } x/C = 1.0 \times \text{length equivalent to } C_p = 1.0}$$

where the area and lengths are expressed in terms of the same linear unit.

A determination of C_L at an angle of attack of $8°$ from pressure measurements gives $C_L = 0.85$, as compared to 0.87 from Eq. (11.9) ($C_L = 2\pi \sin \alpha$). The theoretical value is 0.88. Thus calculations from ideal fluid theory agree well with measured values of lift.

According to the Bernoulli equation, the maximum rise in pressure from the free-stream value p_s is the dynamic pressure. Thus at a stagnation point on an airfoil, the maximum value of $p_x - p_s$ is $\rho u_s^2/2$ and the maximum value of the pressure coefficient C_P in Eq. (11.12) is unity. This is shown in Fig. 11-13.

The low pressures indicated in Fig. 11-13 for the upper, or suction, surface provide most of the lift for this particular foil. Associated with these low pressures are high velocities, so that at high subsonic free-stream velocities, local velocities along the foil surface may become supersonic and shocks may affect the flow. For hydrofoils, the low pressures may cause cavitation, which is undesirable for the profile represented by Fig. 11-13.

The theory of lift and an extensive presentation of the characteristics of subsonic airfoils is given by Abbott and von Doenhoff [7]. Authentic, nontechnical books on drag and lift applied to aerodynamics have been written by Sutton [8] and von Karman [9].

PROBLEMS

11-1. Estimate the wind force on a 3- $\times$ 15-m billboard in standard air blowing at 15 m/s normal to it.

11-2. (a) What is the terminal velocity of a 900-N parachutist-and-parachute combination in standard air for a parachute 4.25 m in diameter having a drag coefficient of $C_D = 1.1$?

(b) From what height would a person have to jump to attain this same velocity in free flight, neglecting air resistance?

11-3. The drag coefficient of a parachute is $C_D = 1.33$. How large a parachute would be required to limit the terminal velocity of a 1000-N parachutist-and-parachute combination to a value no greater than the free-fall velocity from a 3-m height?

11-4. (a) What is the terminal velocity of a parachutist-and-parachute combination weighing 980 N in standard air for a parachute of 4.50 m in diameter with a drag coefficient of $C_D = 1.2$?

(b) From what height would a person have to jump to attain this same velocity?

11-5. A hollow hemisphere has a drag coefficient of about 1.32 with the open side upstream and about 0.34 with the open side downstream. A single-arm anemometer (B–A in Fig. 13-8b) has cups 7.5 cm in diameter on 40-cm centers. Estimate the torque required to hold the arm at rest in a 15-m/s wind normal to the arm. Assume a standard atmosphere.

11-6. A 6-mm-diameter cylinder is used as a pitot cylinder (see Fig. 13-5) to measure the velocity distribution across a diameter in the test section of a 15-cm-diameter cavitation-testing water tunnel. The velocity is very nearly uniform across the entire test section and is 14.25 m/s. Estimate the drag force on the cylinder.

11-7. Explain how the drag on a 30-cm-diameter, 15-m-tall exhaust stack could be the same in a 9.0-m/s crosswind as in a 12.5-m/s crosswind.

11-8. What is the maximum bending moment on a 35-cm-diameter cylindrical piling in 12 m of seawater in an average tidal current of 5 knots (2.57 m/s)?

11-9. Estimate the maximum bending moment on a 50-cm-diameter round piling in 10 m of seawater in a tidal current averaging 2.5 m/s.

11-10. Show that the terminal velocity of a sphere in laminar motion (Stokes' law is considered applicable) falling in a fluid of large extent is

$$u_s = \frac{gd^2(\rho_s - \rho_f)}{18\mu}$$

where g is the gravitational acceleration, d is the sphere diameter, ρ_s and ρ_f are the densities of the sphere and fluid, respectively, and μ is the fluid viscosity. Describe how you could use this equation to determine the viscosity of a fluid.

11-11. The Stokes equation for the drag coefficient of a single sphere in an infinite medium is $C_D = 24/\text{Re}_D$. Oseen's improvement gives

$$C_D = \frac{24}{\text{Re}_D}\left(1 + \frac{3}{16}\,\text{Re}_D\right)$$

Goldstein in 1929 gave an exact solution of Oseen's analysis and obtained

$$C_D = \frac{24}{\text{Re}_D}\left(1 + \frac{3}{16}\,\text{Re}_D - \frac{19}{1280}\,\text{Re}_D^2 + \frac{71}{20,480}\,\text{Re}_D^3 - \cdots\right)$$

Compare these with Eq. (11.5) at $\text{Re}_D = 1$ and with the experimental results of Fig. 11-4.

11-12. Show that for a sphere falling steadily in a fluid

$$C_D u_s^2 = \frac{4Dg}{3}\left(\frac{\rho_s}{\rho_f} - 1\right)$$

The terminal velocity of falling spheres may be obtained from this equation for Re > 0.1 by trial (assume u_s; calculate Re; obtain C_D; compute $C_D u_s^2$ and compare with required value).

11-13. When one is determining the dynamic viscosity of a fluid by measuring the terminal velocity u_s of a sphere falling through the fluid, the Reynolds number may be above that for which Stokes' law is valid. When 0.1 < Re < 100, show that if Eq. (11.5) is used for the drag coefficient, the viscosity may be found from the equation

$$\mu^2 + \left(\frac{3}{16} u_s D \rho \right) \mu - \mu_1^2 = 0$$

where μ_1 is the viscosity if Stokes' law were valid [Eq. (11.6)].

11-14. What is the largest size of spherical sand grains ($s = 2.65$) which will settle according to Stokes' law ($Re_D = 0.1$) in water at (a) 5°C and (b) 25°C?

11-15. Repeat Prob. 11-14 for spherical water droplets in standard air.

11-16. A sphere of aluminum ($s = 2.81$) and a sphere of magnesium ($s = 1.82$) fall freely in water. What should be the ratio of their diameters in order that steady-fluid motion around each should be dynamically similar?

11-17. Repeat Prob. 11-16 for a steel ($s = 7.95$) and an aluminum ($s = 2.81$) sphere.

11-18. A 2.00-mm steel sphere ($s = 7.95$) falls steadily through a vertical distance of 1.00 m in 55 s in an oil with $s = 0.84$. The oil is contained in a vertical cylinder 10.0 cm in diameter. What is the dynamic viscosity of the oil?

11-19. Assume hailstones are smooth spheres with $s = 0.75$. What is the terminal velocity of hailstones in standard air for diameters of
(a) 2 mm,
(b) 10 mm,
(c) 20 mm?
Make a statement regarding the terminal velocity of hailstones as a function of diameter for the range of Reynolds numbers involved.

11-20. Design a sphere for measuring the speed of offshore tidal currents 60 cm from the bottom of the sea by making use of the drag of a sphere. For example, assume the sphere to be mounted on a vertical rod and the bending moment measured with strain gages. Use the largest smooth or rough sphere practicable for water at 10°C flowing at velocities from 0.30 to 1.5 m/s.

11-21. The drag of a golf ball in flight is less than the drag for a smooth sphere of the same size because the golf ball is rough. Is there any advantage in roughening the surface of a spherical water supply tank supported 35 m above the ground in order to reduce the bending moment on the support column in a high wind? Explain.

11-22. Explain why the drag of a rough sphere or cylinder is less than the drag of a smooth sphere or cylinder at Reynolds numbers near 200,000.

11-23. Explain why the drag of a cylinder in a fluid stream normal to its axis and at high Reynolds numbers may be reduced by the addition of a tapered afterbody, forming a short strut.

11-24. A streamlined body falls at a steady velocity in the atmosphere and has a laminar boundary layer over most of its length. Because of an external disturbance, the boundary layer suddenly becomes predominantly turbulent. Will the velocity of fall increase, remain the same, or decrease? Explain.

11-25. Draw a sketch illustrating the effect of top spin on a Ping-Pong ball moving through air.

11-26. The angle of attack at zero lift for a lifting vane is $-3°$. The vane is placed in a standard air stream at 40 m/s at an angle of attack of $+4°$. The chord length is 50 cm. What is the theoretical lift coefficient and the lift per meter of span?

11-27. A lifting vane with a 1.00-m chord has a circulation of 30 m²/s in standard air at a free-stream velocity of 30 m/s. What is
(a) the theoretical lift per meter of span,
(b) the theoretical lift coefficient, and
(c) the angle of attack measured from the angle of zero lift?

11-28. An airplane has a wing loading of 720 Pa. The wing characteristics are shown in Fig. 11-11a and 11-11b. What is the necessary air speed for level flight in standard air for an angle of attack of $4°$.

11-29. A small aircraft with a wing loading of 600 Pa flies at 45 m/s in standard air. The airfoil characteristics are shown in Figs. 11-11a and 11-11b. At what angle of attack should the wing be designed to fly?

11-30. A small aircraft weighs 8000 N and has wings of 9.30 m² total area with characteristics shown in Figs. 11-11a and 11-11b. What is the air speed for level flight in standard air if the wings are designed for a maximum lift–drag ratio?

11-31. Show that the drag on a given aircraft for level flight is independent of the altitude at which level flight occurs.

11-32. Compare the power to maintain level flight for a given aircraft at 3000-m altitude ($\rho = 0.909$ kg/m³) with the power required to maintain level flight at 1500-m altitude ($\rho = 1.058$ kg/m³). Make a qualitative statement regarding the power requirements for maintaining level flight of a given aircraft as a function of altitude.

11-33. Show that the power required to propel a given airfoil supporting a given load is proportional to $C_D/C_L^{3/2}$ and is therefore independent of speed.

11-34. The NACA 23012 airfoil has the characteristics listed in the accompanying table. (from NACA Report 669).

a	-4	0	$+4$	$+8$	$+12$
C_L	-0.281	$+0.122$	$+0.523$	$+0.919$	$+1.32$
C_D	0.0075	0.0061	0.0074	0.0101	0.0164

α	$+16$	$+17$	$+18$	$+20$
C_L	$+1.67$	$+1.74$	$+1.30$	$+1.21$
C_D	0.0317	0.0403	0.153	0.201

(a) Plot C_L, 10 C_D, and L/D as ordinates and α as abscissa.
(b) What is the stall angle?
(c) What is the maximum value of L/D, and at what angle of attack does it occur?
(d) Plot the polar diagram for this airfoil.

11-35. The NACA 2412 airfoil has the measured characteristics at Re = 3×10^6 listed in the table (from NACA Technical Note 404).

α	C_L	C_D	α	C_L	C_D
-3.4	-0.178	0.0096	11.8	1.328	0.0201
-1.7	-0.026	0.0092	13.4	1.457	0.0261
-0.4	$+0.133$	0.0090	15.0	1.566	0.0352
$+1.1$	0.288	0.0090	15.8	1.589	0.0422
2.6	0.439	0.0095	19.8	1.307	0.2269
5.6	0.744	0.0110	26.8	1.003	0.4189
8.7	1.049	0.0143			

(a) Plot C_L, 10 C_D, and L/D as ordinates and α as the abscissa.
(b) What is the stall angle?
(c) What is the maximum value of L/D, and at what angle of attack does it occur?
(d) Plot the polar diagram for this airfoil.

11-36. Measured pressure coefficients on an NACA 4412 airfoil at an $8°$ angle of attack are as listed in the table (from NACA Report 563).

	C_p			C_p	
$100\,\dfrac{x}{C}$	LOWER SURFACE	UPPER SURFACE	$100\,\dfrac{x}{C}$	LOWER SURFACE	UPPER SURFACE
100	0.134		20.0	0.321	-1.239
98.0	0.167	$+0.120$	14.9	0.345	-1.308
94.9	0.180	$+0.079$	10.0	0.402	-1.391
89.9	0.203	-0.009	4.9	0.568	-1.547
74.9	0.231	-0.285	2.9	0.748	-1.647
64.9	0.244	-0.456	1.7	0.916	-1.743
50.0	0.252	-0.690	0.9	1.013	-1.793
40.0	0.265	-0.880	0.4	0.905	-1.740
29.9	0.293	-1.071	0.0	0.157	-1.000

(a) What is the measured lift coefficient according to Eq. (11.14)?

(b) What is the measured lift coefficient according to Eq. (11.13)? From the given data, $C_C = -0.100$.

References

1. H. E. Saunders, Captain, U.S. Navy (retired), *Hydrodynamics in Ship Design*, Vol. 2. (New York: The Society of Naval Architects and Marine Engineers, 1957), p. 100.

2. K. M. S. Davidson, "Resistance and Powering," Chapter 11 of *Principles of Naval Architecture*, edited by H. E. Rossell and L. B. Chapman (New York: The Society of Naval Architects and Marine Engineers, 1941).

3. H. Lamb, *Hydrodynamics* (New York: Cambridge University Press, 1932), pp. 614–616.

4. J. M. Robertson, *Hydrodynamics in Theory and Application* (Englewood Cliffs, N.J.: Prentice-Hall, Inc., 1965), p. 628.

5. S. F. Hoerner, *Fluid-Dynamic Drag* (Midland Park, N.J.: S. F. Hoerner, 1958).

6. A. H. Shapiro, *Shape and Flow* [New York: Doubleday and Company, Inc. (Anchor Books), 1961].

7. I. H. Abbott and A. E. von Doenhoff, *Theory of Wing Sections* (New York: McGraw-Hill Book Company, Inc., 1949).

8. O. G. Sutton, *The Science of Flight* (Baltimore, Md.: Penguin Books, Inc., 1955).

9. T. von Kármán, *Aerodynamics* (Ithaca, N.Y.: Cornell University Press, 1954).

Chapter 12
Open-Channel Flow

Flow in an open channel involves the flow of a liquid (water, in most instances of engineering interest) in which the cross-sectional flow area may change in the direction of flow and for which the surface is always at atmospheric pressure. The forces causing flow are due to gravity, and the forces retarding flow are due to viscous shear along the channel bed. Thus both the Froude number and the Reynolds number are involved. Flow in rivers, flumes, partly full culverts, canals, and irrigation systems are examples of flow in open channels. We will consider steady flow in prismatic channels, in channels made up of successive prismatic channels, and (in Sec. 12-7) flow in channel transitions. A prismatic channel has a constant cross section and constant bed slope, though the flow cross section may vary in the direction of flow because of changes in depth. Analogies exist between flow in open channels and gas flow in pipes and nozzles, and these similarities are discussed in Chapter 15.

12-1. Basic Considerations

Nomenclature for geometry and forces in an open channel is shown in Fig. 12-1.

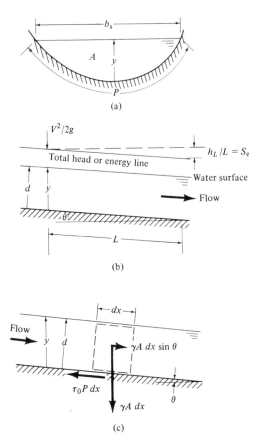

(a)

(b)

(c)

Fig. 12-1.   Nomenclature for an open channel. (a) Cross section: hydraulic radius $= A/P = R_h$; hydraulic depth $= A/b_s = y_h$; actual depth $= y$. (b) Longitudinal section: slope of total head line $= S_e = \sin \theta$; pressure head at channel bed is $h = d \cos \theta = y \cos^2 \theta$. (c) Forces on a fluid elemental length.

The vertical distance from the channel bed to the water surface is designated as y, and the depth normal to the bed as d. For channels which drop less than 14 m per 100 m, y and d differ by less than 1 percent, and y will be used to designate depth. The pressure head at the bottom of the channel for flow with straight streamlines is

$$h = d \cos \theta = y \cos^2 \theta$$

and for slopes of 0.1 or less, the piezometric head for all streamlines in the flow is within 1 percent of y. Open channels are considered quite steep if the bed slope is 0.01, and for this situation $\cos^2 \theta = 0.9999$. Thus for small slopes and straight streamlines (see the discussion of Fig. 12-3 for curved streamlines later in this section) the piezometric head line coincides with the free-water surface.

Mathematically, the slope S_b of the channel bed should be the ratio of change in elevation per horizontal distance. However, for small slopes (less than 0.14), the sine and tangent differ by no more than 1 percent. In open-channel flow analyses, slopes of the bed, water surface, and total head lines are generally expressed as $\sin \theta$, θ being the angle between these and the horizontal; thus, they represent drop per unit length of channel bed.

For steady uniform flow (no fluid acceleration), the gravity and viscous forces parallel to the bed are in equilibrium for any elemental length of channel (Fig. 12-1c).

The momentum theorem applied in a direction parallel to the bed may be expressed as

$$(\gamma A \, dx) \sin \theta - \tau_0 P \, dx = 0$$

since forces owing to pressure cancel out. From this, the average bed shear stress is

$$\tau_0 = \gamma R_h S \tag{12.1}$$

where

R_h = the hydraulic radius, the ratio of flow cross-sectional area A to the wetted perimeter P;

S = slope of the channel bed;

τ_0 = mean bed shear stress, commonly referred to as the average unit tractive force in open-channel hydraulics.

Even though open-channel flow differs from pipe flow in some respects (cross sections are variable, the surface is at constant atmospheric pressure, and gravity forces are involved), there are striking similarities with regard to viscous effects. Recall from Eq. (9.8) that $\tau_0 = f\rho V^2/8$, and thus, from Eq. (12.1),

$$f = \frac{8\gamma R_h S}{\rho V^2} = \frac{8g R_h S}{V^2}$$

where $S = h_f/L$ and represents the slope of the total head line for both pipes and open channels. Measurements on both triangular and rectangular smooth open channels show that the friction factor f is related to the Reynolds number $V4R_h/\nu$ by the *same* curves as for smooth pipes in Fig. 10-6. Typical results are shown in Fig. 12-2.

In rough channels, the friction factor is increased for laminar flow and is influenced by the shape of the channel in the turbulent regime. This is probably owing to secondary flow, which involves cross currents superimposed on the mean flow direction.

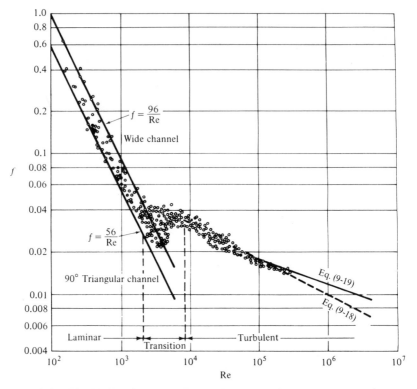

Fig. 12-2. The friction factor vs. Reynolds number from experiments in smooth open channels. (Source: By permission from *Open-Channel Hydraulics* by V. T. Chow. Copyright, 1959. McGraw-Hill Book Company, Inc.)

The gravity effects may be indicated in terms of the Froude number, which is the ratio of the inertia to gravity forces in the system:

$$\mathrm{Fr} = \frac{V}{\sqrt{gy_h}} \tag{12.2}$$

in which V is the mean flow velocity and y_h is the hydraulic depth. In general, the hydraulic depth is the area of the flow cross section divided by the surface width ($y_h = A/b_s$) and is equal to the actual depth for a rectangular channel. The Froude number is also the ratio of the mean flow velocity V to the velocity $\sqrt{gy_h}$ of an elementary surface wave, known as a small gravity wave. Thus, (1) if the Froude number is less than unity, the flow is *subcritical*, or *tranquil*; (2) if the Froude number is equal to unity, the flow is *critical*; and (3) if the Froude number is greater than unity, the flow is *supercritical*, *rapid*, or *shooting*.

Flow regimes of interest include uniform flow, gradually varied flow, and rapidly varied flow (Fig. 12-3). In addition, the flow may be either steady or unsteady. Only steady flow will be considered here.

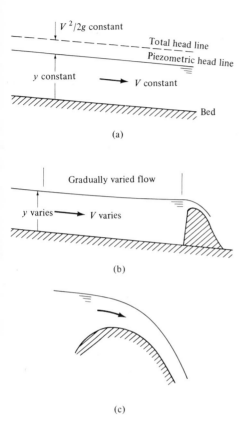

$V^2/2g$ constant

Total head line

Piezometric head line

y constant

V constant

Bed

(a)

Gradually varied flow

y varies ⟶ V varies

(b)

(c)

Fig. 12-3. Types of open-channel flow. (a) Uniform flow. (b) Gradually varied flow. (c) Rapidly varied flow.

Uniform flow exists only in long channels and requires that the depth, cross-sectional area, and hence mean velocity be constant from section to section. In gradually varied flow, these changes take place gradually from section to section, so that the streamlines are essentially parallel and pressures within the flow are hydrostatic. A fundamental assumption in gradually varied flow is that over comparatively short reaches (lengths) of channel, head losses are considered the same as for uniform flow at the average depth. In rapidly varied flow, changes in depth, flow area, and hence average velocity take place in such short reaches of channel that streamlines have pronounced curvature, and pressures within the flow are not hydrostatic. In abrupt changes in channel boundaries and in hydraulic jumps, head losses are high because of separation and intense turbulence; but for curvilinear flow without separation, such as in faired transitions in channel boundaries or in flow over a weir or free overfalls, head losses are small and often may be neglected.

The velocity in an open channel varies from zero along the wetted perimeter to a maximum at or near the surface, depending upon the shape

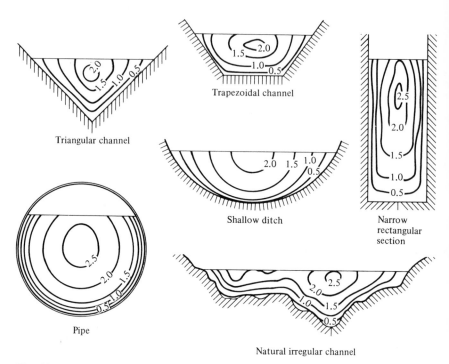

Triangular channel

Trapezoidal channel

Shallow ditch

Narrow rectangular section

Pipe

Natural irregular channel

Fig. 12-4. Typical velocity variations in open channels. (Source: By permission from *Open-Channel Hydraulics* by V. T. Chow. Copyright, 1959. McGraw-Hill Book Company, Inc.)

of the cross section. Some typical velocity variations are indicated in Fig. 12-4. The velocity head in terms of average velocity V should properly be written as $\alpha V^2/2g$ (Sec. 5-1), where α varies from 1.0 to about 1.2 under normal conditions. We will assume one-dimensional flow with $\alpha = 1.0$ in this chapter.

If streamlines are parallel or essentially so (as for uniform flow and gradually varied flow), the pressures within the flow are hydrostatic and determined solely by the depth. For curvilinear streamlines (in rapidly varied flow), the pressures within the flow are less than hydrostatic for convex flow and greater than hydrostatic for concave flow because of centrifugal effects. For channels of large slopes, the pressures within the flow are also less than hydrostatic, the pressure head at any vertical distance y from the surface being

$$h = y \cos^2 \theta$$

where θ is the angle the water surface makes with the horizontal. We will consider in detail only those situations where the pressures within the flow are hydrostatic.

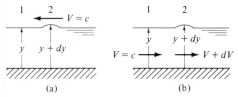

Fig. 12-5.   Elementary surface wave. (a) Observer at rest. (b) Observer moving with wave.

12-2. Velocity of an Elementary Wave

The velocity of a solitary elementary wave on the surface of a liquid may be determined by applying the continuity and Bernoulli equations to a wave with reference axes at rest with respect to the wave (Fig. 12-5). The continuity equation is

$$Vy = (V + dV)(y + dy)$$

or

$$\frac{dV}{V} + \frac{dy}{y} = 0$$

The Bernoulli equation written between sections 1 and 2 is

$$\frac{V^2}{2g} + y = \frac{(V + dV)^2}{2g} + (y + dy)$$

or

$$V\,dV + g\,dy = 0$$

which is equivalent to Eq. (4.7).

Combining the differential forms of the continuity and Bernoulli equations gives (for $V = c$) the wave velocity, or celerity, for waves whose height is small compared with the water depth,[†] for $dy/y \ll 1$

$$c = \sqrt{gy} \tag{12.3a}$$

A more general derivation for nonrectangular channels gives

$$c = \sqrt{gy_h} \tag{12.3b}$$

where y_h is the hydraulic depth.

[†] In Sec. 10-1, the sonic velocity in a gas was derived by using the continuity and momentum equations and was shown to be that of a pressure wave which was very small compared with the ambient pressure.

12-3. Uniform Flow

In uniform flow (Fig. 12-3a), the depth and average velocity remain constant, and thus the channel bed, water surface, and the total head line are parallel ($S_b = S_w = S_e = S$). This type of flow rarely occurs in a natural stream and exists only in relatively long man-made channels. It may be supercritical or subcritical, but uniform critical flow is very unstable and undulations in the water surface permit only an *average* uniform critical flow to occur. Even though uniform flow is not common, its treatment is necessary because many nonuniform flows approach a uniform flow condition, and uniform flow criteria are used in analyzing nonuniform flow.

It would be desirable to have a uniform flow equation which would include all the fluid, geometrical, and flow parameters, but none has been developed as yet. The equation most commonly used is the Manning formula relating the average flow velocity V, the channel slope S, the channel roughness n, and the hydraulic radius R_h of the channel cross section.

In the SI system, Manning's equation is[†]

$$V = \frac{1}{n} R_h^{2/3} S^{1/2} \quad \text{m/s} \tag{12.4a}$$

and

$$Q = \frac{1}{n} R_h^{2/3} S^{1/2} A \quad \text{m}^3/\text{s} \tag{12.4b}$$

The Manning equation may be compared with the Darcy–Weisbach equation for pipe flow [Eq. (9.6)]. The Darcy–Weisbach equation may be written as

$$V = \sqrt{\frac{8g}{f}} \, R_h^{1/2} S^{1/2}$$

so that

$$\frac{1}{\sqrt{f}} = \frac{1}{\sqrt{8g}} \frac{R_h^{1/6}}{n} \tag{12.5}$$

[†] In technical English units

$$V = \frac{1.49}{n} R_h^{2/3} S^{1/2} \quad \text{ft/s}$$

and since $Q = VA$,

$$Q = \frac{1.49}{n} R_h^{2/3} S^{1/2} A \quad \text{ft}^3/\text{s}$$

Manning's experimental work (1889) was in metric units, and in order to use the same value of n in metric and English units, the factor 1.49 (the cube root of the number of feet in a meter) is introduced.

Since Manning's n is a measure of roughness, it would seem illogical that it have dimensions of $T/L^{1/3}$, which results from Eq. (12.5) if the value of 1 is dimensionless. However, if 1 has dimensions of $\sqrt{g}$, then the dimension of n is $L^{1/6}$, and $n/R_h^{1/6}$ then represents a relative roughness similar to $k/D = k/4R_h$ for a pipe. In both instances (pipe flow as well as open-channel flow), flow at high Reynolds numbers for relatively rough surfaces is essentially independent of the Reynolds number. Thus *the Froude number is the significant parameter for open-channel flows of most engineering interest.*

Other uniform flow equations (notably by Ganguillet and Kutter in 1869, Bazin in 1897, and Powell in 1950) have been proposed and used, but they have not received such wide acceptance as Manning's equation.

The determination of an accurate value of Manning's n is difficult, since the value depends on surface roughness, vegetation in the channel bed, channel irregularity, channel alignment, silting and scouring, obstructions, the size and shape of channel, the stage (water surface level) and discharge, seasonal changes, and suspended materials and bed load carried in the stream.

Typical values of n are listed in Table 12-1. We will assume that the same average roughness value n applies to the entire wetted perimeter of a channel. This is not always the case (as for man-made channels with different

Table 12-1. TYPICAL VALUE OF MANNING'S n

Closed conduits flowing partly full	
Smooth brass	0.010
Corrugated metal storm drains	0.024
Concrete culvert	0.013
Unfinished concrete	0.014
Clay drain tile	0.013
Rubble masonry	0.025
Lined or built-up channels	
Unpainted steel	0.012
Planed wood	0.012
Unplaned wood	0.013
Trowel-finished concrete	0.013
Unfinished concrete	0.017
Rough concrete	0.020
Glazed brick	0.013
Brick in cement mortar	0.015
Excavated channels	
Clean earth (straight channel)	0.022
Earth with weeds (winding channel)	0.030
Natural streams	
Clean and straight	0.030
Weedy reaches, deep pools	0.100

SOURCE: Reproduced by permission from *Open-Channel Hydraulics* by V. T. Chow. Copyright, 1959. McGraw-Hill Book Company, Inc.

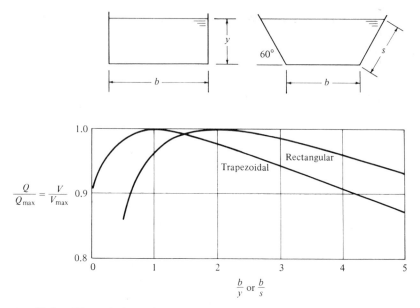

Fig. 12-6. Effect of channel geometry on capacity.

bottom and sides, flooded streams where the flooded surface is usually rougher than the main stream, and ice-covered channels) but the treatment of these other cases is beyond the scope of this text.

Manning's equation indicates that for a given channel slope, roughness, and cross section a shape which has a minimum perimeter will convey the maximum flow. Or a given flow will require a minimum cross section to convey it when the perimeter is also a minimum. Minimizing the perimeter gives the most efficient cross section, the semicircle being the best. Fabricated channels or flumes may be made semicircular, rectangular, or trapezoidal in cross section. In any case, *the most efficient section is one for which the hydraulic radius is one-half the depth.* This condition is automatically met for a semicircle. For a rectangular channel, the width should equal twice the depth ($b = 2y$). For a trapezoidal channel, a half-hexagon is most efficient.

Actually, large departures from these optimum geometries do not affect the flow rate appreciably. Figure 12-6 compares the flow rates for various rectangular and trapezoidal channels of constant cross section in terms of the (maximum) flow rate for the most efficient geometry. It is seen that the ratio of bottom width b to side dimension s may vary from 0 to 4 for the trapezoidal channel, and from 0.6 to more than 5 for the rectangular channel with less than a 10 percent decrease in flow rate for a given area.

The *normal* depth y_n for an open channel is the depth corresponding to uniform flow and may be obtained from Manning's equation [Eq. (12.4)]. The corresponding mean velocity is the *normal velocity* V_n. For most channels, both the normal depth and velocity increase with an increase in total discharge Q. This statement is true for any channel for which the roughness n is considered constant and for which the flow cross section increases at a rate greater than the corresponding increase in wetted perimeter, with an increase in depth. Since

$$V = \frac{1}{n}\left(\frac{A}{P}\right)^{2/3} S^{1/2}$$

differentiation with respect to y_n gives

$$\frac{dV}{dy_n} = \frac{1}{n} S^{1/2}\left(\frac{2}{3}\right)\left(\frac{A}{P}\right)^{-1/3}\left[\frac{P\dfrac{dA}{dy_n} - A\dfrac{dP}{dy_n}}{P^2}\right] \tag{12.6}$$

If the term in brackets is negative, the mean velocity decreases with an increase in depth; if it is zero, the velocity remains constant with an increase in depth; if it is positive—and this is usually the case—the velocity increases with an increase in depth. For a rectangular channel, the term in brackets in Eq. (12.6) is b^2/P^2; for a triangular channel, it is $(1/2)\sin\theta$; and for a trapezoidal channel, it is $(b^2 + 2by_n \tan\theta + 2y_n^2 \tan\theta/\cos\theta)P^2$, where θ is the angle of the sides. In each of these instances, dV/dy_n is positive and both velocity and discharge increase with depth.

Criteria for the design of an open channel for uniform flow depend on whether the channel is nonerodible, erodible, or grass lined. Important criteria in the design of nonerodible channels are lining costs (which depend on the channel shape as well as the lining material), minimum flow velocities in order to avoid deposition of sediment which might be carried in the stream, and free-board requirements. In erodible channels, the maximum flow velocities are limited by the probability of bed scouring. Grass linings are used in erodible channels to prevent bed erosion, and such a channel may be designed on the basis of stability of the bed or for maximum capacity.

Example 12-1

A rectangular trowel-finished concrete channel 5.50 m wide and laid on a slope of 0.002 carries water at a uniform depth of 1.20 m.
(a) What is the average shear stress τ_0 along the channel perimeter?
(b) What is the average flow velocity?
(c) What is the flow regime?
(d) To what value of friction factor f does $n = 0.013$ correspond?

SOLUTION
(a) From Eq. (12.1),

$$\tau_0 = \gamma R_h S = (9810)\left(\frac{6.60}{7.90}\right)(0.002) = 16.4 \text{ Pa}$$

(b) From Table 12-1, $n = 0.013$ and

$$V = \frac{1}{0.013}\left(\frac{6.60}{7.90}\right)^{2/3}(0.002)^{1/2} = 3.05 \text{ m/s}$$

(c)
$$\text{Fr} = \frac{V}{\sqrt{gy}} = \frac{3.05}{\sqrt{(9.81)(1.20)}} = 0.89$$

which is less than 1, and the flow is subcritical.

(d) From Eq. (12.5),

$$f = \frac{8gR_hS}{V^2} = \frac{8gn^2}{R^{1/3}} = \frac{(8)(9.81)(0.013)^2}{(0.835)^{1/3}} = 0.014$$

12-4. Specific Energy and Specific Thrust

SPECIFIC ENERGY

The total energy per unit weight of fluid at any section in a flow stream for an arbitrary streamline in a channel of small slope is the total head—the sum of the velocity head, pressure head, and potential head. If the potential head is taken with respect to the channel bed, the total head or energy is called the *specific energy*. It is expressed in the form of Eq. (5.4) (see Fig. 12-7) as

$$E = \frac{V^2}{2g} + \frac{p}{\gamma} + z$$

and since $(p/\gamma) + z = y$ for any streamline in the section,

$$E = \frac{V^2}{2g} + y \tag{12.7a}$$

For a rectangular channel, the total discharge Q divided by the channel width b is called the unit discharge q. From the continuity equation,

$$V = \frac{Q}{A} = \frac{Q}{yb} = \frac{q}{y}$$

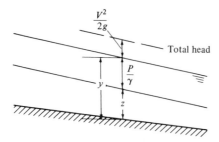

Fig. 12-7. Specific-energy nomenclature.

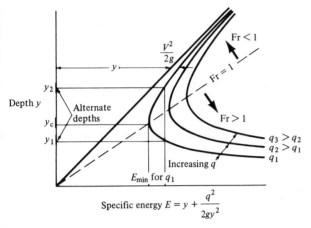

Fig. 12-8. Specific-energy diagram.

so that the specific energy for a rectangular channel is

$$E = \frac{q^2}{2gy^2} + y \tag{12.7b}$$

For a given unit discharge (q constant), the specific energy is a function of the depth of flow [$E = E(y)$]. A plot of this relationship is called a specific-energy diagram, which consists of a family of similar curves each representing a given unit discharge. Examples are shown in Fig. 12-8.

From Eq. (12.7b), as y approaches zero, E approaches an infinite value, and as y becomes very large, E approaches y. Each curve in Fig. 12-8 indicates that there are two depths, y_1 and y_2, for example, for which the specific energy is the same. These depths are called *alternate depths*. Also, each curve has a point of minimum specific energy at which the alternate depths coincide. At this minimum point $dE/dy = 0$, since the slope of the curve is infinite. From Eq. (12.7b), at E_{min},

$$\left(\frac{dE}{dy}\right)_{E_{min}} = -\frac{2q^2}{2gy^3} + 1 = 0$$

so that

$$y_{E_{min}}^3 = \frac{q^2}{g}$$

The velocity head at this point is

$$\frac{V^2}{2g} = \frac{q^2}{2gy^2} = \frac{gy^3}{2gy^2} = \frac{y}{2}$$

or

$$V = \sqrt{gy}$$

This equation corresponds to a Froude number of unity, and thus *flow at minimum specific energy is critical flow.* This statement is often considered to be a definition of critical flow. The minimum specific energy E_{min} *for a given unit discharge* is

$$E_{min} = y_c + \frac{V_c^2}{2g} = y_c + \frac{y_c}{2} = \frac{3}{2} y_c \tag{12.8}$$

The critical depth for a given unit discharge is

$$y_c = \left(\frac{q^2}{g}\right)^{1/3} \tag{12.9a}$$

and the critical velocity for a given unit discharge is

$$V_c = \sqrt{gy_c} \tag{12.10}^\dagger$$

Horizontal distances from the ordinate to the specific energy curve in Fig. 12-8 represent the depth (from the ordinate to the 45° line) plus the velocity head (from the 45° line to the curve). Thus for a given unit discharge at depths greater than critical, the velocities are less than critical so that the Froude number is less than 1 and the flow is called subcritical. For depths less than critical, the velocities are greater than critical so that the Froude number is greater than 1 and the flow is called supercritical.

A *mild* slope is one on which uniform flow occurs at depths greater than critical; a *critical* slope is one on which uniform flow occurs at critical depth; and a *steep* slope is one on which uniform flow occurs at a depth less than critical. Thus, whether a channel slope is mild, critical, or steep does not depend solely on the inclination of the channel bed but on the type of uniform flow it produces. This criterion is similar to that for a hydraulically smooth or rough surface—it depends on the flow as well as the actual surface roughness.

The specific-energy diagram is very useful in analyzing gradually varied flow and is helpful in showing depth variations in rapidly varied flow.

---

† For a nonrectangular channel

$$E = \frac{Q^2}{2gA^2} + y \quad \text{and} \quad \frac{dE}{dy} = -\frac{2Q^2}{2gA^3}\frac{dA}{dy} + 1$$

where $dA/dy = b_s$, the surface width (Fig. 12-1). Thus,

$$\frac{Q^2 b_s}{gA^3} = 1 \tag{12.9b}$$

for *minimum specific energy*, which represents critical flow. From this,

$$\frac{V^2 A^2 b_s}{gA^3} = 1 \quad \text{or} \quad V = \sqrt{\frac{gA}{b_s}} = \sqrt{gy_h}$$

which agrees with Eq. (12.2) for a Froude number of 1.

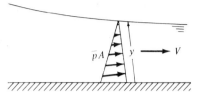

Fig. 12-9.   Specific-thrust nomenclature.

THRUST FUNCTION

The thrust function was defined in Sec. 4-2 as[†]

$$F = pA + \rho V^2 A$$

For a unit width of rectangular channel (Fig. 12-9),

$$pA = \left(\frac{\gamma y}{2}\right)(y)$$

and

$$\rho V^2 A = \left(\frac{\gamma}{g}\right)\left(\frac{q^2}{y^2}\right)(y)$$

The thrust function per specific weight of fluid f is called the *specific thrust function*, or *specific thrust*, and is

$$f = \frac{y^2}{2} + \frac{q^2}{gy} \tag{12.11}$$

For a given unit discharge q, the specific-thrust function depends only on the depth $[f = f(y)]$. A plot of this relationship is called a specific-thrust diagram and consists of a family of similar curves, each applying to a given unit discharge. Examples are shown in Fig. 12-10. As is true for specific energy, here also two depths (y_1 and y_2, for example) exist for which the specific thrust is the same for a given unit discharge. These depths are called *conjugate*, or *sequent*, depths. Each curve has a minimum value of specific thrust at which the conjugate, or sequent, depths coincide. At this minimum point, $df/dy = 0$. From Eq. (12.11),

$$\frac{df}{dy} = y - \frac{q^2}{gy^2} = 0$$

so that

$$y_{f_{min}}^3 = \left(\frac{q^2}{g}\right)$$

[†] By convention, we will use gage pressures here.

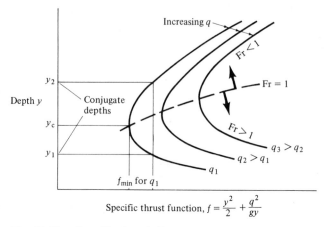

Specific thrust function, $f = \dfrac{y^2}{2} + \dfrac{q^2}{gy}$

Fig. 12-10. Specific-thrust diagram.

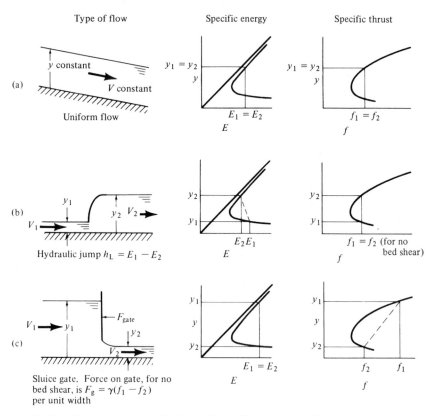

Fig. 12-11. Examples of application of specific-energy and specific-thrust diagrams.

which is the same condition as for minimum specific energy. Thus minimum specific thrust also corresponds to critical flow.

Expressions for the specific-thrust function for nonrectangular channels are more complicated than those for specific energy because the pA term involves the depth to the centroid of the cross section.

Examples of the use of the specific-energy and thrust-function curves are shown in Fig. 12-11. For uniform flow (Fig. 12-11a), the depth of flow and velocity are both constant; from Eqs. (12.7a) and (12.11), the specific energy and specific thrust are both constant. For a hydraulic jump (Fig. 12-11b), the specific thrust is constant for flow on a horizontal bed without bed shear (see Sec. 12-6). The conjugate, or sequent, depths may then be obtained from the specific-thrust curve. When these depths are located on the specific-energy curve, the change in specific energy, which represents the head or energy loss through the jump, may be obtained. For flow under a sluice gate on a horizontal channel (Fig. 12-11c), the energy equation indicates a constant specific energy, and the alternate depths may be obtained from the specific-energy curve. These depths, located on the specific-thrust curve, indicate the change in specific thrust, which represents the force of the sluice gate on the fluid.

Example 12-2

An unfinished concrete trapezoidal channel has a 3.50-m bottom width and sides inclined 45° from the horizontal. It is laid on a slope of 0.001. The flow rate is 14.00 m³/s. (a) What is the normal depth? (b) What is the flow regime? (c) What is the critical depth for this flow? (d) What would be the critical slope for this channel flow?

SOLUTION
(a) From Manning's equation,

$$Q = \frac{1}{0.017} \left(\frac{3.5 y_n + y_n^2}{3.5 + 2\sqrt{2} y_n} \right)^{2/3} (0.001)^{1/2} (3.5 y_n + y_n^2)$$

The value of y_n may be found by a trial solution by inserting various assumed values of y until the calculated flow rate is 14.00 m³/s. From this, $y_n = 1.521$ m. Also, the flow rate obtained from assumed values of y may be plotted and the resulting curve interpolated for the value of y_n, as shown in the following:

Assumed y (m)	1.20	1.30	1.40	1.50	1.60
Calculated Q (m³/s)	9.18	10.57	12.07	13.66	15.35

A plot of y vs. Q indicates that $y_n = 1.52$ m.

(b) The flow regime is determined by the magnitude of the Froude number:

$$Fr = \frac{V}{\sqrt{gy_h}}$$

where $V = Q/A = 14.00/7.637 = 1.83$ m/s and $y_h = A/b_s = 7.637/(3.5 + 2y_n) = 7.637/6.542 = 1.17$ m. Thus

$$Fr = \frac{1.83}{\sqrt{(9.81)(1.17)}} = 0.54$$

which is less than 1, so the flow is subcritical.

(c) The condition which has to be satisfied is that $Q^2/g = A^3/b_s$ from Eq. (12.9b). Thus

$$\frac{(14.00)^2}{9.81} = \frac{(3.5y_c + y_c^2)3}{3.5 + 2y_c}$$

and a trial solution gives $y_c = 1.06$ m.

(d) Uniform flow would occur at critical depth and critical velocity. From Manning's equation

$$S_c = \frac{V^2 n^2}{R_h^{4/3}}$$

where $V_c = Q/A_c = 14.00/4.83 = 2.90$ m/s. The hydraulic depth at critical depth is $(y_h)_c = (A/b_s)_c = 4.83/5.62 = 0.86$ m. The hydraulic radius at critical depth is $(R_h)_c = (A/P)_c = 4.83/5.93 = 0.814$. Thus

$$S_c = \frac{(2.90)^2(0.017)^2}{(0.814)^{4/3}} = 0.00320$$

■

Example 12-3

An unfinished concrete rectangular channel has a bottom width of 3.00 m and a slope of 0.0009. The flow rate is 10 m^3/s. (a) What is the normal depth? (b) What is the flow regime? (c) What is the critical depth? (d) What is the critical slope for this flow?

SOLUTION

(a) From Eq. (12.4d),

$$Q = \frac{1}{0.017}\left(\frac{3y_n}{3 + 2y_n}\right)^{2/3}(0.0009)^{1/2}(3y_n) = 10$$

A trial solution gives $y_n = 2.074$ m.

(b) $Fr = V/\sqrt{gy_h}$, where $V = Q/A = 10/6.222 = 1.607$ m/s, $y_h = y = 2.074$, and $Fr = 1.607/\sqrt{(9.807)(2.074)} = 0.356$ and thus the flow is subcritical.
(c) From Eq. (12.9a), $y_c = (q^2/g)^{1/3} = (3.333^2/9.807)^{1/3} = 1.042$ m.
(d) Uniform flow is at critical depth and velocity.

$$V_c = \sqrt{gy_c} = (9.807)(1.042) = 3.20 \text{ m/s}.$$

From Manning's equation, where $R_h = by_c/(b+2y_c) = (3)(1.042)/(3+2.084) = 0.615$ m,

$$S_c = \frac{V^2 n^2}{R_h^{4/3}} = \frac{(3.20)^2(0.017)^2}{(0.615)^{4/3}} = 0.00565$$

▬▬

12-5. Gradually Varied Flow

In gradually varied flow, the depth varies and changes in depth take place over relatively long reaches of a channel. Whether the depth increases or decreases in the direction of flow is of interest, and this may be determined qualitatively if the sign of dy/dx is known (y is the depth and x is the distance in the flow direction, positive in the direction of flow). If dy/dx is positive, the depth will increase, and if dy/dx is negative, the depth will decrease in the direction of flow. If dy/dx is zero, the depth is constant and the flow is uniform. In addition to finding the change in depth qualitatively, the actual water surface profile may be calculated by any of a number of methods, one of which will be described here.

DEPTH VARIATION

An expression for the rate of change of depth with downstream distance for a wide rectangular channel (two-dimensional flow) is

$$\frac{dy}{dx} = S_b \frac{1 - \left(\dfrac{y_n}{y}\right)^{10/3}}{1 - \left(\dfrac{y_c}{y}\right)^3} \qquad (12.12a)$$

The bed slope S_b is defined as positive when the bed drops in the direction of flow, and the sign of dy/dx is thus determined by the magnitude of the actual depth as compared with the normal depth y_n and the critical depth y_c. The derivation of Eq. (12.12a) involves the combination of the Manning equation with a differential expression for the total head or energy. The total head or energy at a section is referred to an arbitrary datum where $z = 0$ (Fig. 12-12).
This total head or energy H for any channel is

$$H = \frac{V^2}{2g} + y + z_b$$

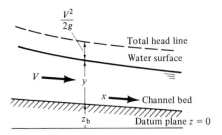

Fig. 12-12. Total head at a section in an open channel: S_e = slope of total head line; S_w = slope of water surface; S_b = slope of channel bed.

and for a rectangular channel

$$H = \frac{q^2}{2gy^2} + y + z_b$$

Differentiation with respect to x gives

$$\frac{dH}{dx} = -\frac{2q^2}{2gy^3}\frac{dy}{dx} + \frac{dy}{dx} + \frac{dz_b}{dx}$$

In this expression dH/dx is the slope of the total head line $-S_e$ and dz_b/dx is the slope of the channel bed $-S_b$. The slope S_e is

$$S_e = \frac{q^2 n^2}{y^{10/3}}$$

since $R_h = y$ for a rectangular channel which is very wide compared to the depth. Thus

$$\frac{dy}{dx} = S_b \frac{1 - \dfrac{q^2 n^2}{y^{10/3}}\dfrac{1}{S_b}}{1 - \dfrac{q^2}{gy^3}} \tag{12.12b}$$

In the numerator

$$\frac{q^2 n^2}{y^{10/3}S_b} = \frac{q^2 n^2}{y^{10/3}}\left[\frac{q^2}{\left(y_n^2 \dfrac{1}{n^2} y_n^{4/3}\right)}\right]^{-1} = \left(\frac{y_n}{y}\right)^{10/3}$$

and in the denominator

$$\frac{q^2}{gy^3} = \left(\frac{y_c}{y}\right)^3$$

Table 12-2. DEPTH VARIATIONS FOR GRADUALLY VARIED FLOW

SLOPE	$\dfrac{y_n}{y}$	SIGN OF NUMER-ATOR	$\dfrac{y_c}{y}$	SIGN OF DENOM-INATOR	SIGN OF $\dfrac{dy}{dx}$	DEPTH	TYPE OF SURFACE PROFILE
Mild	<1	+	<1	+	+	Increases	M-1
$y_n > y_c$	>1	−	<1	+	−	Decreases	M-2
	>1	−	>1	−	+	Increases	M-3
Steep	<1	+	<1	+	+	Increases	S-1
$y_n < y_c$	<1	+	>1	−	−	Decreases	S-2
	>1	−	>1	−	+	Increases	S-3
Critical	<1	+	<1	+	+	Increases	C-1
$y_n = y_c$	>1	−	>1	−	+	Increases	C-3
Horizontal		$-^a$	<1	+	−	Decreases	H-2
$y_n = \infty$		$-^a$	>1	−	+	Increases	H-3
Adverse	<1	$-^b$	<1	+	−	Decreases	A-2
$y_n < 0$	<1	$-^b$	>1	−	+	Increases	A-3

^{a} From Eq. (12.12b).
^{b} $S_b < 0$.

so that

$$\frac{dy}{dx} = S_b \frac{1 - (y_n/y)^{10/3}}{1 - (y_c/y)^3}$$

for a wide rectangular channel.

The sign of dy/dx may be found for various depths on mild, steep, critical, horizontal, and adverse slopes. The procedure and results are shown in Table 12-2. An examination of the magnitude of dy/dx in Eq. (12.12) when the depth y is very large, nearly normal, nearly critical, or very small will indicate the general shapes of the resulting varied-flow water-surface profiles shown in Fig. 12-13. The twelve profile types are the only ones possible for gradually varied flow. Their use is indicated in the following examples.

Example 12-4

What is the water-surface profile for flow passing from a long mild slope to a long steep slope?

SOLUTION

In Fig. 12-14, uniform flow, an M-1 or M-2 curve terminating at a depth greater than critical at the change in bed slope is impossible on the mild

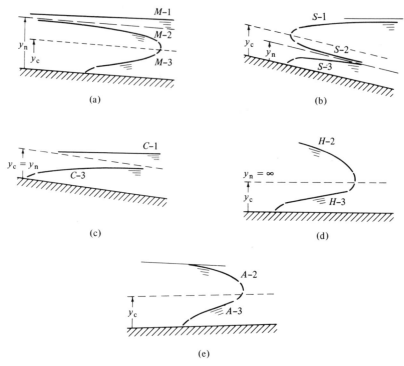

Fig. 12-13. Water surface profiles for gradually varied flow for (a) mild slope, (b) steep slope, (c) critical slope, (d) horizontal slope, and (e) adverse slope.

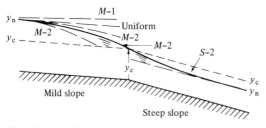

Fig. 12-14. Flow from a long mild slope to a long steep slope.

slope because it would have to be followed by an S-1 curve which would never reach normal depth on the steep slope. An M-2 curve reaching critical depth upstream from the change in bed slope is also impossible because the flow cannot remain critical to the change in bed slope (critical uniform flow cannot occur on a mild slope). Neither can it become less nor more than critical, because if it did the surface would immediately return toward critical along an M-2 or M-3 curve. Therefore, the only possible profile is an M-2 curve passing through critical depth at the location of the change in bed slope,

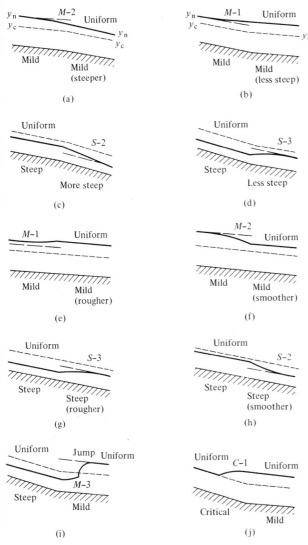

Fig. 12-15. Examples of water surface profiles for gradually varied flow on changes in bed slope and roughness.

followed by an S-2 curve asymptotic to the normal depth line on the steep slope.

Example 12-5

A number of surface profiles are shown in Fig. 12-15. Their correctness and the impossibility of other profiles should be verified.

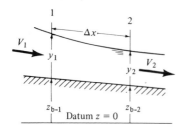

Fig. 12-16. Nomenclature for computing gradually varied flow profiles.

Profiles in Fig. 12-15 indicate that (1) flow on a long steep slope is uniform *upstream* from either a change in bed slope or a change in bed roughness and (2) flow on a long mild slope is uniform *downstream* from either a change in bed slope or a change in bed roughness.

Surface profiles for short channels depend largely on the headwater and tailwater conditions, and generalized results usually cannot be given.

COMPUTATION OF WATER SURFACE PROFILES FOR GRADUALLY VARIED FLOW

Definite points on the surface profiles shown in Fig. 12-13 may be determined by a variety of methods. In the so-called direct-step method,[†] the distance between two sections where depths are assigned may be estimated by assuming that the slope of the total energy line for this distance is (1) the same as for uniform flow at a velocity equal to the average of those at the two sections $(\bar{S}_e = \bar{V}^2 n^2 / \bar{R}^{4/3})$ or (2) equal to the average of the slope of the total energy lines corresponding to uniform flow at the two sections $[\bar{S}_e = (S_{e1} + S_{e2})/2,$ where S_{e1} and S_{e2} are found from Manning's equation].

In Fig. 12-16, the energy equation written for the flow between sections 1 and 2 is

$$\frac{V_1^2}{2g} + y_1 + z_{b1} = \frac{V_2^2}{2g} + y_2 + z_{b2} + h_L$$

Grouping terms gives

$$\left(\frac{V_2^2}{2g} + y_2\right) - \left(\frac{V_1^2}{2g} + y_1\right) = (z_{b1} - z_{b2}) - h_L$$

or

$$E_2 - E_1 = S_b \Delta x - \bar{S}_e \Delta x$$

---

† This method is useful only for prismatic channels.

so that

$$\Delta x = \frac{E_2 - E_1}{S_b - \overline{S}_e} \tag{12.13}$$

Calculations are made either upstream or downstream from a section where the depth is known. For M-1, M-2, H-2, and A-2 curves, calculations generally proceed upstream; for M-3, S-1, S-2, S-3, C-1, C-3, H-3, and A-3 curves, calculations generally proceed downstream. For an assumed flow and channel slope, shape, and roughness, the accuracy obtained depends largely on the size of depth increments chosen. The smaller the depth increments, the more accurate are the results because of the assumptions made regarding head loss. The depth increments should be smaller as the profile approaches uniform depth. If the length of a varied-flow profile between sections where the depths are y_0 and y_k is to be calculated, accuracy would be better if depth increments of $(y_0 - y_k)/20$ were used instead of $(y_0 - y_k)/4$, for example. The calculation effort would be five times greater if 20 rather than 4 intervals were used. This need not be so, however, since the values of Δx increase or decrease monotonically in a well-behaved manner. Thus the value of Δx between y_0 and y_1, y_4 and y_5, y_9 and y_{10}, y_{14} and y_{15}, y_{19} and y_{20} may be calculated and a curve of the various values of Δx vs. interval number plotted. The values of Δx for depth intervals between those calculated may be interpolated from the graph. Maximum accuracy with minimum effort may thus be achieved. The method is illustrated in Example 12-6.

Example 12-6

Water flows in a 15-m-wide rectangular channel at a rate of 112.0 m³/s. The bed slope is 0.001, and the roughness coefficient is $n = 0.025$. A dam increases the depth to 6.20 m immediately upstream. What is the distance upstream to a location where the depth is 3.60 m?

SOLUTION

From the Manning equation, the normal depth is $y_n = 3.37$ m, and from Eq. 12.9a, the critical depth is $y_c = 1.79$ m. Thus $y > y_n > y_c$, so that the water-surface profile is an M-1 curve. Calculation proceeds as indicated in Table 12-3 and Fig. 12-17 (see also Fig. 12-18).

Using 13 intervals of 0.2 m each, the total length is calculated as 4541 m. With two intervals of 1.2 and 1.4 m, the total length is calculated to be 4561 m. With only one depth interval of 2.6 m, the length is calculated to be 4908 m. The use of larger depth increments generally results in an *overestimation* of the length of the varied-flow profile for subcritical flow, and the use of larger depth increments generally results in an *underestimation* of the length of a varied-flow profile for supercritical flow.

Table 12-3 CALCULATION OF VARIED-FLOW PROFILE FOR EXAMPLE 12-6

y	A	V	E	P	$R^{4/3}$	$S_e = \dfrac{V^2 n^2}{R^{4/3}}$	$\overline{S}_e$	ΔE	$S_b - \overline{S}_e$	Δx
Using thirteen intervals										
6.2	93	1.204	6.274	27.4	5.101	0.0001776				
							0.0001860	0.195	0.000814	240
6.0	90	1.244	6.079	27.0	4.979	0.0001943				
5.8										245[a]
5.6										252[a]
5.4										260[a]
5.2										268[a]
5.0	75	1.493	5.114	25.0	4.327	0.0003220				278[a]
							0.0003417	0.191	0.0006583	290
4.8	72	1.556	4.923	24.6	4.187	0.0003614				
4.6										310[a]
4.4										334[a]
4.2										368[a]
4.0										423[a]
3.8	57	1.965	3.997	22.6	3.433	0.0007030				524[a]
							0.0007625	0.178	0.0002375	749
3.6	54	2.074	3.819	22.2	3.271	0.0008219				$\overline{4541}$ m
Using two intervals										
6.2			6.274			0.0001776				
							0.0002498	1.160	0.0007502	1546
5.0			5.114			0.0003220				
							0.0005720	1.293	0.0004280	3025
3.6			3.819			0.0008219				$\overline{4561}$ m
Using one interval										
6.2			6.274			0.0001776				
							0.0004998	2.455	0.0005002	4908 m
3.6			3.819			0.0008219				

[a] These values are obtained from a plot of Δx vs. y in Fig. 12-18.

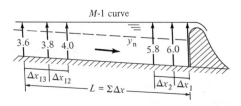

Fig. 12-17. See Example 12-6. Backwater curve.

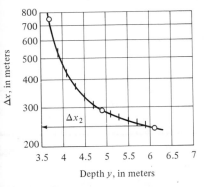

Fig. 12-18. See Example 12-6. Interpolation of distances between assigned depths for backwater curve.

Example 12-7

Water flows from a mild to a steep slope (Fig. 12-14) in an unfinished concrete ($n = 0.020$) rectangular channel 20 m wide at a flow rate of 180 m³/s. The steep channel has a slope of 0.01. What is the distance from the change in grade to a location where the depth is 1.60 m?

SOLUTION

At the change in grade the flow is critical at $y_c = (q^2/g)^{1/3} = (9^2/9.807)^{1/3} = 2.02$ m. The water surface profile is an S-2 curve with an initial depth of 2.02 m, and calculations proceed downstream as indicated in Table 12-4. The normal depth for the steep slope is $y_n = 1.51$ m from a trial solution of the Manning equation.

The distance to $y = 1.60$ m is 57.6 m.

12-6. Hydraulic Jump

A hydraulic jump may occur in channels of any slope and may be found at the foot of spillways and below sluice gates. In all instances, the flow is

Table 12-4 CALCULATION OF VARIED-FLOW PROFILE FOR EXAMPLE 12-7 IN SI UNITS

y	A	V	$\dfrac{V^2}{2g}$	E	P	R	$R^{4/3}$	S_e	$\bar{S}_e$	ΔE	$S_b - \bar{S}_e$	Δx	x
2.02	40.4	4.46	1.014	3.034	24.04	1.680	1.996	0.00404					
									0.00443	0.011	0.00557	1.98	2.0
1.90	38.0	4.74	1.145	3.045	23.80	1.596	1.865	0.00482					
									0.00526	0.029	0.00474	6.12	8.1
1.80	26.0	5.00	1.274	3.074	23.60	1.525	1.755	0.00570					
									0.00627	0.058	0.00373	15.5	23.6
1.70	34.0	5.30	1.432	3.132	23.40	1.453	1.645	0.00683					
									0.00754	0.082	0.00246	33.3	57.6
1.60	32.0	5.62	1.614	3.214	23.20	1.380	1.535	0.00825					

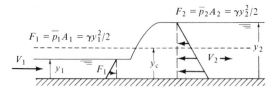

Fig. 12-19. Hydraulic jump.

supercritical upstream of the jump and subcritical downstream of the jump. Equations relating conditions downstream of the jump to conditions upstream may be obtained by writing the continuity, momentum, and energy equations across the jump. If we assume a horizontal channel so that there are no gravity forces, and neglect bed shear for the short reach of channel over which the jump exists, we find that the depth ratio across the jump and the downstream Froude number depend only on the upstream Froude number. Also, the loss of available energy per unit weight of fluid passing through the jump depends only on the depths upstream and downstream.

For the region in which the momentum change takes place (Fig. 12-19) in a unit width of rectangular channel, the continuity equation is

$$V_1 y_1 = V_2 y_2$$

and the momentum equation is

$$\gamma \frac{y_1^2}{2} - \gamma \frac{y_2^2}{2} = \frac{V_1 y_1 \gamma}{g}(V_2 - V_1)$$

so that we get, after combining,

$$\frac{y_1^2}{2} + \frac{q^2}{gy_1} = \frac{y_2^2}{2} + \frac{q^2}{gy_2}$$

which indicates that the thrust function is constant across the jump, in the absence of gravity and bed shear.

The energy equation is[†]

$$E_1 = E_2 + h_L$$

or

$$\frac{q^2}{2gy_1^2} + y_1 = \frac{q^2}{2gy_2^2} + y_2 + h_L$$

The momentum equation may be arranged to yield

$$y_2^2 + y_1 y_2 - \frac{2q^2}{gy_1} = 0$$

[†] The term h_L represents the mechanical energy converted to thermal energy and, in a thermodynamic sense, it is a loss only in available energy.

(since one root of the momentum equation is the trivial case for which $y_1 = y_2$), a quadratic whose solution gives

$$\frac{y_2}{y_1} = \frac{1}{2}\left(\sqrt{1 + \frac{8q^2}{gy_1^3}} - 1\right) = \frac{1}{2}(\sqrt{1 + 8\,\mathrm{Fr}_1^2} - 1) \tag{12.14}$$

Since the momentum equation is symmetrical in y_1 and y_2, it may also be written as

$$y_1^2 + y_2 y_1 - \frac{2q^2}{gy_2} = 0$$

which is another quadratic, whose solution gives

$$\frac{y_1}{y_2} = \frac{1}{2}\left(\sqrt{1 + \frac{8q^2}{gy_2^3}} - 1\right) = \frac{1}{2}(\sqrt{1 + 8\,\mathrm{Fr}_2^2} - 1) \tag{12.15}$$

The depths y_1 and y_2 are *conjugate* depths, since the flow has the same value of thrust function or specific thrust at these depths. Equations (12.14) and (12.15) indicate that for a given unit discharge the smaller the upstream depth, the greater is the downstream depth, and vice versa. As the upstream depth approaches critical ($\mathrm{Fr}_1 \rightarrow 1$), the depth ratio approaches unity. The strength of the jump is expressed in terms of the depth ratio y_2/y_1 or in terms of the upstream Froude number Fr_1.

The head loss through the jump is

$$h_L = E_1 - E_2$$
$$= \frac{q^2}{2gy_1^2} - \frac{q^2}{2gy_2^2} + (y_1 - y_2) = \frac{q^2}{2g}\left(\frac{1}{y_1^2} - \frac{1}{y_2^2}\right) + (y_1 - y_2) \tag{12.16}$$

and since

$$\frac{q^2}{g} = \frac{y_1 y_2(y_1 + y_2)}{2}$$

from the quadratic equation obtained from the momentum equation, the head loss or energy dissipation through the jump is

$$h_L = E_1 - E_2 = \frac{(y_2 - y_1)^3}{4y_1 y_2} \tag{12.17}$$

The hydraulic jump is illustrated on the specific-energy and thrust-function diagrams in Fig. 12-11b.

Equations (12.14), (12.15), and (12.17) are strictly valid only for one-dimensional flow for which the momentum and energy correction factors (α and β in Secs. 5-1 and 4-2) are assumed to be unity. Actually, the velocity profile upstream of the jump is more nearly one dimensional than that downstream. Secondly, the bed shear was assumed zero, and this is less true the longer the jump. The ratio of jump length to downstream depth is about 6 for upstream Froude numbers greater than about 4. Measurements of velocity profiles indicate values of y_2/y_1 from a more exact analysis, which includes the true momentum and the bed shear, to be uniformly less than the approximate value given by Eq. (12.15) [1]. The exact and approximate values agree at $Fr_1 = 1$, and the exact value is about 5 percent less than the approximate value at $Fr_1 = 8$.

In addition to these effects, gravity effects increase as the bed slope increases.

A hydraulic jump can be controlled by a sill, a broad-crested weir, an abrupt rise in the channel floor, or an abrupt drop in the channel floor. In each instance, analysis is made by means of the continuity, momentum, and energy equations. Stilling basins of special design are often used to stabilize the location of a jump and as a means of dissipating energy [2].

Oblique jumps may form in channels where side contractions or expansions occur.

A hydraulic bore is an example of a moving hydraulic jump and may occur in a river emptying into the sea (a tidal estuary), in which case the phenomenon is called a *tidal bore*. The wall of water travels upstream, and if q represents the unit discharge passing through the bore, Eqs. (12.14) and (12.15), as well as those used in their development, apply. The velocities relative to the moving jump are V_1 and V_2.

Example 12-8

Estimate the location of a hydraulic jump downstream of a spillway in the channel of Example 12-6 (for the same flow rate) and calculate the energy and power dissipation in the jump. Assume the dam to be 4.25 m high, that critical flow occurs at its crest, and that there are no head losses on the spillway (Fig. 12-20).

SOLUTION

Here $E_1 + 4.25 = E_2$, writing the Bernoulli equation from the crest of the spillway to its foot. The flow will be supercritical at the foot of the spillway ($y_2 < y_c$) and a M-3 curve describes the water surface downstream to the

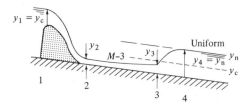

Fig. 12-20.   See Example 12-8. Location of hydraulic jump below a spillway.

hydraulic jump at section 3. From the crest to the spillway foot,

$$E_1 + 4.25 = (\tfrac{3}{2})y_c + 4.25 = (\tfrac{3}{2})(1.79) + 4.25 = 6.93$$

$$= \left(\frac{q^2}{2gy_2^2}\right) + y_2$$

so that $y_2^3 - 6.93y_2^2 + 2.84 = 0$. Roots of this equation are $+6.87$, $+0.673$, and -0.613 m, respectively, and thus $y_2 = 0.673$ m. (The roots $+6.87$ and -0.613 m are physically impossible.)

The depth y_4 is the normal depth of 3.37 m, and this is the downstream depth of a jump for which the upstream depth is y_3. From Eq. (12.15),

$$y_3 = \frac{y_4}{2}\left(\sqrt{1 + \frac{8q^2}{gy_4^3}} - 1\right) = 0.807 \text{ m}$$

The length of the M-3 profile from a depth of 0.67 m to 0.81 m is the distance to the jump. In one step

$$\Delta x = \frac{E_2 - E_3}{|S_b - \bar{S}_e|} = \frac{6.93 - 5.17}{0.001 - 0.114} = 15.6 \text{ m}$$

where $E_2 = 6.93$ m and $E_3 = (q^2/2gy_3^2) + y_3 = 5.17$ m. Then

$$S_{e2} = \frac{(V_2)^2(n)^2}{R_{h_2}^{4/3}} = 0.146$$

$$S_{e3} = \frac{(V_3)^2(n)^2}{R_{h_3}^{4/3}} = 0.082$$

and thus $\bar{S}_e = 0.114$, the average of S_{e_2} and S_{e_3}.

Calculating the distance in two or three steps using smaller depth increments results in a distance about 3 or 4 percent greater than 15.6 m.

The head loss in the jump is $E_3 - E_4 = 5.17 - 3.62 = 1.55$ J/N and there are $(112/15)(9810)$ N/s water flowing per meter width of channel, the power loss is $(1.55)(112/15)(9810) = 113$ kW per meter width of channel, or 1700 kW total power loss.

━━━

Equations for hydraulic jump conjugate depths in nonrectangular channels are very complicated. The general form of the momentum equation for one-dimensional flow for a control volume containing a jump is

$$\rho g h_1 A_1 - \rho g h_2 A_2 = Q\rho(V_2 - V_1) = Q\rho\left(\frac{Q}{A_2} - \frac{Q}{A_1}\right)$$

from which

$$\frac{Q^2}{gh_1 A_1^2} = \frac{(h_2/h_1)(A_2/A_1) - 1}{1 - (A_1/A_2)}$$

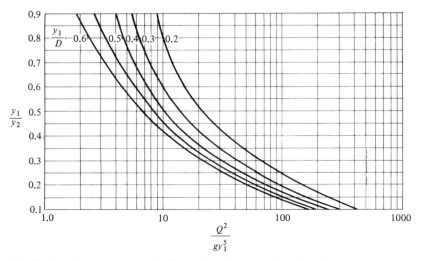

Fig. 12-21. Conjugate depths for hydraulic jumps in a circular-duct open channel. [Source: From A. Thiruvengadam, "Hydraulic Jump in Circular Canals," *Water Power* (Dec. 1961), p. 496. Used with permission of the publisher.]

where h is the depth of the center of gravity of the flow cross section below the water surface; A is the flow cross section; and subscripts 1 and 2 refer to conditions upstream and downstream of the jump, respectively.

Results for a circular-duct open channel have been given in graphical form by Thiruvengadam [3] and are shown in Fig. 12-21, where the depths y_1 and y_2 are measured at the center of the channel. D is the duct diameter.

An analysis for trapezoidal channels gives a fifth-degree equation in y_2/y_1.

12-7. Channel Transitions

The continuity, momentum, and energy equations together with the specific-thrust and specific-energy diagrams may be used to determine the changes in water surface which occur when the channel bottom is raised, when the channel width is increased or decreased, or when a combination of both exists. We will assume that the transitions are gradual so that head losses may be neglected. Flow in channel transitions is rapidly varied flow.

If the channel bottom is raised a height Δz and the width remains constant (Fig. 12-22), then q is constant and the energy equation is

$$E_1 = E_2 + \Delta z$$

A specific-energy curve (q is constant) may be slid along the channel bottom. If the flow upstream of the rise is subcritical, the depth will decrease in order

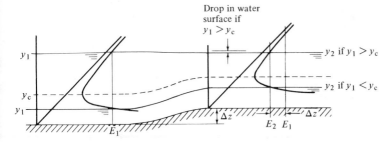

Fig. 12-22.   Effect of rise in channel bed on water surface elevation.

that the specific energy E may decrease; and if the flow upstream of the rise is supercritical, the depth must increase in order that the specific energy may decrease. As the magnitude of Δz is increased, $E_1 - E_2 = \Delta z$ also increases and E_2 eventually corresponds to E_{min}, indicating critical flow over the rise. The Δz which produces critical flow is the maximum rise in the bottom that is possible without causing a backwater (M-1 curve) if the upstream flow is subcritical. Additional increase in Δz beyond $E_1 - E_{min}$ requires an increase in depth, since more specific energy is needed than is available for a depth of the assumed value of y_1. The flow then remains critical over the rise as Δz is increased. If the flow is supercritical upstream and $\Delta z > (E_1 - E_{min})$, a jump will form upstream of the rise, and the flow will then be subcritical and hence the surface will drop slightly as the flow passes over the rise in the channel floor (Fig. 12-23).

In any case the specific-thrust diagram may also be slid along the channel bed, and the net force (per unit width of channel) of the channel transition on the flow stream is the difference in thrust function upstream and downstream of the rise times the specific weight of the water $[F = \gamma(\Delta f)]$.

If the channel sides are contracted, the unit discharge increases but the specific energy remains constant (assuming no losses). As the unit discharge q increases, the depth changes as indicated on successive curves for increasing

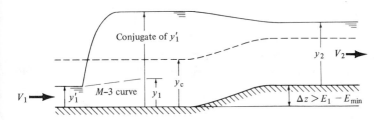

Fig. 12-23.   Supercritical flow upstream of a rise in a channel bed. If $\Delta z > E_1 - E_{min}$, the jump will occur some distance upstream of the rise in the channel bed. On a horizontal bed, the jump will occur where the depth y_1' is less than y_1 (on a M-3 curve) in order that the specific energy E_1' be greater than E_1.

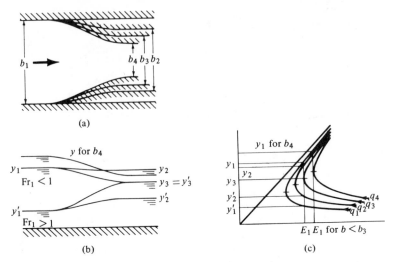

Fig. 12-24. Flow through a constricted channel. (a) Plan view. (b) Elevation. (c) Specific energy.

q on a specific-energy diagram. For a given initial flow, there is a unit discharge q_{max} (q_3 in Fig, 12-24) such that E_{min} for this q_{max} corresponds to the initial specific energy E_1. This condition results in critical flow in the constricted section. Making the channel width even narrower will cause a backwater (M-1 curve) if the upstream flow is subcritical, and the depth will correspond to that value of specific energy on the q_1 curve which is E_{min} for the unit discharge in the constricted section (q_4, for example).

If the upstream flow is supercritical, the depth in the constriction increases as the unit discharge q increases and also becomes critical (for q_3, for example). Further reduction in width causes a jump to form upstream, with a drop in the water surface as for subcritical flow entering the constriction.

12-8. Specialized Examples of Open-Channel Flow

Open-channel flow in culverts may be analyzed by using methods already discussed. Flow over steep spillways often involves an aeration process which results in two-phase flow of air and water. Brief mention of these is made to indicate typical fields of engineering interest and research.

FLOW IN CULVERTS

A culvert should be designed to convey a given discharge with the least head. The flow through a culvert is open-channel flow if the culvert does not flow full and is pipe flow if the culvert does flow full. Factors which determine which type of flow occurs include the size, shape, roughness, ength, and slope of the culvert; the inlet and outlet water surface elevations; and the inlet and outlet culvert geometries.

For a long culvert on a steep slope, the control of flow is considered to be at the inlet for part-full flow because the flow will be critical there and the culvert length, roughness, and outlet conditions will not affect the flow. If the culvert flows full, the control of flow is considered to be owing to culvert friction, although the inlet and outlet losses are also involved but to a lesser degree.

For a long culvert on a mild or horizontal slope with a free outlet, the control of flow is at the outlet for part-full flow because critical flow will occur at the outlet, and the flow will depend on the culvert friction as well as the inlet geometry. If the culvert flows full, control is the same as for a culvert on a long steep slope flowing full.

For a short culvert with a free outlet, part-full flow exists for low inlet heads with both a square and a round inlet (these represent the two extremes of inlet geometry) and for any head on a culvert with a square inlet. A short culvert with a free outlet and a rounded inlet will flow full if the head is about 1.5 diameters above the culvert invert (the bottom of the culvert inlet).

Whether the culvert is on a mild or a steep slope may be determined from the conditions for critical flow [Eq. (12.9b)].

$$\frac{Q^2}{g} = \frac{A^3}{b_s} \qquad\qquad\qquad [12.9b]$$

which, when combined with the Manning equation [Eq. (12.4b)] for critical flow $[Q = (1/n)R_h^{2/3}S_c^{1/2}A$, where S_c is the critical slope] gives the critical slope in dimensionless form. For a circular culvert (subscript o refers to the entire culvert),

$$\frac{S_c}{g(n^2/D^{1/3})} = \frac{4.99(A/A_o)}{(b_s/D)(R_h/R_{ho})^{4/3}} \qquad\qquad (12.18)$$

For a square culvert of sides d and water depth y,

$$\frac{S_c}{g(n^2/d^{1/3})} = \frac{y/d}{\left(\dfrac{y/d}{1 + 2(y/d)}\right)^{4/3}} \qquad\qquad (12.19)$$

Equations (12.18) and (12.19) are plotted in Fig. 12-25, and the uniform flow regime or culvert slope for a given flow may be determined from this figure by calculating the value of $S/g(n^2/D^{1/3})$ or $S/g(n^2/d^{1/3})$ for a given culvert. This is given culvert may be on a mild or steep slope, depending on the depth of flow.

Once the slope of the culvert is determined, the control point may be established for open-channel flow. For pipe flow, the steady-flow energy equation applied between the headwater and culvert exit may be used to determine the flow for a given head or the head required to produce a

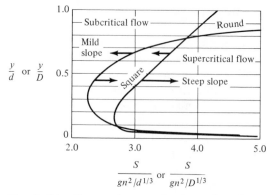

Fig. 12-25. Flow regimes and culvert slope for various depths of flow in square and round culverts.

specified flow rate. The inlet design is of especial importance for long steep culverts and for short culverts when the headwater is more than about one-half the culvert diameter ($D/2$ for a round culvert and $d/2$ for a square culvert) above the culvert crown at inlet [4].

Example 12-9

A 100-cm-diameter concrete culvert ($n = 0.011$) is laid on a slope of 0.010. What will be the flow regime for part-full flow?

SOLUTION

$$\frac{S}{gn^2/D^{1/3}} = \frac{0.01}{(9.81)(0.011)^2(1.00)^{1/3}} = 8.42$$

From Fig. 12-25, the culvert is steep for nearly all depths, and the flow will be supercritical with flow control at the culvert inlet.

■■■

Example 12-10

A 1.2-m-square culvert of unfinished concrete ($n = 0.017$) is 30 m long and has a slope of 0.008. What will be the flow regime for part-full flow?

SOLUTION

$$\frac{S}{gn^2/d^{1/3}} = \frac{0.008}{(9.81)(0.017)^2/(1.2)^{1/3}} = 3.0$$

From Fig. 12-25, the culvert is on a mild slope and the flow will be subcritical for y/d below about 0.05 and above 0.42. Flow control will be at the outlet and, if the outlet is free, critical flow will occur there. For y/d between 0.05

and 0.42 the culvert is steep and the flow will be supercritical. The flow is then controlled at the inlet where the flow is critical.

━━━

AIR-ENTRAINED FLOW ON STEEP SLOPES

When water flows down a steep slope, the boundary layer thickens in the direction of flow. When the boundary layer thickness is equal to the depth of flow, a self-aeration process results in a two-phase flow with a frothy appearance. The aeration process is related to the turbulence conditions within the flow. The mixture consists of bubbles entrained in the water in a lower region near the channel bed, with increasing air concentration in the vertical direction, through a transition region, and to an upper region in which water particles move through air.

There are indications of a bulking effect for aerated flow. The flow velocity, however, is greater for air-entrained flow than for nonaerated flow of water at the same water flow rate. Experiments indicate that the same form of discharge equation may be used for aerated flow as for nonaerated flow with the same value of roughness coefficient if appropriate depths are used in each instance.

PROBLEMS

12-1. Uniform flow of water occurs at an average velocity of 3.50 m/s in a straight open channel whose bed slope is 0.001. What is the energy dissipation (head loss) per kilometer of channel length?

12-2. What is the flow regime in a rectangular channel 4.00 m wide when water flows at 18.20 m^3/s at a 1.20-m depth?

In Probs. 12-3 and 12-4, calculate for uniform flow:
(a) the flow rate;
(b) the Froude number of the flow;
(c) the critical depth for the flow in part (a).

12-3. A rectangular channel is 4.50 m wide, has a bed slope 1/1000, $n = 0.014$, and water flows 1.10 m deep.

12-4. A trapezoidal channel has a bottom width of 4.0 m, sides sloping 1 vertical and 2 horizontal, a bed slope of 1/1681, $n = 0.017$, and a water depth of 2.10 m.

12-5. A smooth steel flume has the shape of a semicircle with a 2-m diameter. What size corrugated metal flume of the same shape conveys the same flow rate at the same slope? Manning's $n = 0.012$ and 0.024, respectively. Both flumes flow full.

12-6. Compare the slopes of a semicircular channel flowing full and a rectangular channel of the same width and cross-sectional area, same surface material, and conveying the same flow rate uniformly.

12-7. Verify the curves for V/V_{max} for
   (a) the rectangular and
   (b) the trapezoidal channels in Fig. 12-6.

12-8. The value of Manning's n is rather insensitive to large variations in the absolute roughness k.
   (a) Show from Eqs. (9.24) and (12.5) that

$$\log_{10} \frac{2R_h}{k} = \frac{1}{4\sqrt{2g}}\left(\frac{R_h^{1/6}}{n}\right) - 0.87$$

   (b) Plot a relation between $2R_h/k$ and $R_h^{1/6}/n$ on log–log paper for a range of $R_h^{1/6}/n$ from 40 to 120, a threefold range in n for a given R_h.
   (c) What is the range in absolute roughness k for the range in n plotted in part (b)?

12-9. What range in absolute surface roughness k is indicated by a range in Manning's n from 0.012 to 0.024 for a hydraulic radius of 1.20 m? Refer to Prob. 12-8.

12-10. What is the unit discharge for uniform flow in a very wide rectangular channel with $n = 0.014$, a bed slope of 1/800, and a flow depth of 1.24 m?

12-11. The depth downstream from a sluice gate in an unfinished concrete rectangular channel 2.50 m wide is 0.50 m. The flow rate is 1.00 m³/s.
   (a) For what channel slope will the depth remain at 0.50 m?
   (b) How will the depth change if the channel slope is 1/600?
   (c) How will the depth change if the slope is 1/1600?

12-12. Uniform flow occurs in an open channel with a 90° V-shape at a rate of 300 L/s. The roughness value is $n = 0.012$. At what channel slope will this flow be critical?

12-13. What is the minimum slope on which a rough concrete rectangular open channel will convey 17 m³/s at an average velocity of 1.50 m/s? *Hint:* The minimum slope is associated with a maximum hydraulic radius.

12-14. What is the minimum slope on which a trapezoidal channel of unfinished concrete will convey 8.50 m³/s at an average velocity of 1.80 m/s?

12-15. It is desired to convey 850 L/s of water in a rectangular channel on a slope of 0.01. What are the dimensions of the channel when a minimum of unplaned timber is used in its construction?

12-16. At what depth will uniform flow occur in a rectangular channel 6.00 m wide, laid on a slope of 0.001, and lined with unfinished concrete for a flow of 28 m³/s? Is the slope of the channel mild or steep?

12-17. What is the normal depth for a unit discharge of 3.70 m³/s per meter width in a wide, rough concrete channel having a bed slope of 0.002?

12-18. A rectangular channel 2.50 m wide has a bed slope of $1/400$, $n = 0.017$, and conveys 9.00 m³/s. What is
(a) the normal or uniform flow depth,
(b) the critical depth for the given flow, and
(c) the flow regime?

12-19. A trapezoidal channel of bottom width 6.0 m has side slopes of 1 vertical and 2 horizontal, and conveys 12.0 m³/s. The bed slope is $1/300$. What is
(a) the normal or uniform flow depth,
(b) the critical depth, and
(c) the flow regime? Here, $n = 0.014$.

12-20. Calculate
(a) the normal or uniform flow depth,
(b) the critical depth, and
(c) the flow regime for a flow of 8.0 m³/s in a trapezoidal channel with a bottom width of 4.0 m and side slopes of 1 vertical and 2 horizontal. The bed slope is $1/1600$ and $n = 0.020$.

12-21. A flow of 30 L/s is conveyed in a 90° triangular channel (sides slope 45°) made of structural aluminum ($n = 0.010$) and at a slope of 0.010.
(a) What is the normal depth?
(b) What is the Froude number?
(c) What is the critical depth?

12-22. Repeat Prob. 12-21 for a flow rate of 50 L/s.

12-23. A channel has vertical side walls 1.20 m apart and a semicircular bottom. What is
(a) the normal or uniform flow depth,
(b) the critical depth, and
(c) the flow regime for a flow rate of 0.6 m³/s, a bed slope of $1/2500$, and $n = 0.017$?

12-24. Derive an expression relating critical depth and critical velocity for a triangular channel.

12-25. Show that for a nonrectangular channel the specific thrust function is

$$f = A\bar{y} + Q^2/Ag$$

where A is the channel cross section and $\bar{y}$ is the depth to the centroid of this section.

12-26. In Eq. (12.7b) there are three variables: E, q, *and* y.
(a) Show that this equation may be written as

$$\frac{E}{y_c} = \frac{1}{2}\left(\frac{y_c}{y}\right)^2 + \left(\frac{y}{y_c}\right)$$

Sketch a curve relating E/y_c as a function of y/y_c. The shape should be like a curve in Fig. 12-8.

(b) Show that Eq. (12.7b) may also be written as

$$\frac{1}{2}\left(\frac{q}{q_c}\right)^2 = \frac{3}{2}\left(\frac{y}{y_c}\right)^2 - \left(\frac{y}{y_c}\right)^3$$

Plot a curve relating y/y_c as a function of q/q_c.

(c) Show that Eq. (12.7b) may also be written as

$$\frac{E}{y} = \frac{1}{2}\left(\frac{q^2}{gy^3}\right) + 1$$

Plot a curve of E/y as a function of $q/\sqrt{g}\,y^{3/2}$.

12-27. Equation (12.7a) is plotted in Fig. 12-8 as E vs. y for various values of constant unit discharge q. If the specific energy E is constant, the unit discharge q is a function of depth y.

(a) Show that for flow at constant specific energy, $q = y\sqrt{2g(E - y)}$.

(b) Show that the flow is critical for maximum q, and that for flow at constant specific energy, $y_c = 2E/3$.

(c) Show that

$$\frac{q}{q_{max}} = \frac{3}{2}\frac{y}{E}\sqrt{3\left(1 - \frac{y}{E}\right)}$$

(d) Plot a dimensionless curve of y/E versus q/q_{max} for flow at constant specific energy.

12-28. Water flows 1.20 m deep at an average velocity of 2.40 m/s in a rectangular channel.

(a) What is the Froude number?

(b) What is the specific energy?

(c) What is the alternate depth?

(d) What is the specific thrust?

(e) What is the conjugate depth?

12-29. Water flows at a rate of 3.40 m³/s in a rectangular channel 2.40 m wide.

(a) What is the maximum depth for supercritical flow?

(b) What is the specific energy for a depth of 1.20 m?

(c) What is the alternate depth for part (b)?

(d) What is the conjugate depth for part (b)?

12-30. Water flows 2.40 m deep in a rectangular channel 3.60 m wide at a rate of 62 m³/s.

(a) What is the critical depth for this flow?

(b) Is the flow subcritical, critical, or supercritical?

(c) At what other depth could the water flow with the same specific energy?

(d) At what other depth could the water flow with the same specific thrust (conjugate depth)?

12-31. Water flows 2.5 m deep in a rectangular channel 4.0 m wide at a rate of 13.5 m^3/s. What is the
  (a) alternate depth and
  (b) conjugate depth for this flow?

12-32. A rectangular channel 5.50 m wide conveys 16.5 m^3/s at a depth of 1.5 m.
  (a) What is the critical depth?
  (b) Is the given flow subcritical or supercritical?
  (c) What is the Froude number of the flow?
  (d) What is the specific energy?
  (e) If Manning's $n = 0.012$, what slope would be required to maintain uniform flow?
  (f) Is this slope a mild or a steep slope?

12-33. Verify the correctness of the water surface profiles and the impossibility of other profiles for each flow situation shown in Fig. 12-15.

12-34. Water flows in a rectangular channel 6.00 m wide on a slope of 0.001 at a rate of 28 m^3/s. The channel is lined with unfinished concrete.
  (a) What is the normal or uniform flow depth?
  (b) A dam increases the depth to 3.60 m. What is the depth 1200 m upstream from the dam?

12-35. Repeat Example 12-6 for a bed slope of 0.002.

12-36. A trapezoidal channel of rough concrete has a bottom width of 3.60 m, a bed slope of 0.0025, and sides which slope 45° from the horizontal. The channel conveys 14.0 m^3/s. A dam increases the depth to 1.80 m.
  (a) What is the normal depth?
  (b) What is the depth 120 m upstream from the dam?
  (c) What is the depth 240 m upstream from the dam?

12-37. A steep rectangular channel 5.0 m wide drains a reservoir whose free-water surface is 2.70 m above the bottom of the channel where it meets the reservoir. What is the flow rate?

12-38. The normal depth for a wide steep spillway is 60 cm, and the critical depth is 120 cm. What type of water surface profile exists downstream of a location where the depth is 120 cm? Plot the profile for a 40-m distance.

12-39. A horizontal rectangular channel 1.50 m wide has $n = 0.020$. The water surface drops from 90 to 60 cm in a distance of 150 m. Estimate the flow rate.

12-40. A wide rectangular channel conveys 2.30 m^3/s per meter width. The upstream bed has a slope of 1/200 with $n = 0.017$, and then the bed slope changes to 1/3000 and $n = 0.020$. Determine the critical and normal depths, and describe the water surface profiles.

12-41. Water flows in a rectangular channel 6.0 m wide on a slope of 0.001 at a rate of 30 m^3/s. The channel is lined with unfinished concrete.

(a) What is the normal or uniform flow depth?

(b) A dam increases the depth to 4.00 m. What is the depth 1000 m upstream from the dam?

12-42. For a hydraulic jump in a rectangular channel, show that the Froude number upstream Fr_1 and the Froude number downstream Fr_2 are related by the equation

$$Fr_2^2 = \frac{8Fr_1^2}{(\sqrt{1 + 8Fr_1^2} - 1)^3}$$

12-43. By equating both unit discharge and specific thrust upstream and downstream of a hydraulic jump, show that

$$\frac{1 + 2Fr_1^2}{Fr_1^{4/3}} = \frac{1 + 2Fr_2^2}{Fr_2^{4/3}} = f(Fr)$$

Plot Fr vs. $f(Fr)$ and compare with Fig. 10-7, a similar plot for normal shocks in a gas.

12-44. Show that for a flow rate Q in a 90° V-notch open channel with the symmetrical axis vertical, the relationship between depths upstream and downstream of a hydraulic jump (y_1 and y_2, respectively) are expressed by

$$\frac{y_1^3}{3} + \frac{Q^2}{gy_1^2} = \frac{y_2^3}{3} + \frac{Q^2}{gy_2^2}$$

and that

$$r^4 + r^3 + r^2 - \tfrac{3}{2}Fr_1^2 r - \tfrac{3}{2}Fr_1^2 = 0$$

where $r = y_2/y_1$.

12-45. The water depth near the foot of a spillway jumps from 1.20 to 2.65 m. What is the power dissipation per meter width of flow in the hydraulic jump?

12-46. At a section in a wide channel the depth is 1.00 m and the velocity is 3.60 m/s. If a hydraulic jump occurs, would it be upstream or downstream of this section? Explain.

12-47. Water flows 1.20 m deep at an average velocity of 4.50 m/s in a rectangular channel 7.5 m wide. What downstream depth is necessary to form a hydraulic jump?

12-48. Water enters a long, wide rectangular channel at the foot of a spillway at a velocity of 7.30 m/s and a depth of 15 cm. The channel has a bed slope of 1/1600 and $n = 0.0165$.

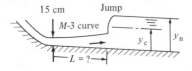

Prob. 12-48

(a) What is the normal depth for the channel? This is the depth downstream of the hydraulic jump which will form.

(b) How far from the foot of the spillway will the jump form?

12-49. Water in a wide rectangular channel flows at an average velocity V_1 and at a depth of y_1. A surge traveling upstream at a velocity V_s produces an increase in depth to y_2 and a velocity V_2 relative to a reference at rest.

(a) Write continuity and momentum equations for a steady-state flow system and show that

$$\frac{(V_1 + V_s)^2}{gy_1} = \frac{y_2}{2y_1}\left(1 + \frac{y_2}{y_1}\right)$$

(b) Show that if $y_2/y_1 > 1$, $(V_1 + V_s) > \sqrt{gy_1}$, indicating a Froude number relative to the surge front greater than unity.

(c) Show that as $y_2/y_1 \to 1$, $(V_1 + V_s) \to \sqrt{gy_1}$. Thus a small surge travels at a velocity $\sqrt{gy_1}$ relative to the water over which it moves.

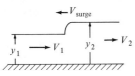

Prob. 12-49

12-50. Water in a wide river flows at an average velocity of 3.60 m/s at a mean depth of 3.60 m. A tidal bore produces a change in depth of 90 cm. At what velocity does the bore travel upstream?

12-51. A tidal bore travels upstream in a wide river at a speed of 2.75 m/s. The river depth upstream of the bore is 3.00 m, and downstream of the bore the river depth is 4.00 m. What is the average velocity of the flow in the river upstream of the bore?

12-52. Water flows from a sluice gate at a depth less than critical; it proceeds downstream on a horizontal channel which extends a great distance, then drops off suddenly. Assume critical flow at the drop-off (actually the flow is critical a very short distance upstream from the drop-off). Identify the water surface profiles, and describe how to determine the location of the hydraulic jump.

12-53. In Prob. 12-52 the depth at the contracted section below the sluice gate (y_2 in Fig. 12-11c) is 30 cm. The horizontal channel is 3.00 m wide and 300 m long, $n = 0.014$, and the unit discharge is $q = 3.00$ m³/s per meter width of channel. Locate the hydraulic jump.

12-54. Repeat Prob. 12-53 for a bed slope of 0.001.

12-55. Repeat Prob. 12-53 for a bed slope of 0.002.

12-56. Through the smooth transition between two rectangular channels the width increases from 3.00 to 3.60 m, but the water surface does not change in elevation. The upstream depth is 1.50 m. How much is the channel bed raised in the transition? Neglect losses.

12-57. Water flows at an average velocity of 3.00 m/s at a depth of 2.00 m in a rectangular channel. What is the change in water depth and the change in stage (water surface level) for
(a) a smooth rise of 0.15 m and
(b) a smooth drop of 0.15 m in the channel bed?

12-58. The bottom of a constant-width rectangular channel is raised gradually. The depth and velocity upstream of the raised bed are 3.00 m and 3.00 m/s, respectively.
(a) For no losses, what is the maximum height the bottom may be raised without affecting the flow depth upstream?
(b) For the condition of part (a), what is the horizontal component of force acting on the curved portion of the raised bed per unit width of channel? Neglect bed shear.

12-59. A rectangular channel 1.20 m wide conveys 1.15 m³/s uniformly at a depth of 90 cm. What is the widest streamlined obstruction which may be placed in the center of the channel and which will not cause a backwater? Increasing the width of the pier will increase the flow velocity adjacent to it until critical flow occurs.

12-60. A river 48 m wide has a horizontal reach where bridge piers restrict the width to 42 m. For a flow rate of 600 m³/s, describe the flow and calculate the depth upstream of the piers.

12-61. Give a verbal definition (use no equations) of the following terms as applied to open-channel flow:
(a) uniform flow;
(b) normal depth;
(c) hydraulic radius;
(d) hydraulic grade line;
(e) specific energy;
(f) mild slope;
(g) critical slope;
(h) steep slope;
(i) critical depth;
(j) hydraulic jump.

References

1. Hunter Rouse, T. T. Siaa, and S. Nagaratnam, "Turbulence Characteristics of the Hydraulic Jump," Hydraulics Division ASCE, Proceedings Paper 1528, Vol. 84, No. HY1 (Feb. 1928), pp. 1–30, and discussion by D. R. F. Harleman, Hydraulics Division ASCE, Proceedings Paper 1856, Vol. 84, No. HY6 (Nov. 1958), pp. 52–55.
2. E. A. Elevatorski, *Hydraulic Energy Dissipators* (New York: McGraw-Hill Book Company, Inc., 1959).

3. A. Thiruvengadam, "Hydraulic Jump in Circular Channels," *Water Power* (Dec. 1961), p. 496.

4. L. G. Straub, A. G. Anderson, and C. E. Bowers, "Importance of Inlet Design on Culvert Capacity," Technical Paper No. 13, Ser. B, University of Minnesota, St. Anthony Falls Hydraulic Laboratory, Aug. 1953; or "Culvert Hydraulics," Research Report 15-B, Highway Research Board, National Academy of Sciences—National Research Council, Washington, D.C., 1953, pp. 53–71.

The reader is referred to the following sources for additional information on open channel flow:

B. A. Bakhmeteff, *Hydraulics of Open Channels* (New York: McGraw-Hill Book Company, Inc., 1932).

V. T. Chow, *Open-Channel Hydraulics* (New York: McGraw-Hill Book Company, Inc., 1959).

F. M. Henderson, *Open Channel Flow* (New York: Macmillan, Inc., 1966).

R. H. J. Sellin, *Flow in Open Channels* (New York: Macmillan, Inc., 1969).

Chapter 13
Flow Measurements

It is both desirable and necessary in many instances to know the magnitude of various flow parameters such as speed or velocity, pressure, temperature, and volumetric, or mass flow rate, in a fluid system. These instances might include (1) manufacturing process control; (2) rating tests of equipment such as pumps, turbines, fans, blowers, propellers, and airfoils; (3) hydrological studies of rainfall and drainage in watershed areas and the apportioning and control of water in irrigation systems; (4) establishment of costs when fluids are sold to a customer, such as water, gas, steam for processing or heating, and liquid fuels; and (5) experimental work in research and development programs.

Fluid-flow measurement techniques are becoming more and more complex because of the greater demand for more detailed information and because of the need for remote measurements using telemetering techniques which involve electronic methods. Measurement systems consist of three component parts: a sensing device, a transmitting medium or system, and an indicating or recording device. This chapter generally will deal with only those components which involve fluids or fluid systems. A treatment of electrical or electronic components is not within the scope of this text.

Included in this chapter are discussions of velocity measurements and flow-rate measurements. Flow measurements might be considered a science of coefficients, because we generally apply the Bernoulli equation for liquid flow and the steady-flow energy equation for isentropic gas flow (in each instance a frictionless fluid is assumed) and then compare the actual behavior of a real (viscous) fluid with the ideal fluid by means of velocity or discharge coefficients.

13-1. Velocity Measurements

The velocity at a point or a number of points throughout a section in a fluid stream is often needed in order to establish the velocity profile. This velocity profile may be of interest in the fundamental studies of boundary layers or wakes, or it may be necessary in order to obtain the average velocity throughout the section from an integration of the velocity profile in order to determine the flow rate. A point velocity is almost impossible to measure, since any sensing device occupies a finite region. However, if the area of flow occupied by the sensing element is very small compared with the total area of the flow stream, we may consider the measured velocity to be essentially a point velocity. For example, a current meter sweeping out an 20-cm circle in a river would measure essentially a point velocity, but in a 30-cm pipe it would not. A 30-cm anemometer measuring wind speeds would indicate essentially a point velocity in the atmosphere, whereas in a 60-cm duct it would not. A 6-mm tube facing upstream in a 15-cm pipe would indicate essentially a point velocity in the central flow area of the pipe, but measurements near the pipe wall, where the velocity gradient is large, would require a tube such as a small hypodermic needle to measure point velocities.

It is essential that the presence of the sensing device in the flow stream not affect the flow being measured, and this requirement also limits the size of instrument which may be used satisfactorily.

Velocities are commonly measured indirectly by measuring the difference between the stagnation and free-stream pressures, by the speed of rotating vaned wheels, by the cooling effect on a thin cylinder in crossflow, and by the angle of oblique shocks in a supersonic gas. In some instances velocities are measured directly by determining the distance traveled by a group of fluid particles in a measured time interval.

PITOT TUBE

If a bent open tube is placed to face upstream in an open liquid stream, the liquid will rise in the tube a height h (capillary effects are assumed absent) as in Fig. 13-1a. The Bernoulli equation written from a point upstream from the submerged end of the tube and the end of the tube itself is

$$\frac{\rho V_1^2}{2} + p_1 = p_0 \tag{13.1}$$

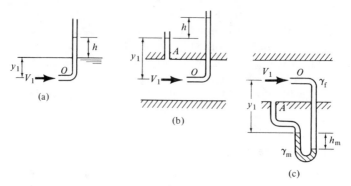

Fig. 13-1. Pitot tube in (a) an open liquid stream, (b) a pipe, and (c) a pipe, connected to a manometer.

since a stagnation condition exists within the tube. Since $p_1 = \gamma y_1$ and $p_0 = \gamma(y_1 + h)$, Eq. (13.1) may be solved for the stream velocity.

$$V_1 = \sqrt{\frac{2\gamma h}{\rho}} = \sqrt{2gh} \tag{13.2}$$

If the stream velocity at a point in a pipe is to be measured (Fig. 13-1b), the same equation is obtained. If the static pressure is high, y_1 is large and reading of the piezometric heights y_1 and $y_1 + h$ may be difficult (y_1 would be about 61 m for water at a pressure of 600 kPa in a pipe).[†] The tubes in Fig. 13-1b may be connected to a simple manometer as in Fig. 13-1c to result in a more convenient system. The difference between stagnation and static (free-stream) pressure is

$$p_0 - p_1 = h_m(\gamma_m - \gamma_f)$$

where h_m is the manometer deflection, and γ_m and γ_f are the specific weights of the manometer and flowing fluids, respectively. The stream velocity becomes

$$V_1 = \sqrt{2gh_m\frac{\gamma_m - \gamma_f}{\gamma_f}} = \sqrt{\frac{2(p_0 - p_s)}{\rho}} \tag{13.3}$$

The total head or impact tube and its opening should be as small as possible in order that essentially a point velocity be measured. If Re = $ur/v < 50$ (r is the radius of the tube opening), laminar flow effects around the tube will cause the indicated stagnation pressure to be in error [1].

[†] The shape of piezometer taps, such as at A in Fig. 13-1b, is critical. They should be made in a smooth region, far removed from internal projections such as flange gaskets, flanges themselves, wavy surfaces, and so forth. No burrs resulting from drilling should project into the flow stream. Good and poor piezometer taps are shown in Fig. 13-2.

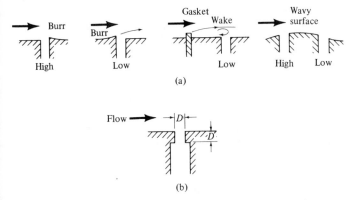

Fig. 13-2. Piezometer tap geometries. (a) Poor taps. Measured pressures are referred to actual pressure. (b) Good tap.

The system shown in Fig. 13-1c may be incorporated into a single instrument known as a *combined pitot-static tube*. This tube consists of two concentric cylinders bent into an L shape and with various head forms, such that the inner cylinder is at the stagnation pressure and the annular region between cylinders is at the free-stream pressure. Sketches of two common head forms for a combined pitot tube are shown in Fig. 13-3. Different head forms have been used to improve accuracy for various angles of yaw (angle between free-stream velocity vector and pitot tube). The indicated static pressure is subject to some uncertainty because the static holes may be exposed to a pressure slightly different from that of the free stream. If the flow is turbulent, the high-frequency velocity and pressure

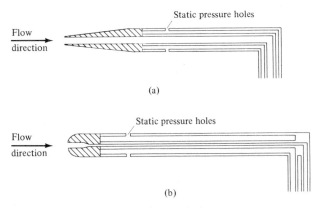

Fig. 13-3. Combined pitot tubes. (a) Brabbee's design. (b) Prandtl's design. Prandtl's design is accurate over a greater range of yaw than Brabbee's design.

variations have the effect of indicating a velocity higher than the time-average velocity, and it is common to use a coefficient C_v, so that

$$V_1 = C_v \sqrt{2gh_m \frac{\gamma_m - \gamma_f}{\gamma_f}}$$

Unless an instrument is carefully calibrated, C_v is usually assumed to be unity. One calibration method consists of towing a pitot tube at a measurable speed through still water or air, in which case the turbulence level of the fluid would be zero, though the instrument may not produce exactly the same manometer deflection in a moving turbulent stream under otherwise identical flow conditions.

It should be pointed out that velocity fluctuations are manifested as pressure fluctuations, and these are generally damped out (depending on their frequency) if a liquid manometer system is used with the pitot or impact tube. A pressure transducer mounted in the instrument and connected to an electronic circuit is necessary if fluctuating pressures are to be measured with any degree of accuracy.

In a gas flow, compressibility must be considered, and the appropriate equation for *subsonic* flow is Eq. (10.9a). This equation may be written as

$$V_1 = \sqrt{\frac{2kRT_1}{k-1}\left[\left(\frac{p_0}{p_1}\right)^{(k-1)/k} - 1\right]} \tag{13.5}$$

or, in terms of the Mach number being measured,

$$M_1 = \sqrt{\frac{2}{k-1}\left[\left(\frac{p_0}{p_1}\right)^{(k-1)/k} - 1\right]} \tag{13.6}$$

For these high-velocity gas flows, the free-stream static pressure p_1 must be known, since it is the value $p_0 - p_1$ which is measured with the manometer. From these values the ratio p_0/p_1 may be obtained.

For a *supersonic* gas stream, the pitot tube equation is given by Eq. (10.56) as

$$\frac{p_0}{p_1} = \left(\frac{k+1}{2}M_1^2\right)^{k/(k-1)}\left(\frac{2k}{k+1}M_1^2 - \frac{k-1}{k+1}\right)^{1/(1-k)} \tag{13.7}$$

and the Mach number corresponding to the measured value of p_0/p_1 may be found from gas tables (Table A-10 in Appendix V, for example). The conventional static tube will not indicate the free-stream pressure, since it is in a subsonic region behind the shock which forms forward of the tube. The static pressure of the free stream must be obtained by other means.

Example 13-1

What water velocity is indicated for a deflection of 9.0 cm on a mercury manometer connected to a combined pitot tube?

SOLUTION

From Eq. (13.3),

$$V_1 = \sqrt{(2)(9.81)(0.090)(13.56 - 1)} = 4.71 \text{ m/s}$$

Example 13-2

A pitot tube in air at a temperature of $-11°C$ at 4000 m ($p = 61.7$ kPa abs) indicates a difference between stagnation and static pressure of 32 kPa. What is the velocity and the Mach number being measured?

SOLUTION

$$\frac{p_0}{p_1} = \frac{61.7 + 32}{61.7} = 1.519 < \frac{1}{0.528}$$

so the flow is subsonic. From Eq. (13.6) or Table A-9 in Appendix V,

$$M_1 = 0.796$$

and

$$V_1 = M_1 c_1 = 0.796\sqrt{(1.4)(287)(262)} = 258 \text{ m/s}$$

A pitot tube may be used in a mixture of gas bubbles and liquid (such as air and water) for gas concentration C no greater than about 0.6 by volume if the concentration is known. The pitot tube, connecting tubing, and manometers must be free of bubbles. The velocity being measured is the mean velocity of the mixture and is

$$V_{\text{mixture}} = \sqrt{\frac{2(p_0 - p_1)}{(1 - C)\rho_{\text{liquid}}}} = \sqrt{\frac{2gh_m}{(1 - C)}\left(\frac{\gamma_m}{\gamma_{\text{liquid}}} - 1\right)} \tag{13.8}$$

where h_m is the manometer deflection and γ_m and γ_{liquid} are the specific weights of the manometer fluid and the flowing liquid, respectively.

Errors due to compressible effects are negligible for high concentrations at low velocities or low concentrations at higher velocities. Equation (13.8) is within the indicated accuracy for conditions as shown in Fig. 13-4. If the gas fraction in a gas–liquid mixture is assumed to flow isentropically to a stagnation point, the steady-flow energy equation applied to the mixture may be shown to result in

$$\frac{V_1^2}{2} = \frac{p_0 - p_1}{\rho_w(1 - C)} + \frac{C}{1 - C}\left(\frac{p_1}{\rho_w}\right)\left[\frac{k}{k - 1}\left(\frac{p_0}{p_1}\right)^{(k-1)/k} - \frac{1}{k - 1} - \left(\frac{p_0}{p_1}\right)\right] \tag{13.9}$$

The last term expresses the effect of gas compressibility, and the error in neglecting it [as in Eq. (13.8)] is the ratio of the second to the first term on

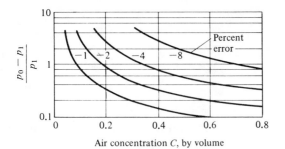

Air concentration C, by volume

Fig. 13-4. Effect of neglecting air compressibility on measurement of velocity of an air–water mixture with a combined pitot tube.

the right-hand side of Eq. (13.9). The expression,

error =

$$\sqrt{1 + \frac{C}{(p_0 - p_1)/p_1}\left[\frac{k}{k-1}\left(\frac{p_0 - p_1}{p_1} + 1\right)^{(k-1)/k} - \frac{1}{k-1} - \left(\frac{p_0 - p_1}{p_1} + 1\right)\right]} - 1$$

(13.10)

is plotted in Fig. 13-4.

PITOT CYLINDER

A small cylinder mounted in a pipe so that it is free to move along the diameter of the pipe and which has a small hole at a forward stagnation point (leading edge of the cylinder) may be used in conjunction with a wall piezometer tap to measure the difference between stagnation and static pressure. A cylinder of this type is known as a pitot cylinder (Fig. 13-5). If velocity measurements are made at a number of points, the velocity profile may be determined and the flow rate calculated.

Equation (13.4), with $C_v = 1.00$, may be used for the pitot cylinder as well as the pitot tube. The flow area obstructed by the cylinder should be kept small, preferably less than about 2 or 3 percent, in order that choking effects may be minimized. The piezometer tap should not be in the plane of the pitot cylinder for the same reason. The static pressure in this plane will be measured too low if the piezometer tap is in the choked section.

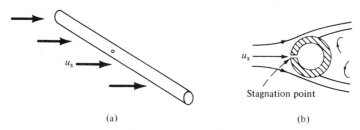

(a)   (b)

Fig. 13-5. Pitot cylinder.

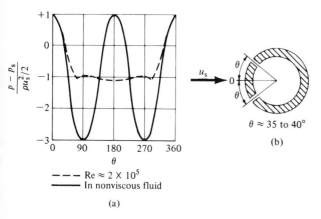

$\theta \approx 35$ to $40°$

(b)

--- Re $\approx 2 \times 10^5$
—— In nonviscous fluid

(a)

Fig. 13-6.   (a) Pressure field around a cylinder. (b) Cylinder used as a direction indicator.

The pressure field around a cylinder, shown in Fig. 13-6a, is such that if the pressure tap or slots in it are placed 35° to 40° above and below the stagnation point and are connected to opposite sides of a differential manometer (or other pressure differential indicator), the cylinder then becomes a sensitive device for determining flow direction. When the manometer deflection is zero, the point 0 faces directly upstream. A slight change in direction of the velocity vector u_s unbalances the manometer, since the pressure at one tap rises and that at the other drops. This device is used in aircraft to indicate rate of climb and in ducts to detect spiral flows.

In order to obtain an average velocity from measurements made at various points throughout a flow cross section, it is often convenient to take the measurements at the centroids of equal subareas. The measurements may then be averaged arithmetically to obtain the average flow velocity. The accuracy of the result depends, of course, on taking sufficient point velocities. In axisymmetric flow in round pipes, point velocities should be taken midway between equal increments of $\Delta(r^2)$. If ten readings are taken, they should be made at distances from the pipe axis equal to 0.05, 0.15, 0.25, ..., and 0.95 R^2, and velocities measured at these points may be averaged directly. If n readings are taken to represent the individual velocities u_i of n subareas, the average velocity V is

$$V = \frac{1}{n} \sum_{i=1}^{i=n} u_i \tag{13.11}$$

A ten-point measurement in laminar flow in a circular tube theoretically gives an exact average velocity based on Eq. (13.11). Similar measurements for turbulent flow give a theoretical error of 0.25 percent for a power-law velocity profile $u/u_m = (y/R)^{1/7}$, and the theoretical error for a logarithmic velocity profile $u/u_m = 1 - 2.5(u/u_m)$, which is equivalent to $V/u_m = 1 - 3.75 \, u/u_m$ [Eq. (9.21)], is a function of the pipe friction factor.

The theoretical error is

0.24 percent for $f = 0.01$
0.33 percent for $f = 0.02$
0.44 percent for $f = 0.04$

Thus a ten-point measurement used with Eq. (13.11) gives an accuracy generally well within the accuracy of measuring point velocities.

If the velocity profile itself is obtained from measurements, the average velocity may be obtained by a graphical integration of the area under the velocity vs. r^2 curve or by the use of Simpson's rule or the trapezoidal rule.

A reasonably good, though necessarily rough, estimate of the average velocity for fully developed flow in a pipe may be made by a measurement of the centerline velocity at a section (preceded by a long section of straight pipe). An estimate of the ratio of maximum to average velocity may be made from Eq. (9.22), but it is at best a close approximation, especially if the pipe roughness is not known or if fully developed flow cannot be assured.

LENGTH–TIME MEASUREMENTS

The distance traversed by a group of fluid particles in a given time interval is a measure of the fluid velocity, averaged over the measured distance. If a salt solution is injected into a stream, its transit time between two sets of electrodes may be measured because of the change in fluid resistivity as the

Fig. 13-7. Surface flow pattern for water approaching spillway of model of a dam. (Courtesy of Dr. Lorenz G. Straub.)

salt slug passes the electrode pairs. This method has been used in large turbine penstocks over rather large distances and in aerated flows where minute cloudlets of salt solution are injected at a rate of 20 injections per second, and the transit time between electrodes 76 mm apart in the flow path are recorded by electronic means [2].

Turbulent flow patterns in liquids may be observed by adding nonsoluble droplets of a fluid whose density is the same as that of the flowing liquid and taking photographs with a known time exposure. If paper punchings (confetti) are spread over the surface of a liquid stream, the surface flow pattern may be obtained from a photograph of known time exposure. Both the magnitude and direction of motion of the surface may be obtained. A typical photograph of this type for flow approaching the spillway of a dam model is shown in Fig. 13-7.

CURRENT METERS AND ANEMOMETERS

Cones on current meters used in water and hemispheres on anemometers used in air are attached to radial arms mounted on a central shaft as shown in Fig. 13-8.

A flow stream in the plane of rotation from any direction will create drag on all cones or cups. The drag on cone A (and cup A) is greater than that on cone B (and cup B), so that the net torque rotates the assemblies as shown. The speed of rotation is generally indicated by means of an electrical contact made once each revolution, and the number of contacts per unit time interval is a direct measure of the average speed of the fluid in the region traversed by the meters. A precalibration by towing the instruments through still water (or air) enables the speed to be determined. The direction of flow is not obtained.

A second type of current meter or anemometer consists of windmill-type vanes mounted in a support so that the fluid flow is parallel to the axis of rotation. Measuring techniques and calibration are similar to those for the cone or cup type of instrument.

HOT-WIRE ANEMOMETERS

A hot-wire anemometer makes use of the convection cooling on a heated cylinder held normal to a gas stream. This cooling is a function of the gas

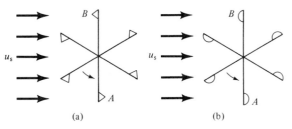

Fig. 13-8.   (a) Current meter used in water. (b) Anemometer used in air.

and wire temperatures and the free-stream velocity. Wires of 0.01 to 0.10 mm in diameter and about 1.5 mm long are held by a forked tube and exposed to the fluid stream. Two methods of measurement are used:

1. The wire resistance is kept constant by adjusting the current flow through it, and the velocity is determined by measuring the current and using a calibration of the instrument.
2. The current flow through the wire is kept constant, and the change in wire resistance from convection cooling is measured in terms of the voltage drop across it. Fluctuations in velocity may be detected and recorded on an oscillograph by suitable circuits.

Hot-wire anemometers have been used for measuring velocity profiles where the velocity gradients are large. This measurement is possible because of the small size of the sensing device. They are also used in measuring turbulence intensities in gas flows.

OBLIQUE SHOCKS IN A GAS

If an infinitesimal disturbance travels in a gas at a speed V, greater than the speed of a sound wave c, a Mach cone will be generated by the wave fronts issuing from the point disturbance P (Fig. 13-9). The spherical sound waves emanating from P as it moves toward the right are indicated. In a time Δt the wave front travels a distance $c \, \Delta t$, while the source P travels a distance $V \, \Delta t$. Thus the sine of the half-angle of the Mach cone is

$$\sin \beta = \frac{c \, \Delta t}{V \, \Delta t} = \frac{1}{M} \tag{13.12}$$

Since a real disturbance is finite, Eq. (13.12) is not exact and the Mach number determined from it will be less than the actual value.

If the point source is at rest with a supersonic gas flowing past it, a Mach cone similar to that in Fig. 13-9 will be generated (it will be opposite in direction to that shown if the gas flows to the right).

It is common practice to place either a wedge (two-dimensional flow) or a cone (three-dimensional flow) at zero incidence in a supersonic gas stream, and the Mach number of the free stream may then be obtained by measuring the wedge half-angle θ and the oblique shock-angle β (Fig. 13-10).

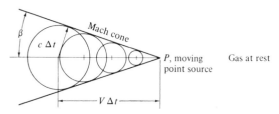

Fig. 13-9.   Mach cone.

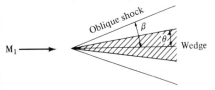

Fig. 13-10.   Mach number measured with wedge.

The wedge or cone half-angle θ must be small enough so that the bow wave remains attached to the wedge or cone tip. For this condition, the continuity and momentum equations together with the system geometry may be written to show that the free-stream Mach number M_1 for a bow wave on a wedge (two-dimensional flow) may be determined from the equation

$$\frac{1}{M_1^2} = \sin^2 \beta - \left(\frac{k+1}{2}\right) \frac{\sin \beta \sin \theta}{\cos(\beta - \theta)}$$

[10.67]

Relationships for a cone are more complicated than for a wedge. However, Eq. (13.12) is a better approximation for an attached shock on a cone than on a wedge. It is difficult to determine the free-stream Mach number from a detached bow wave.

Shock waves are photographed by means of interferometers (these detect density changes across the shock), schlieren systems (these detect density gradients), and shadowgraph systems (these detect the gradient of the density gradient).

Example 13-3

A Mach angle β of 20° on a wedge of half-angle 8° is measured on a schlieren photograph of a two-dimensional wedge in the test section of a supersonic wind tunnel. What is the Mach number of the air flow?

SOLUTION
From Eq. [10-67]

$$\frac{1}{M_1^2} = (0.342)^2 - \frac{(1.4+1)}{2} \frac{(0.342)(0.139)}{(0.978)} = 0.0587$$

Thus, $M_1 = 4.13$.
Equation (13.12) gives $M_1 = 1/0.342 = 2.92$, which is not correct since a wedge rather than a point or line disturbance is used.

$\blacksquare\blacksquare\blacksquare$

13-2. Liquid Flow Rates in Pipes

Liquid flow rates in pipes may be measured by means of a number of devices. Many, such as disk meters of the type used in homes for metering water and the rotameter (which consists of a float in a vertical diverging tube), are direct reading and will not be treated here.

A constriction or an elbow in a pipeline will produce measurable piezometric pressure differences along the pipe walls, and these pressure differences may be used as a means of determining the flow rate. Two methods of approach may be used:

1. The Bernoulli equation (one-dimensional flow with no losses in the form of energy dissipation) and the continuity equation may be written between a section in the pipe and the constricted section, and an expression for the flow rate may be obtained by combining them. An estimate of the ratio of the actual flow rate to the calculated flow rate may be made from previous (published) experiments and applied. From this an estimate of the flow rate in the given system may be obtained.

2. The flow meter may be calibrated, either by the manufacturer or by the user before or after installation. The same fluid and the same range of flows as in the actual installation should be used for the calibration.

Actual calibration (the second method) is necessary for precise measurements, although good estimates (often within 1 or 2 percent) may be possible by using the first method. A given meter may not be exactly geometrically or dynamically similar to one which has been previously calibrated, and thus the flow coefficients may not be the same.

ANALYSIS OF VENTURI, NOZZLE, AND ORIFICE METERS

The Bernoulli and continuity equations applied between sections 1 and 2 of Fig. 13-11 are

$$\frac{V_1^2}{2g} + \frac{p_1}{\gamma} + z_1 = \frac{V_2^2}{2g} + \frac{p_2}{\gamma} + z_2$$

and

$$V_1 A_1 = V_2 A_2$$

Solving for V_2 gives

$$V_2 = \frac{1}{\sqrt{1 - (A_2/A_1)^2}} \sqrt{2g\left[\left(\frac{p_1}{\gamma} + z_1\right) - \left(\frac{p_2}{\gamma} + z_2\right)\right]}$$

where the term in brackets inside the radical represents the change in piezometric head Δh between sections 1 and 2. This change in piezometric head may be measured by means of a differential manometer, for which the deflection is a direct measure of Δh regardless of the inclination of the meter axis (it may be horizontal, vertical, or inclined). The ideal flow rate may be

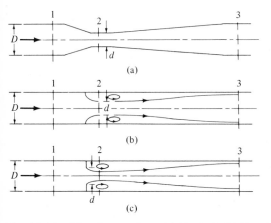

Fig. 13-11. Pipe flow meters. (a) Venturi. (b) Nozzle. (c) Concentric circular orifice.

expressed as

$$Q_{ideal} = \frac{A_2\sqrt{2g\,\Delta h}}{\sqrt{1 - (A_2/A_1)^2}} \qquad (13.13)$$

This equation cannot be exactly correct, because the steady-flow energy equation for liquids should be written as

$$\alpha_1 \frac{V_1^2}{2g} + \frac{p_1}{\gamma} + z_1 = \alpha_2 \frac{V_2^2}{2g} + \frac{p_2}{\gamma} + z_2 + h_L$$

The terms neglected in the Bernoulli equation (α_1, α_2, and h_L) are affected by viscous effects and the boundary roughness.

A flow coefficient K applied to Eq. (13.13) gives a simple form for the actual flow rate Q:

$$Q = K\left(\frac{\pi d^2}{4}\right)\sqrt{2g\,\Delta h} \qquad (13.14)$$

where d is the diameter of the meter throat. This is the diameter at section 2 for the venturi and the nozzle in Fig. 13-11, and at the orifice plate upstream of section 2 for the orifice meter. The value of K is a function of the following:

1. The type of meter.
2. The ratio of meter throat diameter to pipe diameter, d/D. For the venturi and nozzle it includes the so-called approach velocity factor $\sqrt{1 - (A_2/A_1)^2}$. For the orifice it includes this approach velocity factor and, in addition, the contraction of the jet from the orifice plate to the *vena contracta* at section 2 (Fig. 13-11).
3. The viscous effects and the pipe roughness contained in the kinetic energy correction factors α_1 and α_2 and the head loss term h_L, all of

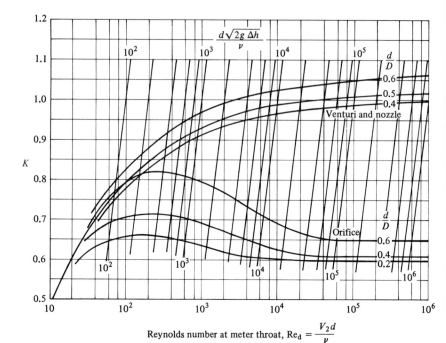

Fig. 13-12. Approximate flow coefficients for pipe meters.

which were neglected in the Bernoulli equation. These effects may be expressed in terms of the Reynolds number at the meter throat (in some instances the Reynolds number for the pipe is used).

Thus we may write

$$K = K\left(\text{meter shape, } \frac{d}{D}, \text{Re}_d\right)$$

Typical values of the flow coefficients are shown in Fig. 13-12. From Eq. 13-14,

$$K = \frac{Q}{\dfrac{\pi d^2}{4}\sqrt{2g\,\Delta h}}$$

and the Reynolds number may be written as

$$\text{Re}_d = \frac{Vd}{\nu} = \frac{Qd}{\dfrac{\pi d^2}{4}\nu} = K\frac{d\sqrt{2g\,\Delta h}}{\nu} \tag{13.15}$$

A family of lines representing constant values of $d\sqrt{2g\,\Delta h}/\nu$ is superimposed on the K vs. Re_d graph of Fig. 13-12. Thus, for a given piezometric head change Δh across a meter of throat diameter d for a fluid of viscosity v the value of $d\sqrt{2g\,\Delta h}/v$ may be calculated, and the intersection of this line with the appropriate meter curve gives the value of K for the meter. The flow rate may then be estimated from Eq. (13.14).

In these meters the throat pressures should not be allowed to become so low that cavitation occurs. In addition, low pressures may cause leakage of air into the manometer lines and cause errors, whereas very slight leakage of liquid out of the lines for high-pressure installations would cause no measurable errors.

Example 13-4

A mercury manometer connected across a 100-mm orifice in a 250-mm pipe in which water flows at 15°C has a deflection of 0.15 m. Estimate the flow rate.

SOLUTION

From Appendix II, $s_m = 13.558$ and $v_w = 1.14 \times 10^{-6}$ m²/s. Then

$$\Delta h = 0.15(s_m - 1) = (0.15)(12.558) = 1.884 \quad m$$

$$d\sqrt{2g\,\Delta h}/v = 0.1\sqrt{(2)(9.807)(1.884)}/1.14 \times 10^{-6} = 5.34 \times 10^5$$

From Fig. 13-12, $K = 0.61$; and from Eq. (13-14),

$$Q = (0.61)(\pi/4)(0.1)^2(6.08) = 0.0291 \quad m^3/s$$

For a venturi meter, $K = 1.00$ and $Q = 0.0477$ m³/s.

■

The total piezometric pressure drop or total head loss across the entire meter (from sections 1 to 3 in Fig. 13-11) is greatest for the orifice and least for the venturi for the same flow rate and meter size. The loss from sections 2 to 3 for the orifice is essentially that of a sudden enlargement from the *vena contracta* to the pipe. For the nozzle, it is also essentially that of a sudden enlargement, but from the lower throat velocity (compared with that for the orifice) to the pipe. For the venturi, it is essentially that of a gradual enlargement (a conical diffuser of about 6° to 7° is generally used).

It should be reemphasized that these meters should be calibrated carefully for precise flow-rate measurements.

ELBOW METER

Flow through an elbow in a pipeline results in a higher pressure at the outer wall surface than at the inner wall surface. This pressure difference is a function of the flow rate. The piezometric or pressure-head difference measured by means of suitable piezometer taps may be indicated on a differential manometer and used as an indication of flow rate. The elbow meter is

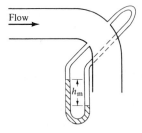

Fig. 13-13. Elbow meter.

inexpensive, since it is already a part of the piping system, and when carefully calibrated it is an accurate meter. A typical calibration consists of a logarithmic plot of manometer deflection vs. volumetric flow rate. A typical installation is shown in Fig. 13-13.

13-3. Liquid Flow Rates in Open Tanks or Open Channels

Flow from open tanks or in open channels is due to gravity, and changes in velocity along a free surface are produced as a result of changes in potential head.

As for pipe meters, estimates of flow rates may be made by using flow coefficients applied to the results of employing the Bernoulli and continuity equations, but precise measurements must be based on careful calibration of the meter.

OPEN TANKS

Flow of a liquid from a circular orifice at the side of an open tank is estimated by writing the Bernoulli equation from a point on the free surface to the contracted jet, where the streamlines are parallel (from point 1 to point 2 in Fig. 13-14). The velocity at 1 is essentially zero, and the pressures at 1 and 2 are both atmospheric. Thus the Bernoulli equation is

$$z_1 = \frac{V_2^2}{2g} + z_2$$

so that

$$V_2 = \sqrt{2g \, \Delta h}$$

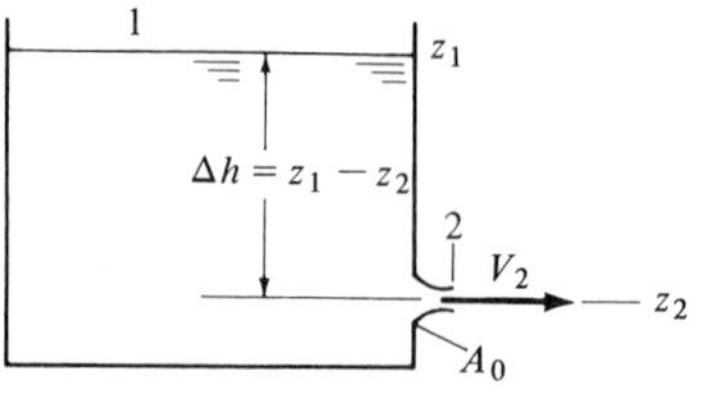

Fig. 13-14. Flow through a free orifice.

and

$$Q_{\text{ideal}} = A_2\sqrt{2g\,\Delta h} \tag{13.16}$$

The steady-flow energy equation contains a term h_L which is not included in the Bernoulli equation. The actual velocity is

$$V_{2\text{ actual}} = \sqrt{2g(\Delta h - h_L)}$$

A simple expression for the actual flow rate is

$$Q = KA_0\sqrt{2g\,\Delta h} \tag{13.17}$$

where A_0 is the area of the orifice hole; and K is a flow coefficient, which depends on the magnitude of h_L, which in turn is affected by the roughness of the inside surface of the tank near the orifice hole and by the flow rate which depends on Δh, and the contraction of the jet from the orifice to section 2.

For an ideal fluid, for a small hole in a large tank, Table 6-2 indicates that

$$\frac{A_2}{A_0} = \frac{\pi}{\pi + 2} = 0.611 \tag{13.18}$$

and for a real fluid the contraction depends on the sharpness of the orifice hole as well as the tank surface roughness just inside the orifice hole. A typical value of K for a sharp-edged orifice is about 0.62.

WEIRS

A weir is a partial obstruction in an open channel over which the liquid accelerates with a free liquid surface. Flow over a sharp-crested weir was illustrated in Fig. 6-4 by means of a flow net. The Bernoulli equation written for a suppressed weir (it extends across the full width of a channel) along a streamline (see Fig. 13-15) is

$$\frac{V_1^2}{2g} + (H + Z) = \frac{V_2^2}{2g} + (H + Z - h)$$

and we get

$$V_2 = \sqrt{2g(h + V_1^2/2g)}$$

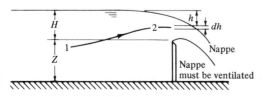

Fig. 13-15. Sharp-crested suppressed weir.

Assumptions made in this derivation include (1) no losses and (2) parallel streamlines without contraction over the weir crest. The flow rate through an element of width b and height dh is

$$dQ = V_2 b\, dh$$
$$= b\sqrt{2g}\sqrt{(h + V_1^2/2g)}\, dh$$

Integration from $h = 0$ to $h = H$ gives

$$Q_{ideal} = \frac{2}{3} b\sqrt{2g}\left[\left(H + \frac{V_1^2}{2g}\right)^{3/2} - \left(\frac{V_1^2}{2g}\right)^{3/2}\right] C_c \tag{13.19}$$

The coefficient of contraction C_c is included because the jet over the weir is less than H, actually $C_c H$.

As before, a simple expression may be written as

$$Q = K\tfrac{2}{3}b\sqrt{2g}\,H^{3/2} \tag{13.20}$$

where K depends on (1) the so-called approach velocity head $V_1^2/2g$, which in turn depends on the ratio of weir head H to weir crest height Z as well as H itself, and (2) the contraction of the streamlines just beyond the weir crest.

Numerous values for K (or a similar coefficient) have been published in hydraulic literature, and a rational analysis by von Mises (see Rouse [3]) indicates K to have the form

$$K = 0.611 + 0.075\frac{H}{Z} \tag{13.21}$$

Example 13-5

Estimate the flow rate over a sharp-crested weir with a ventilated nappe. The weir is a suppressed weir 3.60 m wide, the weir crest is 1.80 m high, and the head is 36 cm.

SOLUTION
From Eq. (13.21),

$$K = 0.611 + 0.075\frac{(0.36)}{(1.80)} = 0.626$$

Thus, from Eq. (13.20),

$$Q = (0.626)(\tfrac{2}{3})(3.60)\sqrt{(2)(9.81)}(0.36)^{1.5} = 1.44 \quad \text{m}^3/\text{s}$$

A weir may be narrower than the channel in which it is installed and is then called a weir with side contractions, or a contracted weir. An empirical equation of Gourley and Crimp for a contracted weir of the type shown in

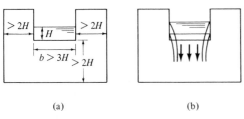

Fig. 13-16.   Contracted weir. (a) Proportions. (b) Jet contraction.

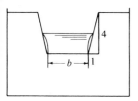

Fig. 13-17.   Cipolletti weir.

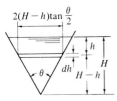

Fig. 13-18.   V-notch weir.

Fig. 13-16 is, in SI units:

$$Q = 1.69b^{1.02}H^{1.47} \quad \text{m}^3/\text{s} \tag{13.22}$$

A Cipolletti weir has sides sloping in order to avoid the varying contraction of the jet with head, as shown in Fig. 13-16b. This weir is shown in Fig. 13-17. An empirical equation for a Cipolletti weir is, in SI units,

$$Q = 1.86bH^{1.5} \quad \text{m}^3/\text{s} \tag{13.23}$$

A V-notch weir is useful for increased accuracy at low flow rates and for higher flow rates as well. From Fig. 13-18, the flow through an element of area dh high and $2(H - h) \tan \theta/2$ wide is

$$dQ = 2\sqrt{2gh}(H - h) \tan \frac{\theta}{2} dh$$

Integration from $h = 0$ to $h = H$ gives

$$Q_{\text{ideal}} = \frac{8}{15}\sqrt{2g} \tan \frac{\theta}{2} H^{5/2}$$

if the velocity of approach is neglected. Applying a flow coefficient K, which depends on the geometry of the system and the roughness of the upstream face of the weir plate, gives

$$Q = K\left(\frac{8}{15}\right)\sqrt{2g}\,\tan\frac{\theta}{2}\,H^{5/2} \tag{13.24}$$

Typical values of K vary from about 0.58 to 0.62 and decrease with H for values of H less than about 6 cm. A typical flow equation for a 90° V-notch weir is, in SI units,

$$Q = 1.40H^{2.5} \quad \mathrm{m^3/s} \tag{13.25}$$

CRITICAL FLOW SECTIONS

Whenever gradually varying flow in an open channel is critical (at a Froude number of 1.0), the flow rate may be determined from Eq. (12.9a) to be

$$q = V_c y_c = \sqrt{gy_c^3} \tag{13.26a}$$

and the total flow rate for a channel of width b is

$$Q = b\sqrt{gy_c^3} \tag{13.26b}$$

It is usually difficult, however, to measure or estimate the critical depth y_c in order to obtain the flow rate Q.

SLUICE GATE

Flow through a sluice gate (Fig. 13-19) is similar to flow through the upper half of a two-dimensional slot or orifice, and the streamline pattern or flow net was shown in Fig. 6-5. The theoretical contraction of the flow is given in Table 6-2 for $\theta = 90°$.

The Bernoulli equation and continuity equation written for one-dimensional flow between sections 1 and 2 of Fig. 13-19 are

$$\frac{V_1^2}{2g} + y_1 = \frac{V_2^2}{2g} + y_2$$

and

$$V_1 y_1 = V_2 y_2$$

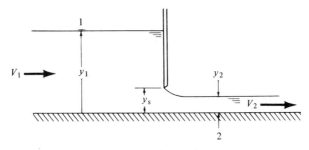

Fig. 13-19. Sluice gate.

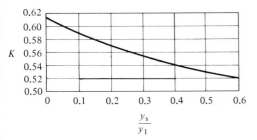

Fig. 13-20. Approximate flow coefficients for a sluice gate. (Source: Reproduced, with permission, from Hunter Rouse, *Elementary Mechanics of Fluids*. Copyright, 1946, John Wiley & Sons, Inc.)

so that

$$V_2 = \frac{1}{\sqrt{1 - (y_2/y_1)^2}} \sqrt{2g(y_1 - y_2)}$$

and for a width b,

$$Q_{ideal} = \frac{by_2}{\sqrt{1 - (y_2/y_1)^2}} \sqrt{2g(y_1 - y_2)} \tag{13.27}$$

A simple form of discharge equation is obtained by introducing the flow coefficient K, so that

$$Q = Kby_s\sqrt{2gy_1} \tag{13.28}$$

where K depends on the system geometry (y_1, y_2, and y_s) and is approximately as shown in Fig. 13-20 as a function of y_s/y_1.

Example 13-6

Estimate the flow rate under a sluice gate 3.0 m wide. The upstream depth is 5.50 m and the gate opening is 60 cm.

SOLUTION

$y_s/y_1 = 0.60/5.50 = 0.11$, and from Fig. 13-20, $K = 0.585$. Thus

$$Q = (0.585)(3.0)(0.60)\sqrt{(2)(9.81)(5.50)} = 10.94 \quad m^3/s$$

▬

Example 13-7

A flow rate of 5.00 m^3/s per meter width of channel is to flow under a sluice gate when the maximum head upstream is 7.0 m. At what opening should the gate be set?

SOLUTION
From Eq. (13.28),

$$y_s = \frac{Q}{K\sqrt{2gy_1}} = \frac{5.00}{K\sqrt{(2)(9.81)(7.0)}} = \frac{0.426}{K}$$

Assume $K = 0.56$; then $y_s = 0.426/0.56 = 0.76$ and $y_s/y_1 = 0.76/7.0 = 1.09$. From Fig. 13-20, $K = 0.585$ for this value of y_s/y_1 and the value of $y_s = 0.426/0.585 = 0.73$ m.

███

Example 13-8

The water depth upstream of a given sharp-crested weir varies, of course, with the discharge. Variations in upstream water level may be reduced by the use of a weir crest longer than the channel width. This weir crest may be in the shape of a V, a W, or square Us connected to inverted square Us, for example. Calculate the variation in water level upstream of a sharp-crested suppressed weir in a wide river for flow rates from 0 to 4.00 m^3/s per meter of channel width for a weir crest length (a) equal to the channel width, (b) twice the channel width, and (c) three times the channel width. Assume $K = 0.63$ at maximum discharge.

SOLUTION
From Eq. (13.20):
(a) $H^{3/2} = 4.00/(0.63)(2/3)\sqrt{(2)(9.81)} = 2.15$ and $H = 1.66$ m, and thus the head varies from 0 to 1.66 m.
(b) For $b = 2$ (2 m of weir width for 1 m of channel width), $H^{3/2} = 1.075$ and $H = 1.05$ m. Thus the head varies from 0 to 1.05 m.
(c) For $b = 3$ (3 m of weir width for 1 m of channel width) $H^{3/2} = 0.54$ and $H = 0.66$ m. Thus the head varies from 0 to 0.66 m.

███

13-4. Subsonic Gas Flow Rates in Pipes

Gas flow rates in pipes may be measured on a mass flow basis. As for liquid flow rates, gas flow meters should be calibrated for accurate measurements, but estimates may be made from published data on flow meters.

In Fig. 13-11, the steady-flow energy equation, the continuity equation, and the isentropic equation may be applied between sections 1 and 2 for the venturi meter and the nozzle meter. These equations are

$$\frac{V_1^2}{2} + h_1 = \frac{V_2^2}{2} + h_2$$

$$V_1 A_1 \rho_1 = V_2 A_2 \rho_2$$

and

$$\frac{T_2}{T_1} = \left(\frac{p_2}{p_1}\right)^{(k-1)/k} \qquad \text{or} \qquad \frac{\rho}{\rho^k} = \text{constant}$$

Combining these equations gives the mass flow rate in terms of quantities at section 2 and at section 1,

$$\dot{m} = \frac{A_2 \sqrt{2kp_1\rho_1\left[\left(\frac{p_2}{p_1}\right)^{2/k} - \left(\frac{p_2}{p_1}\right)^{(k+1)/k}\right]}}{\sqrt{1 - \left(\frac{A_2}{A_1}\right)^2 \left(\frac{p_2}{p_1}\right)^{2/k}}}$$

This equation may be written in a simpler form, comparable to that used for liquid flow, by introduction of the flow coefficient K (the same as for liquid flow) and an expansion factor Y. Then,

$$\dot{m} = KYA_2\sqrt{2\rho_1(p_1 - p_2)} \tag{13.29}$$

where, for the venturi and nozzle meters,

$$Y = \sqrt{\frac{\left(\frac{k}{k-1}\right)\left(\frac{p_2}{p_1}\right)^{2/k}\left[1 - \left(\frac{p_2}{p_1}\right)^{(k-1)/k}\right]\left[1 - \left(\frac{d}{D}\right)^4\right]}{\left[1 - \left(\frac{p_2}{p_1}\right)\right]\left[1 - \left(\frac{d}{D}\right)^4\left(\frac{p_2}{p_1}\right)^{2/k}\right]}} \tag{13.30}$$

and K is obtained from Fig. 13-12. Equation (13.30) indicates that Y is a function of three dimensionless ratios, that is,

$$Y = Y\left(k, \frac{p_2}{p_1}, \frac{d}{D}\right)$$

Thus for a given gas (given k), Y should be a unique function of p_2/p_1 for any specified value of d/D. A graph of the expansion factor Y is shown in Fig. 13-21. The effect of the compressible flow (actually the gas expands in passing to the meter throat) is to *reduce* the flow rate of a gas for given initial conditions and pressure drop, as compared with the flow rate if incompressible flow were assumed. The minimum pressure in the venturi or nozzle throat is the critical pressure of the gas, which, in terms of the stagnation pressure p_0, is

$$\frac{p_c}{p_0} = \left(\frac{2}{k+1}\right)^{k/(k-1)} \tag{10.23}$$

Test results give the expansion factor as

$$Y = 1 - \frac{1}{k}\left[0.41 + 0.35\left(\frac{d}{D}\right)^4\right]\left(1 - \frac{p_2}{p_1}\right) \tag{13.31}$$

for an orifice meter [4]. Values are plotted in Fig. 13-21.

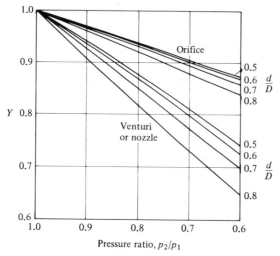

Fig. 13-21. Approximate expansion factors for $k = 1.40$.

PROBLEMS

13-1. What is the minimum velocity which can be measured with a blunt-ended stagnation or total pressure tube 0.6 mm in diameter in
(a) standard air and
(b) water at room temperature in order to avoid laminar flow effects?

13-2. A combined pitot tube is placed in a jet of water and is connected to a mercury manometer which shows a deflection of 40.0 cm. What is the velocity of the water in the jet?

13-3. A combined pitot tube is placed in a stream of standard air and is connected to a differential manometer containing methyl alcohol ($s = 0.80$). The manometer deflection is 9.5 cm. What is the measured air speed?

13-4. A combined pitot tube connected to a micromanometer produces a deflection of 14.5 mm of water when placed in an air stream at atmospheric pressure and 40°C. What is the velocity of the air stream?

13-5. An oil ($s = 0.9$, $\mu = 0.048$ kg/m s) flows in a 15-cm pipe. The centerline velocity is measured with a combined pitot tube with water as the gage fluid in a differential manometer attached to the pitot tube. The manometer deflection is 65.0 cm.
(a) What is the centerline velocity?
(b) Estimate the volumetric flow rate.

13-6. A pitot-static tube on an airplane is connected to a differential manometer which reads 14.0 cm water when the plane flies in standard air. What is the speed of the airplane?

13-7. A combined pitot tube mounted on an airplane is calibrated to read air speed in meters per second when used in standard air. What is the true air speed for an indicated air speed of 80 m/s at

(a) 1500 m, where $\rho = 1.058$ kg/m³, and

(b) 3000 m, where $\rho = 0.909$ kg/m³?

13-8. A combined pitot tube is used to make velocity profile measurements in a boundary layer in air. A manometer indicates a deflection of 4.10 mm water in air of density 1.20 kg/m³.

(a) What is the air velocity?

(b) What is the smallest impact tube diameter which may be used in order to avoid laminar flow effects?

13-9. The difference between stagnation and static pressure measured with an impact tube and a static pressure tap is 8.55 cm alcohol ($s = 0.800$) for air at 25°C and 100 kPa abs. What is the air velocity?

13-10. Repeat Prob. 13-9 for a manometer deflection of 13.1 cm alcohol.

13-11. The stagnation pressure measured with a stagnation tube in a wind tunnel is 20 kPa above the static pressure measured at a static-pressure tap at the wall. The test section is at a pressure of 3 atm. What is the Mach number of the flow?

13-12. A combined pitot tube is mounted on an airplane moving through air at a pressure of 90 kPa abs and -4°C. The stagnation pressure measured by the tube is

(a) 3.00 kPa gage,

(b) 30.0 kPa gage,

(c) 60 kPa gage, and

(d) 90 kPa gage. What are the respective flight velocities and Mach numbers?

13-13. The stagnation hole in a pitot cylinder and a wall pressure tap are connected to a differential manometer containing acetylene tetrabromide. Various point velocities along the pipe radius are to be determined for the flow of water in a round pipe. Manometer readings h_m are taken for various radial locations of the stagnation hole in the pitot cylinder as follows:

r/R	0	0.2	0.4	0.6	0.8	0.9	0.95
h_m (cm)	54.6	52.4	42.2	24.3	10.4	5.56	3.71

What is the average velocity?

13-14. At what radial position in a circular tube should a combined pitot tube be located in order to make a single measurement of the average velocity for

(a) laminar flow and

(b) turbulent flow with a velocity profile given by $u/u_{max} = (y/R)^{1/7}$?

Results for fully developed flow generally are accurate within a few percent when determined in this way, but more accurate results require more detailed measurements.

13-15. Repeat Prob. 13-14 for measurements taken as follows:

r/R	0	0.2	0.4	0.6	0.8	0.9	0.95
h_m (cm)	68.0	65.2	52.6	30.2	12.90	6.93	4.62

13-16. A point disturbance in air produces a Mach cone with a half-angle of 50°. What is the Mach number of the flow?

13-17. Show that Eq. (10.67) is valid for a point source, for which $\theta = 0°$, and becomes equivalent to Eq. (13.12).

13-18. A 20°, two-dimensional wedge produces an oblique shock at an angle of 135° from the free-stream velocity vector in an air stream. What is the free-stream Mach number?

13-19. A 30° wedge in a supersonic air flow produces a wave angle $\beta = 50°$.
(a) What is the Mach number of the flow?
(b) What other wave angle could exist for this Mach number? Refer to Fig. 10-11.

13-20. Water is supplied to a gas water heater through a pipe 35 mm ID and a venturi meter with a throat 21 mm ID.
(a) What is the pressure differential between the venturi inlet and throat for a velocity in the pipe of 30 cm/s? Assume water at 15°C.
(b) The pressure difference operates a gas control which consists of a piston of 40-mm diameter moving in a cylinder. What is the net force on the piston? Neglect friction and the area of the connecting rod.

13-21. A 10-cm-diameter meter is installed in a 25-cm-diameter pipe in which water at 20°C flows at a rate of 56 L/s. What deflection on a mercury manometer connected across the meter would occur for
(a) a venturi meter and
(b) an orifice meter?

13-22. The maximum flow of water at 20°C expected in a 15-cm venturi meter in a 30-cm pipe sloping at 45° from the horizontal is 225 L/s. What deflection of a mercury manometer would occur when the upstream piezometer tap is
(a) 50 cm and
(b) 100 cm from the throat piezometer tap, measured parallel to the meter centerline?

13-23. Water at 20°C flows through a 15-cm-diameter meter in a 25-cm-diameter pipe. A differential manometer connected across the meter shows a deflection of 22.0 cm. What is the flow rate for
(a) a venturi or a nozzle meter and
(b) an orifice meter? The manometer fluid is mercury.

13-24. Crude oil at 20°C flows through a 15-cm meter in a 25-cm-diameter pipeline. A mercury manometer connected across the meter indicates

a deflection of 8.55 cm. What is the flow rate for
(a) a venturi or a nozzle meter and
(b) an orifice meter?

13-25. A 10-cm nozzle meter is installed in a 20-cm water pipeline. What diameter orifice would produce the same deflection on a manometer connected across the meters for the same flow rate of 45 L/s through each meter?

13-26. A mercury manometer connected across a 10-cm orifice in a 20-cm pipe shows a deflection of 30 cm for a certain flow of water. What would be the deflection on the same manometer for the same flow rate through a 10-cm nozzle meter in the pipe?

13-27. Water flows in a 20-cm pipe with a 10-cm nozzle meter. The drop in piezometric head from the pipe to the meter throat is 4.40 m of water.
(a) What is the flow rate?
(b) What is the head loss from the meter inlet to the throat?
(c) What is the head loss from the meter throat to a point downstream of the nozzle? Assume it to be equivalent to a sudden enlargement loss.

13-28. A venturi meter, a nozzle meter, and an orifice meter of the same throat diameter are in series in a pipe, but far apart so there is no mutual interference. Across which meter will the total head loss be the greatest for a given flow? Explain your answer.

13-29. A 10-cm nozzle meter is installed in a 20-cm water pipeline. A differential mercury manometer shows a deflection of 24 cm. Calculate the flow rate through the pipe and estimate the *total* head loss resulting from the meter installation.

13-30. A maximum flow of 100 L/s of water at 20°C in a 30-cm pipe is to be measured with a nozzle or an orifice meter. The maximum deflection of a differential manometer containing acetylene tetrabromide ($s = 2.96$) is limited to 50 cm. What is the smallest diameter
(a) nozzle and
(b) orifice which may be installed?

13-31. Repeat Prob. 13-30 for a flow of 150 L/s.

13-32. A calibration of an elbow meter (Fig. 13-13) is as follows:

Manometer deflection h_m (cm)	7.0	14.0	28.0	56.0	112.0
Flow rate (L/s)		3.68	5.38	7.84	11.46	16.14

What is the calibration equation, assuming it to be of the form $Q = Kh_m^n$ L/s and h is in centimeters?

13-33. Water flows from a circular orifice in a large tank under a head of 12.0 m. The orifice diameter is 10.0 cm. For a flow rate of 13.10 m^3 in 3 min, what is the discharge coefficient?

13-34. Water flows from a circular orifice in a large tank under a head of 11.25 m. The orifice diameter is 9.0 cm. Estimate the flow rate.

13-35. A flow of 6.00 m^3/s of water occurs in a channel nearly horizontal. A rectangular weir the full width of the channel has a crest height of 60 cm above the channel bed. Estimate the water depth a short distance upstream of the weir. The weir nappe is ventilated. Width = 5.00 m.

13-36. In Prob. 13-35, for what crest height would the water upstream of the weir be 1.50 m deep?

13-37. A sharp-crested weir with a crest height $Z = 1.20$ m in a channel 3.0 m wide has a coefficient K which is given by Eq. 13.21. Although K does not vary exponentially with H/Z, an approximate equation for the weir flow rate is $Q = C_1 H^n$. What is a relation between Q in cubic meters per second and H in meters for values of H from 0.10 to 0.20 m?

13-38. Estimate the unit discharge (q in units of cubic meters per second per meter width) for flow over a sharp-crested weir with a well-ventilated nappe for a head of 30 cm and a crest height of
(a) 30 cm,
(b) 60 cm, and
(c) 1.20 m.
Explain the resulting effect of crest height on flow rate on the basis of the variation in the approach velocity V_1 in Eq. (13.19).

13-39. The effect of side contractions on the flow rate over a weir may be illustrated by comparing the flow rate over
(a) a suppressed weir,
(b) a contracted weir, and
(c) a Cipolletti weir, each with the same base width b and a large crest height Z to minimize the effects of approach velocity. Calculate the flow rate over these three weirs for $b = 1.00$ m, $H = 30$ cm, and $Z = 2.00$ m.

13-40. What is the flow rate through a 90°, V-notch weir for a head of
(a) 6 cm and
(b) 36 cm

13-41. A depth recorder for a 90°, V-notch weir is to record flow rates in the range from 30 to 600 L/s. What is the required range of the depth recorder?

13-42. The flow rate in a laboratory model 2.0 m wide reaches 110 L/s.
(a) What is the head on a rectangular weir when the velocity of approach is neglected?
(b) How long a weir (see Example 13-8) would be necessary to limit the head to 2.0 cm?

13-43. The flow rate in a river model is 14 L/s per meter of channel width. How long a sharp-crested, suppressed weir should be installed to limit the head on the weir to

(a) 1.0 cm and

(b) 0.5 cm?

13-44. How many 90°, V-notch weirs are needed at the end of an open channel to limit the depth variation to 6.0 cm for flow rates ranging from 150 to 500 L/s?

13-45. Repeat Prob. 13-44 for a depth variation of 5.0 cm.

13-46. Calculate the flow rate under a sluice gate 2.0 m wide with an upstream depth of 4.00 m and a gate opening of 30 cm.

13-47. A flow rate of 2.00 m/s per meter width of channel issues from a sluice gate with a 60-cm opening. What is the depth upstream of the sluice gate?

13-48. A flow rate of 1.50 m³/s per meter width of channel flows under a sluice gate at a maximum upstream depth of 8.00 m. At what opening should the gate be set?

13-49. Air at 275 kPa gage pressure and 40°C flows in a 20-cm pipe. The pressure drop to the throat of a 10-cm venturi meter is 55 kPa. What is the mass flow rate?

13-50. Air enters a 10-cm venturi throat from a 20-cm pipe at 350 kPa abs and 25°C. The throat pressure is 275 kPa abs. What is the mass flow rate?

13-51. Repeat Prob. 13-50 for an orifice meter.

13-52. Methane at 520 kPa abs and 50°C flows through a 25-cm pipe. The differential pressure between the inlet and the 15-cm throat of a nozzle meter is 125 kPa. What is the mass flow rate? The dynamic viscosity is 1.12×10^{-5} kg/m s.

13-53. Repeat Prob. 13-52 for an orifice meter.

13-54. The flow of methane at 25°C (dynamic viscosity is 1.12×10^{-5} kg/m s) and 125 kPa abs in a 30-cm pipe at an average velocity of 15.0 m/s is to be measured with an orifice meter in the pipeline. The maximum pressure differential across the meter is to be 25 kPa. What is the smallest diameter orifice which may be used?

13-55. Air at 300 kPa and 35°C flows in a 20-cm pipe. The pressure drop to the throat of a 10-cm venturi meter is 65 kPa. What is the mass flow rate?

References

1. C. W. Hurd, K. P. Chesky, and A. H. Shapiro, "Influence of Viscous Effects on Impact Tubes," *Am. Soc. Mech. Engrs., J. Appl. Mech.*, Vol. 20 (1953), p. 253.
2. L. G. Straub, J. M. Killen, and O. P. Lamb, "Velocity Measurements of Air-Water Mixtures," *Trans. Am. Soc. Civil Engrs.*, Vol. 119 (1954), pp. 207–220.
3. H. Rouse, *Elementary Mechanics of Fluids* (New York: John Wiley and Sons, Inc., 1946), p. 93.
4. *Fluid Meters: Their Theory and Application*, 5th ed. (New York: The American Society of Mechanical Engineers, 1959).

See also:

Flow Measurement Symposium (New York: The American Society of Mechanical Engineers, 1966).

E. Ower and R. C. Pankhurst, *Measurement of Air Flow* (Elmsford, N.Y.: Pergamon Press, Inc., 1977).

E. B. Pickett, S. Vigander, J. C. Schuster and L. J. Hooper, "Bibliography on Discharge Measurement Techniques," *Am. Soc. Civil Engrs., J. Hydraulics Div.*, Vol. 103, No. HY 8 (Aug. 1977), pp. 889–903.

Chapter 14
Turbomachines

A rotodynamic machine or turbomachine is one which adds energy to or extracts energy from a fluid by virtue of a rotating system of blades within the machine. The rotating element is called a rotor in a compressor, an impeller in a pump, a rotor in a gas or steam turbine, and a runner in a hydraulic turbine. Centrifugal, axial-flow, and mixed-flow pumps and compressors; fans and blowers; and water, steam, or gas turbines are examples of rotating machines. The general simplified one-dimensional theory of turbomachines will be introduced in this chapter, and then attention will be focused on the operating characteristics of hydraulic machines.

The SI system of units generally is *not* used in this chapter in order to conform with practice of North American pump and turbine manufacturers and to the performance data contained in their literature. Pump flow rates generally are given in gallons per minute (gpm), for example.

14-1. General Theory

Energy is transferred from the rotor or impeller in a compressor or pump and from the fluid to the rotor or runner in a turbine. The fluid leaves the rotor with a resultant average velocity V_2 at a radius R_2 and enters with a resultant average velocity V_1 at a radius R_1. These velocities have components in three mutually perpendicular directions: (1) axial, or parallel to the axis of rotation of the rotor; (2) radial or meridional, which is normal to the axis of rotation and passes through it; and (3) tangential, which is normal to both the axial and the meridional directions. (The tangential component is often called the *whirl component*.) These are designated by subscripts a, m, and t, respectively. Only the tangential components contribute to the rotation of the rotor, and the change in angular momentum, or moment of momentum, per unit mass of the fluid passing through the rotor is $V_{2t}R_2 - V_{1t}R_1$. From Newton's second law, this is equal to the torque on the rotor for steady flow. The energy transfer E per unit weight of fluid equals the product of torque times the angular velocity divided by the acceleration of gravity g. Thus

$$E = \frac{(V_{2t}R_2 - V_{1t}R_1)\omega}{g} \tag{14.1}$$

Since $\omega r = U$, the peripheral velocity of the rotor, Eq. (14.1) may be written as

$$E = \frac{V_{2t}U_2 - V_{1t}U_1}{g} \tag{14.2}$$

Equations (14.1) and (14.2) are known as the Euler equations for a pump or compressor and represent the energy transferred to the fluid by the impeller or rotor. For a turbine, the equations may be written with the subscripts interchanged to represent the energy transferred to the rotor or runner by the fluid.

With reference to Fig. 14-1, the relationships between velocities may be written

$$V_{2m}^2 = V_2^2 - V_{2t}^2 = V_{2r}^2 - (U_2 - V_{2t})^2$$

from which

$$U_2 V_{2t} = \frac{V_2^2 + U_2^2 - V_{2r}^2}{2}$$

where V_{2t} and V_{1r} represent the velocity of the fluid relative to the impeller blades at exit and entrance, respectively. Similarly, at the impeller inlet,

$$U_1 V_{1t} = \frac{V_1^2 + U_1^2 - V_{1r}^2}{2}$$

With these relations, Eq. (14.2) becomes

$$E = \frac{V_2^2 - V_1^2}{2g} + \frac{U_2^2 - U_1^2}{2g} + \frac{V_{1r}^2 - V_{2r}^2}{2g} \tag{14.3}$$

which indicates that E is the sum of the difference in the squares of the absolute fluid velocity, the peripheral rotor velocity, and the relative fluid

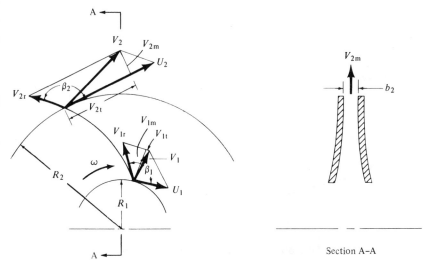

Section A–A

Fig. 14-1.   Flow through a centrifugal pump impeller.

velocity, respectively. The first term represents the change in kinetic energy of the fluid, the second term represents the pressure change from centrifugal effects, and the third term represents the pressure change owing to relative kinetic energy. The second and third terms are considered as static-pressure effects.

In a radial-flow or mixed-flow impeller, all three terms in Eq. (14.3) are effective. In an axial-flow machine, the fluid enters and leaves at the same radius so that $U_2 = U_1$, and only the first and third terms are effective. In a tangential-flow machine, only the first term is effective.

It is common to designate gas turbines or compressors on the basis of the relative importance of the second and third terms in Eq. (14.3) compared with all three terms, though it may be applied to hydraulic machines as well. The ratio of the second and third terms to all three terms is called the *reaction*.

$$\text{reaction} = \frac{\dfrac{U_2^2 - U_1^2}{2g} + \dfrac{V_{1r}^2 - V_{2r}^2}{2g}}{E} \tag{14.4}$$

Rotating machines may be classified according to (a) whether the fluid does work on the rotor or runner (a turbine), or the rotor or impeller transfers energy to the fluid (a compressor or pump); (b) the predominant flow direction through the rotor (axial, radial, tangential, or mixed); or (c) the reaction.

RADIAL-FLOW COMPRESSOR OR CENTRIFUGAL PUMP

From Eq. (14.2), the maximum energy transfer occurs when the angular momentum or whirl at inlet is zero, so that $V_{1t} = 0$. Thus from Fig. 14-1,

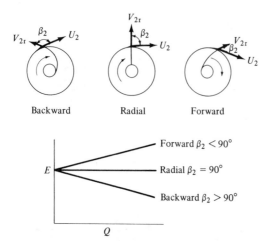

Fig. 14-2. Euler performance curves for a centrifugal pump.

when the inlet whirl is zero, Eq. (14.2) may be written as

$$E = \frac{V_{2t}U_2}{g} = \frac{[U_2 - V_{2m}\cot(180 - \beta_2)]U_2}{g}$$

$$= \frac{U_2^2}{g} + \frac{U_2 V_{2m}}{g}\cot\beta_2 \tag{14.5a}$$

From continuity, the volumetric flow rate through the machine is

$$Q = 2\pi R_2 b_2 V_{2m}$$

and if V_{2m} from this expression is substituted into Eq. (14.5a),

$$E = \frac{U_2^2}{g} + \frac{U_2 \cot\beta_2}{2\pi R_2 b_2 g} Q \tag{14.5b}$$

indicating that for a given impeller at a prescribed peripheral speed E is a function only of the discharge Q.

For a backward-curved impeller blade $\beta_2 > 90°$, $\cot\beta_2$ is negative, and E decreases with an increase in Q. For a radial blade $\beta_2 = 90°$, $\cot\beta_2 = 0$, and E is constant. For a forward-curved blade $\beta_2 < 90°$, $\cot\beta_2$ is positive, and E increases as Q increases. The so-called Euler performance curves for a centrifugal pump or compressor are shown in Fig. 14-2.

AXIAL-FLOW COMPRESSORS OR PUMPS

In an axial-flow machine, the axial component of velocity is generally constant ($U_1 = U_2$) and Eqs. (14.2) and (14.3) become

$$E = \frac{U}{g}(V_{2t} - V_{1t}) = \frac{V_2^2 - V_1^2}{2g} + \frac{V_{1r}^2 - V_{2r}^2}{2g} \tag{14.6}$$

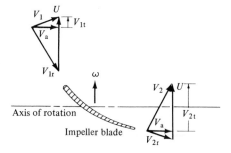

Fig. 14-3. Velocity diagrams for an axial-flow pump.

respectively. If neither V_{1t} nor V_{2t} is zero, the fluid has a whirl component at both inlet and outlet. Equation (14.6) indicates that the absolute velocity of the fluid increases and the relative velocity decreases as the fluid passes through the impeller blades. Figure 14-3 shows axial flow through an impeller and indicates the effect of the blade angle at inlet and outlet on the quantity $V_{2t} - V_{1t}$, which is a direct measure of the energy transfer.

AXIAL-FLOW TURBINE
Equation (14.6) applies with subscripts reversed.

TANGENTIAL-FLOW TURBINE
An impulse turbine involves only a tangential velocity at inlet with a small axial component at the exit of the runner. The fluid enters and leaves at the same radius, so that $U_1 = U_2$, and with essentially the same relative velocity, so that $V_{1r} \approx V_{2r}$. Thus there is no pressure change, and an impulse turbine has zero reaction. The energy transferred is solely due to a jet striking the runner blades, and the change in momentum takes place at constant radius. The energy transfer per unit weight of fluid is, from Eq. (14.2),

$$E = \frac{U}{g}(V_{1t} - V_{2t}) = \frac{U}{g}[(U + V_{1r}) - (U + V_{2r}\cos \beta)]$$

and since $V_{1r} \approx V_{2r}$,

$$E = \frac{U}{g}V_{1r}(1 - \cos \beta) \tag{14.7}$$

Equation (14.7) may be obtained from the momentum theorem directly (Fig. 14-4). The force on the blade is $F = Q\rho(V_{1r} - V_{2r}\cos \beta)$, where $V_{1r} \approx V_{2r}$ as before. The power is $P = FU = Q\rho V_{1r}U(1 - \cos \beta)$. The energy per unit weight of fluid is $E = P/Q\gamma = (U/g)V_{1r}(1 - \cos \beta)$, which is identical to Eq. (14.7).

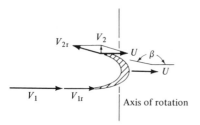

Fig. 14-4. Velocity diagram for a tangential-flow turbine.

14-2. Energy Equation and System Characteristics for Pumps

The total energy added to a fluid by a pump impeller is greater than the net energy or head added to the fluid by the pump itself. Energy or head losses occur owing to secondary flow within the pump impeller and passages, flow separation, disk friction in a centrifugal pump, and leakage. These losses reduce the pump head to a value less than the impeller head.

The energy or head added to a fluid by a pump is obtained from the energy equation [Eq. (5.4b)] applied between the inlet section 1 and the outlet section 2:

$$-w = \frac{V_2^2 - V_1^2}{2g} + \frac{p_2 - p_1}{\gamma} + (z_2 - z_1) \tag{5.4b}$$

or

$$H = H_v + H_p + H_z \tag{14.8}$$

where the increase in total head H results largely from an increase in the pressure head H_p, rather than the velocity head H_v or potential or elevation head H_z. The ratio of the power represented by the total net head H to the power input to the pump is the overall efficiency of the pump,

$$\eta = \frac{\text{output}}{\text{input}} = \frac{Q\gamma H}{(550)(\text{horsepower input})} \tag{14.9}$$

The pump supplies energy to increase the potential head or pressure head of the fluid in the system, to overcome head losses, or both. The system demand increases linearly with discharge for laminar flow and almost parabolically with discharge for turbulent flow (Fig. 14-5). Any pump or combination of pumps operates at the intersection of the system characteristic or demand curve and the characteristic performance curve for the pump or combination of pumps. Typical performance curves are given in Sec. 14-3. The discharge and head corresponding to the point of maximum efficiency are called the *normal*, or *rated*, discharge and head, respectively.

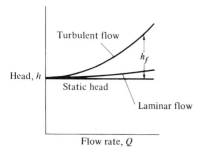

Fig. 14-5.   Piping system demand curve curves.

14-3. Pump Performance Characteristics

The actual performance characteristics of pumps are not obtained from the theory presented in Sec. 14-1 but are obtained by direct measurements made on the particular pump or on a model of the pump. It is customary to express the head, efficiency, and brake horsepower (horsepower input) as functions of capacity or flow rate, either in tabular or graphic form.

The *rated* head, capacity, and power correspond to maximum efficiency at a prescribed speed.

CENTRIFUGAL PUMPS

A centrifugal pump is so named because the head added by the impeller to the fluid is due largely to centrifugal effects. It is generally a high-head, low-capacity type of pump. A centrifugal pump having medium-head and medium-capacity characteristics is shown in Fig. 14-6. Fluid enters the eye of the impeller in an essentially axial direction, with a tangential component (whirl or prerotation) caused by viscous effects. The fluid flows outward through diverging passages between the impeller blades and disks and leaves the impeller periphery at a high pressure and a rather high velocity as it enters the casing, or volute. The flow within the impeller is a combination of a source flow and a forced-vortex flow. The purpose of the volute is to convert the kinetic head, represented by the high discharge velocity, into pressure head before the fluid leaves the pump discharge pipe. If the casing contains fixed guide vanes, the pump is called a diffuser or turbine pump.

A centrifugal pump may have a single suction or a double suction, depending on whether the fluid enters the impeller from one or both axial directions, respectively. In a double-suction pump, end thrust on the pump shaft is essentially eliminated and impeller inlet velocities are reduced for a given impeller size. More than one impeller may be mounted on one shaft, and in this multistage-type pump the fluid discharging from one impeller and its volute enters the eye of the following impeller, and so forth. The total head added by a multistage pump is the sum of the heads added by each stage.

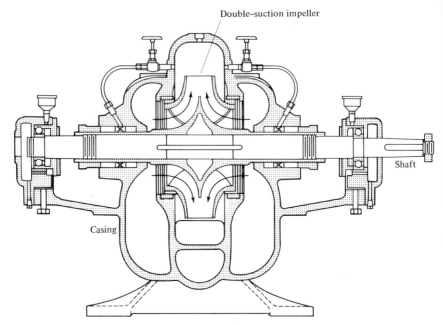

Fig. 14-6. Small-radius, double-suction, single-stage centrifugal pump. (Courtesy Worthington Corporation.)

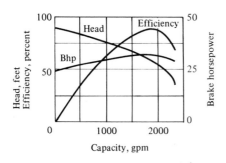

Fig. 14-7. Performance curves for a centrifugal pump.

Typical performance curves for a centrifugal pump are shown in Fig. 14-7. Efficiencies are as high as 85 percent for large pumps.

In Fig. 14-7, the rated discharge is about 1900 gpm and the rated head about 57 ft.

Centrifugal pumps may be classified according to the shape of their head–discharge curves (Fig. 14-8) or their power–discharge curves (Fig. 14-9). The terminology in Fig. 14-8 refers to the change in head for a decreasing discharge. Curves *a*, *c*, and *d* are called stable characteristics because only one head is associated with a given discharge. Conversely, curve *b* is called an unstable characteristic. The stability of operation, however, depends on

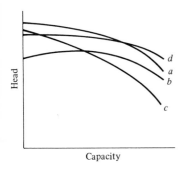

Fig. 14-8. Centrifugal-pump classification based on the shape of the head-capacity curve: (a) drooping curve; (b) rising curve; (c) steep curve; (d) flat curve.

the system characteristic or demand curves (Fig. 14-5) as well as the pump characteristic curve. Pump b of Fig. 14-8 would probably be unstable if used in parallel with a similar pump or when the system characteristic curve is quite flat (head essentially independent of flow rate, as when the head losses are small). It might be stable in a system with low static head compared with higher head losses.

The power–capacity curves labeled a, b, and c in Fig. 14-9 correspond to the head–capacity curves labeled a, b, and c in Fig. 14-8. Curve a is called a *nonoverloading* characteristic, since the power decreases when the head is either increased or decreased. Curve b is called an *overloading* curve with a reduction in head, and curve c is called an *overloading* curve with an increase in head.

AXIAL-FLOW PUMP

An axial-flow pump is a propeller-type pump (Fig. 14-10). It may have three sets of blades: (1) inlet guide vanes to remove any tangential velocity component, (2) impeller blades, and (3) outlet vanes to remove the tangential or

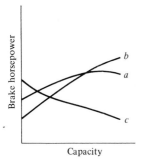

Fig. 14-9. Centrifugal-pump classification based on shape of power-capacity curve: (a) nonoverloading with reduction in head; (b) overloading with reduction in head and increase in capacity; (c) overloading with increase in head and decrease in capacity. Refer to Fig. 14-8.

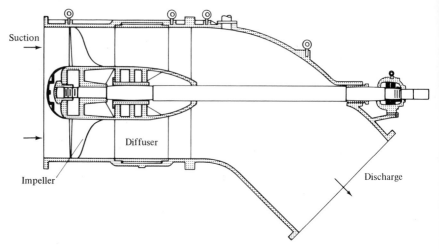

Fig. 14-10. Horizontal axial-flow pump. (Courtesy Worthington Corporation.)

whirl component of velocity at the pump discharge. Larger pumps are often made with controllable-pitch impeller blades so that good efficiency can be achieved over a wide range of flow rates. An axial-flow pump is a high-capacity, low-head type pump. Typical performance curves are shown in Fig. 14-11. Efficiencies are typically as high as 75 percent. The head–capacity curve often has an unstable characteristic, and the power–capacity curve is of type *c* in Fig. 14-9.

MIXED-FLOW PUMP

A mixed-flow pump, shown in Fig. 14-12, has characteristic curves between those of a centrifugal pump and an axial-flow pump.

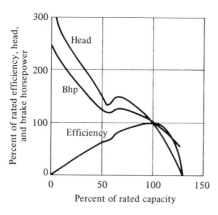

Fig. 14-11. Typical performance curves for an axial-flow pump. (Courtesy Worthington Corporation.)

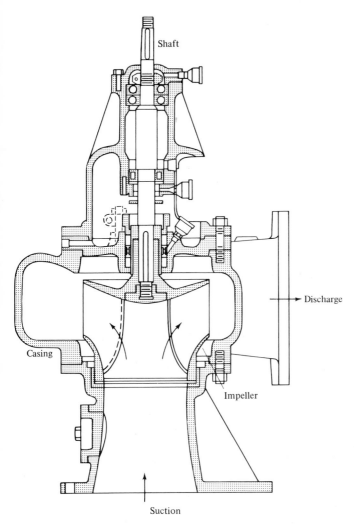

Fig. 14-12.   Pump with mixed-flow impeller. (Courtesy Worthington Corporation.)

14-4. Pump and System Combinations

When two similar pumps are connected in series, the resulting characteristic curves are similar to those obtained by multistaging a number of impellers on a single shaft. For a given capacity, the total head is the sum of the heads added by each individual pump or stage. When two or more similar pumps are in parallel, the total capacity is increased to two or more times the capacity of each individual pump for the same head. The resulting head–capacity operating point will depend on the system characteristic curves as well as the combined pump characteristic curves (Fig. 14-13). One pump will operate at point *A*, and two pumps will operate at point *B*. In neither

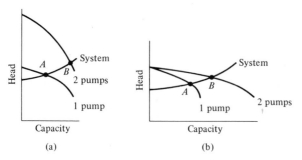

Fig. 14-13. Pump combinations with system characteristics. (a) Series. (b) Parallel.

instance will the head or capacity double for the series or parallel combination of two pumps, respectively.

14-5. Dimensionless Parameters and Dynamic Similarity for Pumps

The dimensional analysis of rotating machines in Sec. 8-6 (which should be reviewed) showed that for dynamically similar or homologous conditions, certain dimensionless parameters remain constant. These are

$$\frac{Q_1}{N_1 D_1^3} = \frac{Q_2}{N_2 D_2^3} \tag{14.10}$$

$$\frac{H_1}{N_1^2 D_1^2} = \frac{H_2}{N_2^2 D_2^2} \tag{14.11}$$

$$\frac{P_1}{\rho_1 N_1^3 D_1^5} = \frac{P_2}{\rho_2 N_2^3 D_2^5} \tag{14.12}$$

and

$$N_{s(P)} = \frac{N_1 \sqrt{Q_1}}{H_1^{3/4}} = \frac{N_2 \sqrt{Q_2}}{H_2^{3/4}} \tag{14.13}$$

These parameters may be used to relate the characteristics of different size but similar pumps operating under dynamically similar conditions (model and prototype units, for example).

For a *given* pump, if the speed N is changed and the efficiency is assumed to remain constant, then Q varies as N, or

$$\frac{Q_1}{Q_2} = \frac{N_1}{N_2} \tag{14.14}$$

H varies as N^2, or

$$\frac{H_1}{H_2} = \left(\frac{N_1}{N_2}\right)^2 \tag{14.15}$$

P varies as N^3, or

$$\frac{P_1}{P_2} = \left(\frac{N_1}{N_2}\right)^3 \tag{14.16}$$

These equations permit an estimate of the pump performance characteristics for speeds other than those for which characteristics are known from direct measurement (see Example 8-6).

The efficiency of large pumps is generally greater than the efficiency of a geometrically similar small pump. The relation between these efficiencies is given approximately by

$$(1 - \eta_1)(N_1 D_1^2)^{1/5} = (1 - \eta_2)(N_2 D_2^2)^{1/5} = \text{constant} \tag{14.17}$$

NONHOMOLOGOUS CONDITIONS

For a pump in which the impeller diameter is reduced in order to obtain slightly different characteristics at a given speed (this is common practice among pump manufacturers), the following relationships apply:

$$\frac{Q_1}{Q_2} = \frac{D_1}{D_2} \tag{14.18}$$

$$\frac{H_1}{H_2} = \left(\frac{D_1}{D_2}\right)^2 \tag{14.19}$$

$$\frac{P_1}{P_2} = \left(\frac{D_1}{D_2}\right)^3 \tag{14.20}$$

From these relations, the characteristic curves for a given impeller in a pump may be converted to characteristic curves for a smaller impeller, so long as the change in diameter is no greater than about 20 percent. Points on a given head–discharge curve for diameter D_1 are transferred to corresponding points on the head–discharge curve for D_2 along parabolas, and points on a given power–discharge curve for diameter D_1 are transferred to corresponding points on the power–discharge curve for diameter D_2 along cubics (Fig. 14-14).

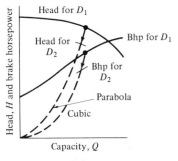

Fig. 14-14. Effect of change in impeller diameter for centrifugal pump.

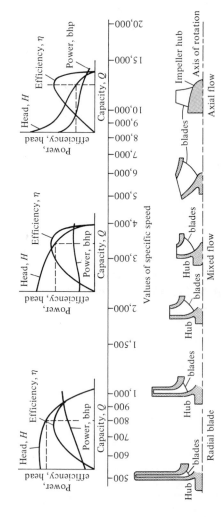

Fig. 14-15. Pump characteristics and specific speed for impellers of various designs. (Source: From Part 4, of *Basic Factors in Centrifugal Pump Application* by Roy Carter and Igor J. Karassik, Report RP-477, Fig. 27. Courtesy Worthington Corporation.)

Specific speeds in North America are generally expressed as a pseudo-dimensionless number with the speed N in revolutions per minute, discharge Q in gallons per minute, and head H in feet at normal, or rated, capacity and head for a single stage. Typical values of specific speeds for various types of pump impellers, together with corresponding pump characteristic curves, are shown in Fig. 14-15.

Example 14-1

What is the specific speed of a pump which delivers 2200 gpm against a head of 90 ft at a speed of 1450 r/min? What type of pump is this?

SOLUTION

From Eq. (14.13)

$$N_{s(P)} = \frac{N\sqrt{Q}}{H^{3/4}} = \frac{1450\sqrt{2200}}{90^{3/4}} = 2330$$

Figure 14-15 indicates a high-specific-speed centrifugal pump or a low-specific-speed mixed flow pump.

Example 14-2

At what head and flow rate would the pump in Example 14-1 operate at 1150 r/min?

SOLUTION

From Eq. (14.14),

$$Q_2 = Q_1\left(\frac{N_2}{N_1}\right) = 2200\left(\frac{1150}{1450}\right) = 1745 \text{ gpm}$$

From Eq. (14.15),

$$H_2 = H_1\left(\frac{N_2}{N_1}\right)^2 = 90\left(\frac{1150}{1450}\right)^2 = 56.6 \text{ ft}$$

The specific speed is the same as in Example 14-1, namely,

$$N_{s(P)} = \frac{1150\sqrt{1745}}{(56.6)^{3/4}} = 2330$$

Example 14-3

A pump is available which has the following characteristics:

H (ft)	120	120	118	113	100	75	43
Q (gpm)	0	800	1200	1600	2000	2400	2800

It is connected to a 12-in. commercial steel pipeline 1200 ft long which contains pipe fittings such that the total head loss is 37 $V^2/2g$ ft. Water is pumped from one open tank to another 50 ft higher than the first. What is the flow rate?

SOLUTION

The pump-piping combination will operate where the pump characteristic curve crosses the piping demand curve. The piping demand curve is obtained from a plot of the following data:

Q (gpm)	0	800	1200	1600	2000	2400	2800
H (ft)	50	53.0	56.6	61.8	68.5	76.6	86.2

A plot of these curves indicates a flow rate of about 2375 gpm.

14-6. Pump Cavitation

It is generally assumed, as a first approximation, that when the absolute pressure at some point within a liquid reaches the vapor pressure corresponding to the liquid temperature, cavities will form. These cavities will contain vapor, undissolved gas, or both. The phenomenon of cavity formation or formation and collapse is known as cavitation. Cavities may exist in a more or less steady state (steady with respect to moving blade element) or as intermittent bubbles. Cavitation, if severe enough, will affect pump performance, cause noisy operation, enhance vibration, and erode metal from the impeller.

There will be some point in the liquid within the pump where the pressure is a minimum, generally in a zone of flow separation, and as the ambient pressure is reduced vapor pressure will be reached and cavitation initiated at this point. Corresponding to this condition will be a fixed absolute pressure at the pump inlet or suction side for a given discharge through the pump. The value of the total head at the centerline of the pump inlet (velocity head plus pressure head) minus the vapor-pressure head of the liquid is called the *net positive suction head* (NPSH) for the pump:

$$\text{NPSH} = \frac{V_1^2}{2g} + \frac{p_1}{\gamma} - \frac{p_v}{\gamma}$$

(14.21)

As the NPSH is decreased for a given pump and flow rate, a value will be reached for which cavitation is considered detrimental; that is, the efficiency drops or the pump loses its prime. This point is called the *minimum* net positive suction head (NPSH_{min}) and is a function of the flow rate through a given pump as well as the type of pump. If a pump is operated above the NPSH_{min}, it will not cavitate detrimentally.

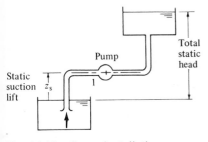

Fig. 14-16. Pump installation.

The cavitation number σ for a pump is defined as the net positive suction head divided by the total head against which the pump operates, or

$$\sigma = \frac{\text{NPSH}}{H} \tag{14.22}$$

The critical cavitation number σ_c corresponds to the minimum net positive suction head. Thus

$$\sigma_c = \frac{\text{NPSH}_{\min}}{H} \tag{14.23}$$

The cavitating characteristics of a pump as designated by its minimum net positive suction head or critical cavitation number determine the required static suction lift for a pump installation. Some terms associated with a pump installation are defined in Fig. 14-16.

The energy equation for flow between the free surface of the lower sump in Fig. 14-16 and the pump inlet (section 1) in terms of absolute pressures is

$$0 + \frac{p_a}{\gamma} + 0 = \frac{V_1^2}{2g} + \frac{p_1}{\gamma} + z_s + h_L$$

where p_a is the atmospheric pressure and h_L is the head loss in the intake. From this equation and the definition of net positive suction head in Eq. (14.21),

$$\text{NPSH} = \frac{V_1^2}{2g} + \frac{p_1}{\gamma} - \frac{p_v}{\gamma} = \frac{p_a}{\gamma} - z_s - h_L - \frac{p_v}{\gamma}$$

and the maximum value of the static suction lift z_s is

$$z_s = \frac{p_a}{\gamma} - \frac{p_v}{\gamma} - \text{NPSH}_{\min} - h_L \tag{14.24a}$$

$$= \frac{p_a}{\gamma} - \frac{p_v}{\gamma} - \sigma_c H - h_L \tag{14.24b}$$

If z_s is negative, the pump must be located below the level of the lower sump.

14-7. Hydraulic Turbines

Hydraulic turbines are used to extract energy from water as it passes through the rotating turbine runner. The momentum of the water results from the velocity gained at the expense of potential energy or head as the water flows from a higher to a lower elevation. Available heads range from several thousand feet to 10 ft or less. Hydraulic turbines are generally connected directly to an electric power generator, although in some instances they may be attached to wheels for grinding wood pulp, for example. As opposed to variable-speed operation of pumps, turbines generally run at a carefully controlled constant speed, depending on the type and design of turbine and generator and the local alternating current power frequency.

The available water supply often varies throughout the year, whereas the power demand remains more uniform. Regulation of the flow resulting from runoff of seasonal rainfall may occur as a result of natural conditions, such as chains of large lakes, or by means of dams to create storage reservoirs.

The *gross*, or *total*, *head* on a hydroelectric plant is the difference between headwater and tailwater elevations (this varies from time to time) when the turbines are not in operation. The overall hydraulic plant efficiency is the ratio of the power output of the turbines to the total available water power input based on the flow rate and gross head. The *net*, or *effective*, *head* is generally considered to be the difference in head between the entrance to the turbine and the tailwater. Thus the turbine itself is not charged with head losses in the penstock or intake passages between the headwater and the turbine. The net head for an impulse turbine is generally assumed to be the total head (velocity plus pressure) at the entrance to the turbine nozzle.

Hydraulic turbines may be classified according to (a) whether they are impulse (zero reaction) or reaction turbines; (b) the predominate flow direction of the water as it passes through the turbine runner; and (c) the specific speed (see Sec. 8-6) $N_{s(T)} = N\sqrt{bhp}/H^{5/4}$, defined at rated or normal capacity.

14-8. Hydraulic Turbine Characteristics

Three general types of hydraulic turbines, based on the flow direction through the runner, are in current use. These are the Pelton turbine (impulse), the Francis turbine (reaction), and the propeller turbine (reaction). Figure 14-17 shows these three types and the flow direction through the runners.

IMPULSE TURBINE

The Pelton wheel is a type of impulse turbine in current use. It is a tangential-flow turbine with a low specific speed (Fig. 14-18). A modern Pelton wheel is shown in Fig. 14-19. Water at a high head is converted into a high-velocity jet in a carefully designed nozzle. The tangential jet strikes a series of vanes

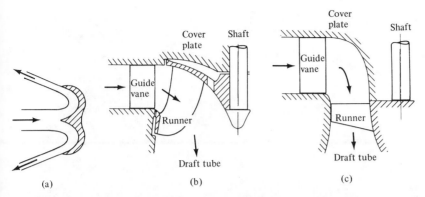

Fig. 14-17. Flow through turbine blades. (a) Impulse turbine. (b) Francis turbine. (c) Kaplan or fixed-blade propeller turbine. (Courtesy Allis–Chalmers Manufacturing Co.)

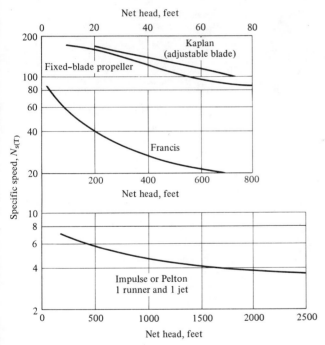

Fig. 14-18. Recommended upper limits of specific speed for turbines. (Courtesy Allis-Chalmers Manufacturing Co.)

or buckets on a rotating wheel, and the momentum change resulting from the jet deflection creates a force over the bucket surface and hence a torque to rotate the wheel (see Fig. 14-4 and the figure for Prob. 5-34). The jet is unconfined and at atmospheric pressure as it passes through the turbine. The theoretical head or energy per unit weight of water is given by Eq. (14.7),

Fig. 14-19.   Impulse turbine wheel. (Courtesy Allis-Chalmers Manufacturing Co.)

and the theoretical power output of the wheel is

$$P_{th} = Q\rho V_{1r} U(1 - \cos \beta) \qquad (14.25a)$$

$$= Q\rho(V_1 - U)U(1 - \cos \beta) \qquad (14.25b)$$

where Q is the flow rate, V_1 is the jet velocity, U is the peripheral blade velocity, and β is the blade angle at exit. The theoretical efficiency is the ratio of the theoretical power output to the power input in the form of kinetic energy of the jet:

$$\eta_{th} = \frac{Q\rho(V_1 - U)U(1 - \cos \beta)}{Q\gamma V_1^2/2g}$$

$$\eta_{th} = 2\left(1 - \frac{U}{V_1}\right)\frac{U}{V_1}(1 - \cos \beta) \qquad (14.26)$$

which is independent of the flow rate and depends only on the speed ratio U/V_1 for a given turbine wheel. It can be shown that the theoretical efficiency is a maximum when $V_1 = 2U$ (see Prob. 5-34). The actual efficiency of the turbine depends on windage losses, mechanical losses in the bearings, and hydraulic losses in the nozzle and on the blades. The power output is based on the flow rate Q and the net head H_1 at the nozzle entrance,

$$P = \frac{Q\gamma H_1}{550}\eta \tag{14.27}$$

A Pelton wheel generally operates under a high head (gross heads over 5000 ft exist in European installations, and up to 3700 ft in American installations) and requires a correspondingly long penstock or supply pipe. Certain relationships exist between the jet and pipe diameter and the pipe length for maximum power. Since the power input to the turbine depends on both the flow rate Q and the jet velocity V_1, it is apparent that too small a jet will provide a small flow rate even though the jet velocity is large, and too large a jet will provide a low jet velocity even though the flow rate is large. In either instance the power input is reduced.

If the losses in the long penstock or supply pipe are considered to be pipe friction h_f only, the net head H at the nozzle entrance is the difference between the total, or gross, head H_t and h_f:

$$H = H_t - f\frac{LV_p^2}{D_p 2g}$$

where V_p is the velocity and D_p is the diameter of the pipe. The power P_1 in the jet is

$$P_1 = \frac{Q\gamma H}{550} = \frac{\gamma A_1 V_1}{550}\left(H_t - f\frac{L}{D_p}\frac{V_1^2}{2g}\frac{A_1^2}{A_p^2}\right) \tag{14.28}$$

which is a maximum when $dP_1/dA_1 = 0$ or when $H_t = 3h_f$. The power is a maximum when the net head H at the nozzle is two-thirds the gross head H_t. For no losses in the nozzle, $H = V_1^2/2g$. Thus, for optimum operation it can be shown that the jet diameter D_1 is related to the penstock diameter D_p by the relation

$$D_1 = \left(\frac{D_p^5}{2fL}\right)^{1/4} \tag{14.29}$$

Hence, for a given penstock or pipe diameter there exists a fixed jet diameter D_1 for maximum power.

REACTION TURBINES

A Francis-type reaction turbine is a radial- or mixed-flow turbine with fixed[†] runner blades and is most efficient at intermediate specific speeds (Figs. 14-17 and 14-18). The radial-flow type has the lowest specific speed and is essentially the reverse of a centrifugal pump. The mixed-flow type has higher

[†] Nonadjustable.

Fig. 14-20. Francis turbine runner. (Courtesy Allis-Chalmers Manufacturing Co.)

specific speeds and is the reverse of a mixed-flow pump. In both instances, however, the efficiency of the Francis turbine (90 percent or more) is higher than the efficiency of the corresponding pump. This is largely owing to the greater losses in the diverging flow through the pump impeller and volute than in the converging flow through the scroll and runner in the turbine. A Francis turbine runner is shown in Fig. 14-20.

The propeller-type reaction trubine is an axial-flow turbine with either fixed or adjustable blades and a high specific speed (Figs. 14-17 and 14-18). It is essentially the reverse of an axial-flow pump. A Kaplan turbine runner has adjustable blades and is shown in Fig. 14-21.

In reaction turbines the water flows from the headwater supply through a penstock (in some instances there is effectively none), through a guide case or scroll, inward past adjustable guide vanes or wicket gates in the guide case, through the turbine runner, and out through a diverging draft tube to the tailrace. In both types of reaction turbine the water is given a whirl

Fig. 14-21. Kaplan turbine runner. (Courtesy Allis-Chalmers Manufacturing Co.)

component before entering the runner and, as the water passes through the runner, energy is absorbed by it in accordance with Eq. (14.3). The whirl component gives the water a desirable angle of attack with respect to the runner blades.

The diverging draft tube is used to reduce the velocity of the water as it enters the tailrace. This in turn reduces the head loss in the system. However, the absolute pressure of the water as it leaves the turbine runner is less with a draft tube than with a constant-area discharge tube. The draft tube requires a lower setting of the turbine in order to avoid cavitation. The cavitation number is often called the *Thoma number*, or the plant sigma, and is defined as

$$\sigma = \frac{H_a - H_v - z_s}{H} \tag{14.30}$$

where H_a is the atmospheric pressure head, H_v is the vapor pressure head, z_s is the static draft head (comparable to the static suction lift for a pump) and is the height from the tailwater to the bottom of a Francis turbine runner

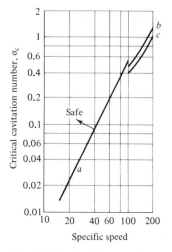

Fig. 14-22. Cavitation limits for reaction turbines. (a) Francis turbine. (b) Fixed-blade propeller turbine. (c) Kaplan turbine.

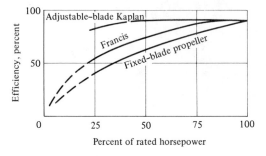

Fig. 14-23. Efficiency curves for reaction turbines.

and to the center of a propeller turbine runner, and H is the net available head. (The value of $H_a - H_v$ is the height of a water barometer.) Typical lower limits of cavitation number for reaction turbines are shown in Fig. 14-22. Figure 14-23 shows typical efficiencies of reaction turbines.

As for pumps, the rated head, capacity, and power for turbines correspond to maximum efficiency at a specified speed.

14-9. Dimensionless Parameters and Dynamic Similarity for Turbines

Dynamic similarity is considered to exist when the dimensionless ratios involving discharge Q, speed N, head H, and size D in Eqs. (14.10), (14.11), and (14.12); the specific speed $N_{s(T)} = N\sqrt{\text{Bhp}}/H^{5/4}$; and the cavitation number of Eq. (14.30) are the same in homologous units. Almost all large turbines are custom built, and model tests are conducted in order that the performance of the prototype may be predicted with reliability. The similarity parameters apply only if the efficiency is the same for similar units. It is well known that larger units are more efficient than small units. An expression by Moody relating efficiencies of models and prototypes is

$$\frac{1 - \eta_m}{1 - \eta_p} = \left(\frac{D_p}{D_m}\right)^{1/5}$$

Example 8-7 illustrates the use of similarity criteria for turbine tests.

PROBLEMS

14-1. A centrifugal pump impeller has a 16-in. diameter, a 1-in. width at exit ($b_2 = 1$ in.), a blade angle $\beta_2 = 150°$, and rotates at 1450 r/min. The flow rate is 1600 gpm.
   (a) Calculate the meridional or radial, the relative, and the absolute fluid velocities at the impeller exit.
   (b) For no inlet whirl, what head is added to water by the impeller?
   (c) What horsepower is required to rotate the impeller?

14-2. Repeat Prob. 14-1 for a speed of 1150 r/min.

14-3. Repeat Prob. 14-1 for an impeller diameter of 14 in.

14-4. Repeat Prob. 14-1 for a blade angle $\beta_2 = 135°$.

14-5. Repeat Prob. 14-1 for a speed of 1750 r/min.

14-6. A centrifugal pump impeller has a 50-cm diameter, a 3-cm width at exit ($b_2 = 3$ cm), a blade angle $\beta_2 = 150°$, and rotates at 1450 r/min. The flow rate is 200 L/s.
   (a) Calculate the meridional or radial, the relative, and the absolute fluid velocities at the impeller exit.
   (b) For no inlet whirl, what head is added to the water by the impeller?
   (c) What power is required to rotate the impeller?

14-7. Repeat Prob. 14-6 for a speed of 1150 r/min.

14-8. Repeat Prob. 14-6 for a blade angle $\beta_2 = 135°$.

14-9. Water flows through the impeller of a centrifugal pump (Fig. 14-1). The flow rate is $Q = 800$ gpm, $R_1 = 2$ in., $R_2 = 8$ in., $b_2 = 0.75$ in., $\beta_2 = 120°$, and $N = 1150$ r/min. Assume no inlet whirl.
   (a) What horsepower is required to rotate the impeller?
   (b) Determine the blade angle β_1 at inlet which will make the relative velocity vector V_{1r} tangent to the blade. This condition provides what is known as a shockless entry.

14-10. Show that the reaction of an axial-flow machine is the average of the *tangential components* of the relative fluid velocities V_{1r} and V_{2r} divided by the peripheral blade velocity U. Show that this may also be expressed as

$$\text{reaction}_{\text{axial flow}} = 1 - \frac{V_{1t} + V_{2t}}{2U}$$

Refer to Fig. 14-3.

14-11. A centrifugal pump with a 4-in.-diameter suction pipe and a 3-in.-diameter discharge pipe has a measured capacity of 300 gpm, a suction-side vacuum of 8 in. mercury, and a discharge pressure of 35 psig. The suction and discharge pipes are at the same elevation. The measured power input is 9.1 hp. What is the pump efficiency?

14-12. A centrifugal pump has a 6-in. suction pipe and a 5-in. discharge pipe. It is tested at 490 gpm with a measured input of 13.4 hp. The suction pressure is measured with a mercury manometer (Fig. 2-7) with a deflection $h_m = 8.00$ in. and the upper mercury meniscus is 12.0 in. below the suction pipe centerline ($y = 12.0$ in.). The discharge pressure is 31 psig. What is the pump efficiency?

14-13. A centrifugal pump with a 30-cm suction pipe and a 25-cm discharge pipe pumps 140 L/s. The suction-side vacuum is 20 cm mercury and the discharge pressure is 300 kPa gage. Inlet and discharge pipes are at the same elevation. For an efficiency of 81 percent, what is the power input to the pump?

14-14. A centrifugal pump delivers 600 gpm against a head of 100 ft at 1750 r/min. What are the corresponding values of
(a) discharge and
(b) head at a speed of 1650 r/min?

14-15. Under what conditions would two similar pumps in parallel in a system deliver essentially twice the discharge of a single pump connected in the same system?

14-16. A 14.125-in.-diameter impeller in a centrifugal pump has the following characteristics at 1750 r/min:

Q (gpm)	200	300	400	500
H (ft)	205	201	191	167

(a) Plot the characteristic curve for a speed of 1450 r/min for this pump.
(b) What impeller diameter would produce a head of 160 ft at 365 gpm at a speed of 1750 r/min?
(c) What brake horsepower would be required in part (b) for an efficiency of 70 percent?

14-17. A 14.5-in.-diameter impeller in a centrifugal pump has the following characteristics at 1750 r/min:

Q (gpm)	200	300	350	400
H (ft)	177	165	156	147

(a) Plot characteristic curves for a speed of 1150 r/min for this pump.

(b) What diameter impeller would produce a head of 140 ft at 300 gpm at a speed of 1750 r/min?

(c) What brake horsepower would be required in part (b) for an efficiency of 72 percent?

14-18. A 12-in.-diameter impeller in a centrifugal pump delivers 450 gpm at a head of 100 ft at 1750 r/min and requires 14 hp to drive it.

(a) What would be the corresponding capacity and head if the impeller diameter is reduced to 11 in.?

(b) What horsepower would be required in part (a)? Assume no change in efficiency.

14-19. A pump with a 25-in. impeller is to deliver 13,000 gpm at a head of 102 ft at 860 r/min. A model pump with a 6-in. impeller is run at 1750 r/min.

(a) What are the corresponding capacity and head for the model, assuming homologous conditions?

(b) For an assumed efficiency of 80 percent for both model and prototype, what horsepower will be required to drive each?

(c) Calculate the specific speed and determine what type of pump this is.

14-20. A pump with a 65-cm impeller is to deliver 800 L/s at a head of 31 m at 850 r/min. A model pump with a 16-cm impeller is run at 1750 r/min.

(a) What are the corresponding flow rate and head for the model for dynamically similar conditions?

(b) For an efficiency of 82 percent for both model and prototype pumps, what power will be required to drive each?

14-21. A 6-in. model of a centrifugal pump has an efficiency of 66 percent. Estimate the efficiency of similar pumps

(a) 9 in. and

(b) 12 in. in diameter at the same speed.

14-22. A pump for which $\sigma_c = 0.08$ pumps against a total head of 200 ft. The water temperature is 100°F, the barometric pressure is 14.5 psia, and the head loss in the suction pipe is 3 ft. What is the maximum value of the static suction lift?

14-23. Repeat Prob. 14-22 for a water temperature of 180°F ($\gamma = 60.6$ lb$_f$/ft^3 and $p_v = 7.51$ psia).

14-24. Select the type of pump appropriate for the following operating conditions:

(a) 1450 r/min, 200 gpm, 50-ft head;

(b) 720 r/min, 10,000 gpm, 8-ft head; and

(c) 1150 r/min, 1000 gpm, 30-ft head.

14-25. Plot a curve of the theoretical efficiency for an impulse turbine as a function of speed ratio U/V_1 for a blade angle $\beta_2 = 165°$.

14-26. In Eq. 14-28, let $A_1 V_1 = Q$ and $dP_1/dA_1 = 0$. Show that $H_t = 3h_f$ for maximum power.

14-27. Verify Eq. (14.29).

14-28. A water jet 2.5 in. in diameter at a velocity of 200 ft/s drives an impulse wheel whose diameter is 60 in. at a speed of 360 r/min. The jet is deflected through an angle of 165°. What horsepower is delivered to the turbine runner?

14-29. A water jet 8 cm in diameter at a velocity of 70 m/s drives an impulse wheel whose diameter is 150 cm at a speed of 360 r/min. The jet is deflected through an angle of 165°. What power is delivered to the turbine?

14-30. A turbine develops 8000 hp under a head of 100 ft at 450 r/min. Estimate the
(a) speed and
(b) power developed under a head of 80 ft.
(c) What type of turbine is this?

14-31. A penstock for an impulse turbine is 4000 ft long, 2 ft in diameter, has a total head of 1200 ft, and $f = 0.024$. What is the optimum diameter of the jet? Neglect losses in the nozzle.

14-32. Plot the jet velocity, jet horsepower, and flow rate as ordinates vs. jet diameter from 0 to 2 ft as abscissa for the conditions of Prob. 14-31. Verify that maximum power occurs when $H_t = 3h_f$.

14-33. A reaction turbine develops 35,000 hp at 514 r/min under a head of 850 ft. For a barometer reading of 28.00 in. mercury and a maximum water temperature of 60°F ($p_v = 0.26$ psia), what is the maximum height of the bottom of the turbine runner above the tailwater?

14-34. A fixed-blade propeller turbine develops 45,000 hp at 85.7 r/min under a head of 48 ft. For the water and atmospheric conditions of Prob. 14-33, what is the maximum height of the turbine runner with respect to the tailwater?

14-35. A Kaplan turbine with a 22-ft-diameter propeller develops 36,000 hp at 75 r/min under a head of 36 ft.
(a) At what speed should a 2-ft-diameter model be run under a head of 18 ft for homologous conditions?
(b) What flow rate should exist for the model, assuming an efficiency of 93 percent for both prototype and model?
(c) What horsepower will be developed by the model?
(d) What is the specific speed?

14-36. Verify Eqs. (14.10), (14.11), and (14.12) on the basis of similarity of vector diagrams (Fig. 14-1) for dynamically similar or homologous machines.

14-37. A 6-in. model of a Francis turbine has an efficiency of 90 percent. Plot a curve showing expected efficiencies for homologous units 10, 20, and 30 times the size of the model.

14-38. Tests on a 16-in.-diameter turbine runner under a head of 24 ft at 400 r/min for a flow rate of 17.8 ft³/s show a power output of 40.0 hp.
(a) What is the specific speed?

(b) What type of turbine runner is this?

(c) If a 36-in. runner is used under a head of 120 ft, what is the speed, discharge, and horsepower for a homologous condition? Include effect of a slightly greater efficiency for the larger turbine.

14-39. Select the type of hydraulic turbine appropriate for the following conditions:

(a) 100 r/min, 35,000 hp, 92-ft head;

(b) 360 r/min, 20,000 hp, 2245-ft head; and

(c) 125 r/min, 40,000 hp, 70-ft head.

14-40. A 36-cm-diameter impeller in a centrifugal pump has the following characteristics at 1750 r/min:

Q (L/s)	10	15	20	25	30
H (m)	54.6	52.7	49.4	44.8	39.3

(a) Plot characteristic curves for a speed of 1150 r/min for this pump.

(b) What diameter impeller would produce a head of 42 m at 20 L/s at a speed of 1750 r/min?

(c) What power input would be required in part (b) for an efficiency of 72 percent?

14-41. A 15-cm model of a centrifugal pump has an efficiency of 65 percent. Estimate the efficiency of similar pumps

(a) 45 cm and

(b) 90 cm at the same speed.

14-42. A pump with $\sigma_c = 0.10$ pumps against a head of 60 m. The water temperature is 50°C, the barometric pressure is 100 kPa abs, and the head loss in the suction piping is 1.2 m. What is the maximum value of the static suction lift?

References

The reader is referred to the following sources for additional information on turbomachinery:

H. Addison, *Centrifugal and Rotodynamic Pumps* (London: Chapman and Hall, Ltd., 1948).

T. Baumeister, "Turbomachinery," Sec. 19 of *Handbook of Fluid Dynamics*, edited by V. L. Streeter (New York: McGraw-Hill Book Company, Inc., 1961).

J. W. Daily, "Hydraulic Machinery," Chap. 13 of *Engineering Hydraulics*, edited by H. Rouse (New York: John Wiley and Sons, Inc., 1950).

D. H. Norrie, *An Introduction to Incompressible Flow Machines* (New York: American Elsevier Publishing Company, Inc., 1963).

D. G. Shepherd, *Principles of Turbomachinery* (New York: Macmillan Inc., 1956).

Chapter 15
Varied Flow in Open Channels and Gas Flow in Pipes and Nozzles

One-dimensional flow of water in an open channel is analogous in many respects to the one-dimensional flow of a gas in pipes and in nozzles. The trivial situation is uniform flow in an open channel and the flow of a gas in a pipe at such low velocities that the gas flow is essentially incompressible, and thus uniform as well. In uniform open-channel flow, the slope of the water surface is constant in the direction of flow, and for uniform gas flow in a pipe, the pressure gradient dp/dx is constant. These water-surface and pressure gradients may be expressed quantitatively by means of the Manning equation [Eq. (12.4)] for open-channel flow and by the Darcy–Weisbach equation [Eq. (9.5)] for pipe flow.

Analogies of greater interest exist between nonuniform flow of water in open channels (see Chapter 12) and (1) isentropic gas flow in nozzles where friction is neglected (Sec. 10-3), (2) adiabatic gas flow in pipes with friction (Sec. 10-5), (3) diabatic flow of gases in pipes without friction (Sec. 10-4), and (4) shocks in gas flow (Sec. 10-6). Gradually varying and rapidly varying open-channel flow in prismatic channels, in otherwise prismatic channels but with bottom slopes in the direction of flow, and in channels with bottom and side constrictions have analogous counterparts in high-velocity gas flows in pipes and in nozzles. In all instances, the same basic equations of

512

continuity, energy, and momentum are applicable. The flow patterns or behavior depend in all instances on the ratio of the local flow velocity to that of a weak disturbance. This weak disturbance consists of an elementary surface wave in open-channel flow and of an acoustic or sound wave in gas flow. Thus the controlling parameters are the local Froude number and the local Mach number, respectively. We will consider only rectangular channels in order to use the actual depth rather than the hydraulic depth (cross-section area of flow divided by the surface width) for nonrectangular channels.

15-1. Froude and Mach Numbers

In Chapter 8 the Froude number was defined as the ratio of inertia to gravity forces in a flow system, and on this basis the resulting expression for the Froude number is

$$\text{Fr} = \frac{V}{\sqrt{gL}} \tag{15.1}$$

where V is the average flow velocity at a section and L is a characteristic length. In Sec. 12-2, the velocity of an elementary wave on the free surface of water (a wave whose height Δy is much smaller than the water depth y) was shown to be

$$c_{\text{wave}} = \sqrt{gy} \tag{15.2}$$

The Froude number, then, in a rectangular channel is the ratio of the average flow velocity at a section to that of an elementary (small) surface wave at that section. Thus, when a small pebble is dropped into flowing water, the upstream portions of the circular surface wave will travel upstream if the Froude number is less than unity, it will remain essentially at rest if the Froude number is equal to unity, and it will travel only downstream if the Froude number is greater than unity. In any case, the downstream portion of the circular surface wave will travel downstream superimposed on the stream flow. Flows at Froude numbers less than unity, equal to unity, and greater than unity are called subcritical, critical, and supercritical flows, respectively.

In Chapter 8, the Mach number was defined as the ratio of inertia to compressible forces in a flow system, and on this basis, the resulting expression is

$$\text{M} = \frac{V}{\sqrt{K/\rho}} \tag{15.3}$$

where V is the local flow velocity, K is the bulk compression modulus, and ρ is the fluid density. In Sec. 10-1, the velocity of a very weak pressure disturbance (an acoustic wave in which the change in pressure Δp is much smaller than the ambient pressure p) was shown to be

$$c = \sqrt{K/\rho} = \sqrt{kp/\rho} \qquad (15.4)$$
$$= \sqrt{kRT} \qquad \text{for a perfect gas}$$

The Mach number in a gas is, then the ratio of the fluid velocity to that of a very weak pressure wave. Thus a sound wave will travel upstream if the local Mach number is less than unity, it will not travel upstream if the local Mach number is unity, and it will travel only downstream if the local Mach number is greater than unity. In each instance, the sound wave will travel downstream with the gas and will be superimposed on the gas flow. Flows at Mach numbers less than unity, equal to unity, and greater than unity are called subsonic, sonic, and supersonic flows, respectively.

15-2. Specific Energy and the Fanno Line; Specific Thrust and the Rayleigh Line

The continuity, energy, and momentum equations applied to both open-channel flow and to gas flow, together with gas relations which give entropy as a function of pressure and density, and enthalpy as a function of entropy and density, may be used to produce some useful graphic comparisons between these flows.

SPECIFIC ENERGY

The specific energy E in an open channel was defined in Sec. 12-4 as the total energy of an open channel with respect to the channel bed. This relation is indicated in Fig. 15-1 as

$$E = \frac{V^2}{2g} + \frac{p}{\gamma} + z = \frac{V^2}{2g} + y \qquad (15.5a)$$

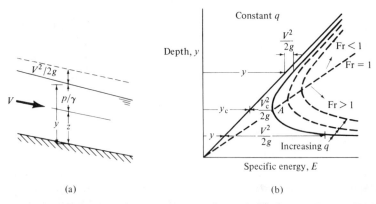

Fig. 15-1.   Specific energy in an open channel. (a) Nomenclature. (b) Diagram.

For a unit width of channel, the unit discharge is $q = Vy$ and may be considered as the constant flow-rate intensity (in volumetric flow rate per unit width). If this term is substituted in Eq. (15.5a), the specific energy may be expressed in terms of the depth y for a constant unit discharge:

$$E = \frac{q^2}{2gy^2} + y \qquad (15.5b)$$

For a constant unit discharge, $E = E(y)$ from the energy and continuity equations, and a plot of this relation is shown in Fig. 15-1b. At any depth, the horizontal distance to the $45°$ line ($y = E$) represents the depth, and the horizontal distance from the line $y = E$ to the specific-energy curve represents the velocity head. At the minimum specific energy (point A), $dE/dy = 0$, and from Eqs. (15.5) this is equivalent to

$$\frac{V^2}{2g} = \frac{y}{2} \qquad \text{or} \qquad \frac{V}{\sqrt{gy}} = 1$$

Thus the minimum specific energy represents critical flow at a Froude number of unity. Also, flow at depths greater than the critical depth (along the upper part of a specific-energy curve) is subcritical with $\text{Fr} < 1$, and flow at depths less than critical (along the lower part of a curve) is supercritical with $\text{Fr} > 1$.

THE FANNO LINE

For adiabatic flow of a gas in a pipe with friction, the continuity and energy equations, together with gas relations, may be combined and plotted graphically to give the Fanno line, which is similar to the specific-energy curve for open-channel flow.

The continuity equation is $V\rho = G$, a constant mass-flow intensity (in kilograms per square meter-second, for example). The energy equation, in the absence of heat transfer, work, and potential energy changes, is

$$h + \frac{V^2}{2} = h_0 \qquad (15.6a)$$

where h is the enthalpy and h_0 the stagnation or total enthalpy which is constant for this adiabatic flow. Substituting the value of V from the continuity equation into Eq. (15.6a) gives

$$h = h_0 - \frac{G^2}{2} \left(\frac{1}{\rho}\right)^2 \qquad (15.6b)$$

which is the equation of the Fanno line shown in Fig. 15-2a. This Fanno line is a constant (h_0) minus a parabolic quantity $(G^2/2)(1/\rho)^2$. Also shown in Fig. 15-2a is a family of constant entropy curves for which $s = s(p,\rho)$. If the curve is replotted with enthalpy h vs. entropy s, the result is as shown in Fig. 15-2b.

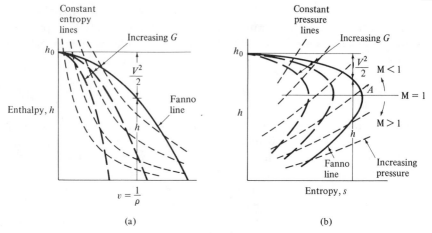

Fig. 15-2. Fanno line: (a) h vs. $1/\rho$; (b) h vs. s.

At point A, $ds/dh = 0$, and from the second law of thermodynamics as expressed by Eq. (1.6) ($T\,ds = dh - dp/\rho$), at point A, $dh = dp/\rho$. Substituting this value of dh, at point A, in the differential form of the energy equation ($dh + V\,dV = 0$), we get

$$\frac{dp}{\rho} + V\,dV = 0$$

This equation corresponds to the equation of motion in a frictionless fluid in the absence of gravitational effects, which is true for point A. From continuity, $\rho\,dV + V\,d\rho = 0$, so that

$$\frac{dp}{\rho} + V\,dV = \frac{dp}{\rho} - \frac{V^2\,d\rho}{\rho} = 0$$

and thus at point A,

$$V^2 = \frac{dp}{d\rho} \qquad \text{at constant entropy} \tag{15.7}$$

This equation shows that the velocity at point A is sonic and the Mach number at point A is unity. This condition corresponds to a Froude number of unity at point A in Fig. 15-1b. Since from Eq. (10.14) the velocity and Mach number either both increase or both decrease, and since the velocity along the upper part of the Fanno line on the h–s (enthalpy–entropy) diagram in Fig. 15-2b is less than sonic, the upper part of the Fanno line in Fig. 15-2b represents subsonic flow with M < 1 and the lower part of the Fanno line represents supersonic flow with M > 1. From the second law of thermodynamics, the entropy of the gas must increase for adiabatic flow with friction. Therefore, subsonic flow is in a direction of increasing Mach number and supersonic flow is in a direction of decreasing Mach number, both flows approaching M = 1 as a limit (see Sec. 10-5). From the family of pressure lines in Fig. 15-2b, subsonic flow is accompanied by a decreasing pressure, whereas supersonic flow is accompanied by an increasing pressure.

Similarly, from Fig. 15-1b, subcritical flow in an open channel and supercritical flow in an open channel approaching point A, where Fr = 1, are accompanied by decreasing and increasing depths, respectively.

SPECIFIC-THRUST FUNCTION IN OPEN-CHANNEL FLOW

The continuity and momentum (thrust function) equations applied to an open channel yield a graphical representation of the specific-thrust function. From Sec. 12-4, these equations are

$$Vy = q$$

indicating a constant unit discharge, or volumetric flow-rate intensity, and

$$F = \frac{\gamma y^2}{2} + \rho V^2 y$$

The specific-thrust function is the thrust function divided by the specific weight of the fluid and is

$$f = \frac{y^2}{2} + \frac{q^2}{gy} \tag{15.8}$$

For a constant unit discharge, $f = f(y)$ from the continuity and thrust function, and a plot of this specific-thrust function is shown in Fig. 15-3b.

At the minimum specific thrust (point B), $df/dy = 0$, and from Eq. (15.8), this is equivalent to

$$y_B^3 = \frac{q^2}{g} \qquad \text{or} \qquad y_B = \frac{V^2}{g}$$

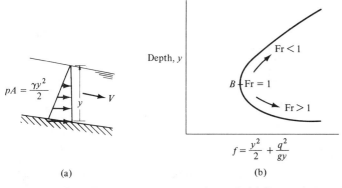

(a)   (b)

Fig. 15-3. Specific thrust in an open channel. (a) Pressure term in thrust function. (b) Specific thrust function.

Thus $V/\sqrt{gy_B} = 1$, which is the same condition as for minimum specific energy and corresponds to critical flow at which Fr = 1. As for specific energy, the upper part of the specific-thrust function curve (above B) represents subcritical flow for which Fr < 1, and the lower part of the specific-thrust function curve (below B) represents supercritical flow for which Fr > 1.

THE RAYLEIGH LINE

For diabatic flow in a constant area duct without friction, the continuity and momentum equation (thrust function), together with gas relations, give an equation whose graph is called the Rayleigh line. Diabatic flow without friction implies flow with heat transfer which is reversible.

From continuity, $V\rho = G$, a constant mass flow rate intensity. From the momentum theorem, the thrust function is constant in the absence of external drag or friction. Thus

$$p + \rho V^2 = \text{constant}$$

Combining the continuity and thrust-function equations gives

$$p + \frac{G^2}{\rho} = \text{constant} \tag{15.9}$$

Both entropy and enthalpy are functions of pressure and density, and therefore Eq. (15.9) may be plotted as shown in Fig. 15-4.

Point B in Fig. 15-4 corresponds to a sonic velocity and a Mach number of unity, just as point B on the specific-thrust curve of Fig. 15-3b corresponds to critical flow with a Froude number of unity. The sonic condition at point B may be shown by differentiating Eq. (15.9) to obtain

$$dp = \frac{G^2}{\rho^2} d\rho = 0$$

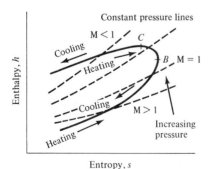

Fig. 15-4.   Rayleigh line (for reversible flow in a constant area duct with heat transfer).

so that

$$\frac{dp}{d\rho} = \frac{G^2}{\rho^2} = V_B^2$$

Then $V_B = \sqrt{dp/d\rho}$, and since $ds = 0$ (entropy constant) at point B, the velocity at point B is sonic and the Mach number is unity. The upper part of the Rayleigh curve corresponds to subsonic flow, and the lower part to supersonic flow. The flow is without friction and thus is reversible, so that heat addition (heating) is associated with an entropy increase and heat removal (cooling) is associated with an entropy decrease. (For reversible flow, $ds = dQ/T$.) Thus, flow with heat addition is towards a sonic condition, and heat removal is away from a sonic condition, for both subsonic and supersonic flow. This conclusion is compatible with the conclusions reached in Sec. 10-4 for diabatic flow. Note from Fig. 15-4 that for heating at points between the highest point on the Rayleigh line (point C) and point B, the enthalpy of the gas decreases. This means that heating in this region will cool the gas. Conversely, heat removal in this region between points B and C will warm the gas. From Sec. 10-4, point C corresponds to a Mach number of $1/\sqrt{k}$. These conclusions are coincident with those reached in Sec. 10-4 in the analysis of the differential equations for diabatic flow in a pipe without friction which were developed there.

15-3. Hydraulic Jump and Normal Shock

A hydraulic jump normal to a stream is similar to a normal shock in a gas flow, and hydraulic jumps oblique to a stream flow direction are similar to oblique shocks in a gas flow.

HYDRAULIC JUMP

In Sec. 12-6, the hydraulic jump was shown to occur with constant specific thrust and with a decrease in specific energy and is the only way in which supercritical flow may become subcritical without passing through a channel transition (Sec. 12-7). The conditions on a specific-energy and a specific-thrust diagram were shown in Fig. 12-11b and are shown again in Fig. 15-5.

The energy equation written across a jump on a horizontal bed is

$$E_1 = E_2 + h_L$$

so that the head loss (energy dissipation per unit weight of fluid) is

$$h_L = E_1 - E_2$$

which is indicated in Fig. 15-5. The Froude number changes from a value greater than unity to a value less than unity. The depths on each side of the jump were shown in Sec. 12-6 to be a function of the upstream Froude

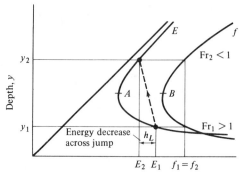

Fig. 15-5. Hydraulic jump.

number Fr_1:

$$\frac{y_2}{y_1} = \frac{1}{2}(\sqrt{1 + 8Fr_1^2} - 1) \qquad (15.10)$$

NORMAL SHOCK

In Sec. 10-6 the normal shock was shown to occur at a constant thrust function (constant momentum flux or transport). It can also be shown that the entropy change across the shock is positive in the direction of flow for upstream Mach numbers M_x greater than unity. This increase in entropy is required by the second law of thermodynamics. (Actually, for weak shocks—Mach numbers no greater than about 1.2—the increase in entropy is extremely small, so that the flow is nearly isentropic.) Since the mass-flow rate, energy, and thrust function or momentum flux are all constant across a normal shock (see Sec. 10-6), conditions on each side of the shock must lie on *both* the Fanno line and the Rayleigh line. Thus for a given mass flow intensity G, the intersections of the Fanno and Rayleigh lines locate upstream and downstream conditions for a normal shock (Fig. 15-6).

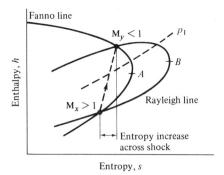

Fig. 15-6. Normal shock.

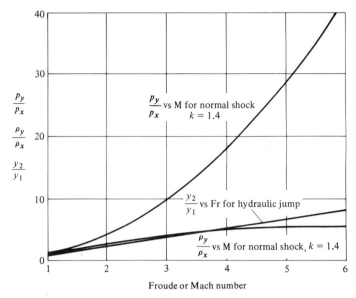

Fig. 15-7.   Depth ratio for hydraulic jump for various Froude numbers, and pressure and density ratio across normal shock for various Mach numbers ($k = 1.40$).

Either adiabatic flow in a pipe with friction or diabatic flow (with heat transfer) in a pipe without friction may be initially subsonic and may then proceed towards the sonic condition along a Fanno line or a Rayleigh line towards point A or B, respectively. These are limiting points, and they provide choking effects which may change the initial assumed conditions in the pipe if the gas tries to reach these limiting conditions upstream from the end of the pipe. Different Fanno and Rayleigh lines would then apply, since the mass flow intensity G would change. If the gas starts out supersonic (on the lower part of the lines), the gas may proceed to A or B as a limit if the back pressure beyond the end of the pipe does not prevent this. If the back pressure is at p_1, for example, a shock will form within the pipe and the gas may proceed towards A or B at subsonic flow ($M < 1$).

The pressure across a normal shock were shown in Sec. 10-6 to be a function of the upstream Mach number M_x [Eq. (10-54)].

$$\frac{p_y}{p_x} = \frac{2k}{k+1} M_x^2 - \frac{k-1}{k+1} \tag{15.11}$$

Figure 15-7 shows the depth ratios for a hydraulic jump as a function of the upstream Froude number, together with the pressure and density ratios for a normal shock as a function of the upstream Mach number. It may be noted that the depth ratio across a hydraulic jump increases almost linearly with the Froude number, whereas the pressure ratio across a normal shock increases as the square of the Mach number.

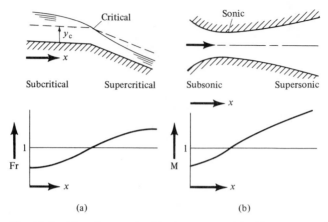

Fig. 15-8. Flow from subcritical to supercritical in an open channel and from subsonic to supersonic in a nozzle. (a) Flow from a mild to a steep slope in an open channel. (b) Gas flow in a converging–diverging nozzle with low back pressure.

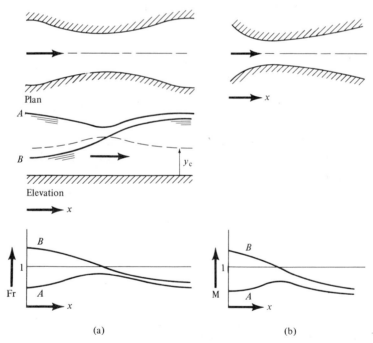

Fig. 15-9. Flow through channel constriction to a high tailwater and gas flow through a converging–diverging nozzle with a high back pressure. (a) Constriction in channel of constant bed slope, with high tailwater. Curve *A* represents initially subcritical flow; curve *B*, supercritical flow. (b) Converging–diverging nozzle, with high back pressure. Curve *A* represents initially subsonic flow; curve *B*, supersonic flow.

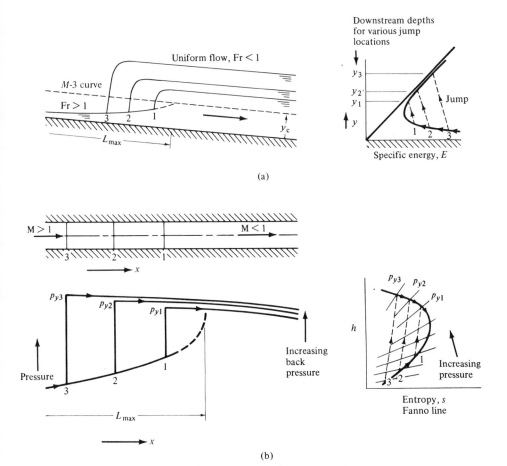

Fig. 15-10. Hydraulic jump in open channel for various tailwater elevations, and shock in pipe for various back pressures. (a) Supercritical open channel flow on a mild slope with various tailwater elevations. Location of jump is dependent upon tailwater elevation; L_{max} refers to length of M-3 curve for supercritical flow. (b) Supersonic gas flow entering pipe with friction, adiabatic flow, with various back pressures. Location of shock in pipe is dependent upon back pressure; L_{max} refers to maximum length for supersonic flow.

15-4. Some Examples of Similar Flows

There are various instances in which open-channel flows are similar to gas flows. Some examples are illustrated in Figs. 15-8, 15-9, and 15-10.

Reference

E. Preiswerk, "Application of the Methods of Gas Dynamics to Water Flows with Free Surface: Part I, Flows with no Energy Dissipation," NACA TM 934 (1940); "Part II, Flows with Momentum Discontinuities (Hydraulic Jumps)," NACA TM 935 (1940).

Chapter 16
Convective Heat Transfer

The momentum equation was used in Chapter 7 to make calculations for hydrodynamic boundary layers. Useful results are obtained. When a fluid at one temperature flows along a surface at a different temperature, there is a temperature variation as well as a velocity variation near that surface. Equations giving energy balances for control volumes, taking into account these variations, also give useful results which are applied in making heat transfer calculations. This is introduced in this chapter.

16-1. The Thermal Boundary Layer

When a fluid at one temperature flows past a surface at another temperature, heat transfer takes place from one to the other. The fluid temperature T varies from the free-stream temperature T_s to the wall temperature T_w, the variation taking place in a region known as the *thermal boundary layer* (Fig. 16-1). Heat flows by conduction at the wall surface, either from the wall to the fluid or from the fluid to the wall, according to the well-known relation known as Fourier's heat conduction equation

$$q = -k\frac{\partial T}{\partial y} \tag{16.1}$$

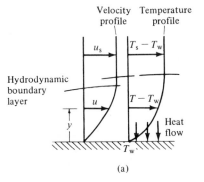 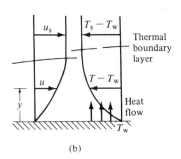

Fig. 16-1. Hydrodynamic and thermal boundary layers. (a) Wall temperature below free–stream temperature. (b) Wall temperature above free–stream temperature.

where q is the heat transferred per unit time and unit area; k is the thermal conductivity of the fluid in energy per unit area, unit time, and unit temperature difference for a unit thickness of fluid[†]; and $\partial T/\partial y$ is the temperature gradient at the wall surface. Heat flows in a direction of decreasing temperature (negative temperature gradient) and thus the negative sign is required in Eq. (16.1). If the wall temperature is higher than the free-stream temperature of the fluid, the heat transferred from the wall is carried away within the thermal boundary layer of the fluid. If the wall temperature is lower than the free-stream temperature of the fluid, the heat transferred to the wall is carried to it within the thermal boundary layer. Heat transfer of this type is called *heat transfer by convection*. If the fluid motion is produced by external means (by a pump or blower, for example), the heat transfer is due to *forced convection*. If the fluid is set in motion because of buoyancy effects owing to changes in fluid density which result from the heat transfer, the heat transfer is due to *free convection*. In either case, it is customary to define a heat-transfer coefficient h by the equation

$$q = h(T_w - T_s) \tag{16.2}$$

where h has the dimensions of energy per unit time, unit area, and unit temperature difference. The heat-transfer coefficient h depends on the fluid property values, such as specific heat capacity, thermal conductivity, viscosity, and density; the fluid flow characteristics, such as pressure gradient and whether the flow is laminar or turbulent; and the geometry of the system.

Of primary interest is the determination or prediction of the magnitude of the heat-transfer coefficient h or its dimensionless equivalent, Nu, the *Nusselt number* hL/k, for various conditions (where L is a characteristic length similar to that used for the Reynolds number). In addition, the thickness of the thermal boundary layer δ_t is also of interest. The skin-friction

---

[†] In this chapter, k is *not* the ratio of specific-heat capacities.

coefficient and the thickness of the hydrodynamic boundary layer are analogous to the heat-transfer coefficient and the thickness of the thermal boundary layer. A thermal boundary layer is always accompanied by a hydrodynamic boundary layer, whereas a hydrodynamic boundary layer need not be accompanied by a thermal boundary layer. The ratio of the thickness of the thermal to the hydrodynamic boundary layer, δ_t/δ, depends inversely on a dimensionless fluid property known as the *Prandtl number* ($\Pr = c_p\mu/k$). The Prandtl number is much less than unity for liquid metals, about unity for gases, and up to 10,000 or more for some viscous oils. Therefore, the thermal boundary layer is much thicker than the hydrodynamic boundary layer for liquid metals, about equal to it for gases, and much thinner than the hydrodynamic boundary layer for viscous oils.

A dimensional analysis would show that the forced-convection heat-transfer coefficient for flow past a flat plate depends on the Reynolds number of the flow and on the Prandtl number of the fluid:

$$\text{Nu} = f(\text{Re}, \Pr) \tag{16.3}$$

Also, the thickness of the thermal boundary layer for this flow may be shown to depend on the thickness of the hydrodynamic boundary layer and on the Prandtl number for a plate heated over its entire length:

$$\frac{\delta_t}{\delta} = f(\Pr) \tag{16.4}$$

For fully developed flow in a tube, the Nusselt number is a constant, and the thermal and the hydrodynamic boundary layers are both equal to the tube radius.

Exact solutions of the boundary-layer momentum and energy equations may be made, but a simpler approximate method, similar to that employed in Chapter 7 for the hydrodynamic boundary layer, can be used to obtain useful results.

In order to illustrate the interrelationships between the thermal and hydrodynamic boundary layers, an approximate boundary-layer analysis for laminar flow over a flat plate and in a tube and the Reynolds analogy for turbulent flow over a flat plate and in a tube will be given.

16-2. Flow over a Flat Plate with no Pressure Gradients

The boundary-layer heat-flow equation is obtained by making a heat balance on a region of unit width, of infinitesimal length dx, and height y_2 (y_2 is greater than either δ or δ_t). Assumptions include steady two-dimensional flow, constant fluid properties, and velocities in the boundary layer small so that temperature changes owing to dissipation from viscous shear may

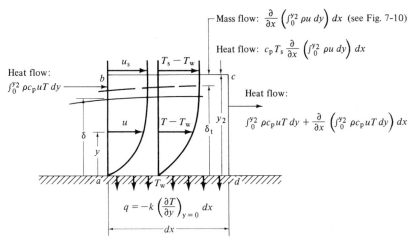

Fig. 16-2.   Infinitesimal length of boundary layer with heat balance.

be considered negligible. It should be noted that for a constant property fluid, the heat transfer does not change or affect the fluid-flow pattern.

The net heat flow (in minus flow out) must be zero. Thus, from Fig. 16-2

$$\rho c_p T_s \frac{\partial}{\partial x}\left[\int_0^{y_2} u\, dy\right]dx - \rho c_p \frac{\partial}{\partial x}\left[\int_0^{y_2} uT\, dy\right]dx - k\left(\frac{\partial T}{\partial y}\right)_{y=0} dx = 0$$

so that the heat flow equation for the boundary layer is

$$\frac{\partial}{\partial x}\int_0^{y_2}(T_s - T)u\, dy = \frac{k}{\rho c_p}\left(\frac{\partial T}{\partial y}\right)_{y=0} \tag{16.5}$$

LAMINAR FLOW

The thermal boundary-layer thickness δ_t and the local (h) or average $(\bar{h})$ heat-transfer coefficient may be obtained for any prescribed shape of temperature and velocity profiles within the boundary layer. The velocity profile (see Sec. 7-5) may be assumed to be

$$\frac{u}{u_s} = \frac{3}{2}\left(\frac{y}{\delta}\right) - \frac{1}{2}\left(\frac{y}{\delta}\right)^3 \tag{7.18}$$

Temperature profiles may be assumed to be of the form

$$T - T_w = a + by + cy^2 + dy^3$$

with appropriate boundary conditions similar to those used for the various velocity profiles in Sec. 7-5. The results are tabulated in Table 16-1. The condition $\partial^2 T/\partial y^2 = 0$ at $y = 0$ is obtained from a simplification of the

Table 16.1 TEMPERATURE PROFILES IN LAMINAR BOUNDARY LAYER

BOUNDARY CONDITION		EQUATION
AT $Y = 0$	AT $Y = \delta_t$	
$T = T_w$	$T = T_s; \dfrac{dT}{dy} = 0$	$\dfrac{T - T_w}{T_s - T_w} = 2\left(\dfrac{y}{\delta_t}\right) - \left(\dfrac{y}{\delta_t}\right)^2$ (16.6)
$T = T_w; \dfrac{\partial^2 T}{\partial y^2} = 0$	$T = T_s; \dfrac{dT}{dy} = 0$	$\dfrac{T - T_w}{T_s - T_w} = \dfrac{3}{2}\left(\dfrac{y}{\delta_t}\right) - \dfrac{1}{2}\left(\dfrac{y}{\delta_t}\right)^3$ (16.7)

boundary-layer energy equation for steady flow:

$$u\frac{\partial T}{\partial x} + v\frac{\partial T}{\partial y} = \frac{k}{\rho c_p}\frac{\partial^2 T}{\partial y^2} \tag{16.8}$$

The tangential velocity u and the normal velocity v are both zero at the wall and thus at $y = 0$, $\partial^2 T/\partial y^2 = 0$.

The integral of Eq. 16-5 need be carried out only from $y = 0$ to $y = \delta_t$, since beyond that point the fluid temperature is the free-stream temperature $(T = T_s)$ and the integrand is zero whether $u = u_s$ or not. The value of the integral of Eq. (16.5) becomes

$$I_{\text{Eq. (16.6)}} = u_s(T_s - T_w)\delta\left[\frac{1}{8}\left(\frac{\delta_t}{\delta}\right)^2 - \frac{1}{120}\left(\frac{\delta_t}{\delta}\right)^4\right] \tag{16.9}$$

using Eq. 16.6, and

$$I_{\text{Eq. (16.7)}} = u_s(T_s - T_w)\delta\left[\frac{3}{20}\left(\frac{\delta_t}{\delta}\right)^2 - \frac{3}{280}\left(\frac{\delta_t}{\delta}\right)^4\right] \tag{16.10}$$

using Eq. (16.7). If $\delta_t < \delta$, the $(\delta_t/\delta)^4$ terms in Eqs. (16.9) and (16.10) may be neglected, since they will be small compared to the $(\delta_t/\delta)^2$ terms. Equation (16.9) or (16.10) may be differentiated and the results, together with the values of $\delta(d\delta/dx)$ and δ^2 obtained from Eq. (7.19), inserted into Eq. (16.5) to obtain

$$\left(\frac{\delta_t}{\delta}\right)^3 + \frac{4}{3}x\frac{d\left(\frac{\delta_t}{\delta}\right)^3}{dx} = \frac{52}{35}\frac{k}{c_p\mu} = \frac{52}{35}\frac{1}{\text{Pr}}$$

if the temperature profile of Eq. (16.6) is used. A solution of this differential equation in δ_t/δ is

$$\left(\frac{\delta_t}{\delta}\right)^3 = \frac{52}{35\,\text{Pr}} + Cx^{-3/4}$$

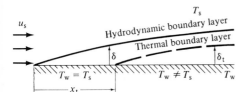

Fig. 16-3.   Plate heated beyond $x = x_t$ to a uniform temperature T_w.

If the plate is at the same temperature as the fluid up to a point x_t from the leading edge (Fig. 16-3), the constant of integration is

$$C = -\frac{52}{35 \, \text{Pr} \, x_t^{3/4}}$$

Therefore the ratio of thermal to hydrodynamic boundary-layer thickness is

$$\frac{\delta_t}{\delta} = \frac{1}{0.88 \, \text{Pr}^{1/3}} \left[1 - \left(\frac{x_t}{x}\right)^{3/4} \right]^{1/3} \qquad \text{for} \quad \delta_t < \delta \qquad (16.11)$$

Note that $x_t = 0$ if the plate is heated or cooled over its entire length.

To get the heat-transfer coefficient, Eqs. (16.1) and (16.2) are equated to give

$$h = \frac{k}{T_s - T_w} \left(\frac{\partial T}{\partial y}\right)_{y=0} = \frac{2k}{\delta_t} = \frac{2k}{\delta \left(\dfrac{\delta_t}{\delta}\right)}$$

Inserting Eq. (7.19) for δ and Eq. (16.11) for δ_t/δ gives the local heat-transfer coefficient,

$$h = \frac{0.376 \, k \, \text{Pr}^{1/3}}{x \left[1 - \left(\dfrac{x_t}{x}\right)^{3/4} \right]^{1/3}} \, \text{Re}_x^{1/2} \qquad (16.12a)$$

The dimensionless local heat-transfer coefficient, the Nusselt number, is

$$\text{Nu}_x = \frac{hx}{k} = \frac{0.376 \, \text{Pr}^{1/3} \, \text{Re}_x^{1/2}}{\left[1 - \left(\dfrac{x_t}{x}\right)^{3/4} \right]^{1/3}} \qquad (16.12b)$$

Equations (16.12) apply for a uniform wall temperature T_w beyond x_t. If the temperature profile given by Eq. (16.7) is used, the coefficient for δ_t/δ in Eq. (16.11) becomes $1/1.03$ instead of $1/0.88$, and the coefficient for Nu_x in Eq. (16.12b) becomes 0.332 instead of 0.376. Exact solutions of the energy equation [Eq. (16.8)] agree quite well with the value 0.332 for a range of Prandtl numbers from 0.6 to 15 [1].

For a plate heated or cooled over its entire length, the average value of the heat-transfer coefficient is twice the local value at the downstream end.

Using the value of h from Eq. (16.12a),

$$\bar{h} = \frac{1}{x} \int_0^x h\, dx = 2h \qquad (16.12c)$$

TURBULENT FLOW

An analogy between momentum transfer and heat transfer by turbulent motion of masses of fluid particles was derived by Reynolds in 1874. If a mass G of fluid moves through a unit area per unit time from one layer at a temperature T_1 and velocity u_1 to another layer at a temperature T_2 and velocity u_2, the heat transported is

$$q_{turb} = Gc_p(T_1 - T_2)$$

and the apparent shear stress, obtained from equating the force per unit area to the change of momentum flux, is

$$\tau_{turb} = G(u_1 - u_2)$$

Eliminating G gives

$$q_{turb} = \tau_{turb} c_p \frac{T_1 - T_2}{u_1 - u_2}$$

which may be written in differential form as

$$q_{turb} = \tau_{turb} c_p \frac{dT}{du} \qquad (16.13)$$

This is known as *Reynolds' analogy.*

Within the laminar sublayer the shear stress is, by definition,

$$\tau_{lam} = \mu \frac{du}{dy}$$

and the heat flow from the boundary (at the wall) is, from Eq. (16.1),

$$q_w = -k\left(\frac{dT}{dy}\right)_{y=0}$$

Combining these equations (using the shear stress at the wall, τ_w) gives

$$q_w = -\tau_{lam} \frac{k}{\mu}\left(\frac{dT}{du}\right) \qquad (16.14)$$

If $k/\mu = c_p$ (true for a fluid for which the Prandtl number is unity), Eqs. (16.13) and (16.14) indicate that the heat flow in the turbulent and laminar regions may be given by the same equation. Then from Eq. (16.14),

$$\frac{q_w}{\tau_w c_p}\, du = -dT \qquad (16.15a)$$

and this equation may be integrated from the wall ($u = 0$ and $T = T_w$) to the free stream ($u = u_s$ and $T = T_s$) to give

$$\frac{q_w}{\tau_w c_p} u_s = (T_s - T_w) \tag{16.15b}$$

From Eq. 16.2, the magnitude of $h = q/(T_w - T_s)$ and from Eq. (7.3), $c_f = \tau_w/(\rho u_s^2/2)$, so that

$$h = \frac{q}{T_w - T_s} = \frac{\tau_w c_p}{u_s} = \frac{c_f}{2} \rho u_s c_p \tag{16.16}$$

This equation may be written in terms of the dimensionless heat-transfer coefficient, the Nusselt number, as

$$Nu_x = \frac{hx}{k} = Re_x \, Pr \, \frac{c_f}{2} \tag{16.17a}$$

which gives the local value of the Nusselt number in terms of the local Reynolds number, the Prandtl number, and the local skin-friction coefficient. The value of the local skin-friction coefficient c_f is given by Eq. (7.27b). The Nusselt number may thus be written as

$$Nu_x = \frac{hx}{k} = 0.030 \, Pr\left(\frac{u_s x}{\nu}\right)^{4/5} \tag{16.17b}$$

for a Prandtl number approximately equal to unity. Comparison with Eq. (16.12b) for the laminar case with $x_t = 0$ indicates (since $h = k \, Nu_x/x$): (1) a higher heat-transfer coefficient for turbulent flow than for laminar flow at the same Reynolds number; (2) the heat-transfer coefficient decreases with $x^{-1/2}$ for laminar flow, and less rapidly with $x^{-1/5}$ for turbulent flow.

For fluids with a Prandtl number other than unity, the equations are slightly different. For a range of Pr from 0.6 to about 50, Eq. (16.17a) and Eq. (16.17b) are quite valid if Nu_x is replaced by $Nu_x Pr^{2/3}$[2]. Equation (16.17b) then becomes

$$Nu_x = 0.030 \, Pr^{1/3} \, Re_x^{4/5} \tag{16.17c}$$

For a plate heated or cooled over its entire length, the average value of the heat-transfer coefficient is 1.25 times the local value at the downstream end. Using the value of h from Eq. (16.17c),

$$\bar{h} = \frac{1}{x} \int_0^x h \, dx = 1.25h \tag{16.17d}$$

16-3. Flow in a Pipe with a Uniform Wall Temperature

As in the case of flow past a flat plate, velocity and temperature profiles may be used to analyze laminar flow in pipes and the Reynolds' analogy used to analyze turbulent flow.

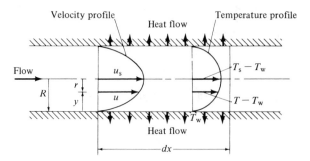

Fig. 16-4. Laminar flow in a pipe with heat transfer.

LAMINAR FLOW

For fully developed laminar flow, the velocity profile is given by Eq. (9.11b) as a parabola

$$\frac{u}{u_s} = 2\left(\frac{y}{R}\right) - \left(\frac{y}{R}\right)^2 \tag{16.18}$$

The heat flow by conduction through a peripheral area $2\pi r\, dx$ (Fig. 16-4) is

$$Q = -k2\pi r\, dx \frac{dT}{dy} \tag{16.19}$$

If a suitable expression for the temperature profile is obtainable, the heat-transfer coefficient h, defined by

$$q = h(T_b - T_w) \tag{16.20}$$

may be derived by combining Eqs. (16.19) and (16.20). The temperature T_b is called the *bulk temperature*, defined by

$$T_b = \frac{\int Tu\, dA}{\int u\, dA} \tag{16.21}$$

and represents the temperature obtained by mixing the fluid passing a section in a given time interval.

As in the case of a flat plate, a temperature profile of the form

$$T - T_w = a + by + cy^2 + dy^3$$

may be assumed, with boundary conditions $T - T_w = 0$ at $y = 0$, $T - T_w = T_s - T_w$, and $dT/dy = 0$ at $y = R$. In addition, if $r = R - y$ is substituted in Eq. (16.19) and the equation solved for dT/dy and differentiated, it is found that

$$\frac{d^2T}{dy^2} = \frac{1}{R}\left(\frac{dT}{dy}\right)_{y=0}$$

This equation indicates that the curvature of the temperature profile is *not* zero at the wall, as was true for the flat plate. These conditions yield a temperature profile whose dimensionless form is invariant in the direction of flow:

$$\frac{T - T_w}{T_s - T_w} = \frac{6}{5}\left(\frac{y}{R}\right) + \frac{3}{5}\left(\frac{y}{R}\right)^2 - \frac{4}{5}\left(\frac{y}{R}\right)^3 \tag{16.22}$$

When this equation is differentiated and introduced into Eq. (16.19), the heat flow per unit area of pipe wall is shown to be

$$q = -k\left(\frac{dT}{dy}\right)_{y=0} = \frac{6}{5}k\frac{(T_s - T_w)}{R} \tag{16.23}$$

From Eqs. (16.18), (16.21), and (16.22) the bulk temperature is expressible as

$$T_b - T_w = 0.583(T_s - T_w)$$

Thus the local heat-transfer coefficient is

$$h = \frac{q}{T_b - T_w} = \frac{6}{5}\frac{k(T_s - T_w)}{R(T_b - T_w)} = \frac{12}{5}\left(\frac{k}{0.583D}\right) = 4.12\frac{k}{D} \tag{16.24}$$

The dimensionless heat-transfer coefficient (the Nusselt number) based on the diameter as the characteristic length is constant and does not vary in the direction of flow for a fully developed velocity and temperature profile for laminar flow in a pipe:

$$Nu_D = \frac{hD}{k} = 4.12 \tag{16.25a}$$

Exact calculations by Graetz and Nusselt give

$$Nu_D = 3.65 \tag{16.25b}$$

Equations (16.25) are for a uniform wall temperature T_w.

TURBULENT FLOW

The Reynolds analogy for fully developed turbulent flow or fluids having a Prandtl number near unity gives the Nusselt number in terms of the pipe Reynolds number, the Prandtl number, and the friction factor. This relationship is

$$Nu_D = Re_D \, Pr\frac{f}{8} \tag{16.26}$$

This relation is obtained by integrating Eq. (16.15a) from the pipe wall ($u = 0$ and $T = T_w$) to the bulk conditions ($u = V$ and $R = T_b$) to get

$$\frac{q_w}{\tau_w c_p}V = T_b - T_w \tag{16.27}$$

From the definition of the heat-transfer coefficient for a pipe given in Eq. (16.20),

$$q_w = h(T_b - T_w) \qquad\qquad\qquad [16.20]$$

we obtain from Eqs. (16.20) and (16.27),

$$Vh = \tau_w c_p$$

The wall shear stress in a pipe is defined in Eq. (9.8) as

$$\tau_w = \frac{f\rho V^2}{8}$$

so that

$$\frac{h}{\rho V c_p} = \frac{f}{8}$$

This equation may be expanded to the form

$$\left(\frac{hD}{k}\right)\left(\frac{\mu}{D\rho V}\right)\left(\frac{k}{\mu c_p}\right) = \frac{f}{8}$$

which is equivalent to Eq. (16.26).

Agreement with experimental data for gases is rather good if Nu_D in Eq. (16.26) is replaced by $Nu_D\, Pr^{2/3}$. Equation (16.26) then becomes

$$Nu_D = Re_D\, Pr^{1/3}\, \frac{f}{8} \qquad\qquad (16.28)$$

PROBLEMS

16-1. Show that the Prandtl number ($Pr = c_p\mu/k$) is dimensionless.

16-2. Derive the temperature profiles given by Eqs. (16.6) and (16.7) from $T - T_w = a + by + cy^2 + dy^3$ and the boundary conditions given in Table 16-1.

16-3. Show that the integral of Eq. (16.5) is given by Eqs. (16.9) and (16.10) for the temperature profiles of Eqs. (16.6) and (16.7), respectively, and the velocity profile of Eq. (7.18).

16-4. Derive the differential equation in δ_t/δ following Eq. (16.10).

$$\left(\frac{\delta_t}{\delta}\right)^3 + \frac{4}{3}\, x\, \frac{d\left(\frac{\delta_t}{\delta}\right)^3}{dx} = \frac{52}{35}\, \frac{k}{c_p\mu} = \frac{53}{35}\, \frac{1}{Pr}$$

by differentiating Eq. (16.9), using values of $\delta(d\delta/dx)$ and δ^2 from Eq. (7.19) the value of $(\partial T/\partial y)_{y=0}$ from Eq. (16.6), and inserting into Eq. (16.5).

16-5. Show that Eq. (16.11) is a solution of the differential equation in δ_t/δ derived in Prob. 16-4.

16-6. For a linear velocity distribution ($u/u_s = y/\delta$) within a laminar boundary layer for a fluid flowing past a flat plate, the boundary layer thickness is, from Table 7-2,

$$\frac{\delta}{x} = \frac{3.46}{Re_x^{1/2}}$$

Assume a linear temperature profile as well

$$\left(\frac{T - T_w}{T_s - T_w}\right) = \frac{y}{\delta_t}$$

with a constant wall temperature T_w.

(a) Show that

$$\frac{\delta_t}{\delta} = \frac{1}{Pr^{1/3}}$$

(b) Show that

$$Nu_x = \frac{1}{3.46} Re_x^{1/2} Pr^{1/3}$$

16-7. Show that the average heat-transfer coefficient $\bar{h}$ for laminar flow along a flat plate heated or cooled over its entire length is twice the local value of the heat-transfer coefficient h at any arbitrary distance x from the leading edge; that is, verify Eq. (16.12c).

16-8. Show that from Eq. (16.16), $Nu_x = hx/k = Re_x Pr\, c_f/2$ [Eq. (16.17a)].

16-9. Assume that a boundary layer is turbulent from the leading edge of a flat plate and that heat transfer takes place according to Eq. (16.17b). Show that the average value of the heat-transfer coefficient over a distance x from the leading edge is 1.25 times the local value of the heat-transfer coefficient at x; that is, show that

$$\overline{Nu_x} = 1.25\, Nu_x \qquad \text{or} \qquad \bar{h} = 1.25\, h$$

16-10. Derive the temperature profile for fully developed laminar flow in a pipe given by Eq. (16.22) from $T - T_w = a + by + cy^2 + dy^3$ and the given boundary conditions.

16-11. Verify the expression for the bulk temperature

$$T_b - T_w = 0.583(T_s - T_w)$$

using the laminar velocity and temperature profiles given by Eqs. (16.18) and (16.22), respectively.

16-12. For a parabolic velocity profile for laminar flow in a pipe given by Eq. (16.18), and a linear temperature profile given by

$$\frac{T - T_w}{T_s - T_w} = \frac{y}{R}$$

show that the bulk temperature T_b is given by the expression

$$T_b - T_w = \frac{7}{15}(T_s - T_w)$$

This means that the heat content of the fluid is the same as though the temperature were uniform across the section at a value equal to

$$\frac{7}{15}(T_s - T_w) + T_w$$

References

1. E. R. G. Eckert and R. M. Drake, *Heat and Mass Transfer* (New York: McGraw-Hill Book Company, Inc., 1959), p. 188.
2. A. P. Colburn, "A Method of Correlating Forced Convection Heat Transfer Data and a Comparison with Fluid Friction," *Trans. Am. Inst. Chem. Engrs.*, Vol. 29 (1933), pp. 174–210.
3. J. P. Holman, *Heat Transfer*, 4th ed. (New York: McGraw-Hill Book Company, Inc., 1976).

Appendix I
Nomenclature

Letter Symbols

a	linear acceleration	C_c	contraction coefficient
A	area	C_D	drag coefficient
A	constant	C_f	average skin friction coefficient
b	breadth	C_L	lift coefficient
b_s	surface breadth	C_p	pressure coefficient
B	breadth	C_v	coefficient of velocity
B	constant	d	depth normal to flow in an open channel
c	wave velocity		
c_f	local skin friction coefficient	d	diameter
c_p	specific heat capacity at constant pressure	D	diameter
		D	drag force
c_v	specific heat capacity at constant volume	D_h	hydraulic diameter
		e	base of natural logarithms
C	chord length	e	eccentricity
C	concentration	e	energy per unit mass
C	constant	E	energy transfer per unit weight in a rotating machine
°C	degrees Celsius		
C	Hazen–Williams pipe roughness coefficient	E	specific energy in an open channel

f	force	N	newton force
f	friction factor	N	rotational speed
f	function	N_s	specific speed
f	specific thrust function	p	pressure (p_a, atmospheric; p_v,
°F	degrees Fahrenheit		vapor; p_g, gage; p_c, critical;
F	force		p_0, stagnation; p_s, free stream)
F	thrust function	P	perimeter
g	acceleration of gravity	P	power
g_c	factor of proportionality	q	heat transfer per unit mass
G	mass flow intensity	q	heat transfer per unit time and
h	convection heat transfer		area
	coefficient	q	unit discharge in an open
h	enthalpy per unit mass		channel
h	head	Q	heat transferred
h	height	Q	volumetric flow rate
H	head	r	radius
i	$\sqrt{-1}$	°R	degrees Rankine
$\mathbf{i}$	unit vector in x direction	R	gas constant
I	moment of inertia	R	radius
$\mathbf{j}$	unit vector in y direction	R_h	hydraulic radius
ID	inside diameter	s	arc length
J	mechanical equivalent of heat	s	entropy per unit mass
k	ratio of specific heat capacities	s	specific gravity relative to
	($k = c_p/c_v$)		water
k	roughness height	s	streamline direction
k	thermal conductivity	s	surface
$\mathbf{k}$	unit vector in z direction	S	slope (S_c, critical; S_b bed;
k_L	loss coefficient		S_e, energy or total head line)
K	compressibility (or elastic	t	thickness
	modulus)	t	time
K	consistency index for	T	temperature (T_0, stagnation;
	non-Newtonian fluids		T_b, bulk; T_s, free stream;
°K	degrees Kelvin		T_w, wall)
K	flow coefficient	T	time (dimensional
l	length		representation only)
l	mixing length	T	torque
L	length	u	internal energy per unit mass
L	lift force	u	variable velocity at a section
L_e	entrance length		(u_s, free stream; u_b, edge of
m	mass		laminar sublayer)
$\dot{m}$	mass flow rate	u	x component of velocity
M	mass (dimensional	U	peripheral velocity of a rotor
	representation)		or runner
n	direction normal to streamline	v	specific volume
n	flow behavior index for	v	y component of velocity
	non-Newtonian fluids	v_*	shear velocity
n	Manning roughness coefficient	V	average velocity at a section
n	polytropic exponent	V	velocity in general (V_c, critical;
			V_n, normal velocity for

	uniform flow in an open channel)
Ψ	volume
w	$\phi + i\psi$ (in $\phi - \psi$ plane)
w	work per unit mass (or per unit weight)
w	z component of velocity
W	total weight
x	flow direction
X	body force
y	coordinate normal to flow, or depth (y_c, critical; y_h, hydraulic; y_n, normal)
Y	body force
Y	expansion factor
z	elevation above datum plane
z	$x + iy$ (in $x-y$ plane)
Z	body force
Z	compressibility factor for gases
Z	weir crest height

Greek Letters

α	angle
α	angle of attack
α	angular acceleration
α	energy correction factor
β	blade angle
β	momentum correction factor
β	wave angle
γ	specific weight
Γ	circulation
δ	boundary layer thickness (δ^*, displacement; δ_i, momentum; δ_b, sublayer; δ_t, thermal)
Δ	increment of a quantity
∂	partial derivative
ϵ	eddy viscosity
η	dimensionless coordinate
η	efficiency
θ	angle
θ	temperature (dimensional representation only)
θ	deflection angle
κ	proportionality factor in turbulent flow
μ	dynamic viscosity
v	kinematic viscosity

π	dimensionless group
π	3.1416
ρ	mass density
Σ	summation
σ	cavitation number or index
σ	normal stress
σ	surface tension
τ	shear stress (τ_0, at boundary; τ_w, at wall for convective heat transfer)
ϕ	velocity potential
ψ	stream function
ω	angular velocity
∇	vector operator del

Subscripts

0	at boundary ($y = 0$)
0	stagnation condition
a	atmospheric
a	axial
b	barometer
b	bulk
b	sublayer
c	cavity
c	critical
CG	center of gravity
D	diameter
D	drag
e	energy
e	entrance
f	flowing
f	force
f	friction
F	force
g	gas
g	gravity
h	hydraulic
H	horizontal
L	lift
l	liquid
L	loss
m	manometer
m	mass
m	meridional
m	mixture
m	model
n	normal
N	normal

o	oil
p	pressure
p	prototype
P	pump
r	radial
R	radial
s	in direction parallel to streamline
s	free stream
s	shear
s	solid
s	specific (speed)
t	tangential
t	thermal
T	tangential
T	turbine
v	vapor
V	vertical
w	wall
w	water
x	in x direction

x	upstream of normal shock
y	downstream of normal shock
y	in y direction
z	in z direction
θ	normal to radial direction

Dimensionless Numbers or Groups

Fr	Froude number
M	Mach number
$N_{s(P)}$	specific speed of a pump
$N_{s(T)}$	specific speed of a turbine
Nu	Nusselt number
Pr	Prandtl number
Re	Reynolds number
Re'	pseudo-Reynolds number
We	Weber number
σ	cavitation number or index

Appendix II
Conversion Factors and
Property Values

Table A-1 UNIT CONVERSION FACTORS

MULTIPLY	BY	TO GET
LENGTH		
feet	0.3048	meters
inches	2.54	centimeters
meters	3.281	feet
AREA		
square feet	0.0929	square meters
square inches	6.452	square centimeters
square meters	10.764	square feet
VOLUME		
cubic feet	0.02832	cubic meters
U.S. gallons	3.785	liters
U.S. gallons	0.1336	cubic feet
cubic meters	35.31	cubic feet
MASS		
slugs	14.59	kilograms
kilograms	0.0686	slugs
pounds mass	0.4536	kilograms
pounds mass	0.03108	slugs
slugs	32.174	pounds mass

Table A-1. UNIT CONVERSION FACTORS (*continued*)

MULTIPLY	BY	TO GET
DENSITY		
slugs/ft^3	515.4	kg/m^3
lb$_m$/ft^3	16.019	kg/m^3
kg/m^3	0.001940	slugs/ft^3
VELOCITY		
ft/s	0.3048	m/s
miles per hour	0.4470	m/s
knots	0.5148	m/s
knots	1.689	ft/s
FORCE		
pounds	4.448	newtons
newtons	0.2248	pounds
TORQUE		
pound feet	1.356	newton meters
ENERGY, WORK, HEAT		
ft lb$_f$	1.356	N m (joule)
Btu	778.2	ft lb$_f$
POWER		
horsepower	550	ft lb$_f$/s
ft lb$_f$/s	1.356	watts
joules/second	1	watts
PRESSURE, STRESS		
lb$_f$/in.^{2}	6895	N/m^2 (pascal, Pa)
lb$_f$/ft^2	47.88	N/m^2 (Pa)
bars	10^5	N/m^2 (Pa)
TEMPERATURE		
degrees Kelvin	1.80	degrees Rankine
GAS CONSTANT, SPECIFIC HEAT		
ft lb$_f$/slug °R	0.1673	J/kg °K
DYNAMIC VISCOSITY		
slug/ft s	47.88	kg/m s
poises	0.10	kg/m s
KINEMATIC VISCOSITY		
ft^2/s	0.0929	m^2/s
ft^2/s	929	cm^2/s (Stokes)
HEAT TRANSFER COEFFICIENT		
Btu/hr ft^2 °F	5.678	J/s m^2 °C (W/m^2 °C)

TEMPERATURE
$T(°K) = T(°C) + 273.15$
$T(°R) = T(°F) + 459.67$

ACCELERATION OF GRAVITY
$32.174 \text{ ft/s}^2 = 9.8066 \text{ m/s}^2$

Table A-2 TYPICAL PROPERTIES OF
GASES AT ROOM TEMPERATURE

		SI UNITS (m N/kg °K)	
GAS	k	R	c_p
Air	1.40	287.1	1,005
Helium	1.66	2077	5,224
Hydrogen	1.40	4124	14,434
Methane	1.31	518	2,190
Xenon	1.66	63.3	159

Table A-3 DENSITY OF SOME LIQUIDS

TEMPERATURE $(°C)$	DENSITY (kg/m^3)			
	WATER	MERCURY	BENZENE	GLYCERINE
0	999.8	13,595		1275
5	999.9	13,583		
10	999.7	13,571		
15	999.1	13,558	880	1270
20	998.2	13,546		
25	997.0	13,534		
30	995.6	13,521		
35	994.0	13,509		
40	992.2	13,497	866	1250

Table A-4 TYPICAL SPECIFIC GRAVITY
OF SOME LIQUIDS AT 20°C

LIQUID	SPECIFIC GRAVITY
Gasoline	0.66–0.69
Denatured alcohol	0.80
Kerosene	0.80–0.84
Crude oil	0.80–0.92
Castor oil	0.97
Sea water	1.025
Carbon tetrachloride	1.594
Acetylene tetrabromide	2.962
Mercury	13.546

Table A-5 DENSITY OF WATER

TEMPERATURE (°C)	DENSITY ρ (kg/m³)
0	999.8
5	999.9
10	999.7
15	999.1
20	998.2
25	997.1
30	995.6
35	994.1
40	992.2
45	990.2
50	988.1
55	985.7
60	983.2
65	980.6
70	977.8
75	974.9
80	971.8
85	968.6
90	965.3
95	961.9
100	958.4

SOURCE: From *Handbook of Chemistry and Physics* (West Palm Beach, Fla.: CRC Press. 59th edition, 1978–1979).

Table A-6 SURFACE TENSION OF LIQUIDS IN CONTACT WITH AIR, WATER, OR THEIR OWN VAPORS

	SURFACE TENSION	
SUBSTANCE	(lb_f/ft)	(N/m)
Benzene–air	0.0020	0.029
Carbon tetrachloride–air	0.0018	0.027
Water–air	0.0050	0.073
Mercury–air	0.030	0.435
Benzene–water	0.0024	0.035
Carbon tetrachloride–water	0.0031	0.045
Mercury–water	0.026	0.375
Benzene vapor	0.0020	0.029
Carbon tetrachloride vapor	0.0018	0.027
Water vapor	0.0050	0.073

Table A-7 VAPOR PRESSURE OF SOME LIQUIDS

TEMPERATURE		WATER		MERCURY	KEROSENE	METHYL ALCOHOL
(°F)	(°C)	$(lb_f/in.^2)$	(kPa)	(kPa)	(kPa)	(kPa)
32	0	0.088	0.61	0.000025		
41	5	0.126	0.87		2.28	5.10
50	10	0.178	1.23	0.000066	2.55	7.10
59	15	0.247	1.70			9.65
68	20	0.339	2.34	0.00016	3.52	12.6
77	25	0.459	3.16		4.14	16.4
86	30	0.615	4.24	0.00037	4.8	21.4
104	40	1.070	7.38	0.00081	6.1	
122	50	1.789	12.34		7.1	
140	60	2.89	19.92			
158	70	4.52	31.2			

Table A-8 VISCOSITY OF WATER
(See Figs. A-1 and A-2)

TEMPERATURE (°C)	DYNAMIC μ (kg/m s)	KINEMATIC v (m^2/s)
0	1.787×10^{-3}	1.787×10^{-6}
5	1.519	1.519
10	1.307	1.307
15	1.139	1.140
20	1.002	1.004
25	0.890	0.893
30	0.798	0.801
35	0.719	0.723
40	0.653	0.658
45	0.596	0.602
50	0.547	0.553
55	0,504	0.511
60	0.466	0.474
65	0.433	0.441
70	0.404	0.409

SOURCE: From *Handbook of Chemistry and Physics* (West Palm Beach, Fla.: CRC Press, 59th edition, 1978–1979).

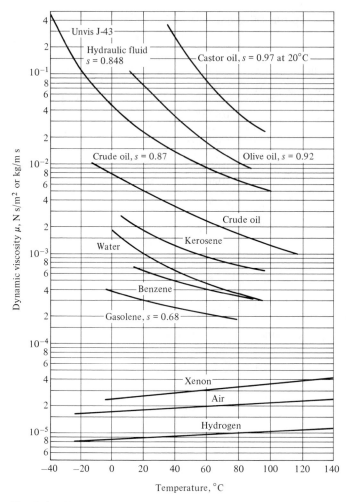

Fig. A-1. Dynamic viscosity of some liquids and gases. Values of specific gravity apply at about 20°C.

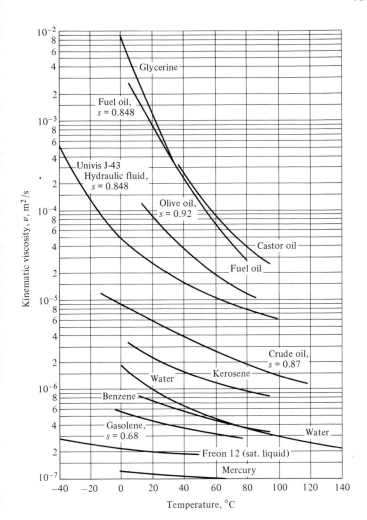

Fig. A-2. Kinematic viscosity of some liquids. Values of specific gravity apply at about 20°C.

Appendix III
Laminar Flow of a
Power-Law Fluid

From Fig. 9-3 the shear stress for steady, fully developed flow in a circular tube given by Eq. (9.7) may be equated to the shear stress for a power-law fluid given by Eq. (1.30). From this may be obtained a general expression for the shape of the velocity profile, the average velocity and volumetric flow rate in terms of the pressure gradient, and the friction factor. The results may then be compared with those for laminar flow of a Newtonian fluid in a circular tube given in Sec. 9-3. When $n = 1$, the general results for a power-law fluid are those for a Newtonian fluid.

Equating Eq. (1.30) to (9.7) gives

$$\tau = -K\left(\frac{du}{dr}\right)^n = \frac{\Delta p}{L}\frac{r}{2}$$

so that

$$du = -\left(\frac{\Delta p}{2LK}\right)^{1/n}\int r^{1/n}\, dr$$

Integration with boundary conditions $u = 0$ at $r = R$ gives

$$u = \left(\frac{\Delta p}{2LK}\right)^{1/n}\left(\frac{n}{n+1}\right)R^{1+1/n}\left[1 - \left(\frac{r}{R}\right)^{1+1/n}\right] \tag{AIII.1}$$

The average velocity V is

$$V = \frac{2\pi \int ur\,dr}{\pi R^2} = R^{1+1/n}\left(\frac{\Delta p}{2LK}\right)^{1/n}\left(\frac{n}{3n+1}\right) \tag{AIII.2}$$

The maximum velocity u_m occurs at $r = 0$, and is

$$u_m = R^{1+1/n}\left(\frac{n}{n+1}\right)\left(\frac{\Delta p}{2LK}\right)^{1/n} \tag{AIII.3}$$

Thus we have the following ratios of point to average velocity

$$\frac{u}{V} = \frac{3n+1}{n+1}\left[1 - \left(\frac{r}{R}\right)^{(n+1)/n}\right] \tag{AIII.4}$$

of average to maximum velocity

$$\frac{V}{u_m} = \frac{n+1}{3n+1} \tag{AIII.5}$$

and of point to maximum velocity

$$\frac{u}{u_m} = 1 - \left(\frac{r}{R}\right)^{(n+1)/n} \tag{AIII.6}$$

For $n = \frac{1}{2}$, $u/u_m = [1 - (r/R)^3]$, $V/u_m = \frac{3}{5}$, and the velocity profile is a cubic equation.

For $n = \frac{1}{2}$, $u/u_m = [1 - (r/R)^3]$, $V/u_m = \frac{3}{5}$, and the velocity profile is a cubic equation.

For $n = 1$, $u/u_m = [1 - (r/R)^2]$, $V/u_m = \frac{1}{2}$, and the velocity profile is a parabola, as was obtained in Sec. 9-3.

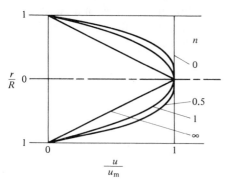

Fig. A-3. Velocity profiles for power-law fluids for fully developed flow in circular tubes.

For $n \to \infty$, $u/u_m = 1 - (r/R) = y/R$, $V/u_m = \frac{1}{3}$, and the velocity profile consists of straight lines. These velocity profiles are shown in Fig. A-3.

The volumetric flow rate Q is

$$Q = VA = \frac{\pi n}{3n + 1}\left(\frac{R^{3n+1}}{2K}\frac{\Delta p}{L}\right)^{1/n} \tag{AIII.7}$$

The friction factor, f, is

$$f = \frac{8\tau_0}{\rho V^2} = \frac{8}{\rho V^2}K\left(\frac{du}{dy}\right)^n$$

where

$$\frac{du}{dy} = -\frac{du}{dr} = \frac{3n + 1}{n}\left(\frac{2V}{D}\right) = \frac{3n + 1}{4n}\left(\frac{8V}{D}\right)$$

Thus

$$f = \frac{8}{\rho V^2}K\left(\frac{3n + 1}{4n}\right)^n\left(\frac{8V}{D}\right)^n$$

$$= \frac{64}{D^n V^{2-n}\rho/\mu'} \tag{AIII.8a}$$

$$= \frac{64}{Re'} \tag{AIII.8b}$$

where $\mu' = 8^{n-1}K[(3n + 1)/4n]^n$.

For a Newtonian fluid, $n = 1$, $\mu' = K = \mu$, and the value of f is the same as that given in Eq. (9.14).

Appendix IV
The Prandtl Boundary
Layer Equations

Navier–Stokes Equations

The Navier–Stokes equations were simplified by Prandtl in 1904. He stated that viscous effects were concentrated in a thin layer of fluid along solid boundaries. By estimating the order of magnitude of the terms in the Navier–Stokes equations, he simplified them by keeping only terms of magnitude 1 and dropping those of smaller magnitude, δ or less. His method will be shown for two-dimensional steady boundary-layer flow. The boundary layer thickness is assumed to be small as compared with any characteristic dimension of the boundary—the size of a body immersed in the fluid, for example. Distances in the x direction and velocities u in that direction (including the free-stream velocity u_s) are assumed to be of order 1, whereas distances in the y direction and velocities v in that direction are assumed to be of order δ. The continuity equation with the order of magnitude of each term written beneath it is

$$\frac{\partial u}{\partial x} + \frac{\partial v}{\partial y} = 0$$

$$\frac{1}{1} \qquad \frac{\delta}{\delta}$$

Both terms are seen to be of order 1.

Outside the boundary layer the Bernoulli equation applies, and the differential form of this equation, with the order of magnitude of each term written beneath it, is

$$\frac{\partial p}{\partial x} + \rho\, u_s \frac{\partial u_s}{\partial x} = 0$$

$$1 \quad 1 \quad \frac{1}{1}$$

Thus $\partial p/\partial x$ is of order 1 when the density is assumed to be of order 1.

The Navier–Stokes equation for the x direction [Eq. 4.21] is, neglecting gravity,

$$u\frac{\partial u}{\partial x} + v\frac{\partial u}{\partial y} = -\frac{1}{\rho}\frac{\partial p}{\partial x} + v\left(\frac{\partial^2 u}{\partial x^2} + \frac{\partial^2 u}{\partial y^2}\right)$$

$$1\frac{1}{1} \quad \delta\frac{1}{\delta} \qquad 1\ 1 \qquad 1 \qquad \frac{1}{\delta^2}$$

The first term in parentheses is negligible as compared with the second term. The term containing viscosity must be of the same order of magnitude as the other terms in the equation, and thus the viscosity v is of the order δ^2. This means that a low-viscosity fluid is required in order to obtain boundary layer flow.

The Navier–Stokes equation for the y direction [Eq. (4.21)] is, again neglecting gravity,

$$u\frac{\partial v}{\partial x} + v\frac{\partial v}{\partial y} = -\frac{1}{\rho}\frac{\partial p}{\partial y} + v\left(\frac{\partial^2 v}{\partial x^2} + \frac{\partial^2 v}{\partial y^2}\right)$$

$$1\frac{\delta}{1} \quad \delta\frac{\delta}{\delta} \qquad 1 \qquad \delta^2\left(\frac{\delta}{1} \qquad \frac{\delta}{\delta^2}\right)$$

The term $\partial p/\partial y$ must be of the order of magnitude of δ and thus may be neglected as compared with the terms in the x component Navier–Stokes equation. This indicates that the pressure gradients across the boundary layer may be neglected and that the free-stream pressure is impressed on the boundary layer.

The simplified Navier–Stokes equations thus become

$$u\frac{\partial u}{\partial x} + v\frac{\partial u}{\partial y} = -\frac{1}{\rho}\frac{\partial p}{\partial x} + v\frac{\partial^2 u}{\partial y^2}$$

$$\frac{\partial p}{\partial y} = 0$$

$$\frac{\partial u}{\partial x} + \frac{\partial v}{\partial y} = 0$$

These are the Prandtl boundary-layer equations for steady, two-dimensional incompressible flow.

Appendix V
Tables for Compressible Gas Flow

Table A-9   ISENTROPIC FLOW, $k = 1.4$

M	$\dfrac{p}{p_0}$	$\dfrac{T}{T_0}$	$\dfrac{\rho}{\rho_0}$	$\dfrac{A}{A^*}$	$\dfrac{V}{V^*}$
0	1.0000,0[a]	1.00000	1,0000,0	∞	0
.01	.9999,3	.99998	.9999,5	5,7.874	.01096
.02	.9997,2	.99992	.9998,0	2,8.942	.02191
.03	.9993,7	.99982	.9995,5	1,9.300	.03286
.04	.9988,8	.99968	.9992,0	14.,482	.04381
.05	.9982,5	.99950	.9987,5	11.5,915	.05476
.06	.9974,8	.99928	.9982,0	9.6,659	.06570
.07	.9965,8	.99902	.9975,5	8.2,915	.07664
.08	.9955,3	.99872	.9968,0	7.2,616	.08758
.09	.9943,5	.99838	.9959,6	6.4,613	.09851

SOURCE: The values in this table were originally calculated and reported in M.I.T. Meteor Report No. 14, Bureau of Ordnance, U.S. Navy Department, December, 1947, by A. H. Shapiro, W. R. Hawthorne, and G. M. Edelman.

[a] *Note:* Linear interpolation may be made (a) for values without a comma and (b) to the left of comma for values containing a comma.

Table A-9 (*Continued*)

M	$\dfrac{p}{p_0}$	$\dfrac{T}{T_0}$	$\dfrac{\rho}{\rho_0}$	$\dfrac{A}{A^*}$	$\dfrac{V}{V^*}$
.10	.9930,3	.99800	.9950,2	5.8,218	.10943
.11	.9915,7	.99758	.9939,8	5.2,992	.12035
.12	.9899,8	.99714	.9928,4	4.8,643	.13126
.13	.9882,6	.99664	.9916,0	4.4,968	.14216
.14	.9864,0	.99610	.9902,7	4.18,24	.15306
.15	.9844,1	.99552	.9888,4	3.91,03	.16395
.16	.9822,8	.99490	.9873,1	3.67,27	.17483
.17	.9800,3	.99425	.9856,9	3.46,35	.18569
.18	.9776,5	.99356	.9839,8	3.27,79	.19654
.19	.9751,4	.99283	.9821,7	3.11,22	.20738
.20	.9725,0	.99206	.9802,7	2.96,35	.21822
.21	.9697,3	.99125	.9782,8	2.82,93	.22904
.22	.9668,5	.99041	.9762,1	2.70,76	.23984
.23	.9638,3	.98953	.9740,3	2.59,68	.25063
.24	.9607,0	.98861	.9717,7	2.49,56	.26141
.25	.9574,5	.98765	.9694,2	2.40,27	.27216
.26	.9540,8	.98666	.9669,9	2.31,73	.28291
.27	.9506,0	.98563	.9644,6	2.23,85	.29364
.28	.9470,0	.98456	.9618,5	2.16,56	.30435
.29	.9432,9	.98346	.9591,6	2.09,79	.31504
.30	.9394,7	.98232	.9563,8	2.035,1	.32572
.31	.9355,4	.98114	.9535,2	1.976,5	.33638
.32	.9315,0	.97993	.9505,8	1.921,8	.34701
.33	.9273,6	.97868	.9475,6	1.870,7	.35762
.34	.9231,2	.97740	.9444,6	1.822,9	.36821
.35	.9187,7	.97608	.9412,8	1.778,0	.37879
.36	.9143,3	.97473	.9380,3	1.735,8	.38935
.37	.9097,9	.97335	.9347,0	1.696,1	.39988
.38	.9051,6	.97193	.9312,9	1.658,7	.41039
.39	.9004,4	.97048	.9278,2	1.623,4	.42087
.40	.8956,2	.96899	.9242,8	1.590,1	.43133
.41	.8907,1	.96747	.9206,6	1.558,7	.44177
.42	.8857,2	.96592	.9169,7	1.528,9	.45218
.43	.8806,5	.96434	.9132,2	1.500,7	.46256
.44	.8755,0	.96272	.9094,0	1.474,0	.47292
.45	.8702,7	.96108	.9055,2	1.448,7	.48326
.46	.8649,6	.95940	.9015,7	1.424,6	.49357
.47	.8595,8	.95769	.8975,6	1.401,8	.50385
.48	.8541,3	.95595	.8934,9	1.380,1	.51410
.49	.8486,1	.95418	.8893,6	1.359,4	.52432

Table A-9 (*Continued*)

M	$\dfrac{p}{p_0}$	$\dfrac{T}{T_0}$	$\dfrac{\rho}{\rho_0}$	$\dfrac{A}{A^*}$	$\dfrac{V}{V^*}$
.50	.8430,2	.95238	.8851,7	1.339,8	.53452
.51	.8373,7	.95055	.8809,2	1.321,2	.54469
.52	.8316,6	.94869	.8766,2	1.303,4	.55482
.53	.8258,9	.94681	.8722,7	1.286,4	.56493
.54	.8200,5	.94489	.8678,8	1.270,3	.57501
.55	.8141,6	.94295	.86342	1.255,0	.58506
.56	.8082,2	.94098	.85892	1.240,3	.59508
.57	.8022,4	.93898	.85437	1.226,3	.60506
.58	.7962,1	.93696	.84977	1.213,0	.61500
.59	.7901,2	.93491	.84513	1.200,3	.62491
.60	.78400	.93284	.84045	1.188,2	.63480
.61	.77784	.93074	.83573	1.176,6	.64466
.62	.77164	.92861	.83096	1.165,6	.65448
.63	.76540	.92646	.82616	1.155,1	.66427
.64	.75913	.92428	.82132	1.145,1	.67402
.65	.75283	.92208	.81644	1.1356	.68374
.66	.74650	.91986	.81153	1.1265	.69342
.67	.74014	.91762	.80659	1.1178	.70307
.68	.73376	.91535	.80162	1.1096	.71268
.69	.72735	.91306	.79662	1.1018	.72225
.70	7.2092	.91075	.79158	1.0943,7	.73179
.71	.71448	.90842	.78652	1.0872,9	.74129
.72	.70802	.90606	.78143	1.0805,7	.75076
.73	.70155	.90368	.77632	1.0741,9	.76019
.74	.69507	.90129	.77119	1.0681,4	.76958
.75	.68857	.89888	.76603	1.0624,2	.77893
.76	.68207	.89644	.76086	1.0570,0	.78825
.77	.67556	.89399	.75567	1.0518,8	.79753
.78	.66905	.89152	.75046	1.0470,5	.80677
.79	.66254	.88903	.74524	1.0425,0	.81597
.80	.65602	.88652	.74000	1.0382,3	.82514
.81	.64951	.88400	.73474	1.0342,2	.83426
.82	.64300	.88146	.72947	1.0304,6	.84334
.83	.63650	.87890	.72419	1.0269,6	.85239
.84	.63000	.87633	.71890	1.0237,0	.86140
.85	.62351	.87374	.71361	1.0206,7	.87037
.86	.61703	.87114	.70831	1.0178,7	.87929
.87	.61057	.86852	.70300	1.0153,0	.88817
.88	.60412	.86589	.69769	1.0129,4	.89702
.89	.59768	.86324	.69237	1.0108,0	.90583

Table A-9 (*Continued*)

M	$\dfrac{p}{p_0}$	$\dfrac{T}{T_0}$	$\dfrac{\rho}{\rho_0}$	$\dfrac{A}{A^*}$	$\dfrac{V}{V^*}$
.90	.59126	.86058	.68704	1.0088,6	.91460
.91	.58486	.85791	.68171	1.0071,3	.92333
.92	.57848	.85523	.67639	1.0056,0	.93201
.93	.57212	.85253	.67107	1.0042,6	.94065
.94	.56578	.84982	.66575	1.0031,1	.94925
.95	.55946	.84710	.66044	1.0021,4	.95781
.96	.55317	.84437	.65513	1.0013,6	.96633
.97	.54691	.84162	.64982	1.0007,6	.97481
.98	.54067	.83887	.64452	1.0003,3	.98325
.99	.53446	.83611	.63923	1.0000,8	.99165
1.00	.52828	.83333	.63394	1.0000,0	1.00000
1.01	.52213	.83055	.62866	1.0000,8	1.00831
1.02	.51602	.82776	.62339	1.0003,3	1.01658
1.03	.50994	.82496	.61813	1.0007,4	1.02481
1.04	.50389	.82215	.61288	1.0013,0	1.03300
1.05	.49787	.81933	.60765	1.0020,2	1.04114
1.06	.49189	.81651	.60243	1.0029,0	1.04924
1.07	.48595	.81368	.59722	1.0039,4	1.05730
1.08	.48005	.81084	.59203	1.0051,2	1.06532
1.09	.47418	.80800	.58685	1.0064,5	1.07330
1.10	.46835	.80515	.58169	1.0079,3	1.08124
1.11	.46256	.80230	.57655	1.0095,5	1.08914
1.12	.45682	.79944	.57143	1.0113,1	1.09699
1.13	.45112	.79657	.56632	1.0132,2	1.10480
1.14	.44545	.79370	.56123	1.0152,7	1.11256
1.15	.43983	.79083	.55616	1.0174,6	1.1203
1.16	.43425	.78795	.55112	1.0197,8	1.1280
1.17	.42872	.78507	.54609	1.0222,4	1.1356
1.18	.42323	.78218	.54108	1.0248,4	1.1432
1.19	.41778	.77929	.53610	1.0275,7	1.1508
1.20	.4123,8	.77640	.53114	1.0304,4	1.1583
1.21	.4070,2	.77350	.52620	1.0334,4	1.1658
1.22	.4017,1	.77061	.52129	1.0365,7	1.1732
1.23	.3964,5	.76771	.51640	1.0398,3	1.1806
1.24	.3912,3	.76481	.51154	1.0432,3	1.1879
1.25	.3860,6	.76190	.50670	1.0467,6	1.1952
1.26	.3809,4	.75900	.50189	1.0504,1	1.2025
1.27	.3758,6	.75610	.49710	1.0541,9	1.2097
1.28	.3708,3	.75319	.49234	1.0581,0	1.2169
1.29	.3658,5	.75029	.48761	1.0621,4	1.2240

Table A-9 (*Continued*)

M	$\dfrac{p}{p_0}$	$\dfrac{T}{T_0}$	$\dfrac{\rho}{\rho_0}$	$\dfrac{A}{A^*}$	$\dfrac{V}{V^*}$
1.30	.3609,2	.74738	.48291	1.0663,1	1.2311
1.31	.3560,3	.74448	.47823	1.0706,0	1.2382
1.32	.3511,9	.74158	.47358	1.0750,2	1.2452
1.33	.3464,0	.73867	.46895	1.0795,7	1.2522
1.34	.3416,6	.73577	.46436	1.0842,4	1.2591
1.35	.3369,7	.73287	.45980	1.0890,4	1.2660
1.36	.3323,3	.72997	.45527	1.0939,7	1.2729
1.37	.3277,4	.72707	.45076	1.0990,2	1.2797
1.38	.3231,9	.72418	.44628	1.1042,0	1.2865
1.39	.3186,9	.72128	.44183	1.1095,0	1.2932
1.40	.3142,4	.71839	.43742	1.1149	1.2999
1.41	.3098,4	.71550	.43304	1.1205	1.3065
1.42	.3054,9	.71261	.42869	1.1262	1.3131
1.43	.3011,9	.70973	.42436	1.1320	1.3197
1.44	.2969,3	.70685	.42007	1.1379	1.3262
1.45	.2927,2	.70397	.41581	1.1440	1.3327
1.46	.2885,6	.70110	.41158	1.1502	1.3392
1.47	.2844,5	.69823	.40738	1.1565	1.3456
1.48	.2803,9	.69537	.40322	1.1629	1.3520
1.49	.2763,7	.69251	.39909	1.1695	1.3583
1.50	.2724,0	.68965	.39498	1.1762	1.3646
1.51	.2684,8	.68680	.39091	1.1830	1.3708
1.52	.2646,1	.68396	.38687	1.1899	1.3770
1.53	.2607,8	.68112	.38287	1.1970	1.3832
1.54	.2570,0	.67828	.37890	1.2042	1.3894
1.55	.2532,6	.67545	.37496	1.2155	1.3955
1.56	.2495,7	.67262	.37105	1.2190	1.4016
1.57	.2459,3	.66980	.36717	1.2266	1.4076
1.58	.2423,3	.66699	.36332	1.2343	1.4135
1.59	.2387,8	.66418	.35951	1.2422	1.4195
1.60	.23527	.66138	.35573	1.2502	1.4254
1.61	.23181	.65858	.35198	1.2583	1.4313
1.62	.22839	.65579	.34826	1.2666	1.4371
1.63	.22501	.65301	.34458	1.2750	1.4429
1.64	.22168	.65023	.34093	1.2835	1.4487
1.65	.21839	.64746	.33731	1.2922	1.4544
1.66	.21515	.64470	.33372	1.3010	1.4601
1.67	.21195	.64194	.33016	1.3099	1.4657
1.68	.20879	.63919	.32664	1.3190	1.4713
1.69	.20567	.63645	.32315	1.3282	1.4769

Table A-9 (*Continued*)

M	$\dfrac{p}{p_0}$	$\dfrac{T}{T_0}$	$\dfrac{\rho}{\rho_0}$	$\dfrac{A}{A^*}$	$\dfrac{V}{V^*}$
1.70	.20259	.63372	.31969	1.3376	1.4825
1.71	.19955	.63099	.31626	1.3471	1.4880
1.72	.19656	.62827	.31286	1.3567	1.4935
1.73	.19361	.62556	.30950	1.3665	1.4989
1.74	.19070	.62286	.30617	1.3764	1.5043
1.75	.18782	.62016	.30287	1.3865	1.5097
1.76	.18499	.61747	.29959	1.3967	1.5150
1.77	.18220	.61479	.29635	1.4071	1.5203
1.78	.17944	.61211	.29314	1.4176	1.5256
1.79	.17672	.60945	.28997	1.4282	1.5308
1.80	.17404	.60680	.28682	1.4390	1.5360
1.81	.17140	.60415	.28370	1.4499	1.5412
1.82	.16879	.60151	.28061	1.4610	1.5463
1.83	.16622	.59888	.27756	1.4723	1.5514
1.84	.16369	.59626	.27453	1.4837	1.5564
1.85	.16120	.59365	.27153	1.4952	1.5614
1.86	.15874	.59105	.26857	1.5069	1.5664
1.87	.15631	.58845	.26563	1.5188	1.5714
1.88	.15392	.58586	.26272	1.5308	1.5763
1.89	.15156	.58329	.25984	1.5429	1.5812
1.90	.14924	.58072	.25699	1.5552	1.5861
1.91	.14695	.57816	.25417	1.5677	1.5909
1.92	.14469	.57561	.25138	1.5804	1.5957
1.93	.14247	.57307	.24862	1.5932	1.6005
1.94	.14028	.57054	.24588	1.6062	1.6052
1.95	.13813	.56802	.24317	1.6193	1.6099
1.96	.13600	.56551	.24049	1.6326	1.6146
1.97	.13390	.56301	.23784	1.6461	1.6193
1.98	.13184	.56051	.23522	1.6597	1.6239
1.99	.12981	.55803	.23262	1.6735	1.6285
2.00	.12780	.55556	.23005	1.6875	1.6330
2.01	.12583	.55310	.22751	1.7017	1.6375
2.02	.12389	.55064	.22499	1.7160	1.6420
2.03	.12198	.54819	.22250	1.7305	1.6465
2.04	.12009	.54576	.22004	1.7452	1.6509
2.05	.11823	.54333	.21760	1.7600	1.6553
2.06	.11640	.54091	.21519	1.7750	1.6597
2.07	.11460	.53850	.21281	1.7902	1.6640
2.08	.11282	.53611	.21045	1.8056	1.6683
2.09	.11107	.53373	.20811	1.8212	1.6726

Table A-9 (*Continued*)

M	$\dfrac{p}{p_0}$	$\dfrac{T}{T_0}$	$\dfrac{\rho}{\rho_0}$	$\dfrac{A}{A^*}$	$\dfrac{V}{V^*}$
2.10	.10935	.53135	.20580	1.8369	1.6769
2.11	.10766	.52898	.20352	1.8529	1.6811
2.12	.10599	.52663	.20126	1.8690	1.6853
2.13	.10434	.52428	.19902	1.8853	1.6895
2.14	.10272	.52194	.19681	1.9018	1.6936
2.15	.10113	.51962	.19463	1.9185	1.6977
2.16	.09956	.51730	.19247	1.9354	1.7018
2.17	.09802	.51499	.19033	1.9525	1.7059
2.18	.09650	.51269	.18821	1.9698	1.7099
2.19	.09500	.51041	.18612	1.9873	1.7139
2.20	.09352	.50813	.18405	2.0050	1.7179
2.21	.09207	.50586	.18200	2.0229	1.7219
2.22	.09064	.50361	.17998	2.0409	1.7258
2.23	.08923	.50136	.17798	2.0592	1.7297
2.24	.08784	.49912	.17600	2.0777	1.7336
2.25	.08648	.49689	.17404	2.0964	1.7374
2.26	.08514	.49468	.17211	2.1154	1.7412
2.27	.08382	.49247	.17020	2.1345	1.7450
2.28	.08252	.49027	.16830	2.1538	1.7488
2.29	.08123	.48809	.16643	2.1734	1.7526
2.30	.07997	.48591	.16458	2.1931	1.7563
2.31	.07873	.48374	.16275	2.2131	1.7600
2.32	.07751	.48158	.16095	2.2333	1.7637
2.33	.07631	.47944	.15916	2.2537	1.7673
2.34	.07513	.47730	.15739	2.2744	1.7709
2.35	.07396	.47517	.15564	2.2953	1.7745
2.36	.07281	.47305	.15391	2.3164	1.7781
2.37	.07168	.47095	.15220	2.3377	1.7817
2.38	.07057	.46885	.15052	2.3593	1.7852
2.39	.06948	.46676	.14885	2.3811	1.7887
2.40	.06840	.46468	.14720	2.4031	1.7922
2.41	.06734	.46262	.14557	2.4254	1.7957
2.42	.06630	.46056	.14395	2.4479	1.7991
2.43	.06527	.45851	.14235	2.4706	1.8025
2.44	.06426	.45647	.14078	2.4936	1.8059
2.45	.06327	.45444	.13922	2.5168	1.8093
2.46	.06229	.45242	.13768	2.5403	1.8126
2.47	.06133	.45041	.13616	2.5640	1.8159
2.48	.06038	.44841	.13465	2.5880	1.8192
2.49	.05945	.44642	.13316	2.6122	1.8225

Table A-9 (Continued)

M	$\dfrac{p}{p_0}$	$\dfrac{T}{T_0}$	$\dfrac{\rho}{\rho_0}$	$\dfrac{A}{A^*}$	$\dfrac{V}{V^*}$
2.50	.05853	.44444	.13169	2.6367	1.8258
2.51	.05763	.44247	.13023	2.6615	1.8290
2.52	.05674	.44051	.12879	2.6865	1.8322
2.53	.05586	.43856	.12737	2.7117	1.8354
2.54	.05500	.43662	.12597	2.7372	1.8386
2.55	.05415	.43469	.12458	2.7630	1.8417
2.56	.05332	.43277	.12321	2.7891	1.8448
2.57	.05250	.43085	.12185	2.8154	1.8479
2.58	.05169	.42894	.12051	2.8420	1.8510
2.59	.05090	.42705	.11918	2.8689	1.8541
2.60	.05012	.42517	.11787	2.8960	1.8572
2.61	.04935	.42330	.11658	2.9234	1.8602
2.62	.04859	.42143	.11530	2.9511	1.8632
2.63	.04784	.41957	.11403	2.9791	1.8662
2.64	.04711	.41772	.11278	3.0074	1.8692
2.65	.04639	.41589	.11154	3.0359	1.8721
2.66	.04568	.41406	.11032	3.0647	1.8750
2.67	.04498	.41224	.10911	3.0938	1.8779
2.68	.04429	.41043	.10792	3.1233	1.8808
2.69	.04361	.40863	.10674	3.1530	1.8837
2.70	.04295	.40684	.10557	3.1830	1.8865
2.71	.04230	.40505	.10442	3.2133	1.8894
2.72	.04166	.40327	.10328	3.2440	1.8922
2.73	.04102	.40151	.10215	3.2749	1.8950
2.74	.04039	.39976	.10104	3.3061	1.8978
2.75	.03977	.39801	.09994	3.3376	1.9005
2.76	.03917	.39627	.09885	3.3695	1.9032
2.77	.03858	.39454	.09777	3.4017	1.9060
2.78	.03800	.39282	.09671	3.4342	1.9087
2.79	.03742	.39111	.09566	3.4670	1.9114
2.80	.03685	.38941	.09462	3.5001	1.9140
2.81	.03629	.38771	.09360	3.5336	1.9167
2.82	.03574	.38603	.09259	3.5674	1.9193
2.83	.03520	.38435	.09158	3.6015	1.9220
2.84	.03467	.38268	.09059	3.6359	1.9246
2.85	.03415	.38102	.08962	3.6707	1.9271
2.86	.03363	.37937	.08865	3.7058	1.9297
2.87	.03312	.37773	.08769	3.7413	1.9322
2.88	.03262	.37610	.08674	3.7771	1.9348
2.89	.03213	.37448	.08581	3.8133	1.9373

Table A-9 (Continued)

M	$\dfrac{p}{p_0}$	$\dfrac{T}{T_0}$	$\dfrac{\rho}{\rho_0}$	$\dfrac{A}{A^*}$	$\dfrac{V}{V^*}$
2.90	.03165	.37286	.08489	3.8498	1.9398
2.91	.03118	.37125	.08398	3.8866	1.9423
2.92	.03071	.36965	.08308	3.9238	1.9448
2.93	.03025	.36806	.08218	3.9614	1.9472
2.94	.02980	.36648	.08130	3.9993	1.9497
2.95	.02935	.36490	.08043	4.0376	1.9521
2.96	.02891	.36333	.07957	4.0763	1.9545
2.97	.02848	.36177	.07872	4.1153	1.9569
2.98	.02805	.36022	.07788	4.1547	1.9593
2.99	.02764	.35868	.07705	4.1944	1.9616
3.00	.027,22	.357,14	.076,23	4.23,46	1.964,0
3.10	.023,45	.342,23	.068,52	4.65,73	1.986,6
3.20	.020,23	.328,08	.061,65	5.12,10	2.007,9
3.30	.0174,8	.314,66	.055,54	5.6,287	2.027,9
3.40	.0151,2	.301,93	.050,09	6.1,837	2.046,6
3.50	.0131,1	.289,86	.045,23	6.7,896	2.064,2
3.60	.0113,8	.278,40	.040,89	7.4,501	2.080,8
3.70	.0099,0	.267,52	.0370,2	8.1,691	2.096,4
3.80	.0086,3	.257,20	.0335,5	8.9,506	2.111,1
3.90	.0075,3	.247,40	.0304,4	9.7,990	2.125,0
4.00	.0065,8	.238,10	.0276,6	10.7,19	2.138,1
4.10	.0057,7	.229,25	.0251,6	11.7,15	2.150,5
4.20	.0050,6	.2208,5	.0229,2	12.7,92	2.162,2
4.30	.0044,5	.2128,6	.0209,0	13.9,55	2.173,2
4.40	.0039,2	.2052,5	.0190,9	15.2,10	2.183,7
4.50	.0034,6	.1980,2	.0174,5	16.5,62	2.193,6
4.60	.0030,5	.1911,3	.0159,7	18.0,18	2.203,0
4.70	.0027,0	.1845,7	.0146,3	19.5,83	2.211,9
4.80	.0024,0	.1783,2	.0134,3	21.2,64	2.220,4
4.90	.0021,3	.1723,5	.0123,3	23.0,67	2.228,4
5.00	.00189	.16667	.01134	25.000	2.2361
6.00	$.0_3633^b$	.12195	.00519	53.180	2.2953
7.00	$.0_3242$	.09259	.00261	104.143	2.3333
8.00	$.0_3102$	.07246	.00141	190.109	2.3591
9.00	$.0_4474$	.05814	$.0_3815$	327.189	2.3772
10.00	$.0_4236$	.04762	$.0_3495$	535.938	2.3904
∞	0	0	0	∞	2.4495

^{b} Note: 0_3495 signifies .000495.

Table A-10 NORMAL SHOCK, $k = 1.4$

M_x	M_y	$\dfrac{p_y}{p_x}$	$\dfrac{\rho_y}{\rho_x}$	$\dfrac{T_y}{T_x}$	$\dfrac{p_{0y}}{p_{0x}}$	$\dfrac{p_{0y}}{p_x}$
1.00	1.0000,0[a]	1.0000,0	1.0000,0	1.0000,0	1.00000	1.8929
1.01	.9901,3	1.0234,5	1.0166,9	1.0066,5	.99999	1.9152
1.02	.9805,2	1.0471,3	1.0334,4	1.01325	.99998	1.9379
1.03	.9711,5	1.0710,5	1.0502,4	1.01981	.99997	1.9610
1.04	.9620,2	1.0952,0	1,0670,9	1.02634	.99994	1.9845
1.05	.9531,2	1.1196	1.0839,8	1.03284	.99987	2.0083
1.06	.9444,4	1.1442	1.10092	1.03931	.99976	2.0325
1.07	.9359,8	1.1690	1.11790	1.04575	.99962	2.0570
1.08	.9277,2	1.1941	1.13492	1.05217	.9994,4	2.0819
1.09	.9196,5	1.2194	1.15199	1.05856	.9992,1	2.1072
1.10	.9117,7	1.2450	1.1691	1.06494	.9989,2	2.1328
1.11	.9040,8	1.2708	1.1862	1.07130	.9985,8	2.1588
1.12	.8965,6	1.2968	1.2034	1.07764	.9982,0	2.1851
1.13	.8892,2	1.3230	1.2206	1.08396	.9977,6	2.2118
1.14	.8820,4	1.3495	1.2378	1.09027	.9972,6	2.2388
1.15	.8750,2	1.3762	1.2550	1.09657	.9966,9	2.2661
1.16	.8681,6	1.4032	1.2723	1.10287	.9960,5	2.2937
1.17	.8614,5	1.4304	1.2896	1.10916	.9953,4	2.3217
1.18	.8548,8	1.4578	1.3069	1.11544	.9945,5	2.3499
1.19	.8484,6	1.4854	1.3243	1.12172	.9937,1	2.3786
1.20	.8421,7	1.5133	1.3416	1.1280	.9928,0	2.4075
1.21	.8360,1	1.5414	1.3590	1.1343	.9918,0	2.4367
1.22	.8299,8	1.5698	1.3764	1.1405	.9907,3	2.4662
1.23	.8240,8	1.5984	1.3938	1.1468	.9895,7	2.4961
1.24	.8183,0	1.6272	1.4112	1.1531	.9883,5	2.5263
1.25	.8126,4	1.6562	1.4286	1.1594	.9870,6	2.5568
1.26	.8070,9	1.6855	1.4460	1.1657	.9856,8	2.5876
1.27	.8016,5	1.7150	1.4634	1.1720	.9842,2	2.6187
1.28	.7963,1	1.7448	1.4808	1.1782	.9826,8	2.6500
1.29	.7910,8	1.7748	1.4983	1.1846	.9810,6	2.6816
1.30	.7859,6	1.8050	1.5157	1.1909	.9793,5	2.7135
1.31	.7809,3	1.8354	1.5331	1.1972	.9775,8	2.7457
1.32	.7760,0	1.8661	1.5505	1.2035	.9757,4	2.7783
1.33	.7711,6	1.8970	1.5680	1.2099	.9738,2	2.8112
1.34	.7664,1	1.9282	1.5854	1.2162	.9718,1	2.8444

SOURCE: The values in this table were originally calculated and reported in M.I.T. Meteor Report No. 14, Bureau of Ordnance, U.S. Navy Department, December, 1947, by A. H. Shapiro, W. R. Hawthorne, and G. M. Edelman.

[a] *Note:* Linear interpolation may be made (a) for values without a comma and (b) to the left of comma for values containing a comma.

Table A-10 (*Continued*)

M_x	M_y	$\dfrac{p_y}{p_x}$	$\dfrac{\rho_y}{\rho_x}$	$\dfrac{T_y}{T_x}$	$\dfrac{p_{0y}}{p_{0x}}$	$\dfrac{p_{0y}}{p_x}$
1.35	.7617,5	1.9596	1.6028	1.2226	.9697,2	2.8778
1.36	.7571,8	1.9912	1.6202	1.2290	.9675,6	2.9115
1.37	.7526,9	2.0230	1.6376	1.2354	.9653,4	2.9455
1.38	.7482,8	2.0551	1.6550	1.2418	.9630,4	2.9798
1.39	.7439,6	2.0874	1.6723	1.2482	.9606,5	3.0144
1.40	.7397,1	2.1200	1.6896	1.2547	.9581,9	3.0493
1.41	.7355,4	2.1528	1.7070	1.2612	.9556,6	3.0844
1.42	.7314,4	2.1858	1.7243	1.2676	.9530,6	3.1198
1.43	.7274,1	2.2190	1.7416	1.2742	.9503,9	3.1555
1.44	.7234,5	2.2525	1.7589	1.2807	.9476,5	3.1915
1.45	.7195,6	2.2862	1.7761	1.2872	.9448,3	3.2278
1.46	.7157,4	2.3202	1.7934	1.2938	.9419,6	3.2643
1.47	.7119,8	2.3544	1.8106	1.3004	.9390,1	3.3011
1.48	.7082,9	2.3888	1.8278	1.3070	.9360,0	3.3382
1.49	.7046,6	2.4234	1.8449	1.3136	.9329,2	3.3756
1.50	.7010,9	2.4583	1.8621	1.3202	.9297,8	3.4133
1.51	.6975,8	2.4934	1.8792	1.3269	.9265,8	3.4512
1.52	.6941,3	2.5288	1.8962	1.3336	.9233,1	3.4894
1.53	.6907,3	2.5644	1.9133	1.3403	.9199,9	3.5279
1.54	.6873,9	2.6003	1.9303	1.3470	.9166,2	3.5667
1.55	.6841,0	2.6363	1.9473	1.3538	.9131,9	3.6058
1.56	.6808,6	2.6725	1.9643	1.3606	.9097,0	3.6451
1.57	.6776,8	2.7090	1.9812	1.3674	.9061,5	3.6847
1.58	.6745,5	2.7458	1.9981	1.3742	.9025,5	3.7245
1.59	.6714,7	2.7828	2.0149	1.3811	.8988,9	3.7645
1.60	.66844	2.8201	2.0317	1.3880	.8952,0	3.8049
1.61	.66545	2.8575	2.0485	1.3949	.8914,4	3.8456
1.62	.66251	2.8951	2.0652	1.4018	.8876,4	3.8866
1.63	.65962	2.9330	2.0820	1.4088	.8838,0	3.9278
1.64	.65677	2.9712	2.0986	1.4158	.8799,2	3.9693
1.65	.65396	3.0096	2.1152	1.4228	.87598	4.0111
1.66	.65119	3.0482	2.1318	1.4298	.87201	4.0531
1.67	.64847	3.0870	2.1484	1.4369	.86800	4.0954
1.68	.64579	3.1261	2.1649	1.4440	.86396	4.1379
1.69	.64315	3.1654	2.1813	1.4512	.85987	4.1807
1.70	.64055	3.2050	2.1977	1.4583	.85573	4.2238
1.71	.63798	3.2448	2.2141	1.4655	.85155	4.2672
1.72	.63545	3.2848	2.2304	1.4727	.84735	4.3108
1.73	.63296	3.3250	2.2467	1.4800	.84312	4.3547
1.74	.63051	3.3655	2.2629	1.4873	.83886	4.3989

Table A-10 (*Continued*)

M_x	M_y	$\dfrac{p_y}{p_x}$	$\dfrac{\rho_y}{\rho_x}$	$\dfrac{T_y}{T_x}$	$\dfrac{p_{0y}}{p_{0x}}$	$\dfrac{p_{0y}}{p_x}$
1.75	.62809	3.4062	2.2791	1.4946	.83456	4.4433
1.76	.62570	3.4472	2.2952	1.5019	.83024	4.4880
1.77	.62335	3.4884	2.3113	1.5093	.82589	4.5330
1.78	.62104	3.5298	2.3273	1.5167	.82152	4.5783
1.79	.61875	3.5714	2.3433	1.5241	.81711	4.6238
1.80	.61650	3.6133	2.3592	1.5316	.81268	4.6695
1.81	.61428	3.6554	2.3751	1.5391	.80823	4.7155
1.82	.61209	3.6978	2.3909	1.5466	.80376	4.7618
1.83	.60993	3.7404	2.4067	1.5542	.79926	4.8083
1.84	.60780	3.7832	2.4224	1.5617	.79474	4.8511
1.85	.60570	3.8262	2.4381	1.5694	.79021	4.9022
1.86	.60363	3.8695	2.4537	1.5770	.78567	4.9498
1.87	.60159	3.9130	2.4693	1.5847	.78112	4.9974
1.88	.59957	3.9568	2.4848	1.5924	.77656	5.0453
1.89	.59758	4.0008	2.5003	1.6001	.77197	5.0934
1.90	.59562	4.0450	2.5157	1.6079	.76735	5.1417
1.91	.59368	4.0894	2.5310	1.6157	.76273	5.1904
1.92	.59177	4.1341	2.5463	1.6236	.75812	5.2394
1.93	.58988	4.1790	2.5615	1.6314	.75347	5.2886
1.94	.58802	4.2242	2.5767	1.6394	.74883	5.3381
1.95	.58618	4.2696	2.5919	1.6473	.74418	5.3878
1.96	.58437	4.3152	2.6070	1.6553	.73954	5.4378
1.97	.58258	4.3610	2.6220	1.6633	.73487	5.4880
1.98	.58081	4.4071	2.6369	1.6713	.73021	5.5385
1.99	.57907	4.4534	2.6518	1.6794	.72554	5.5894
2.00	.57735	4.5000	2.6666	1.6875	.72088	5.6405
2.01	.57565	4.5468	2.6814	1.6956	.71619	5.6918
2.02	.57397	4.5938	2.6962	1.7038	.71152	5.7434
2.03	.57231	4.6411	2.7109	1.7120	.70686	5.7952
2.04	.57068	4.6886	2.7255	1.7203	.70218	5.8473
2.05	.56907	4.7363	2.7400	1.7286	.69752	5.8997
2.06	.56747	4.7842	2.7545	1.7369	.69284	5.9523
2.07	.56589	4.8324	2.7690	1.7452	.68817	6.0052
2.08	.56433	4.8808	2.7834	1.7536	.68351	6.0584
2.09	.56280	4.9295	2.7977	1.7620	.67886	6.1118
2.10	.56128	4.9784	2.8119	1.7704	.67422	6.1655
2.11	.55978	5.0275	2.8261	1.7789	.66957	6.2194
2.12	.55830	5.0768	2.8402	1.7874	.66492	6.2736
2.13	.55683	5.1264	2.8543	1.7960	.66029	6.3280
2.14	.55538	5.1762	2.8683	1.8046	.65567	6.3827

Table A-10 (*Continued*)

M_x	M_y	$\dfrac{p_y}{p_x}$	$\dfrac{\rho_y}{\rho_x}$	$\dfrac{T_y}{T_x}$	$\dfrac{p_{0y}}{p_{0x}}$	$\dfrac{p_{0y}}{p_x}$
2.15	.55395	5.2262	2.8823	1.8132	.65105	6.4377
2.16	.55254	5.2765	2.8962	1.8219	.64644	6.4929
2.17	.55114	5.3270	2.9100	1.8306	.64185	6.5484
2.18	.54976	5.3778	2.9238	1.8393	.63728	6.6042
2.19	.54841	5.4288	2.9376	1.8481	.63270	6.6602
2.20	.54706	5.4800	2.9512	1.8569	.62812	6.7163
2.21	.54572	5.5314	2.9648	1.8657	.62358	6.7730
2.22	.54440	5.5831	2.9783	1.8746	.61905	6.8299
2.23	.54310	5.6350	2.9918	1.8835	.61453	6.8869
2.24	.54182	5.6872	3.0052	1.8924	.61002	6.9442
2.25	.54055	5.7396	3.0186	1.9014	.60554	7.0018
2.26	.53929	5.7922	3.0319	1.9104	.60106	7.0597
2.27	.53805	5.8451	3.0452	1.9194	.59659	7.1178
2.28	.53683	5.8982	3.0584	1.9285	.59214	7.1762
2.29	.53561	5.9515	3.0715	1.9376	.58772	7.2348
2.30	.53441	6.0050	3.0846	1.9468	.58331	7.2937
2.31	.53322	6.0588	3.0976	1.9560	.57891	7.3529
2.32	.53205	6.1128	3.1105	1.9652	.57452	7.4123
2.33	.53089	6.1670	3.1234	1.9745	.57015	7.4720
2.34	.52974	6.2215	3.1362	1.9838	.56580	7.5319
2.35	.52861	6.2762	3.1490	1.9931	.56148	7.5920
2.36	.52749	6.3312	3.1617	2.0025	.55717	7.6524
2.37	.52638	6.3864	3.1743	2.0119	.55288	7.7131
2.38	.52528	6.4418	3.1869	2.0213	.54862	7.7741
2.39	.52419	6.4974	3.1994	2.0308	.54438	7.8354
2.40	.52312	6.5533	3.2119	2.0403	.54015	7.8969
2.41	.52206	6.6094	3.2243	2.0499	.53594	7.9587
2.42	.52100	6.6658	3.2366	2.0595	.53175	8.0207
2.43	.51996	6.7224	3.2489	2.0691	.52758	8.0830
2.44	.51894	6.7792	3.2611	2.0788	.52344	8.1455
2.45	.51792	6.8362	3.2733	2.0885	.51932	8.2083
2.46	.51691	6.8935	3.2854	2.0982	.51521	8.2714
2.47	.51592	6.9510	3.2975	2.1080	.51112	8.3347
2.48	.51493	7.0088	3.3095	2.1178	.50706	8.3983
2.49	.51395	7.0668	3.3214	2.1276	.50303	8.4622
2.50	.51299	7.1250	3.3333	2.1375	.49902	8.5262
2.51	.51204	7.1834	3.3451	2.1474	.49502	8.5904
2.52	.51109	7.2421	3.3569	2.1574	.49104	8.6549
2.53	.51015	7.3010	3.3686	2.1674	.48709	8.7198
2.54	.50923	7.3602	3.3802	2.1774	.48317	8.7850

Table A-10 (*Continued*)

M_x	M_y	$\dfrac{p_y}{p_x}$	$\dfrac{\rho_y}{\rho_x}$	$\dfrac{T_y}{T_x}$	$\dfrac{p_{0y}}{p_{0x}}$	$\dfrac{p_{0y}}{p_x}$
2.55	.50831	7.4196	3.3918	2.1875	.47927	8.8505
2.56	.50740	7.4792	3.4034	2.1976	.47540	8.9162
2.57	.50651	7.5391	3.4149	2.2077	.47155	8.9821
2.58	.50562	7.5992	3.4263	2.2179	.46772	9.0482
2.59	.50474	7.6595	3.4376	2.2281	.46391	9.1146
2.60	.50387	7.7200	3.4489	2.2383	.46012	9.1813
2.61	.50301	7.7808	3.4602	2.2486	.45636	9.2481
2.62	.50216	7.8418	3.4714	2.2589	.45262	9.3154
2.63	.50132	7.9030	3.4825	2.2693	.44891	9.3829
2.64	.50048	7.9645	3.4936	2.2797	.44522	9.4507
2.65	.49965	8.0262	3.5047	2.2901	.44155	9.5187
2.66	.49883	8.0882	3.5157	2.3006	.43791	9.5869
2.67	.49802	8.1504	3.5266	2.3111	.43429	9.6553
2.68	.49722	8.2128	3.5374	2.3217	.43070	9.7241
2.69	.49642	8.2754	3.5482	2.3323	.42713	9.7932
2.70	.49563	8.3383	3.5590	2.3429	.42359	9.8625
2.71	.49485	8.4014	3.5697	2.3536	.42007	9.9320
2.72	.49408	8.4648	3.5803	2.3643	.41657	10.0017
2.73	.49332	8.5284	3.5909	2.3750	.41310	10.0718
2.74	.49256	8.5922	3.6014	2.3858	.40965	10.1421
2.75	.49181	8.6562	3.6119	2.3966	.40622	10.212
2.76	.49107	8.7205	3.6224	2.4074	.40282	10.283
2.77	.49033	8.7850	3.6328	2.4183	.39945	10.354
2.78	.48960	8.8497	3.6431	2.4292	.39610	10.426
2.79	.48888	8.9147	3.6533	2.4402	.39276	10.498
2.80	.48817	8.9800	3.6635	2.4512	.38946	10.569
2.81	.48746	9.0454	3.6737	2.4622	.38618	10.641
2.82	.48676	9.1111	3.6838	2.4733	.38293	10.714
2.83	.48607	9.1770	3.6939	2.4844	.37970	10.787
2.84	.48538	9.2432	3.7039	2.4955	.37649	10.860
2.85	.48470	9.3096	3.7139	2.5067	.37330	10.933
2.86	.48402	9.3762	3.7238	2.5179	.37013	11.006
2.87	.48334	9.4431	3.7336	2.5292	.36700	11.080
2.88	.48268	9.5102	3.7434	2.5405	.36389	11.154
2.89	.48203	9.5775	3.7532	2.5518	.36080	11.228
2.90	.48138	9.6450	3.7629	2.5632	.35773	11.302
2.91	.48074	9.7127	3.7725	2.5746	.35469	11.377
2.92	.48010	9.7808	3.7821	2.5860	.35167	11.452
2.93	.47946	9.8491	3.7917	2.5975	.34867	11.527
2.94	.47883	9.9176	3.8012	2.6090	.34570	11.603

Table A-10 (*Continued*)

M_x	M_y	$\dfrac{p_y}{p_x}$	$\dfrac{\rho_y}{\rho_x}$	$\dfrac{T_y}{T_x}$	$\dfrac{p_{0y}}{p_{0x}}$	$\dfrac{p_{0y}}{p_x}$
2.95	.47821	9.9863	3.8106	2.6206	.34275	11.679
2.96	.47760	10.055	3.8200	2.6322	.33982	11.755
2.97	.47699	10.124	3.8294	2.6438	.33692	11.831
2.98	.47638	10.194	3.8387	2.6555	.33404	11.907
2.99	.47578	10.263	3.8479	2.6672	.33118	11.984
3.00	.47519	10.333	3.8571	2.6790	.32834	12.061
3.50	.45115	14.125	4.2608	3.3150	.21295	16.242
4.00	.43496	18.500	4.5714	4.0469	.13876	21.068
4.50	.42355	23.458	4.8119	4.8751	.09170	26.539
5.00	.41523	29.000	5.0000	5.8000	.06172	32.654
6.00	.40416	41.833	5.2683	7.9406	.02965	46.815
7.00	.39736	57.000	5.4444	10.469	.01535	63.552
8.00	.39289	74.500	5.5652	13.387	.00849	82.865
9.00	.38980	94.333	5.6512	16.693	.00496	104.753
10.00	.38757	116.500	5.7143	20.388	.00304	129.217
∞	.37796	∞	6.0000	∞	0	∞

Answers to Selected Problems

1-5. (a) $0.0315 \text{ m}^3/\text{s}$ (b) $0.0143 \text{ m}^3/\text{s}$ (c) 95.8 kPa (d) 26.8 m/s
(e) 801 kg/m^3

1-11. (a), (b), and (c) 50 J/kg

1-12. (a) Work is added to fluid. (b) Work is removed from fluid. (e) Heat is removed from water in radiator and added to it in water jackets around cylinders. (g) Heat is added to air as it passes through radiator, and work is done on air by the fan.

1-14. For helium, $v = 5.91 \text{ m}^3$

1-16. (a) 160 kPa gage (b) 2150 kPa gage

1-24. 907 m^3

1-26. (a) $278°\text{K}$ (b) 1.111 kg/m^3

1-28. (a) 17.4 kg/m^3 (b) 17.8 kg/m^3

1-34. $R_{140} = 190.7$ and $R_{1400} = 195.5 \text{ J/kg °K}$. Thus the gas is not a perfect gas.

1-38. 10.25 kN/m^3

1-39. (a) $2.56 \times 10^9 \text{ Pa}$ (b) 4.7 percent

1-42. (a) 312 m/s (b) 101 m/s

1-52. 49 Pa

1-60. 0.90 N/m^2

1-64. $4.71 \times 10^{-5} \text{ m}^2/\text{s}$

1-66. 6.91 N

1-68. 31.9 N

1-70. (a) Curve C (b) Curve A
1-72. 0.214 kg/m s
1-74. 105 N m
1-76. 11.6 N m

2-4. (a) 55.3 MPa (b) 55.9 MPa
2-5. (a) 100 kPa (b) 10.19 m
2-8. 0.68 cm
2-12. 0.750
2-17. (a) 38.0 kPa (b) 3.35 m
2-19. (a) 0.50 m (b) 0.075 m
2-24. (a) 30.1 kPa (b) 12.45 kPa
2-29. (a) 106 kN (b) 212 kN (c) 2.00 m
2-34. (a) 217 kN (b) 1.125 m
2-38. (a) 33.1 kN (b) 8.28 kN at top, 16.56 kN at bottom
2-40. W/4
2-45. (a) 27.4 N (b) 44.0 N
2-52. 678 m
2-57. 1600 m
2-62. 11.5 m maximum error

3-4. (a) $25\mathbf{j} + 45\mathbf{k}$
3-10. (d) -37.5 m/s^2
3-16. (a) 1.27 m/s^2 (b) 5.33 m/s^2 (c) 40.5 m/s^2
3-19. 8 m^3/s per meter
3-21. (a) $v = -2xy + f(x)$ (c) $v = -2xy - y + f(x)$

(e) $v = \dfrac{x}{x^2 + y^2} + f(x)$

3-23. (a) and (c) satisfy continuity for a rotational fluid,
(b) and (e) for an irrotational fluid; (d) does not satisfy continuity
3-30. Zero circulation
3-36. 3.18 m/s
3-39. 1.60 m/s
3-41. 78 m/s
3-43. (c) 30.1 kg/s
3-45. 0.50
3-49. 14.00 m/s

4-1. (a) 1130 N (b) 2220 N (c) 407 N
4-7. 3530 N
4-9. 1660 N downstream
4-11. 9190 N
4-16. (a) and (b) 66.4 kN (c) 71.5 kN
4-18. (a) 104 kN (b) 87 kN
4-19. 1.18 for Prob. 3-50
4-21. $\rho u_s^2 \delta / 6$
4-24. (a) $(2/3)\rho u_s^2 D$ (b) 4/3
4-27. 13.2 rad/s
4-29. (a) and (b) 44.2 m

4-30. (c) $\omega^2 R/g$

4-34. (a) $1.125 \, \rho u_s^2$ (b) 10.1 kPa

4-36. $0.093 \text{ m}^3/\text{s}$

4-42. 16.8 m/s

4-46. $0.878 \text{ m}^3/\text{s}$

4-48. (a) 0.134 m   (b) 61.4 kN

4-51. 4.42 kN

4-53. $45.6 \text{ m}^3/\text{s}$

4-56. 16.0 m/s

4-61. $u = -\dfrac{1}{2\mu}\dfrac{dp}{dx}(b^2 - y^2)$

4-62. $u = U\dfrac{y}{h} - \dfrac{h^2}{2\mu}\dfrac{dp}{dx}\dfrac{y}{h}(1 - y/h)$

4-65. $u_2 = U\left(\dfrac{\mu_1}{\mu_1 + \mu_2}\right)\left(\dfrac{y}{b} + \dfrac{\mu_2}{\mu_1}\right)$ and $u_1 = U\left(\dfrac{\mu_2}{\mu_1 + \mu_2}\right)\left(\dfrac{y}{b} + 1\right)$

5-1. 1.556 for Prob. 3-50

5-2. (b) 1.058

5-6. 428 m/s

5-10. (a) 21°C   (b) 116 kPa   (c) 18.7 kN

5-12. (a) 75.1 m/s   (b) 5.1 kN   (c) 16.9 kJ/kg

5-15. (a) 380 m/s   (b) 388 m/s

5-17. 9.1 m

5-20. (a) 147 kPa gage   (b) 0.49 m

5-27. 0.133 m

5-28. (a) 0.75   (b) $\dfrac{2}{9}\dfrac{V_{pipe}^2}{2g}$

5-30. $15.51 \text{ m}^3/\text{s}$

5-35. (a) -25.9 Pa gage   (b) 388 Pa gage   (c) 3.04 kW

6-1. (a), (c), and (h)

6-5. (a) Continuity not satisfied   (b) $\phi = x^3/3 - x^2 - xy^2 + y^2$
(c) $\phi = xy^2 - x^3/3 - x^2/2 + y^2/2$

6-11. (a) $-\dfrac{1}{3}\dfrac{b^2}{\mu}\dfrac{dp}{dx}$

6-14. (a) Zero   (b) Infinite

6-19. $\dfrac{Q}{4\pi}\cos\theta$

6-22. (a) $\psi = -u_s y - \dfrac{q}{2\pi}\arctan\dfrac{y}{x+6} + \dfrac{q}{2\pi}\arctan\dfrac{y}{x-6}$

(b) $v = 0$ along x and y axes and $x_0 = \pm 6.164$ m, $y_0 = 0$

(e) $\psi = -u_s r \sin\theta - \dfrac{q}{2\pi}(\theta_{source} - \theta_{sink})$

(f) 0.992 m   (g) along x and y axes   (h) 1.20 kPa

6-25. 808 Pa

6-37. (a) 33.3 m/s (b) 93.3 m/s (c) 213.7° and 326.3°
(d) 5.33 kPa (e) 2.31 kN/m

6-44. At $y = \pm a$

6-49. 4.32 kN

7-4. (a) 2.69 (b) 2.66 (c) 1.29

7-9. $\dfrac{\delta}{x} = \dfrac{0.398}{\mathrm{Re}_x^{1/5}}$ and $C_f = \dfrac{0.071}{\mathrm{Re}_x^{1/5}}$

7-13. 3.9

7-16. (a) 8.34 m/s (b) 7.35 m/s

7-19. (a) 0.707 (b) 0.574

7-24. (a) 3.9 kN (b) 97 kW

7-26. (a) 355 kN (b) 440 kN

7-28. (a) 0.029 m (b) 0.010 m (c) 1.34 m/s

8-3. $\dfrac{F}{\rho V^2 L^2} = f\left(\dfrac{VL\rho}{\mu}, \dfrac{k}{L}, \dfrac{V}{\sqrt{gL}}\right)$

8-4. $\dfrac{P}{\rho N^3 D^5} = f\left(\dfrac{Q}{ND^3}, \dfrac{\rho ND^2}{\mu}\right)$

8-10. $\dfrac{u}{u_s} = f\left(\dfrac{u_s x\rho}{\mu}, \dfrac{\tau_0}{\rho u_s^2}, \dfrac{y}{x}\right)$

8-13. $\dfrac{x_R}{s} = f\left(\dfrac{u_s s\rho}{\mu}, \dfrac{\delta}{s}\right)$

8-16. $\dfrac{Q}{g^{1/2} H^{5/2}} = f\left(\dfrac{\mu}{\rho g^{1/2} H^{3/2}}, \dfrac{\sigma}{H^2 g\rho}, \theta, \dfrac{H}{Z}\right)$

8-17. $\dfrac{F}{\rho D^2 V^2} = f\left(\dfrac{VD\rho}{\mu}, \dfrac{V}{ND}, \dfrac{gD}{V^2}\right)$

8-19. $\dfrac{u}{\sqrt{\tau_0/\rho}} = f\left(\dfrac{y\rho}{\mu}\sqrt{\dfrac{\tau_0}{\rho}}, \dfrac{R\rho}{\mu}\sqrt{\dfrac{\tau_0}{\rho}}, \dfrac{k}{R}\right)$

8-21. (a) 38 m/s (b) 2.4 m/s

8-24. 4.4 atmospheres

8-29. Max $L_m/L_p = 1/30.2$

8-33. 15 min, 31 s

8-34. 1/29.5

8-37. (a) 1.90 m/s (b) 343 kN

8-40. (a) 60 N (b) 49 N (c) 12.1 m/s (d) 510 kN (e) 424 kN
(f) 11.3 MW and 283 W, respectively

8-43. 1.86 min

8-50. 7 percent

8-56. (a) 0.315 m³/s (b) 6.36 m (c) 23.1 kW

8-60. 1.48 MW

9-2. (a) 0.20 m/s (b) 3.2 kPa

9-5. 60 mm

9-11. (a) 1.25

9-16. 164 kPa
9-17. 2.00
9-23. 830 kPa
9-31. (a) 0.0142   (b) 0.00014   (c) Smooth cast iron
9-37. (a) 408 kPa   (b) 0.132 m
9-50. (a) $V = 0.783\ u_m$ (b) 0.0147
9-52. (a) 1.63   (b) Greater than 1.63
9-53. 7.87   for the 4:1 aspect ratio
9-60. 0.43 kPa
9-68. (a) 0.49 L/s   (b) 0.033 m   (c) No
9-73. $D_2 = D_1 D_3 \sqrt{2/(D_1^2 + D_2^2)}$
9-78. $V_{\text{small}} = 0.77\ V_{\text{large}}$
9-84. 1660 kW
9-92. 410 L/s

10-1. 175.5 m/s for xenon
10-3. (b) 295 m/s
10-5. 1.95
10-8. (a) 288 m/s   (b) 349 m/s   (c) 2.20
10-9. 2.54 for methane
10-11. (c) 248°C
10-13. (c) 0.245
10-19. (a) 1.118   (c) 0.471 or 1.763   (f) 2.076
10-21. (a) 0.805   (b) 270 m/s   (c) 41°C
10-27. (a) 370 kPa abs   (b) 248°K   (c) 2.50 cm
10-33. 0.590 and 1.232, respectively
10-39. 229 kJ/kg
10-45. (b) 1.01   (f) 1.31
10-56. (a) $M_x = 1.523$ (b) $M_y = 0.692$ (c) $M_x = 2.299$ (d) $M_x = 2.58$
10-64. (a) $p_y = 400$ kPa abs   (b) $p_{0y} = 546$ kPa abs
10-69. (a) 0.303   (b) $p_{\text{exit}}/p_{0x} = 0.758$ (c) $p_{\text{rec}} = 0.0640\ p_0$
10-70. 7.0 s
10-76. 0.872 atmospheres
10-74. (a) 1.62

11-1. 7.4 kN
11-16. $D_{\text{mg}}/D_{\text{al}} = 1.30$
11-18. 0.82 kg/m s
11-19. (b) 14 m/s
11-26. $C_L = 0.766$ and lift = 375 N/m
11-35. (b) 15.8°   (c) 73.4 at 8.7°
11-36. (a) 1.004   (b) 1.018

12-2. Supercritical
12-4. (a) 29.2 m³/s   (b) 0.46   (c) 1.39 m
12-6. $S_{\text{circ}}/S_{\text{rect}} = 0.84$
12-8. (c) 5170:1
12-10. 3.61 m³/s per meter
12-12. 0.00348
12-18. (a) 1.56 m   (b) 1.10 m   (c) Subcritical
12-22. (a) 0.18 m   (b) 1.69   (c) 0.22 m

12-31. (a) 0.53 m   (b) 0.33 m
12-37. 68.2 m^3/s
12-41. (a) 2.27 m   (b) 3.25 m
12-45. 18.2 kW
12-51. 4.02 m/s
12-51. (a) 0.32 m drop   (b) 0.04 m rise

13-2. 9.94 m/s
13-7. (a) 86.1 m/s
13-8. (a) 8.19 m/s   (b) 0.18 mm
13-18. 1.77
13-24. (a) 92 L/s   (b) 57 L/s
13-30. (a) 0.167 m   (b) 0.208 m
13-33. 0.604
13-35. 1.30 m
13-37. $Q = 5.70 \, H^{1.52} \, m^3/s$
13-44. 37
13-46. 3.16 m^3/s
13-49. 4.04 kg/s
13-55. 4.63 kg/s

14-1. (a) $V_{2m} = 10.2$ ft/s, $V_{2r} = 20.4$ ft/s, $V_2 = 84.1$ ft/s   (b) $E = 262$ ft   (c) 106 hp
14-11. 76.5 percent
14-16. (b) 12.92 in.   (c) 21.1 hp
14-20. (a) 24.6 L/s and 7.96 m   (b) 297 kW and 2.34 kW, respectively
14-22. 12.5 ft
14-30. (a) 402 r/min   (b) 5700 hp   (c) $N_{s(T)} = 127$, Kaplan
14-33. 8.1 ft

Index

Acceleration
  convective, 84, 85, 126, 128
  local, 84, 85, 126, 128
Adiabatic
  ellipse, 340
  flow in pipes with friction, 356−363
  flow of gases, 162, 167, 169, 337, 364
  flow or system, definition, 16
Aerostatics, 56
Airfoil characteristics, 394−400
Alternate depths, 419
Anemometers, 461
Annulus
  laminar flow, 280, 302, 323, 324
  turbulent flow, 299, 303
Answers to selected problems, 571
Atmosphere
  adiabatic, 58

isothermal, 57
lapse rate, 58−59
polytropic, 57
U. S. standard, 58−59

Barometer, 48
Bernoulli equation
  gases, 137, 169
  liquids, 126, 130, 137, 168, 196, 233,
      300, 413, 453, 464, 468, 469,
      472
Blasius
  boundary layer analysis, 230, 236
  law of pipe friction, 291
Body forces, 46, 114, 125, 136, 139
Borda mouthpiece, 155, 205, 304, 306

Boundary layer
  definition, 218
  description, 223−229
  equation of Prandtl, 229, 553
  growth in pipes, 227, 282
  hydrodynamic, 226−244
  instabilities, 224
  laminar, 224, 225, 230, 232, 234,
      243
  laminar sublayer, 242
  momentum equation, 124, 232, 241
  separation, 226
  thermal, 524, 527, 529
  thickness, 228, 236, 239
  transition, 225
  turbulent, 238
Bulk temperature, 532
Buoyancy, 55

Capillarity, 22−26
Cavitation, 24
  number or index, 265
  pumps, 498−499
  turbines, 505−506
Circulation
  airfoil, 395−397
  cylinder, 195
  definition, 91, 191−192
Complex variables, 202−206
Compressibility factor, 341
Compressible flow, 336−373
  in ducts without friction, 341−349
  in pipes, 352−363
Conformal transformations, 201
Conjugate depths, 421
Conservation
  of energy, 12, 159
  of mass, 95−102
  of momentum, 113−124
Continuity equations, 99
Continuum, 17, 221
Contraction coefficients, 205
Contraction in pipes, 306
Control surface, 11, 39
Control volume, 11, 77−78, 232,
      527
Conversion factors, 541
Critical depth, 420, 422, 515

Critical flow
  gases, 83, 345
  open channels, 420, 422, 517
Critical pressure of gas, 345
Critical slope, 420
Critical temperature of a gas, 345
Culvert flow, 440−443
Curved surfaces, forces on, 54
Cylinder
  in nonviscous fluid, 194
  in viscous fluid, 389
  with circulation, 195

Darcy-Weisbach equation, 284
Deformation of fluid, 88
  angular, 89, 172
  linear, 88, 172
  rotation, 90, 93
  translation, 88
Density
  definition, 17
  of liquids, 543−544
Diabatic flow, 349−352, 518
Diffuser, subsonic, 308
Dimensional analysis, 255
Dimensionless parameters, 253−273
  propellers, 270
  pumps, 270−272, 494−495
  turbines, 270−273, 506
Dimensions, 255
Dissipation function, 172
Dissipation of energy, 163, 167, 168,
      170
Distorted models, 266
Doublet, 192
Drag
  cylinder, 196, 389
  definition, 244, 384−385
  induced, 397
  pressure, 387
  profile, 384, 388
  smooth flat plate, 243, 386
  sphere, 389−393
  wave, 385, 393
Drag coefficient, 228, 263
  infinite cylinder, 389
  plane surfaces, 243, 386
  plate normal to flow, 387

ship hull, 386
spheres, 389
Dynamic pressure, 126
Dynamic similitude, 261
  rectilinear flow systems, 268
  rotating systems, 270
Dynamic viscosity, 26, 545, 546

Eddy viscosity, 222
Efficiency
  diffuser, 308
  pump, 488, 490, 492, 495
  turbine, 503, 506
Elasticity, bulk modulus, 19
Elbow meter, 467
Energy
  conservation, 12, 159−169
  content, 12, 159
  displacement, 160
  internal, 12, 14, 160
  kinetic, 159
  potential, 160
Energy equation, 159
  gases, 161, 168, 337, 339, 340, 342,
    350, 352, 356, 364
  liquids, 163, 168, 307, 311, 430, 434,
    465, 488, 499
Enthalpy, 13, 15, 337
Entrance region of a pipe, 227, 282, 300
Entropy, 13, 354, 369, 516, 518, 520
Equations of motion
  nonviscous fluids, 126−138
  viscous fluids, 138−141
Equipotential lines, 183
Equivalent sand-grain roughness, 293,
  294
Euler equation
  nonviscous fluids, 126, 129, 137
  turbomachines, 484−487
Euler performance curves for pumps, 486
Expansion factors for gas flow, 475−476
Expansions in pipes, 307

Fanno line, 515−516, 520
First law of thermodynamics, 12, 159
Flow
  culverts, 440−443

measurements, 452−476
  open channels, 407−443
  pipes, 281−322, 349−363
  steep slopes, 420, 443
Flow coefficients
  pipe meters, 466
  sluice gates, 473
  weirs, 470−472
Flow nets, 189−204
Fluid, definition, 3−4
Fluid mechanics, history, 5
Fluid statics, 43−65
Forced convection, 525
Forces in flow systems, 258
Francis turbine, 272, 501, 504
Free convection, 525
Friction factor
  definition, 257, 284−285
  laminar flow, 287, 288, 295
  open channels, 409−410, 414
  turbulent flow, 291, 292, 294, 295
Froude number, 259, 261, 264, 268, 410,
  435, 513, 521
Fully developed flow in a duct, 227, 282

Gas constants, 543
Gas dynamics, 336−373, 515−518,
  520−523
Gas flow
  hypersonic, 83, 340
  pipes, 352−363, 474, 523
  subsonic, 93, 340
  supersonic, 93, 340
  tables, 555−569
  transonic, 93, 340
Gases
  air as imperfect gas, 14−15
  perfect, 14
  properties, 543, 546
Gradually varied flow, open channels,
  411, 425

Half-body, 192
Hazen-Williams equation, 315−317
Head
  loss, 163
  piezometric, 45

Head *(continued)*
  potential, 45, 127
  pressure, 45, 127
  total, 127
  velocity, 127
Heat, 11
Heat-flow equation, boundary layer, 527
Heat transfer by convection, 524−534
Heat-transfer coefficient, 525
Horsepower, 542
Hydraulic bore, 436
Hydraulic depth, 408, 410
Hydraulic diameter, 281, 286, 287
Hydraulic jump, 422, 429, 433, 519,
    520, 521, 523
Hydraulic radius, 408, 414
Hydrodynamic entrance length, 300
Hydrostatics, 44
Hypersonic flow, 83, 340

Ideal fluid, 4
Images, method of, 198
Impulse turbine, 271, 487, 501−503
Incomplete similarity, 267
Incompressible flow, 83, 340
Induced drag, 385, 397
Internal energy, 12−14, 160
Irrotational flow, 82, 91, 94
Isentropic flow, 16, 127
  ducts, 341−349
  nozzles, 344−349
  perfect gas, 340−349
  tables, 555−563
Isothermal flow, 16
  in pipes with friction, 353, 356,
    360−363

Kaplan turbine, 501, 505−506
Kinematic viscosity, 27, 545, 547
Kinetic energy, 159
  correction factor, 160, 308, 435, 465
Kutta-Joukowski theorem, 395

Laminar flow, 82, 219−223
  boundary layer, 223, 225, 230, 234
  entrance region of ducts, 227, 282, 300

fully developed flow in ducts, 286
non-Newtonian fluid, 549
sublayer, 224, 242
Laplace equation, 183, 186
Lift, 393
  airfoils, 395−401
  coefficient, 264, 394, 395, 398,
    400
  cylinder with circulation, 196
  definition, 385
  measurement, 400
Limiting condition, gas flow in pipes,
    355−360

Mach cone, 462
Mach number, 254, 259, 260, 261, 264,
    268, 399, 521
  limiting value in pipes, 355, 358
Magnus effect, 393
Manning equation, 414
Manning roughness coefficient, 415
Manometers, 47−51
  tubing, 24
Measurements, 452−476
  flow rates in open channels, 469−474
  gases in pipes, 474−476
  liquid flow rates in pipes, 463−468
  static pressure, 455
  velocity, 453−463
Metacentric height, 62
Micromanometer, 50
Mild slope, 420, 428
Mixing length, 290
Mixtures in pipes, 320
Modeling ratios, 264
Models
  Froude number, 268
  Mach number, 268
  pumps, 494
  Reynolds number, 267
  ship, 267
  turbines, 506
Moment of inertia, plane surfaces, 53
Moment of momentum, 484
Momentum correction factor, 122, 283,
    435
Momentum equation, 113−124
  boundary layer, 124, 232

hydraulic jump, 434
  normal shock, 364
Momentum flux, 114, 122
Momentum transport, 114
Moody diagram for pipe flow, 295

Natural coordinate system, 85, 126
Navier-Stokes equations 138−141, 260
  Prandtl's simplification, 229, 553
Net positive suction head, 498
Newton force, 8, 542
Newtonian fluid, 29
Nomenclature, 537
Noncircular ducts, 287, 298
Non-Newtonian fluids, 29, 549
Nonuniform flow, 82, 411, 425−433
Normal depth, open channels, 416
Normal shocks, 364−369, 520
  tables, 564−569
Normal stress, 138, 140, 141
Nozzles
  converging, gas flow in, 346
  converging-diverging, gas flow,
    347, 522
  pipe meters
    gas flow, 474−476
    liquid flow, 464−467
Numerical relaxation method, 206
Nusselt number, 526, 529, 531, 533, 534

Oblique shock in a gas, 369−373
One-dimensional analysis, 80, 337
One-dimensional flow, 80
Open-channel flow, 407−443
Orifice
  gas flow in pipe, 474−476
  liquid flow in pipe, 464−467
  nonviscous fluid, 205

Paper pulp in pipes, 320
Pathline, 86
Pelton wheel, 271, 500−503
Perfect gas, 14
  properties, 543
Pi theorem, 256
Piezometric head, 45

Piezometric head line, 314
Pipe fittings, losses in, 309
Pipe networks, 317
Pipes
  entrance region, 282, 300
  parallel, 312
  series, 312
Pitot cylinder, 458
Pitot tubes, 453−458
Polar diagrams for airfoils, 398
Polytropic
  atmosphere, 57
  flow or process, 16
Potential
  energy, 160
  head 45, 127
Potential flow, 181−211
Potential functions, 182−186
  cylinder, 194
    with circulation, 195
  doublet, 192
  half-body, 192
  rectilinear flow, 189
  sink, 191
  source, 190
  tornado, 196
  vortex, 191
Pound force, 8, 542
Power, 169, 542
Prandtl, 7
  boundary layer equations, 229, 553
  law of pipe friction, 292
  mixing length theory, 290
  number, 260, 526
Pressure, 47, 140, 542
  coefficient, 259, 400
  distribution on airfoil, 400
  drag, 387
  gradient, 130
    negative, 226
    pipes, 256, 283, 287, 303, 354,
      358
    positive, 226
  head, 45
  on steep slopes, 408
Profile drag, 384, 388
Propeller thrust and torque, 270,
  399
Pumps, 271, 483−499

Rankine-Hugoniot equation, 368, 372
Rapidly varied flow in open channels, 411, 433−440
Rayleigh line, 518
Reaction of turbomachines, 485
Reaction turbines, 503−506
Rectilinear flow, 189, 268
Relative roughness of pipes, 257, 292, 294, 295
Repeating variables in pi theorem, 256
Reynolds analogy, 530, 533
Reynolds number, 220, 257, 259, 260, 267
  critical
    for flat plate boundary layer, 236, 243
    for pipes, 281, 287
    for spheres and cylinders, 388−393
Reynolds transport theorem, 77−78, 114, 159
Rotation, 82, 90, 93
Rotational flow, 82, 93
Rough surface, definition of, 225

Second law of thermodynamics, 12
Separation, 226, 392
Sequent depths, 421
Shear stress, 26, 114, 138, 139, 141, 222, 223
Shear velocity, 285, 290−293
Ship modeling, 267
Shocks
  normal, 364−369, 520, 523
    tables, 564−570
  oblique, 369−373
  strength, 368
Sink, 191
Skin-friction drag coefficients, 236−243
  boundary layer, 243, 386
  ship hull, 386
Slug, definition, 8, 541
Sluice gate, 133, 188, 422, 472
Smooth surface, definition, 225, 243
Source, 190
Specific energy, 418, 422, 434, 438, 514
Specific gravity, 19, 543
Specific heat capacity, 14−16, 543
Specific speed, 271, 494, 506

Specific thrust function, 421, 517, 520
Specific volume, 18
Specific weight, 18
Sphere, drag coefficients, 389
Stability
  floating bodies, 61
  submerged bodies, 61−65
Stagnation enthalpy, diabatic flow, 352
Stagnation points on cylinder, 196
Stagnation pressure in a gas, 341, 456
Stagnation temperature, 340, 352
Standard air, 18, 59
Starting vortex, airfoil, 396
Steady flow, definition, 79, 85, 86
Steel slope, definition, 420
Stokes' hypothesis, 140
Stokes' law for spheres, 389
Streakline, 86
Stream function, 86, 182−186
  cylinder, 194, 195
  doublet, 194
  half-body, 192
  rectilinear flow, 190
  sink, 191
  source, 190
  tornado, 197
  vortex, 191
Streamlines, 85, 183
Streamtube, 98
Submerged surfaces, forces on, 51−56
Subsonic flow, 83, 340
Supersonic flow, 83, 340
Surface tension, 22−25, 544
System, definition, 10, 77

Temperature, definition, 11
Temperature profiles
  flat plates, 528
  pipes, 533
Thermal boundary layer, 524−534
Thermodynamics 10−16
  first law, 12, 159−169
  second law, 12
Thoma number, 505
Three-dimensional flow, 81
Thrust function, 121, 421, 517, 520
Tornado, 196
Total head, 127, 311

Total head line, 311, 314
Total pressure, 126
Transition, 82, 225, 243
Transonic flow, 80, 340
Turbines, 271, 483−488, 500−507
Turbomachines, 483
Turbulent flow, 82, 219−222
    boundary layers, 223−229, 238−243
    entrance region, pipes, 227, 301
    fully developed, pipes, 289−294
    Prandtl mixing length, 290
    rough pipes, 291−295
    shear stress, 222, 223
Two-dimensional flow, 80−81

Uniform flow, 82, 85
    in open channels, 411, 414
Unit discharge, 418
Units
    conversion factors, 541−542
    SI system, 8
    table of symbols, 10
    technical English, 8
Unsteady flow, 80, 86

Vapor pressure, 25
    in pumps, 498−499
    table, 545
Velocity
    critical
        gas flow in pipes, 355, 358
        isentropic gas flow, 345
        open channels, 420
    elementary wave on free surface, 413
    gradient, 26

head, 127, 164
measurements, 453−463
potential, 182
profile, 26, 101, 122
sonic, 21, 337−338
Velocity distribution
    laminar boundary layer, 222, 231,
        234, 236, 237
    laminar flow in pipes, 222, 286
    laminar sublayer, 239, 242
    measurement in pipes, 459−460
    turbulent boundary layer, 222, 239,
        241, 242
    turbulent flow in pipes, 222, 290−291
Velocity potential, 182
Vena contracta, 465
Venturi meter
    gas flow, 474−476
    liquid flow, 464−467
Viscosity
    definition, 25
    dynamic, 26, 545, 546
    eddy or turbulent, 222−223
    kinematic, 27, 545, 547
von Kármán, mixing length theory, 290
Vortex
    irrotational, 94
    on airfoil, 396
    rotational, 94
Vorticity, 91

Wake, 219, 226, 391−392
Wave drag, 267, 385, 393
Weber number, 259, 261, 264
Weirs, 188, 469−472
Work, 11, 169, 542